AF906249

Nanomembranes

Nanomembranes

Materials, Properties, and Applications

Edited by

Yongfeng Mei
Gaoshan Huang
Xiuling Li

WILEY-VCH

Editors

Prof. Yongfeng Mei
Fudan University
Department of Materials Science
220 Handan Rd.
Yangpu District
200433 Shanghai
China

Prof. Gaoshan Huang
Fudan University
Department of Materials Science
220 Handan Rd.
Yangpu District
200433 Shanghai
China

Prof. Xiuling Li
The University of Texas at Austin
Electrical and Computer Engineering
10100 Harry Ransom Trail
78758 Austin TX
United States

Cover Image: © Ana Babii/Shutterstock

All books published by **WILEY-VCH** are carefully produced. Nevertheless, authors, editors, and publisher do not warrant the information contained in these books, including this book, to be free of errors. Readers are advised to keep in mind that statements, data, illustrations, procedural details or other items may inadvertently be inaccurate.

Library of Congress Card No.: applied for

British Library Cataloguing-in-Publication Data
A catalogue record for this book is available from the British Library.

Bibliographic information published by the Deutsche Nationalbibliothek
The Deutsche Nationalbibliothek lists this publication in the Deutsche Nationalbibliografie; detailed bibliographic data are available on the Internet at <http://dnb.d-nb.de>.

© 2022 WILEY-VCH GmbH, Boschstr. 12, 69469 Weinheim, Germany

All rights reserved (including those of translation into other languages). No part of this book may be reproduced in any form – by photoprinting, microfilm, or any other means – nor transmitted or translated into a machine language without written permission from the publishers. Registered names, trademarks, etc. used in this book, even when not specifically marked as such, are not to be considered unprotected by law.

Print ISBN: 978-3-527-34446-8
ePDF ISBN: 978-3-527-81390-2
ePub ISBN: 978-3-527-81392-6
oBook ISBN: 978-3-527-81393-3

Typesetting Straive, Chennai, India
Printing and Binding CPI Group (UK) Ltd, Croydon, CR0 4YY

C082843_080822

Contents

Preface

With the development of nanoscience and nanotechnology, "nanomembrane" becomes a new research direction of nanomaterials. The nanomembrane is basically the thinnest part of a film. As a rapidly developing field, the definition of nanomembranes may vary in different literatures, and there is no commonly accepted definition. In this book, the term "nanomembrane" is defined as a structure with its thickness limited to about one to several hundred nanometers and much larger (typically at least two orders of magnitude large, or even macroscopic scale) in lateral dimensions. The nanomembranes are generally isolated from their environment on both sides (e.g. by air, vacuum, or other deliberately introduced dissimilar materials). The vertical dimension of the nanomembrane has a characteristic size between atomic and micro scale, so it perfectly fills the gap between nanoscale and macroscopic levels. This, together with the large surface area, makes nanomembranes quite different from its macro counterpart. Traditional theories for macro scale need to be re-interpreted in nanomembranes, and this brings new physical phenomena. For example, nanomembranes have much stronger deformation capability. This, combined with a large number of surface/interface states, makes the nanomembrane exhibit interesting and extraordinary electronic, optical, thermal, mechanical, and chemical properties. It is worth noting that compared with the commonly used definition of nanomaterials, the upper thickness limit of nanomembranes covered in this book is slightly expanded. This is because the nanomembrane with a larger thickness may still have properties different from the macroscopic reference, and a clear boundary is unlikely to be established. In addition, due to the similarity in structure and research methodology, this book also includes discussion of monolayer two-dimensional materials such as graphene and transition metal dichalcogenides.

The unique nature of the nanomembrane leads to new technologies and applications, and relevant studies are booming in recent years. The nanomembrane can be prepared and integrated into functional devices and systems by using conventional solid film technology. In addition, the flexibility of nanomembrane makes it highly suitable for assembling three-dimensional structures. Inspired by origami/kirigami, a variety of three-dimensional structures can be constructed by strain engineering. The combination of material property and three-dimensional geometry brings combined value to the nanomembrane structure, and the application is thus further expanded. Interesting applications in areas including flexible electronics, nanophotonics, robotics, biology, microelectromechanical system, lab on a chip, etc. have already been demonstrated.

The editors of this book attempt to assemble a series of chapters reviewing recent advances by leading researchers in the nanomembrane field. In this book, there are totally 14 chapters, which broadly focus on the fabrications, properties, and applications of nanomembranes and assembled structures made from semiconductor, metal, insulator, polymer, and composite materials. Chapter 1 summarizes the fabrication and applications of buckling-induced origami, and the design concepts and challenges are specifically stressed upon. Chapter 2 focuses on the materials, fabrication, and self-destruction of nanomembrane-based transient electronics. In Chapter 3, nanomembranes made from polymeric materials are discussed, and their applications in environmental and energy field are reviewed. Chapter 4 gives an overview of the recent development of inorganic flexible electronics together with the discussion about challenges and future directions. Chapter 5 is dedicated to magnetic nanomembranes, and the fundamentals of magnetic nanomembranes and their applications in microbots are reviewed. Chapter 6 focuses on the discussion of mechanical forces during deformation and assembly of nanomembrane, both theoretically and experimentally. Chapter 7 describes the bio-applications of three-dimensional scaffolds produced by assembling bio-compatible nanomembranes. Chapter 8 summarizes recent applications of nanomembranes in batteries and supercapacitors, and the design strategy and underneath mechanism are discussed in detail. Chapter 9 deals with nanomembrane-based robots, and the basic principles and potential tasks performed by the robots are reviewed. In Chapter 10, three-dimensional origami is upgraded to four-dimensional by incorporating stimuli-dependent materials into the nanomembrane. Chapter 11 emphasizes on passive electronic components based on self-rolled-up nanomembranes, while Chapter 12 deals with optics of rolled-up nanomembranes. Chapter 13 gives a broad overview of the two-dimensional materials and corresponding assembled structures like origami and kirigami structures. Chapter 14 highlights two-dimensional transition metal dichalcogenides, and the strain tuning as well as its effect on material properties are summarized.

The book consolidates the knowledge concerning nanomembranes and provides a coherent viewpoint of the field that is difficult to obtain solely by reading individual journal papers. The book would act as a resource to those who aspire to further extend either the science or the technological developments in the field of nanomembranes. The nanomembrane platform is inherently an integration scheme with additional degrees of freedoms provided by the curvature and interfaces for multidimensional manipulation of physical properties that are otherwise impossible. We hope that this book can attract the attention of graduate students, young scientists, and experienced researchers across many disciplines to this ever-growing field.

Shanghai, China
30 September 2021

Yongfeng Mei
Gaoshan Huang
Xiuling Li

1

Buckling-Induced Origami Assembly of 3D Micro/Nanostructures: Designs, Materials, and Applications

Xu Cheng and Yihui Zhang

Tsinghua University, Applied Mechanics Laboratory, Department of Engineering Mechanics; Center for Flexible Electronics Technology, No. 1, Qinghua Yuan, Haidian District, Beijing 100084, People's Republic of China

1.1 Introduction

With the rapid development of micro/nano-fabrication technologies in recent years, the fundamental research on nanomembranes [1–8] and their applications in various fields, such as electronics/optoelectronics [3, 8, 9], micro/nano-electromechanical systems (MEMS/NEMS) [6, 10] and optics [11, 12], has become an area with enormous potentials and opportunities. Meanwhile, nanomembrane technologies are gradually deepening into people's daily life in extensive ways and play an increasingly important role in the development of modern intelligent society. For example, the microelectronics based on nanomembrane technologies (e.g. cell phones, laptops, and wearable devices), which serve as physical carriers of vital signal acquisition and transmission in artificial intelligence technologies, are integrated with more and more functionalities, with constantly decreasing sizes [13, 14]. However, the applications of the nanomembrane in microelectronics mostly focused on the two-dimensional (2D) micro/nanostructures and planar devices. The 2D layouts of nanomembranes may not be conducive to realizing further performance improvement or fulfilling specific crucial requirements in some scenarios [13, 15–17], such as the spatial light modulation [18], unconventional near field communication (NFC) with high Q factors [19, 20], and high-efficiency energy harvesters [21]. Development of techniques to transform nanomembranes into 3D micro/nanostructures could bypass some of the challenges encountered in the planar designs, providing a feasible route to achieve more diversity in device designs, better performance, and more advanced functionalities [22, 23]. However, many technological challenges existed in the fabrication of 3D micro/nanostructures [24]. Extensive efforts have been devoted to the development of new manufacturing methods in the past decades, and significant progress has been made in 3D nanomembrane fabrications. Of these methods, the 2D-to-3D assembly methods stand out and have received broad attention due to their inherent advantages, such as the excellent compatibility with modern planar fabrication

Nanomembranes: Materials, Properties, and Applications, First Edition.
Edited by Yongfeng Mei, Gaoshan Huang, and Xiuling Li.
© 2022 WILEY-VCH GmbH. Published 2022 by WILEY-VCH GmbH.

technologies (e.g. lithography, deposition, and etching) and the extensive material applicability [22, 23]. Most of the existing 3D assembly methods can be categorized into four main classes based on their different loading/deformation characteristics, including the rolling, folding, curving, and buckling methods [23].

The rolling and folding methods rely on intrinsic gradient stresses (e.g. the heteroepitaxial crystalline bilayers [25, 26], nonepitaxially deposited nanomembranes [27], and ion–solid interactions [28, 29]) or extrinsic driving forces (e.g. capillary forces [30–32], magnetic forces [33, 34], and cell traction forces [35]) to assemble 2D nanomembranes into 3D micro/nanostructures, through global bending deformations and bendable hinge deformations, respectively. The feature of bending-dominated mode facilitates the formation of 3D micro/nanostructures with cylindrical and polyhedral geometries and their variants [36, 37] (e.g. tubular, helical, hexahedral, and dodecahedral), but it also limits the applications to other complex configurations. The curving methods mainly focus on integrating the 2D membranes onto nonplanar substrates [38–41] (e.g. convex/concave paraboloid surfaces, cylindrical surface, and spherical surfaces) through efficient transfer-printing methods (e.g. prestretching methods [42, 43], hydroprinting methods [44, 45], vacuum-assisted transfer printing [46], punch methods [47], and hierarchical stamp methods [48]). This type of method requires an accurate positioning of the nanomembrane components with respect to the target nonplanar substrate. The buckling methods introduce the soft elastomer substrates to serve as the assembly platform, which provides driving forces to trigger mechanical buckling deformations of 2D membrane precursors [22, 23].

A schematic illustration in Figure 1.1 shows the buckling-induced assembly process, which is composed of three main steps, including the fabrication of 2D precursors, transfer-printing process, and compressive buckling assembly [23, 49, 50]. Specifically, the nanomembrane structures were first fabricated on a source wafer

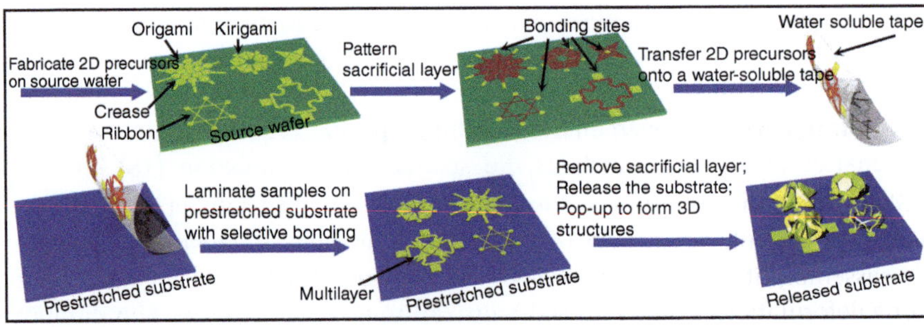

Figure 1.1 Schematic illustrations of origami assembly process of micro/nanostructures guided by controlled buckling. The process begins with transferring the 2D precursor structures with sacrificial layers and bonding sites, fabricated by advanced planar processing technologies, onto a water-soluble tape, followed by laminating the samples on a prestretched substrate with selective bondings. Removing the sacrificial layer and popping up the 3D microstructures by releasing the prestretched substrate completes the buckling assembly process. Source: Adapted with permission from Cheng and Zhang [23]. Copyright 2019, Wiley-VCH.

(e.g. sheet of monocrystalline silicon or glass) by modern planar manufacturing technologies, including lithography, deposition, and etching generally. Then, the bonding sites and the sacrificial layers were defined selectively through the electron beam (E-beam) evaporation technologies or magnetron sputtering technologies. In the transfer-printing process, a soft polydimethylsiloxane (PDMS) stamp or a water-soluble tape was exploited to transfer the 2D precursors onto the top surface of a prestretched elastomer substrate, followed by the ultraviolet (UV) light and heating treatments to activate the robust covalent bondings between the silicon dioxide layers on the bonding sites and the elastomer substrate. In the last step, releasing the prestretched substrate induces driving forces at the bonding sites, transforming the 2D precursors through spatial folding or coupled bending/twisting deformations and complex translational/rotational motions into target 3D geometries.

By introducing various design strategies on the patterned 2D precursors [19, 49, 51, 52], elastomer substrates [53–55], and paths of strain release [56, 57], a diversity of complex 3D geometries with wide-spanning length scales (e.g. from nanometers to millimeters) have been achieved. Furthermore, the integration of broad-ranging advanced functional materials (e.g. from typical rigid materials to newly emerging soft active materials) with the 3D micro/nanostructures has spawned compelling applications not only in electronics [58, 59], robotics [60], and sensors [61–63], but also in biomedical devices [64, 65], efficient energy harvesters [66, 67], and optics [55, 68].

This chapter aims to offer a comprehensive review of the latest progress of the aforementioned buckling-induced origami assembly methods, highlighting the design strategies, material integrations, and applications. It begins with the discussions on the buckling-induced folding assembly methods in Section 1.2, followed by the introduction to buckling-induced bending/twisting assembly in Section 1.3. The integration of advanced functional materials with 3D micro/nanostructures and the typical applications are summarized in Sections 1.4 and 1.5, respectively. Finally, an outlook on the existing challenges and future opportunities is provided.

1.2 Buckling-Induced Folding Assembly

The 3D micro/nanostructures induced by the folding methods typically contain creases with much lower stiffness than the other regions. In the buckling-induced folding assembly, it is essential to develop design strategies that can enable localized bending deformations at the crease regions without failure. This section discusses three representative folding strategies and their principles, including nonuniform thickness strategy [51], plasticity strategy [69], and loading path strategy [56].

1.2.1 Nonuniform Thickness Strategy

The thickness strategy realized folding creases in the 2D precursor structures through the spatial variation of thickness [51]. As shown in Figure 1.2a, a straight ribbon, consisting of two panels (marked blue), two bonding sites (marked red), and three creases (marked gray), serves as an example to illustrate the design concepts and principles. The thickness (t_2) of creases is smaller than that (t_1) of panels, leading to negligible deformations in the panel parts, in the case of sufficiently small thickness ratio (t_2/t_1), because of the cubic downscaling of the bending stiffness with the nanomembrane thickness. Furthermore, another two key parameters controlling the folding angle are the prestretching strain (ε_{pre}) and the length ratio (L_2/L), where the L_2 and L are the lengths of creases and ribbons, respectively. The experimental and computational results in Figure 1.2a show that the folding angle of the straight ribbon increases with the increase of the length ratio and the prestrain (the middle and the bottom images in Figure 1.2a). In the folding assembly, it is important to avoid the fracture of the creases that accommodate most of the overall compressive deformation. Based on the finite element analysis (FEA) and the experimental verification, the maximum material strain in the creases can be effectively reduced by decreasing the thickness ratio and increasing the crease length. In addition to the accurate prediction of the folding assembly process through FEA methods, a theoretical solution for multi-segment thickness-varying structures can be obtained based on the classical elastica theory, which provides guidelines for the optimal design of folding structures [70].

The nonuniform thickness strategies can realize a broad set of versatile 3D mesostructures by elaborately designing the 2D precursors and crease patterns [51]. Figure 1.2b shows five representative mesostructures assembled through bidirectional or even hierarchical folding, including a "pyramid," a "windmill," a "cylindrical shell," a "car," and a "three-floor building with textured steps." It is noteworthy that these mesostructures are capable of reversible and deterministic control over the 2D-to-3D transformation by tuning the applied stretching strain on the elastomer substrate, due to the elastic nature of the assembly scheme.

1.2.2 Plasticity Strategy

The plasticity strategy provides an alternative and complementary method to 3D mesostructures with folding geometries, which relies on controlled plastic deformations of the 2D precursor structures [69]. Figure 1.2c illustrates the design concepts and principles through a straight ribbon structure with spatial variation of widths. According to the linear downscaling of the bending stiffness with the width of the straight ribbon, the sections with smaller widths are prone to out-of-plane bending deformation in the initial compression of the ribbon, causing plastic yielding of the metal. Furthermore, the plastic deformation exacerbates the stress concentration at the sections with smaller widths, leading to the well-designed creases as the valley and mountain folds during the 2D-to-3D transformation. In general, the reduction of widths and the strain localization-induced plastic deformations can dramatically decrease the bending stiffness at desired regions. As shown in the four experimental images of Figure 1.2c, the straight ribbon structure made of the metallic copper shows increasingly evident folding deformations as the prestrain (ε_{pre}) increases, while the straight ribbon structure made of the polyethylene glycol

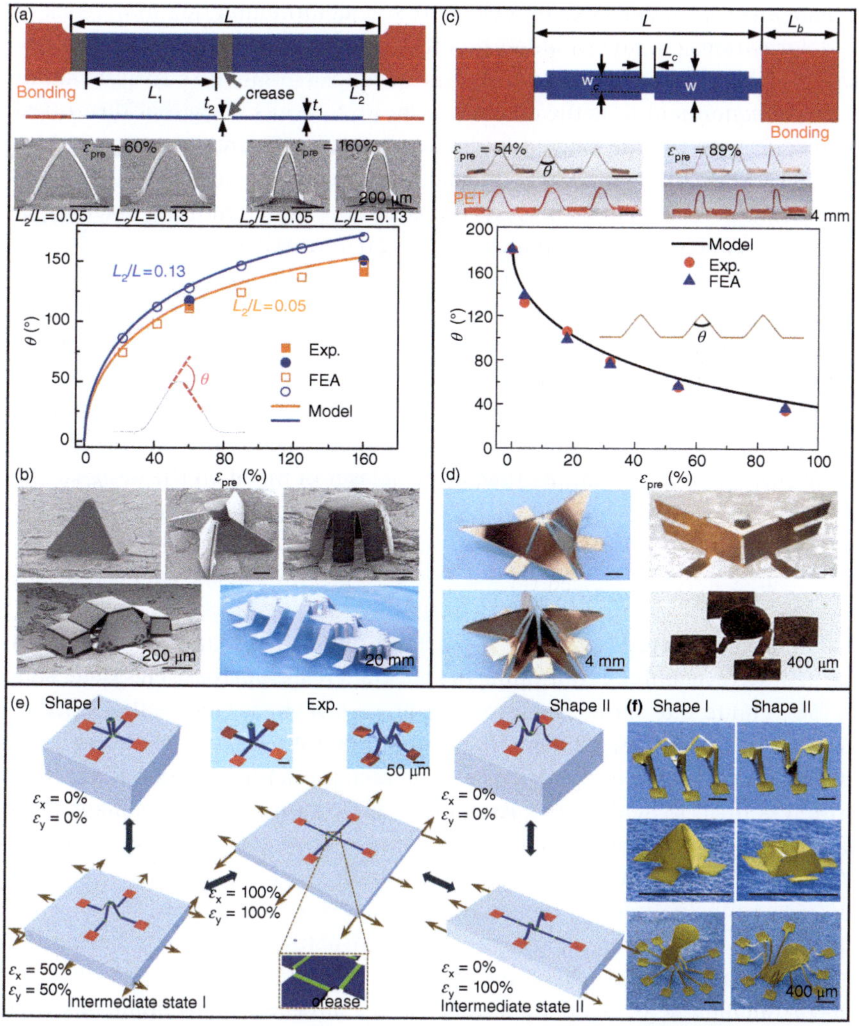

Figure 1.2 The 2D precursor design strategies and loading strategies of buckling-induced folding assembly. (a) Schematic illustrations of the nonuniform thickness strategy (top), SEM images of two deformed ribbons (made of bilayers of Au and SU8) with fixed thickness ratio and different length ratios under different prestrain levels (middle), and the folding angle (θ) versus the prestrain (ε_{pre}) for two straight ribbons with different length ratios (bottom). (b) Five representative examples of folding assembly of 3D mesostructures with nonuniform thicknesses. (c) Schematic illustrations of the plasticity strategy with nonuniform widths (top), optical images of two deformed ribbons (made of copper and PET, respectively) under two different levels of prestrain (middle), and the folding angle (θ) versus the prestrain (ε_{pre}) for the straight ribbon with a fixed thickness (bottom). (d) Four representative origami mesostructures based on the plasticity strategy. (e) Illustration of the reconfigurable folding assembly based on the loading path strategies through a sequence of FEA results and two SEM images. (f) SEM images of three representative reconfigurable origami mesostructures. Source: (a and b) Yan et al. [51]/with permission of John Wiley & Sons, Inc. (c and d) Shi et al. [69]/with permission of Elsevier. (e and f) Fu et al. [56]/with permission of Springer Nature.

terephthalate (PET) undergoes no obvious folding deformation due to the lack of plastic deformations. A dimensionless quantity (ρ_c/L_c) was defined to characterize the extent of folding deformations, where ρ_c is the ratio of minimum curvature radius in the crease region and L_c is the crease length. FEA results show that the critical prestrain (ε_c) triggering the evident folding deformations decreases under a given threshold (ρ_c/L_c) with the membrane thickness increasing. Besides, the folding angle (θ) decreases as the prestrain increases (Figure 1.2c, bottom), indicating that a larger prestrain (ε_{pre}) enables a more significant folding deformation.

Figure 1.2d highlights the use of the aforementioned folding strategy in several metallic 3D mesostructures, such as a "rotated table" and an "aircraft" [69]. In terms of the failure of metallic structures, the quantitative analyses relying on FEA show that the maximum material strain of the crease is in the range of 5–8%, which is well below the fracture threshold (~10%) of copper. It should be pointed out that the plasticity strategy is applicable not only to the 3D mesostructure made of metals, but also to diverse material systems by integrating a layer of metal on the designed 2D precursor structures.

1.2.3 Loading Path Strategy

The loading path strategy provides a practical route to reshape the folded 3D mesostructures reversibly by controlling the prestrain release paths of elastomer substrates [56]. Figure 1.2e shows the reversible process of shape change that involves the assembly of a 2D cross ribbon structure into two different stable 3D origami mesostructures with well-designed creases (marked green). The center of the cross structure was lifted to achieve the assembly of the column structure (Shape I) by synchronous release of the biaxial prestrain ($\varepsilon_x = \varepsilon_y = 100\,\% \rightarrow \varepsilon_x = \varepsilon_y = 0\%$). The sequential release of the x-direction prestrain ($\varepsilon_x = \varepsilon_y = 100\,\% \rightarrow \varepsilon_x = 0$, $\varepsilon_y = 100\%$) and the y-direction prestrain ($\varepsilon_x = 0$, $\varepsilon_y = 100\,\% \rightarrow \varepsilon_x = 0\,\%$, $\varepsilon_y = 0\%$) reshaped the column structure into the socket structure (Shape II). During the x-direction release process, the center of the cross ribbon was constrained by the ribbon along the y-direction, leading to the downward buckling mode of the final configuration. The fabrication of creases exploited the nonuniform thickness strategy and the creases played an important role in increasing the energy barrier between different stable configurations during the folding assembly. To analyze the bistability of the representative cross-shaped ribbon structure, a finite-deformation model based on the elastica theory and the perturbation method was established, providing guidelines for the designs of morphable folding structures [71].

Figure 1.2f shows three representative morphable mesostructures that can be reshaped between a "mountain" and a "valley," between a "pyramid" and a "swimming pool," and between an "octopus" and a "spider" [56]. It is important to point out that this strategy can also realize multiple stable geometries (>2) with well-designed release sequences by introducing various bistable elements (e.g. cross-shaped ribbons with creases) in the 2D precursor structures.

1.3 Buckling-Induced Bending/Twisting Assembly

The 3D micro/nanostructures induced by the bending/twisting method typically involve global bending and twisting deformations. The mode of bending and twisting deformations is mainly determined by the patterns of 2D precursor structures and the strain distributions of elastomer substrates. This section first introduces the deformation mode ratio [49] to characterize the bending and twisting deformation, and then discusses the principles of three representative design strategies, including kirigami design strategy [52, 72], multilayer design strategy [19], and engineered substrate strategies [53–55].

1.3.1 Deformation Mode Ratio

Filamentary structures typically consisting of slender thin ribbons (with the thickness and width much smaller than the width and arc length, respectively) represent a class of 3D mesostructures with combined bending and twisting deformations. The buckling of filamentary structures always involves considerable out-of-plane bending deformations. One practical approach to characterize the deformation components relies on a dimensionless quantity R, which is defined by the average twisting curvature (κ_{twist}) ratio to the average bending curvature (κ_{bend}) [49]. The deformation mode ratio varies drastically as the patterns of 2D precursor structure and bonding sites change. As shown in Figure 1.3a, the two representative mesostructures assembled from the same 2D precursor structures possess totally different deformation mode ratios, due to the different alignment of bonding sites [49]. The circular helix structure experiences similar levels of bending and twisting deformations ($R \approx 1.09$), while the flower-shaped structure mainly undergoes bending deformation ($R \approx 0.11$). By assuming the deformation mode of the helical structures and the frame structures, the analytical solutions can be established based on the principle of energy minimization [73, 74]. Furthermore, the deformed configurations for more extensive filamentary structures can be predicted by a double perturbation theory of postbuckling analyses [75].

Figure 1.3b shows two representative filamentary structures made of monocrystalline silicon [49]. Multilevel filamentary networks with combined bending and twisting deformations can be achieved by interconnecting multi-ribbons based on the mechanisms of hierarchical buckling.

1.3.2 Kirigami Design Strategy

The kirigami strategy relies on the strategically designed patterns of cuts to form the diverse 3D mesostructures with curved surfaces, achieving high levels of topographical complexity [52]. One of the advantages of the kirigami strategy is that the cuts change the load-transfer paths in the membrane during the compressive buckling process, with capabilities to reduce the stress concentration evidently and avoid the formation of folding creases. The top panel in Figure 1.3c serves as an example to

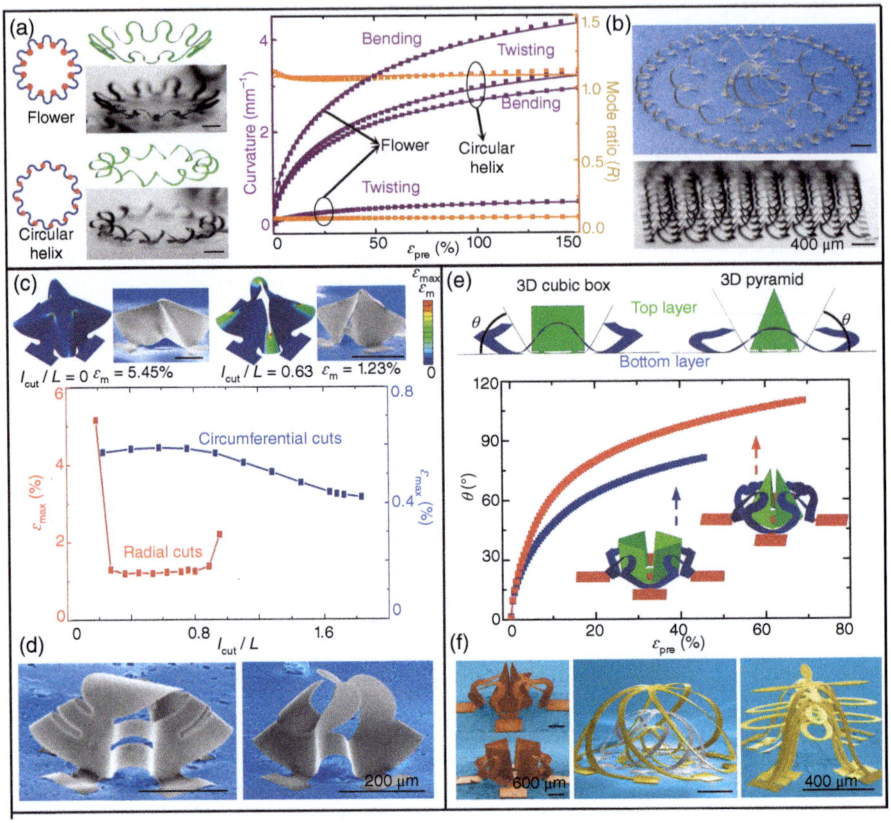

Figure 1.3 The 2D precursor and elastomer substrate design strategies of buckling-induced bending and twisting assembly. (a) Analyses of curvature components (bending and twisting) and mode ratio (*R*) for two representative filamentary mesostructures with the same 2D layout and different patterns of bonding sites. (b) SEM images of a complex circular helical structure (top) and a network of double-floor helices (bottom). (c) FEA results and SEM images of two mesostructures without and with radial cuts, and maximum material strain (ε_{max}) versus the dimensionless cut lengths (l_{cut}/L) for circular membranes. (d) Two representative kirigami mesostructures assembled with different kirigami patterns. (e) Computational results for the folding angle as a function of prestrain for the assembly of 3D cube and pyramid assisted by the lower ribbons. (f) SEM images of four unique multilayer 3D mesostructures. (g) Schematic illustrations of the buckling and twisting assembly of a morphable cross ribbon structure based on the kirigami substrates, and the rotation angles at the center of an individual square unit versus the applied biaxial strain. (h) SEM images of two representative mesostructures assembled on a typical kirigami elastomer substrate. Source: (a and b) Xu et al. [49]/with permission of American Association for the Advancement of Science – AAAS. (c and d) Zhang et al. [52]/with permission of National Academy of Sciences. (e and f) Yan et al. [19]/with permission of American Association for the Advancement of Science – AAAS. (g and h) Zhao et al. [55]/with permission of National Academy of Sciences.

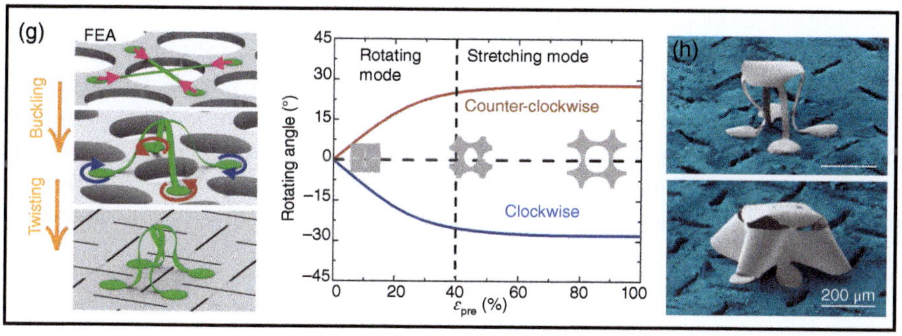

Figure 1.3 *(Continued)*

illustrate this advantage by comparing two mesostructures made of monocrystalline silicon with and without radial cuts. The corresponding FEA results show that the maximum material strain ($\varepsilon_m = 5.45\%$) in the mesostructure without cuts is much larger than that ($\varepsilon_m = 1.23\%$) of the structure with radial cuts due to the evident folding deformations. The bottom panel in Figure 1.3c shows the maximum material strain exhibits a nonlinear change with the increase of length ratio (l_{cut}/L), due to the influences of the cutting length on the deformation mode, where the l_{cut} and L are the length of cuts and the membrane, respectively.

Another advantage through use of this strategy is that the introduction of cuts enhances the design flexibility of 3D assembly in realizing 3D mesostructures with complex curved geometry. Figure 1.3d provides two representative mesostructures assembled from the same circular precursor geometries but tailored with different patterns of cuts [52]. The first symmetric kirigami mesostructure was achieved with the aid of cuts along the radial or circumferential directions, while the second antisymmetric mesostructure was induced by the cuts with serpentine patterns.

1.3.3 Multilayer Design Strategy

The multilayer design strategy provides a practical route to assemble multilayer 3D mesostructures with complex 3D topologies through the use of laminated and multilayer 2D precursors [19]. In the fabrication process of multilayer 3D mesostructures, releasable 2D precursor structures need to be transferred separately and stacked together on the prestretched elastomer substrate, thereby requiring precise positioning between different layers. Figure 1.3e shows two representative multilayer mesostructures leveraging the combined bending and twisting deformations of the bottom layer (e.g. ribbon structures) to drive the 3D folding assembly of the top layer (e.g. 3D cube and pyramid-like structure). The mechanical interactions between the two overlaying layers exerting contact forces at strategic locations can be well designed quantitatively by means of FEA. The computational results show that the folding angle and the contact forces increase as the prestrain increases, and gradually converge to a stable value. Besides, the driving forces can be transmitted over a relatively long distance by introducing a domino structure.

In addition to the above advantage, the multilayer strategy is capable of realizing densely distributed 3D architectures, which can be exploited to improve the functional and integration density at the device level. The mesostructures on the right of Figure 1.3f are examples (e.g. 3D trilayer nested cages and bionic trees) of densely distributed 3D structures. Other complex multilayer structures can be formed by entangling a plurality of ribbons and arranging the bonding sites of the upper layers on the underlying structures.

1.3.4 Engineered Substrate Strategies

The engineered substrate strategies rely precisely on tailored strain distributions of the elastomer platform to control the bending and twisting deformations of 3D mesostructures during buckling assembly. Three representative strategies have been proposed, including the engineered thickness strategy [54], engineered modulus strategy [53], and kirigami strategy [55].

The engineered thickness strategy tailors the distribution of substrate thickness through the casting and curing process of soft lithography [54]. Based on the linear downscaling of the tensile stiffness with the substrate thickness, the deformations in thick regions are much smaller than the thin regions, leading to much smaller out-of-plane deformations of 3D mesostructures. The inverse designs of the distributions of substrate strain and thickness relying on the iterative calculations of FEA can guide the assembly of unique 3D mesostructures (e.g. a 3D concave mirror-like structure and a 3D elevated helical coil).

The engineered modulus strategy is suitable for achieving strain distributions with larger strain ratios and larger strain gradients through the use of heterogeneous materials, in comparison to the engineered thickness strategy [53]. The heterogeneous substrates, typically prepared through multi-material 3D printing methods or casting and curing techniques, undergo much larger deformations in lower modulus regions and sharp strain variation in the material transition zone. Based on a finite deformation model, the strain distributions on the uniaxial substrate can be predicted and inversely designed. A variety of 3D mesostructures have been realized, including some biomimetic examples (e.g. a scorpion-like structure and a manta ray-like structure), which are not accessible through use of uniform substrates.

The kirigami designs of substrates introduce local twisting deformations into the assembly of 3D mesostructures through the use of interconnected, rotatable units [55]. The kirigami patterns fabricated by laser-cutting technology undergo large rotational motions with the applied prestretching strain. The cross-shaped ribbon structure in Figure 1.3g serves as an example to illustrate the deformation process of kirigami substrates. At the beginning stage of strain release in the substrate (stretching mode), the structure mainly undergoes out-of-plane buckling deformation ($40\% < \varepsilon_{pre} < 100\%$), followed by the twisting deformations when the substrate changes into the rotating dominated mode ($\varepsilon_{pre} < 40\%$). According to the FEA results, the rotation angle at the center of a square unit converges to a fixed value (e.g. ~30° for the square kirigami patterns) when the applied biaxial strain exceeds ~40%. Increasing the length of cut can effectively increase the rotation angle of the square units, and, at the same time, increase the maximum material strain of the substrate. Therefore, a relatively large rotation angle can be obtained without fracture at the cutting regions by optimizing the ratio of the cut length to the length of the unit. Based on the nature of rotating features, a broad set of 3D mesostructures with local chiralities can be obtained. Figure 1.3h shows two representative 3D mesostructures based on the kirigami substrate. Besides, by incorporating the folding crease design and the loading path strategy with kirigami substrates, morphable 3D mesostructures with local chiralities can be constructed, further enhancing the design flexibility.

1.4 Functional Materials Integrated with 3D Mesostructures

The buckling-guided 3D assembly strategies apply very well to a wide range of advanced functional material systems, including but not limited to metallic materials, semiconductor materials, piezoelectric materials, 2D materials, and phase-change materials. The compatibility with diverse range of materials suggests a great potential in developing high-performance devices.

1.4.1 Metallic Conductors

The metallic materials (e.g. gold, copper, aluminum, nickel, and titanium) typically possess good ductility and good conductivity of heat and electricity, which can be integrated with 2D precursor structures conveniently through the E-beam evaporation technologies or magnetron sputtering technologies (Figure 1.4a) [49, 52]. Metallic materials integrated with the 3D mesostructures can function either as electrical interconnects [57, 65] or as coils to achieve electromagnetic signal transmissions [19, 79, 80] in two ways. Taking the recently proposed 3D strain sensor as an example, the two separate ribbons of the metallic spiral structure contact with each other at designed tensile strain, indicating the value of strain visually through the connected light-emitting diodes (LEDs) [57]. The 3D NFC device, assembled from a two-layered coil structure made of copper, has an enhanced Q factor and improved functional angle in comparison to the 2D counterparts [19].

1.4.2 Semiconductors

Semiconductor materials (e.g. silicon and GaAs) are widely used in microelectronic devices, which play an important role in the development of modern intelligent society. The buckling-induced origami assembly methods work well for monocrystalline silicon with a very small brittle fracture strain ($\sim$1%), by exploiting optimized geometrical parameters [22, 49]. The 3D high-density microelectronic devices can be realized by integrating large-scale chips on the 2D silicon precursors. Figure 1.4b shows two representative multilevel mesostructures made of monocrystalline silicon, where the material strain can be tailored by the FEA. Besides, the piezo-resistive

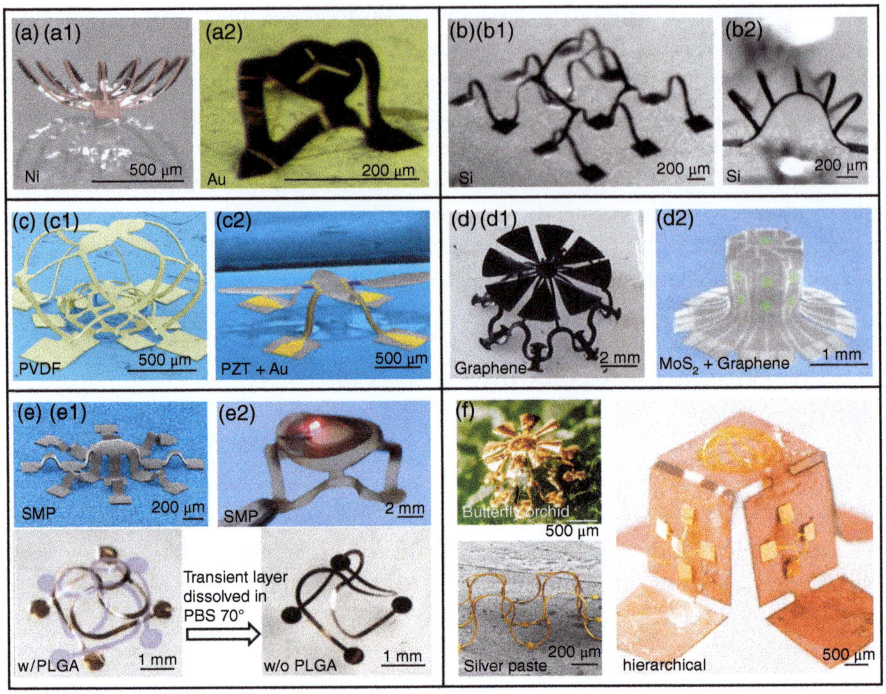

Figure 1.4 The integration strategies of advanced functional materials in the buckling-induced origami assembly. Representative 3D mesostructures integrating metallic conductors (e.g. Au, Ni in (a)), monocrystalline silicon (b), piezoelectric materials (e.g. PZT, PVDF in (c)), 2D materials (e.g. graphene, MoS_2 in (d)), and phase-change materials (e.g. SMP, PLGA in (e)). (f) Transfer printing of the buckled 3D origami mesostructures onto diverse substrates (e.g. butterfly orchid and silver paste) and hierarchical designs. Source: (a1and b1) Xu et al. [49]/with permission of American Association for the Advancement of Science - AAAS. (a2) Zhang et al. [52]/with permission of National Academy of Sciences. (c1) Han et al. [21]/with permission of Springer Nature. (c2) Ning et al. [61]/with permission of American Association for the Advancement of Science - AAAS. (d1) Ling et al. [76]/with permission of American Chemical Society. The right image in panel (d2) Lee et al. [68]/with permission of Springer Nature. (e1) Wang et al. [77]/with permission of John Wiley & Sons, Inc. (e2) Park et al. [78]/with permission of John Wiley & Sons, Inc. (f) Yan et al. [60]/with permission of National Academy of Sciences.

effect of silicon materials can also be exploited to design sensitive pressure sensors with multimodal responses (e.g. normal force, shear force, bending, and temperature) and relative wide dynamic range [81].

1.4.3 Piezoelectric Materials

There are two types of piezoelectric materials commonly used in buckling-induced origami assembly, including polyvinylidene difluoride (PVDF) [21] and $Pb(Zr_{0.52}Ti_{0.48})O_3$ (PZT) [61]. Their piezoelectric properties have a great potential for applications as 3D sensors and actuators. The left panel in Figure 1.4c shows an overlapping network made of the PVDF membrane [21]. The electrical signals generated by the deformations of PVDF membranes can be collected by integrating metallic electrodes (e.g. Au) at strategic regions of the 3D mesostructure. Spin coating an eccentric layer (e.g. a polyimide [PI] layer) on the PVDF membrane can offset the sensing layer from the neutral mechanical plane, leading to an increase of the open-circuit output voltage by 2 orders of magnitude. The right panel in Figure 1.4c shows a fly-shaped mesostructure with a pair of mechanical actuators made of PZT nanomembrane on the wings [61]. The layouts of the PZT and metal electrodes are optimized by the quantitative FEA to ensure the material strain below the fracture strain (~0.6%). The deterministic control over the dynamical behavior and flapping resonate mode (resonant frequency ≈1.15 kHz) of 3D fly-shaped mesostructure relies on the strategically placed PZT microactuators on 3D geometries and the in-phase driving voltages applied to the actuators.

1.4.4 2D Materials

2D materials typically refer to the materials with electrons moving freely in a 2D plane at the nanometer scale (1–100 nm), such as the graphene and molybdenum disulfide (MoS_2) [68]. The left panel in Figure 1.4d shows a jellyfish-like mesostructure made of the CO_2 laser-induced cellular graphene [76]. Incorporation of the cellular graphene can form hierarchical mesostructures and greatly enhance the stretchability of cellular graphene while maintaining its excellent conductivity, which can hardly be achieved by its planar counterparts due to the irregularly distributed, porous microstructures. The 3D interdigital supercapacitors made of cellular graphene and solid-state electrolytes exhibit excellent electromechanical properties under the cyclic loading conditions. The right panel in Figure 1.4d shows an octagonal prism incorporating monolayer graphene (light gray) and MoS_2 (green), each of which provides an excellent combination of electrical, optical, and mechanical properties [68]. Specifically, graphene materials are suitable for use as flexible, transparent conductors and MoS_2 materials are of great interest for the excellent photovoltaic conversion properties. Such a heterogeneous material system can be used to fabricate high-performance 3D photodetectors with the aid of well-designed 3D geometries.

1.4.5 Phase-Change Materials

Introduction of phase-change materials into the 3D assembly methods can realize further actuations of as-assembled 3D mesostructures, allowing quick response to the environmental stimuli. The top panel in Figure 1.4e shows two representative mesostructures made of shape memory polymers (SMPs), a kind of soft active material, which can recover the initial 3D configurations under an external stimulus [77]. Cooling the heated mesostructures (~57 °C) to room temperature and maintaining the applied forces on the strategic locations of structures can fix the temporary geometries [77, 82]. This type of actuation can be realized either by directly heating the mesostructures or indirectly by light exposure of the mesostructures with embedded nanoparticles. Besides, with the use of SMPs, freestanding

3D mesostructures can be obtained by dissolving a sacrificial layer at bonding sites after the shape fixation (see the image at the top right of Figure 1.4e). The bottom panel in Figure 1.4e shows a 4D assembly process in which a 3D interwoven multilayer structure made of Cu/PI and poly(lactic-*co*-glycolic acid) (PLGA) bilayer was reshaped to a monolayer 3D mesostructure [78]. Exposing the assembled 3D mesostructures to specific environmental conditions, such as immersing the structures into phosphate-buffered saline solution (PBS) at 70 °C, can dissolve the transient layer and trigger the 4D assembly process, leading to the reconfiguration of geometries over time.

1.4.6 Transfer Printing of 3D Mesostructures

The elastomeric substrate provides supporting and driving forces for the assembly of 3D mesostructures, but it also limits the applications of these structures due to poor heat resistance, easy degradation, and other disadvantages [60]. Transfer-printing technologies allow the integration of assembled 3D mesostructures with nearly any class of platforms, greatly facilitating the practical applications in extensive environmental conditions. The transfer printing of 3D structures relies on a meltable wax, which starts with embedding the assembled 3D mesostructures in the wax, instead of the typically used soft stamps (e.g. PDMS) in planar fabrication process. Melting the wax (raising the ambient temperature) after printing on the desired substrate coated with a thin adhesive layer completes the process without changing 3D geometries of mesostructures. A set of 3D mesostructures (e.g. filamentary, kirigami, origami, multilayer, and hierarchical designs) on different target substrates (e.g. butterfly orchid, silver paste, quartz, chicken breast, and silicon) have been obtained (Figure 1.4f), exhibiting the versatility and flexibility of this transfer-printing strategy [60, 83].

1.5 Applications

1.5.1 Electronics

Figure 1.5a shows a concealable electromagnetic device capable of switching the working mode through shape change. The device consists of three planar antennas on the middle pad for wireless communication and an elastomeric substrate for assembly and support [56]. Based on the loading path strategy introduced in Section 1.2, the simultaneous release of the biaxial prestrain yields the working mode (Shape I) with antennas elevated to the top and high radiant efficiency due to the off state of electromagnetic shielding. Sequential release of the substrate strain leads to a concealing mode (Shape II) with low radiant efficiency, because of the metallic membrane structure shielding the three antennas. Figure 1.5b shows a deformable hemispherical electrically small antenna (ESA) with the meander line conducting paths, allowing large frequency bandwidths and relatively small physical sizes simultaneously [79]. The optimized inner interconnects under the ESA enable the approximation of non-developable hemisphere surface by eight circumferentially distributed ribbons with variable widths. The 3D configuration can be easily reshaped by stretching or releasing the elastomeric substrate (ε_{appl}), providing a practical route to tune the frequency and normalized quality factor (Q/Q_{lb}) continuously and reversibly. Combined FEA results and experimental measurements showed that the best normalized quality factor occurs at strain-free state ($\varepsilon_{appl} = 0\%$) due to its hemispherical geometry. In addition to the 3D antennas, the 3D silicon electronic systems can also be fabricated by directly incorporating the high-performance n-channel silicon nanomembrane (Si NM) MOSFETs (n-channel metal oxide semiconductor [nMOS]) and p-channel Si NM MOSFETs (p-channel metal oxide semiconductor [pMOS]) into the assembled the 3D mesostructures [58].

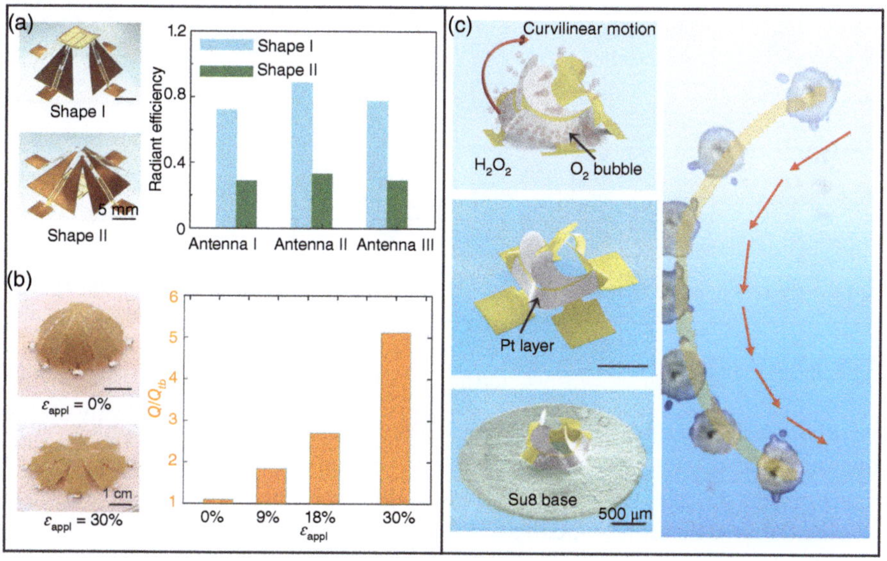

Figure 1.5 Applications of 3D mesostructures in electronics, robotics, sensors, and biomedical devices. (a) Optical images of the concealable electromagnetic device and the computational results of radiant efficiency at two different stable shapes (Shapes I and II) of the integrated three antennas. Source: Adapted with permission from Fu et al. [56]. Copyright 2018, Springer Nature. (b) Optical images of the hemispherical electrically small antenna under two different strain levels and the normalized quality factor (Q/Q_{lb}) versus applied stretching strain (ε_{appl}). Source: Liu et al. [79]/with permission of John Wiley & Sons, Inc. (c) Illustrations, SEM images, and super-imposed images of the 3D micro-swimmer with controlled motion modes and trajectories (the designed curvilinear motion). Source: Adapted with permission Yan et al. [60]. Copyright 2017, National Academy of Sciences. (d) The optical image of the impact-based 3D piezoelectric sensor and the time-domain output voltage by bending the encapsulated sensor along two vertical directions. Source: Adapted with permission from Han et al. [21]. Copyright 2019, Springer Nature. (e) A 3D fluid property sensor with five independently addressable PZT micro-actuators and the simulated resonant modes of the double-floor mesostructure (left). Measured and calculated viscosity of the water–glycerol mixtures with multiple mixing ratios (right). Source: Adapted with permission from Ning et al. [61]. Copyright 2018, AAAS. (f) Optical images of a health-monitoring device composed of helical network interconnects and the measured time-domain output signals of the electrophysiological recordings (e.g. EEG, EOG). Source: Jang et al. [65]/with permission of Springer Nature.

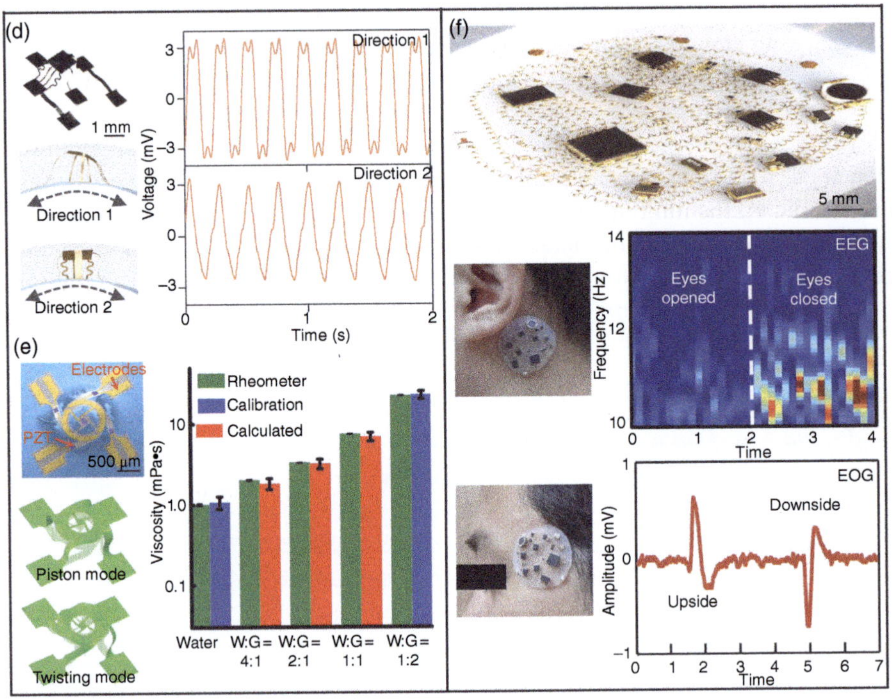

Figure 1.5 (*Continued*)

Systematic experimental and simulation studies on the 3D silicon electronic systems illustrated that the interconnected 3D bridges or coils ensure that the commercial transistors are not affected by external mechanical deformations (e.g. bending and shearing deformations) with negligible changes in transfer curves.

1.5.2 Robotics

Figure 1.5c shows 3D self-propelled micro-swimmers with controlled swimming modes and motion trajectories based on freestanding kirigami mesostructures [60]. The isolation of propeller-shaped mesostructure from the assembling substrate relies on the photocrosslinkable bases, which begins with dropping polymer liquids (e.g. SU8 droplets) onto the 3D mesostructures. The bonding sites can be constrained by a photodefined base with UV light radiation through a photomask. Besides, the freestanding mesostructures can also be obtained through the aforementioned SMPs [77], controlled plastic deformations, and mechanical interlocking systems [78]. Specifically, the plastic deformations of the thin metal films deposited in the 2D precursor fabrication process can maintain a certain amount of deformations instead of fully restoring the original 2D geometries. In this case, it is essential to take the springback effects into considerations to achieve precise geometric control of isolated mesostructures. The mechanical interlocking approach relies on well-designed mechanical interlocking elements consisting of female-type lugs and male-type hooks. Irreversibly "lock-in" of the hooks and corresponding lugs along different directions (e.g. x-axis and y-axis) can be realized by controlling the loading paths. The strategically patterned nanomembranes of platinum (Pt, 100 nm in thickness) on the swimming robots are capable of catalyzing the hydrolysis of hydrogen peroxide (H_2O_2, 30% by weight), generating bubbles of O_2 at room temperature to propel the micro-swimmer in a controlled mode. Based on the multibody dynamics simulations and experimental results, the curvilinear motions rely on the Pt nanomembrane integrated on four petals and one side of the structure.

1.5.3 Sensors

Integrating piezoelectric materials (e.g. PVDF and PZT) with the 3D mesostructures implies attractive applications in MEMS/NEMS and sensors. Figure 1.5d shows impact-based 3D piezoelectric sensors encapsulated in a soft silicone elastomer, protecting the 3D device physically and maintaining its flexibility simultaneously [21]. One of the advantages of the 3D sensor is that the output voltage responses vary with the magnitude, applied location, and the applied direction of the mechanical stimuli. Under the bending mode of operation, the output voltage increases linearly as the curvature of the encapsulation elastomer increases. Besides, different bending directions (e.g. Directions 1 and 2) lead to distinguishable electrical signals with different voltage peaks. Figure 1.5e shows a fluid property sensor integrated with five independently addressable PZT nanomembranes [61]. Based on the well-tailored rotated-table geometry, the mesostructure has two qualitatively distinguished resonant modes (twisting motion and piston motion), the resonant frequencies of which have decoupled sensitivities to the viscosity and density of the fluids. A theoretical model of 3D vibrations has been established to offer guidelines for the measurement of fluid properties once the resonant frequencies of these working modes are known [84–86]. Compared with the experimental results based on a commercial rheometer, the calculated results based on the abovementioned principles exhibit good consistency for the water–glycerol mixtures. Besides, another 3D vibratory sensor driven by the Lorentz forces has been fabricated to measure the nanomembrane properties (e.g. modulus and density) based on the partially decoupled sensitivities of resonant frequencies in two distinct vibration modes [62].

1.5.4 Biomedical Devices

Figure 1.5f shows a stretchable health-monitoring device consisting of more than 50 electronic components (e.g. data storage/processing chips, capacitances, and sensors) interconnected by 3D open-mesh networks of helical microcoils [65]. Compared with the planar interconnects widely used in stretchable electronics (e.g. serpentine or fractal interconnects), the 3D helical mesostructures offer much higher levels of stretchability while maintaining the mechanical and electrical robustness, due to the reduction of stress concentration. Here, a two-stage encapsulation strategy was introduced to maximize the system-level stretchability of the 3D health-monitoring device and protect the electronic components from the corrosions of water and oxygen [87]. Based on the detailed experimental results, this demonstrated device is capable of implementing reliable multifunctional operations for health monitoring, including 3D motion tracking, respiration monitoring, and electrophysiological signal monitoring (e.g. electrocardiograph [ECG], electromyography [EMG], electrooculogram [EOG], and electroencephalograph [EEG]). In addition to its application to health monitoring, the 3D micro/nanostructures can also be used as cell scaffolds to study the interactions of cell growth with the external environments. Systematic experimental results showed that the fibroblasts could adjust their alignments according to the geometries of 3D cell scaffolds. The fibroblasts exhibited highly ordered networks alignments on the 3D helical structures, while random and messy alignments on the 3D table-shaped structures [64]. Besides, introducing addressable microelectrodes into 3D cellular scaffolds enables the recording of electrophysiological signals of dorsal root ganglion (DRG) neural cells [60].

1.5.5 Energy Harvesters

Figure 1.6a shows a thermoelectric energy harvester composed of an array of compliant and stretchable helical microcoils [66]. The 3D helical geometries, integrating silicon ribbons with p-type and n-type segments at the coil legs, allow large temperature gradients across the thermoelectric device and low thermal impedance interfaces to the ambient flowing air. Introducing an encapsulation layer on the coil legs can enhance the heat flow through the legs and increase the total surface cooling capacity, exemplified by 9.7 nW increase in silicon heat flow. The experimentally measured output characteristics of the thermoelectric harvester satisfy the design expectations with good power outputs ($\approx$2 nW). Figure 1.6b shows a 3D piezoelectric energy harvester with an ultralow-stiffness serpentine design [21, 88]. To avoid the influence of interface adhesion [89] on the configuration of the ultralow-stiffness serpentine structure, a sacrificial structure was introduced in the buckling assembly process. The property of ultralow stiffness ensures particularly high sensitivity to low-frequency vibrations at small scale (in-plane size $\approx$1 mm). Implanting the piezoelectric energy harvester into the hind leg of a mouse, the bioenergy generated by mouse activity, including but not limited to trotting and climbing, can be converted into electrical energy, inducing more than 1 mV output voltages. The third recently reported energy harvester based on the buckling-induced 3D assembly is a graphene hygroelectric generator [67], which is capable of harvesting the chemical potential energy. By exposing the 3D pyramid-shaped hygroelectric generator to a moisture environment, the stable electricity output can be realized even under extreme deformations, such as uniaxially/biaxially stretching and out-of-plane compression.

1.5.6 Optical Devices

Figure 1.6c shows a mechanically tunable optical chiral metamaterial [55]. Controlled release of the kirigami substrate allows access to two reconfigurable 3D

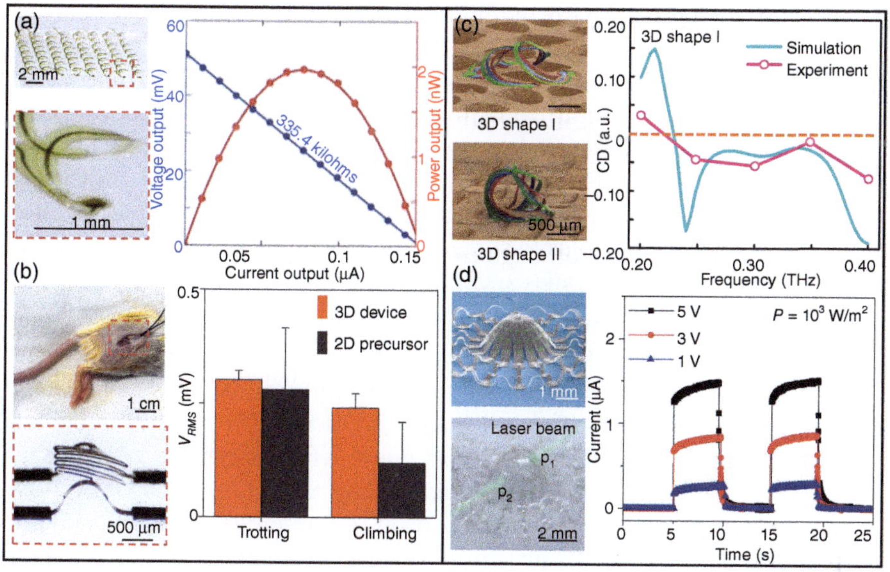

Figure 1.6 Applications of 3D mesostructures in energy harvesters and optics. (a) A 3D thermoelectric energy harvester with an array of 8 × 8 coils and the measured power output characteristics. Source: Nan et al. [66]/with permission of American Association for the Advancement of Science - AAAS. (b) A 3D piezoelectric energy harvester implanted into the hind leg of a mouse and the root-mean-square output voltages under different behaviors of the animal (e.g. trotting, climbing) in comparison to its 2D counterpart. Source: Han et al. [21]/with permission of Springer Nature. (c) SEM images of the 3D tunable optical chiral metamaterials with two unique 3D shapes based on the kirigami substrate, and the experimental and computational optical circular dichroism (CD) versus the frequency. Source: Zhao et al. [55]/with permission of National Academy of Sciences. (d) SEM and optical images of a 3D photodetector integrating 2D materials (graphene and MoS_2), and the photocurrent responses under the green laser illumination at different power densities. Source: Lee et al. [68]/with permission of Springer Nature.

geometries of the trilayer cage-like mesostructure (Shapes I and II), both of which have strongly chiral features. The distinct microscale configurations and the suitable physical scales (e.g. in-plane size <300 μm, ribbon width = 40 μm) trigger the mutually detuned resonant responses in the terahertz (THz) range. Simulation results and electromagnetic measurements under left-handed circularly polarized light (LCP) and right-handed circularly polarized light (RCP) show the phenomena of wavelength-dependent chirality and asymmetric absorption. Figure 1.6d shows a 3D photo-detecting device constructed with graphene and MoS_2 membranes [68]. The 3D hemispherical mesostructure with 16 ribbon components allows simultaneous detection of direction and intensity of the laser light. The time-resolved measurements at different bias voltages showed that the photo-detecting device can quickly switch between the "on" and "off" states as the laser changes. Based on the principles of geometric optics, the positions of entry and the exit points (P1, P2) of the laser on the 3D hemispherical surfaces can be determined by interpolating the photocurrent intensity.

1.6 Concluding Remarks

This chapter reviews the buckling-induced origami assembly methods, highlighting various design concepts, fabrication methods, applied materials, and device applications. The sophisticated 3D micro/nanostructures, determined by not only the layouts of 2D precursors, but also the deformable substrates and loading schemes, have compelling applications in different research fields, such as electronics, robotics, sensors, and optics. Although tremendous progress has been achieved in the assembly and application of 3D mesostructures, intractable challenges remain before realizing the full potential of the buckling assembly methods, thus providing fruitful opportunities for future explorations. One major challenge lies in the inverse design of 3D mesostructures due to the complicated nature of bifurcation and steric deformation of mechanical buckling process. The recently proposed topology optimization approaches, such as the moving morphable components method (MMC) [90] and the multiparameter genetic algorithm method (GA) [91], provide practical routes to the inverse designs of certain kirigami and filamentary mesostructures. The complex 3D structures with curved membrane features remain to be solved due to the much higher computational costs and the lack of straightforward mechanics models/algorithms. Besides, it is very important to further improve the manufacturing accuracy (e.g. positions, geometries, and dimensions) in the assembly process, as a significant step to the large-scale manufacturing of 3D mesostructures. With the further progress in the buckling methods and the applications, industrial-level and multifunctional devices can be expected in recent decades.

Acknowledgments

Y.Z. acknowledges support from the National Natural Science Foundation of China (#11672152 and #11722217), the Tsinghua National Laboratory for Information Science and Technology, and the State Key Laboratory of Digital Manufacturing Equipment and Technology, Huazhong University of Science and Technology (#DMETKF2019005).

References

1 Kim, D.-H. and Rogers, J.A. (2009). Bend, buckle, and fold: mechanical engineering with nanomembranes. *ACS Nano* 3 (3): 498–501.

2 Huang, G. and Mei, Y. (2012). Thinning and shaping solid films into functional and integrative nanomembranes. *Adv. Mater.* 24 (19): 2517–2546.

3 Liu, Z., Cui, A., Li, J., and Gu, C. (2019). Folding 2D structures into 3D configurations at the micro/nanoscale: principles, techniques, and applications. *Adv. Mater.* 31 (4): 1802211.

4 Ning, X., Wang, X., Zhang, Y. et al. (2018). Assembly of advanced materials into 3D functional structures by methods inspired by origami and kirigami: a review. *Adv. Mater. Interfaces* 5 (13): 1800284.

5 Xu, C., Wu, X., Huang, G., and Mei, Y. (2019). Rolled-up nanotechnology: materials issue and geometry capability. *Adv. Mater. Technol.* 4 (1): 1800486.

6 Rogers, J., Huang, Y., Schmidt, O.G., and Gracias, D.H. (2016). Origami MEMS and NEMS. *MRS Bull.* 41 (2): 123–129.

7 Huang, G. and Mei, Y. (2018). Assembly and self-assembly of nanomembrane materials-from 2D to 3D. *Small* 14 (14): 1703665.

8 Shulaker, M.M., Hills, G., Park, R.S. et al. (2017). Three-dimensional integration of nanotechnologies for computing and data storage on a single chip. *Nature* 547 (7661): 74.

9 Sharma, R., Bufon, C.C.B., Grimm, D. et al. (2014). Large-area rolled-up nanomembrane capacitor arrays for electrostatic energy storage. *Adv. Energy Mater.* 4 (9): 1301631.

10 Li, J., de Avila, B.E.-F., Gao, W. et al. (2017). Micro/nanorobots for biomedicine: delivery, surgery, sensing, and detoxification. *Sci. Rob.* 2 (4): eaam6431.

11 Li, X. (2011). Self-rolled-up microtube ring resonators: a review of geometrical and resonant properties. *Adv. Opt. Photonics* 3 (4): 366–387.

12 Randhawa, J.S., Gurbani, S.S., Keung, M.D. et al. (2010). Three-dimensional surface current loops in terahertz responsive microarrays. *Appl. Phys. Lett.* 96 (19): 191108.

13 Waldrop, M.M. (2016). The chips are down for Moore's law. *Nature* 530: 144–147.

14 Lee, B.-H., Kang, M.-H., Ahn, D.-C. et al. (2015). Vertically integrated multiple nanowire field effect transistor. *Nano Lett.* 15 (12): 8056–8061.

15 Gao, Y. (2016). The 'more Moore' and 'more than Moore' in microelectronics technology and IC industry. *Natl. Sci. Rev.* 3 (2): 158–158.

16 De Marchi, M., Sacchetto, D., Zhang, J. et al. (2014). Top-down fabrication of gate-all-around vertically stacked silicon nanowire FETs with controllable polarity. *IEEE Trans. Nanotechnol.* 13 (6): 1029–1038.

17 Kirihata, T., Golz, J., Wordeman, M. et al. (2016). Three-dimensional dynamic random access memories using through-silicon-vias. *IEEE J. Emerging Select. Topic Circuit Syst.* 6 (3): 373–384.

18 Cho, J.-H., Keung, M.D., Verellen, N. et al. (2011). Nanoscale origami for 3D optics. *Small* 7 (14): 1943–1948.

19 Yan, Z., Zhang, F., Liu, F. et al. (2016). Mechanical assembly of complex, 3D mesostructures from releasable multilayers of advanced materials. *Sci. Adv.* 2 (9): e1601014.

20 Zhang, F., Liu, F., and Zhang, Y. (2019). Analyses of mechanically-assembled 3D spiral mesostructures with applications as tunable inductors. *Sci. China-Technol. Sci.* 62 (2): 243–251.

21 Han, M., Wang, H., Yang, Y. et al. (2019). Three-dimensional piezoelectric polymer microsystems for vibrational energy harvesting, robotic interfaces and biomedical implants. *Nat. Electron.* 2 (1): 26–35.

22 Zhang, Y., Zhang, F., Yan, Z. et al. (2017). Printing, folding and assembly methods for forming 3D mesostructures in advanced materials. *Nat. Rev. Mater.* 2 (4): 17019.
23 Cheng, X. and Zhang, Y. (2019). Micro/nanoscale 3D assembly by rolling, folding, curving, and buckling approaches. *Adv. Mater.* 31 (36): e1901895.
24 Guo, X., Xue, Z., and Zhang, Y. (2019). Manufacturing of 3D multifunctional microelectronic devices: challenges and opportunities. *NPG Asia Mater.* 11: 29.
25 Schmidt, O.G. and Jin-Phillipp, N.Y. (2001). Free-standing SiGe-based nanopipelines on Si (001) substrates. *Appl. Phys. Lett.* 78 (21): 3310–3312.
26 Prinz, V.Y. (2003). A new concept in fabricating building blocks for nanoelectronic and nanomechanic devices. *Microelectron. Eng.* 69 (2–4): 466–475.
27 Mei, Y., Huang, G., Solovev, A.A. et al. (2008). Versatile approach for integrative and functionalized tubes by strain engineering of nanomembranes on polymers. *Adv. Mater.* 20 (21): 4085–4090.
28 Cui, A., Liu, Z., Li, J. et al. (2015). Directly patterned substrate-free plasmonic "nanograter" structures with unusual Fano resonances. *Light-Sci. Appl.* 4: e308.
29 Mao, Y., Pan, Y., Zhang, W. et al. (2016). Multi-direction-tunable three-dimensional meta-atoms for reversible switching between midwave and long-wave infrared regimes. *Nano Lett.* 16 (11): 7025–7029.
30 Legrain, A., Janson, T.G., Berenschot, J.W. et al. (2014). Controllable elastocapillary folding of three-dimensional micro-objects by through-wafer filling. *J. Appl. Phys.* 115 (21): 214905.
31 Cho, J.-H., Azam, A., and Gracias, D.H. (2010). Three dimensional nanofabrication using surface forces. *Langmuir* 26 (21): 16534–16539.
32 Py, C., Reverdy, P., Doppler, L. et al. (2007). Capillary origami: spontaneous wrapping of a droplet with an elastic sheet. *Phys. Rev. Lett.* 98 (15): 156103.
33 Iwase, E. and Shimoyama, I. (2005). Multistep sequential batch assembly of three-dimensional ferromagnetic microstructures with elastic hinges. *J. Microelectromech. Syst.* 14 (6): 1265–1271.
34 In, H.J., Lee, H., Nichol, A.J. et al. (2008). Carbon nanotube-based magnetic actuation of origami membranes. *J. Vac. Sci. Technol., B* 26 (6): 2509–2512.
35 Kuribayashi-Shigetomi, K., Onoe, H., and Takeuchi, S. (2012). Cell origami: self-folding of three-dimensional cell-laden microstructures driven by cell traction force. *PLoS One* 7 (12): e51085.
36 Pandey, S., Ewing, M., Kunas, A. et al. (2011). Algorithmic design of self-folding polyhedra. *Proc. Natl. Acad. Sci. U.S.A.* 108 (50): 19885–19890.
37 Huang, W., Koric, S., Yu, X. et al. (2014). Precision structural engineering of self-rolled-up 3D nanomembranes guided by transient quasi-static FEM modeling. *Nano Lett.* 14 (11): 6293–6297.
38 Choi, C., Choi, M.K., Liu, S. et al. (2017). Human eye-inspired soft optoelectronic device using high-density MoS_2-graphene curved image sensor array. *Nat. Commun.* 8: 1664.
39 Kim, D.-H., Lu, N., Ghaffari, R. et al. (2011). Materials for multifunctional balloon catheters with capabilities in cardiac electrophysiological mapping and ablation therapy. *Nat. Mater.* 10 (4): 316–323.

40 Song, Y.M., Xie, Y., Malyarchuk, V. et al. (2013). Digital cameras with designs inspired by the arthropod eye. *Nature* 497 (7447): 95–99.

41 Xue, Z., Song, H., Rogers, J.A. et al. (2019). Mechanically-guided structural designs in stretchable inorganic electronics. *Adv. Mater.* 32: e1902254.

42 Ko, H.C., Shin, G., Wang, S. et al. (2009). Curvilinear electronics formed using silicon membrane circuits and elastomeric transfer elements. *Small* 5 (23): 2703–2709.

43 Ko, H.C., Stoykovich, M.P., Song, J. et al. (2008). A hemispherical electronic eye camera based on compressible silicon optoelectronics. *Nature* 454 (7205): 748–753.

44 Saada, G., Layani, M., Chernevousky, A., and Magdassi, S. (2017). Hydroprinting conductive patterns onto 3D structures. *Adv. Mater. Technol.* 2 (5): 1600289.

45 Le Borgne, B., De Sagazan, O., Crand, S. et al. (2017). Conformal electronics wrapped around daily life objects using an original method: water transfer printing. *ACS Appl. Mater. Interfaces* 9 (35): 29424–29429.

46 Xu, X., Davanco, M., Qi, X., and Forrest, S.R. (2008). Direct transfer patterning on three dimensionally deformed surfaces at micrometer resolutions and its application to hemispherical focal plane detector arrays. *Org. Electron.* 9 (6): 1122–1127.

47 Fan, D., Lee, B., Coburn, C., and Forrest, S.R. (2019). From 2D to 3D: strain- and elongation-free topological transformations of optoelectronic circuits. *Proc. Natl. Acad. Sci. U.S.A.* 116 (10): 3968–3973.

48 Park, H., Cho, H., Kim, J. et al. (2014). Multiscale transfer printing into recessed microwells and on curved surfaces via hierarchical perfluoropolyether stamps. *Small* 10 (1): 52–59.

49 Xu, S., Yan, Z., Jang, K.-I. et al. (2015). Assembly of micro/nanomaterials into complex, three-dimensional architectures by compressive buckling. *Science* 347 (6218): 154–159.

50 Yan, Z., Han, M., Yang, Y. et al. (2017). Deterministic assembly of 3D mesostructures in advanced materials via compressive buckling: a short review of recent progress. *Extreme Mech. Lett.* 11: 96–104.

51 Yan, Z., Zhang, F., Wang, J. et al. (2016). Controlled mechanical buckling for origami-inspired construction of 3d microstructures in advanced materials. *Adv. Funct. Mater.* 26 (16): 2629–2639.

52 Zhang, Y., Yan, Z., Nan, K. et al. (2015). A mechanically driven form of kirigami as a route to 3D mesostructures in micro/nanomembranes. *Proc. Natl. Acad. Sci. U.S.A.* 112 (38): 11757–11764.

53 Luan, H., Cheng, X., Wang, A. et al. (2019). Design and fabrication of heterogeneous, deformable substrates for the mechanically guided 3D assembly. *ACS Appl. Mater. Interfaces* 11 (3): 3482–3492.

54 Nan, K., Luan, H., Yan, Z. et al. (2017). Engineered elastomer substrates for guided assembly of complex 3D mesostructures by spatially nonuniform compressive buckling. *Adv. Funct. Mater.* 27 (1): 1604281.

55 Zhao, H., Li, K., Han, M. et al. (2019). Buckling and twisting of advanced materials into morphable 3D mesostructures. *Proc. Natl. Acad. Sci. U.S.A.* 116: 201901193.

56 Fu, H., Nan, K., Bai, W. et al. (2018). Morphable 3D mesostructures and microelectronic devices by multistable buckling mechanics. *Nat. Mater.* 17 (3): 268–276.

57 Guo, X., Wang, X., Ou, D. et al. (2018). Controlled mechanical assembly of complex 3D mesostructures and strain sensors by tensile buckling. *NPJ Flexible Electron.* 2 (1): 14.

58 Kim, B.H., Lee, J., Won, S.M. et al. (2018). Three-dimensional silicon electronic systems fabricated by compressive buckling process. *ACS Nano* 12 (5): 4164–4171.

59 Kim, B.H., Liu, F., Yu, Y. et al. (2018). Mechanically guided post-assembly of 3D electronic systems. *Adv. Funct. Mater.* 28 (48): 1803149.

60 Yan, Z., Han, M., Shi, Y. et al. (2017). Three-dimensional mesostructures as high-temperature growth templates, electronic cellular scaffolds, and self-propelled microrobots. *Proc. Natl. Acad. Sci. U.S.A.* 114 (45): E9455–E9464.

61 Ning, X., Yu, X., Wang, H. et al. (2018). Mechanically active materials in three-dimensional mesostructures. *Sci. Adv.* 4 (9): eaat8313.

62 Nan, K., Wang, H., Ning, X. et al. (2019). Soft three-dimensional microscale vibratory platforms for characterization of nano-thin polymer films. *ACS Nano* 13 (1): 449–457.

63 Fu, H., Nan, K., Froeter, P. et al. (2017). Mechanically-guided deterministic assembly of 3D mesostructures assisted by residual stresses. *Small* 13: 1700151.

64 McCracken, J.M., Xu, S., Badea, A. et al. (2017). Deterministic integration of biological and soft materials onto 3D microscale cellular frameworks. *Adv. Biosyst.* 1 (9): 1700068.

65 Jang, K.-I., Li, K., Chung, H.U. et al. (2017). Self-assembled three dimensional network designs for soft electronics. *Nat. Commun.* 8: 15894.

66 Nan, K., Kang, S.D., Li, K. et al. (2018). Compliant and stretchable thermoelectric coils for energy harvesting in miniature flexible devices. *Sci. Adv.* 4 (11): eaau5849.

67 Yang, C., Huang, Y., Cheng, H. et al. (2019). Rollable, stretchable, and reconfigurable graphene hygroelectric generators. *Adv. Mater.* 31 (2): 1805705.

68 Lee, W., Liu, Y., Lee, Y. et al. (2018). Two-dimensional materials in functional three-dimensional architectures with applications in photodetection and imaging. *Nat. Commun.* 9: 1417.

69 Shi, Y., Zhang, F., Nan, K. et al. (2017). Plasticity-induced origami for assembly of three dimensional metallic structures guided by compressive buckling. *Extreme Mech. Lett.* 11: 105–110.

70 Zhang, F., Fan, Z., and Zhang, Y. (2018). A theoretical model of postbuckling in straight ribbons with engineered thickness distributions for three-dimensional assembly. *Int. J. Solids Struct.* 147: 254–271.

71 Luo, G., Fu, H., Cheng, X. et al. (2019). Mechanics of bistable cross-shaped structures through loading-path controlled 3D assembly. *J. Mech. Phys. Solids* 129: 261–277.

72 Humood, M., Shi, Y., Han, M. et al. (2018). Fabrication and deformation of 3D multilayered kirigami microstructures. *Small* 14: 1703852.

73 Liu, Y., Yan, Z., Lin, Q. et al. (2016). Guided formation of 3D helical mesostructures by mechanical buckling: analytical modeling and experimental validation. *Adv. Funct. Mater.* 26 (17): 2909–2918.

74 Liu, Y., Xu, Z., Hwang, K.C. et al. (2019). Postbuckling analyses of frame mesostructures consisting of straight ribbons for mechanically guided three-dimensional assembly. *Proc. Math. Phys. Eng. Sci.* 475 (2225): 20190012.

75 Fan, Z., Hwang, K.-C., Rogers, J.A. et al. (2018). A double perturbation method of postbuckling analysis in 2D curved beams for assembly of 3D ribbon-shaped structures. *J. Mech. Phys. Solids* 111: 215–238.

76 Ling, Y., Zhuang, X., Xu, Z. et al. (2018). Mechanically assembled, three-dimensional hierarchical structures of cellular graphene with programmed geometries and outstanding electromechanical properties. *ACS Nano* 12 (12): 12456–12463.

77 Wang, X., Guo, X., Ye, J. et al. (2019). Freestanding 3D mesostructures, functional devices, and shape-programmable systems based on mechanically induced assembly with shape memory polymers. *Adv. Mater.* 31 (2): 1805615.

78 Park, Y., Luan, H., Kwon, K. et al. (2019). Transformable, freestanding 3D mesostructures based on transient materials and mechanical interlocking. *Adv. Funct. Mater.* 29: 1903181.

79 Liu, F., Chen, Y., Song, H. et al. (2019). High performance, tunable electrically small antennas through mechanically guided 3D assembly. *Small* 15 (1): 1804055.

80 Liu, F., Cheng, X., Zhang, F. et al. (2019). Design and assembly of reconfigurable 3D radio-frequency antennas based on mechanically triggered switches. *Adv. Electron. Mater.* 5 (6): 1900256.

81 Won, S.M., Wang, H., Kim, B.H. et al. (2019). Multimodal sensing with a three-dimensional piezoresistive structure. *ACS Nano* 13: 10972.

82 Guo, X., Xu, Z., Zhang, F. et al. (2018). Reprogrammable 3D mesostructures through compressive buckling of thin films with prestrained shape memory polymer. *Acta Mech. Solida Sin.* 31 (5): 589–598.

83 Shi, Y., Pei, P., Cheng, X. et al. (2018). An analytic model of two-level compressive buckling with applications in the assembly of free-standing 3D mesostructures. *Soft Matter* 14 (43): 8828–8837.

84 Wang, H., Ning, X., Li, H. et al. (2018). Vibration of mechanically-assembled 3D microstructures formed by compressive buckling. *J. Mech. Phys. Solids* 112: 187–208.

85 Li, H., Wang, X., Zhu, F. et al. (2018). Viscoelastic characteristics of mechanically assembled three-dimensional structures formed by compressive buckling. *J. Appl. Mech.* 85 (12): 121002.

86 Ning, X., Wang, H., Yu, X. et al. (2017). 3D tunable, multiscale, and multistable vibrational micro-platforms assembled by compressive buckling. *Adv. Funct. Mater.* 27: 1605914.

87 Li, K., Cheng, X., Zhu, F. et al. (2019). A generic soft encapsulation strategy for stretchable electronics. *Adv. Funct. Mater.* 29 (8): 1806630.

88 Li, S., Han, M., Rogers, J.A. et al. (2019). Mechanics of buckled serpentine structures formed via mechanics-guided, deterministic three-dimensional assembly. *J. Mech. Phys. Solids* 125: 736–748.

89 Liu, Y., Wang, X., Xu, Y. et al. (2019). Harnessing the interface mechanics of hard films and soft substrates for 3D assembly by controlled buckling. *Proc. Natl. Acad. Sci. U.S.A.* 116 (31): 15368–15377.

90 Xue, R., Li, R., Du, Z. et al. (2017). Kirigami pattern design of mechanically driven formation of complex 3D structures through topology optimization. *Extreme Mech. Lett.* 15: 139–144.

91 Xu, Z., Fan, Z., Fu, H. et al. (2019). Optimization-based approach for the inverse design of ribbon-shaped three-dimensional structures assembled through compressive buckling. *Phys. Rev. Appl.* 11 (5): 054053.

2

Design and Realization of Transient Electronics Enabled by Nanomembranes

Chunyu You, Bofan Hu, Ziyu Zhang, Borui Xu, Gaoshan Huang, and Yongfeng Mei

Fudan University, Department of Materials Science, 220 Handan Road, Shanghai 200433, People's Republic of China

2.1 Introduction

Transient electronics refer to a class of electronics that can dissolve, disintegrate, or disappear physically or chemically in a relatively short period of time [1, 2]. Unlike conventional electronics, where reliability simply means the longest possible lifetime, transient electronics have a designed life and failure process and are therefore suitable for many novel applications [3, 4]. For example, self-disappearing biomedical implants could save much time to take them out after the temporally functional period for medical staff and relieve pain for patients [5–7]. And hardware-level chip security could be welcomed by military or business applications for the process of confidential electronic document elimination is eased [8, 9]. Moreover, transient electronics enable a possibility of disposable electronics [10, 11], aiming at the difficulties in the recycling of current integrated circuits, which is becoming a heavy burden on our earth.

Combined uses of natural or synthesized biodegradable polymers with non-transient electronic materials were examples of early research direction of water-soluble electronic devices, which allowed to be partially disintegrated into fragmentation due to collapse of the substrates [2]. The study was limited to a passive resolving mode [10, 12, 13], where the self-destruction process could not be modulated precisely and started timely.

In addition, like the case in modern flexible electronics, a much higher carrier mobility and robust mechanical property are still essential for their out-of-lab application [14–16], which, considering the current situation, still have a long way to go.

The first inorganic transient electronic device was pioneered by Rogers and coworkers in 2009 [17], featuring a silicon nanomembrane (SiNM) device on a silk substrate. Three years later, a milestone of transient electronics, which indeed drew much attention and interest, was presented by the same group [18]. Figure 2.1a,b shows images and schematic diagrams of the prototype device containing nearly all the components needed in modern electronics, including passive components,

Nanomembranes: Materials, Properties, and Applications, First Edition.
Edited by Yongfeng Mei, Gaoshan Huang, and Xiuling Li.
© 2022 WILEY-VCH GmbH. Published 2022 by WILEY-VCH GmbH.

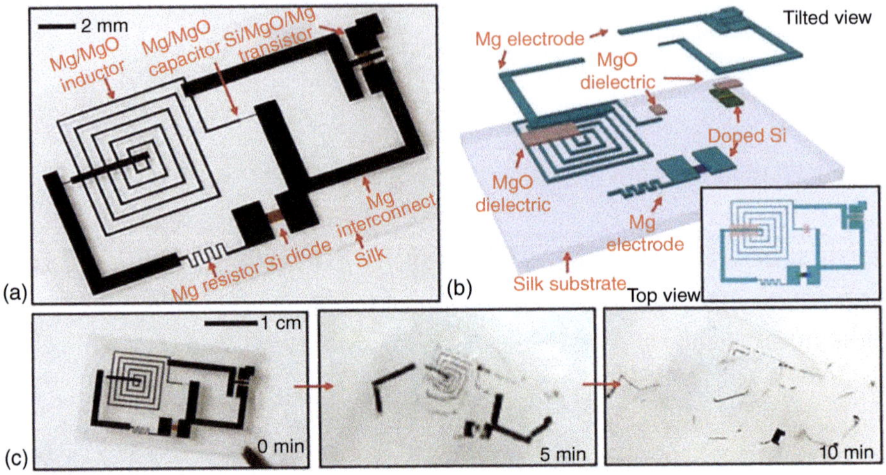

Figure 2.1 Typical example of a passive transient electronic device based on nanomembranes. (a) Image of a device that includes various components all on a thin silk substrate. (b) Schematics of the device. The inset shows the top view. (c) Optical images showing dissolution of the device in DIW. Source: Reprinted from Hwang et al. [18]. Copyright 2012, The American Association for the Advancement of Science.

such as inductors, capacitors, and resistors, as well as active components, such as diodes and transistors. Once it contacts deionized water (DIW) the whole device starts to disintegrate and dissolve (Figure 2.1c).

The key factor in the novel form of electronics above is the utilizing of nanomembranes. Compared to their bulk alternative, the dissolution and reaction are speeded up, owing to their high surface-to-volume ratio [4, 19]. The ultrathin character also makes nanomembranes vulnerable to break, if an out-of-plane force is applied properly [20]. In the case shown above, the former factor plays the role. Another important point is that the nanomembrane form of materials is much easier to transfer and combine with those organic-based dissolvable substrates, compared to other novel forms of materials, e.g. nanowire or carbon nanotube [21–23].

Since the first creation of nanomembrane-based transient electronics, great efforts have been taken to realize many brilliant conceives of this new form of electronics. More useful devices, e.g. *in vivo* pressure sensors [3, 6, 7, 24], antennas [24–29], and optical waveguides [30], have been demonstrated to broaden the family, signifying the great potential. To achieve a desired transient property, materials selection, active and passive device design, circuit layout, and packaging process should all be taken into consideration as well as the triggering methods and dysfunction strategies.

In this chapter, we will first explain the material characters of the nanomembranes for transient electronics, covering the basic electric performance, degradable properties as well as their biocompatibility. Then the fabrication methods for this novel type of electronics and the pros and cons of different processes will be presented. After that, as the core of transient electronics, the typical dysfunction mechanism compatible with nanomembrane setup will be covered. Since the on-demand destruction property is preferred in many ways and has its special significance, various triggering methods and designs will be discussed. Last but not least, we will talk about the opportunities and challenges in the field, and a brief outlook of the development of transient electronics will be demonstrated.

2.2 Material Selection for Transient Electronics

All electronics consist of three different kinds of materials, which are conductors, semiconductors, and dielectrics. Turning the materials to nanomembrane form largely increases the surface-to-volume and reduces the amount of the substance, thus making the dissolution of materials begin to take hold. In modern integrated circuit industry, the conductors and dielectrics are readily obtained in their nanomembrane form. While for the cases of semiconductors, only the surface part is functional, so the transformation of bulk counterparts to nanomembranes does not overdeteriorate the electronic properties. The main factors to be considered are dissolution rates in different environments, process compatibility with integrated circuits, and biocompatibility in applications such as medical implantations

2.2.1 Conductors

Conducting materials are indispensable in constructing complete circuits for their ability to connect electronic components. Among various approaches, metals are prominent due to their high thermal stability and low resistivity. Several metals, such as magnesium (Mg), zinc (Zn), iron (Fe), molybdenum (Mo), and tungsten (W) in their nanomembrane forms, have been widely investigated and can be particularly appealing options in transient electronics because of their relatively rapid rates of dissolution and biocompatibility [31, 32]. For cases of metal nanomembranes, due to the smaller change of width or length than that of thickness during dissolution, normally it is easier to measure the electrical properties than thicknesses. The effective thickness h can be obtained by converting the electrical resistance R with equation $R \times h = R_0 \times h_0$, where R_0 and h_0 refer to the initial resistance and thickness, respectively. The electrical dissolution rate (EDR) is therefore used to describe the dissolution velocity to obtain quantitative analysis of experimental observations, which is defined by the equation: $v_{\mathrm{EDR}} = -\frac{\partial h}{\partial t}$.

Hwang et al. introduced a presentative example of transient electronic devices, which can completely disintegrate and dissolve in DIW [18]. In this case, Mg was applied to fabricate interconnects and electrodes. Mg and Zn are among the most extensively used metals in transient electronics. Research had illustrated that the hydrolysis kinetics of Mg is similar to that of Zn, which mainly accounts for their dissolution behavior [32]. The chemical reaction $X + 2H_2O \rightarrow X(OH)_2 + H_2$, where X is for Mg and Zn, happens during their dissolution.

Dissolution manners of Mg and Zn traces with 300 nm thickness, as a function of immersion time in DIW and Hanks' solution in different conditions, are shown in Figure 2.2a,d, respectively. Metal nanomembranes are fabricated into serpentine traces with two top probing pads at the ends of line. Considerably faster dissolution

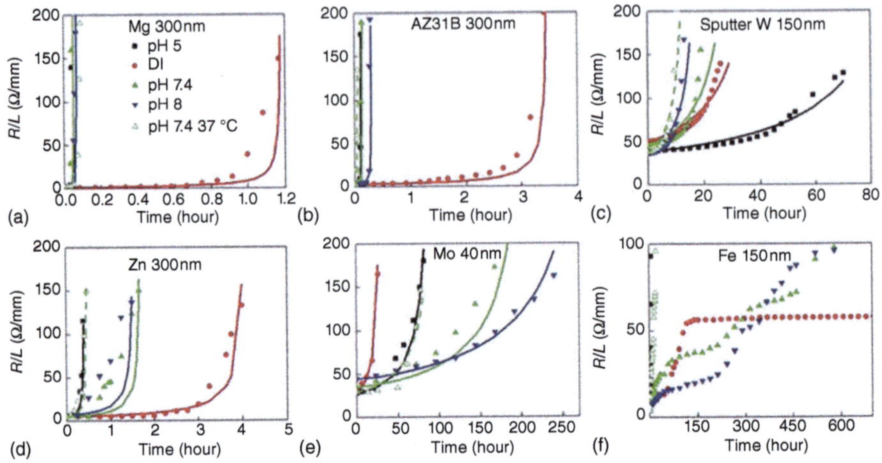

Figure 2.2 The variation of resistance of different metal nanomembranes in the form of serpentine traces with time immersed in DIW and Hanks' solution with different temperatures or pH. (a) Mg with thickness of 300 nm; (b) Mg alloy (AZ31B, containing 3 wt% Al and 1 wt% Zn) with thickness of 300 nm; (c) W formed by sputter deposition with thickness of 150 nm; (d) Zn with thickness of 300 nm; (e) Mo with thickness of 40 nm; (f) Fe with thickness of 150 nm; (a–f) share the same legend. Source: Adapted from Yin et al. [32]. Copyright 2013, Wiley-VCH Verlag GmbH & Co. KGaA, Weinheim.

behavior can be observed in Hanks' solution, which contains inorganic salts and glucose. According to the related studies on the corrosion of bulk Mg and Zn, chlorides (Cl^-) that exist in Hanks' solution can promote rapid chemical attack, which accounts for the accelerated dissolution [33, 34] while temperature and pH levels do not exert significant influence in EDRs for Mg and Zn.

As for Mo and W, the dissolution of which involves the presence of oxygen, their hydrolysis reaction can be described as follows, $2X + 2H_2O + 3O_2 \rightarrow 2H_2XO_4$, where X is for Mo and W. Mo dissolves considerably slower in Hanks' solution than in DIW, as indicated in Figure 2.2e. The descending oxygen solubility in solution accounts for this phenomenon. Ions in aqueous solution, such as Na^+ and Cl^-, can reduce the oxygen solubility, leading to decreased concentration of oxygen in Hanks' solution than in DIW [35]. Figure 2.2c shows that W exhibits higher dissolution rate in acidic salt solution than in basic solution, which is consistent with the related study on the corrosion of bulk W [36]. The dissolution manner of W is more sensitive to fabrication process than other metals. A lower EDR can be found in chemical vapor deposited (CVD) W than in sputtered W [32].

Due to the presence of oxygen, surface oxides form on the nanomembranes before immersion in aqueous solution. A two-stage dissolution process has been proposed. The metal degrades rapidly in the first stage and subsequently the residue oxides dissolve relatively slowly [32]. Take Mg as an example;$Mg(OH)_2$ accounts for a large part of surface during dissolution, with a low level of MgO. Mg nanomembrane serpentine traces tested (300 nm thick) are found to be electrically discontinuous within 1.5 hours in DIW. However, there are still some remaining base metals. Over time, they will react into residual $Mg(OH)_2$ and MgO, which thoroughly dissolve within 1–4 days because of their relatively high solubility [32].

Figure 2.2f suggests that the EDR for Fe is relatively high in an acidic solution or at elevated temperature, while lower in other cases. This arises from extremely low solubility of Fe_2O_3 and Fe_3O_4, the hydrolysis products of Fe, which form on the surfaces, leading to irregular and inhomogeneous dissolution [32]. A plateau can be found in the variation of resistance of Fe with time immersed in DIW, meaning that the dissolution apparently suspends after 120 hours.

2.2.2 Semiconductors

Semiconductor is another main type of materials applied in electronics [37–39]. It plays the most critical role in various scenarios, such as integrated circuits, communication systems, photovoltaic power generation, and other fields. Single crystalline silicon has been extensively used as semiconductors in traditional integrated circuit technologies. The solubility of single crystalline silicon establishes the foundation for high-performance nanomembrane-based transient electronics. Hydrolysis of Si NM mainly accounts for the dissolution of Si NM in aqueous solutions at near-neutral conditions [39]. Factors such as crystal structure and morphology of silicon, the properties of aqueous solution, temperature, and doping condition can considerably exert influence on the dissolution behavior of Si NMs. Related studies had illustrated that Si NM reacts into silicic acid $Si(OH)_4$, with oxidizing into intermediate product SiO_2 in some cases, in phosphate-buffered saline (PBS, pH = 7.4) at physiological temperatures (37 °C) [40, 41].

A linear relationship between thickness of Si NM and time immersed in the solution can be observed in Figure 2.3a, which arises from the hydrolysis that only happens on the surface of Si NM, with no significant contribution of reactive diffusion processes. Increased dissolution rate with elevated pH appears in Figure 2.3b, arising from the increase of the concentration of OH^-. Previous study had revealed that the dissolution rate (R) of silicon in alkaline solution with high pH, described as $R = \frac{\partial h}{\partial t}$ where h refers to the thickness of silicon, depends on the concentrations of H_2O and OH^- according to the formula $R = k_0[H_2O]^4[OH^-]^{0.25}e^{-\frac{E_A}{k_B T}}$, where k_0, E_A, k_B, and T are the reaction constant, the activation energy, the Boltzmann constant, and the Kelvin temperature, respectively [41]. For silicon dissolution in potassium hydroxide (KOH) solution with high pH, the reaction constant k_0 is 2480 μm/h × (mol/l)$^{-4.25}$ and 4500 μm/h × (mol/l)$^{-4.25}$ for <100> surface and <110> surface, respectively [41]. By changing the power-law exponent for OH^-, the formula can be applied in the cases of dissolution in solutions with lower pH [19].

Based on simulations with density functional theory (DFT), Figure 2.3c represents a schematic illustration of the adsorption process for ions (OH^-, HPO_4^{2-}, Cl^-, and K^+) on the silicon surfaces. These ions can assist in dissolution of silicon. Two ions first attack the silicon surface. Nucleophilic ions bonded to the silicon surface can elongate the silicon interior bonds. The strength of interior bonds between silicon surface atoms is then weakened, being inducible to be broken when there is a third ion.

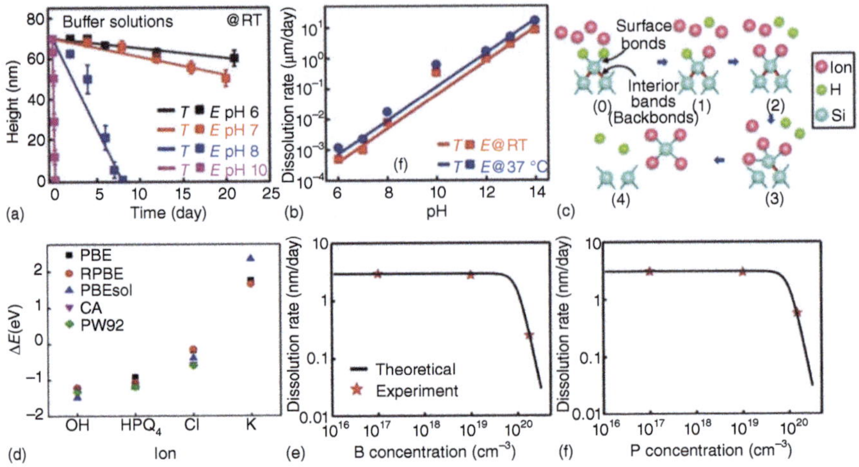

Figure 2.3 Analysis of dissolution behavior of Si NMs. (a) The theoretical (*T*, lines) and experimental (*E*, symbols) dissolution manner of Si NMs at room temperature in buffer solutions with different pH (black: 6; red: 7; blue: 8; purple: 10). A linear relationship can be found. (b) The variation of dissolution rates of Si NM in buffer solutions with pH at room temperature (red) and 37 °C (blue). The dissolution rates increase with higher pH and temperature. Source: (a, b) Adapted from Hwang et al. [19]. Copyright 2014, Wiley-VCH Verlag GmbH & Co. KGaA, Weinheim. Schematic illustration of dissolution of Si NM based on density function theory (DFT). (c) Atomic configurations for ion adsorption; (d) The calculated energy change for various ions based on exchange-correlation (XC) functions. The results suggest that OH^- inclines to form bonding with Si more than HPO_4^{2-} and Cl^-. And Si prefers to form bonding with H than K. Source: (c, d) Adapted from Yin et al. [42]. Copyright 2015, Wiley-VCH Verlag GmbH & Co. KGaA, Weinheim. The calculated (black) and experimental (red) variation of dissolution rate in buffer solution of (e) boron-doped and (f) phosphorus-doped Si NM with different doping levels. Lower rate of dissolution can be found when doping concentration exceeds a certain level. Source: (e, f) Adapted from Hwang et al. [43]. Copyright 2014, American Chemical Society.

Figure 2.3d shows the total change in energy for each ion ΔE_{Ion} ($\Delta E_{\text{Ion}} = E_{\text{Bonded}} - E_{\text{Nonbonded}}$, where E_{Bonded} and $E_{\text{Nonbonded}}$ are the energies of the system when an ion is bonded and nonbonded, respectively) computed using different exchange-correlation (XC) functionals. The results suggest that OH^- inclines to form bonding with Si more than HPO_4^{2-} and Cl^-, and Si prefers to form bonding with H than K.

As mentioned before, the doping condition of the Si NMs can be important. To investigate the influence of dopants, Si NMs are doped with boron at three different levels (10^{17} cm^{-3}, 10^{19} cm^{-3}, 10^{20} cm^{-3}). Significantly lower rate of dissolution occurs in Figure 2.3e,f when dopant concentration exceeds a certain level (10^{20} cm^{-3} in this case). Previous studies on silicon etching in alkaline solutions indicated that the dissolution rate remains unchanged (R_i) until the dopant concentration exceeds a critical level (C_0), and then the rate decreases dramatically [44]. The relationship between dissolution rate ($R_{\text{dissolution}}$) and dopant concentration (C) can be described as $R_{\text{dissolution}} = R_i/[1+(C/C_0)^4]$ [44]. Similarly, phosphorus-doped Si NM exhibits lower dissolution rate as the dopant concentration increases. However, the reduction of dissolution rate in phosphorus-doped Si NM is smaller than that of boron-doped one, which can be attributed to the absence of electrons in the conduction band at high boron concentration [44].

Biocompatibility is another distinguishing feature of Si NM, which makes it one of the best candidates for bioresorbable electronics. The *in vitro* cytotoxicity studies on Si NM have been carried out [43]. An array of Si NMs was made as culture substrate, on which metastatic breast cancer cells were cultured. Then the cell viability as a function of time was examined. The results revealed that no measurable toxicity could be found after the dissolution of Si NMs. *In vivo* studies are performed as well; a transient device including Si NM was implanted in the subdermal dorsal region of mice [43]. The studies suggested the Si NMs to be biocompatible and their potential application in long-term implantation.

Apart from single crystalline silicon, other semiconducting materials, such as polycrystalline silicon (poly-Si), amorphous silicon (a-Si), germanium (Ge), and alloys of silicon and germanium (SiGe), are of great interest and value as well due to their extensive use in electronic technologies.[38] In PBS with pH 7.4 at physiological temperatures (37 °C), the dissolution rates of poly-Si, a-Si, Ge, and SiGe are 2.8, 4.1, 3.1, and 0.1 nm/day, respectively [38]. Poly-Si and a-Si share qualitatively similar dissolution behaviors with single crystalline silicon. Compared to poly-Si (0.524 eV), a-Si has lower activation energy, E_A (0.518 eV), and lower density, which contribute to higher rate of dissolution for a-Si than that of poly-Si [38]. The bandgap of Ge is smaller than that of Si, while the minority-carrier mobility of Ge is larger than that of Si. The dissolution of Ge in aqueous solutions can be described as $Ge + O_2 + H_2O \rightarrow H_2GeO_3$. Stemming from band-bending effect, SiGe alloy has a large activation energy (0.76 eV), leading to a comparatively low rate of dissolution [45].

Zinc oxide (ZnO) is another appealing candidate because of its direct bandgap (3.37 eV) with large exciton-binding energy (60 meV) and piezoelectric properties. Prior studies found out that different thicknesses of ZnO and volume of the aqueous solutions lead to distinct hydrolysis reactions [46, 47]. Typically, the chemical reaction applied in dissolution of ZnO nanomembranes can be described as $ZnO + H_2O \rightarrow Zn^{2+} + 2OH^-$. While in the cases of bulk ZnO, the hydrolysis product is moderately soluble hydroxide Zn $(OH)_2$, and the reaction can be described as $ZnO + H_2O \rightarrow Zn(OH)_2$. It was reported that the dissolution rate sputtered ZnO is ~12 nm/h in DIW at room temperature [47]. A transient thin-film transistor (TFT) with ZnO (200 nm thick) as active layer has been fabricated, being stable up to 3 hours, and degrades within 45 minutes in DIW. The electrical properties of the fabricated transient devices are similar to those of non-transient ones [47].

2.2.3 Dielectrics

Dielectrics are the last building block of electronics. For modern scenarios where the working voltage is usually less than 2 V, dielectrics with several nanometer thicknesses could adequately avoid electrical breakdown and current leakage. Silicon dioxide (SiO_2) and silicon nitride (Si_3N_4) have long been used as interlayers,

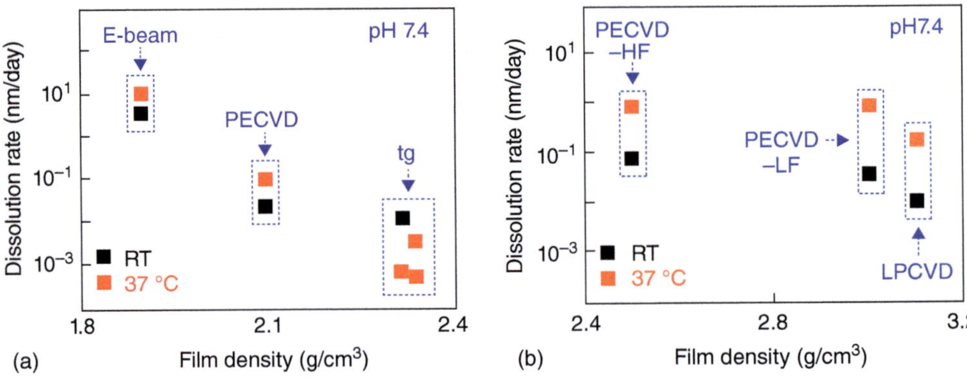

Figure 2.4 The variation of dissolution manners for different types of (a) silicon dioxides and (b) silicon nitrides with film density in buffer solution (pH 7.4) at room temperature (black) and 37 °C (red). Source: Adapted from Kang et al. [48]. Copyright 2014, Wiley-VCH Verlag GmbH & Co. KGaA, Weinheim.

gate dielectrics, and passivation coating in commercial electronics. Excellent electric properties and ability of dissolution in aqueous solutions make them stand out and related studies revealed their promising applications as insulation, passivation, and encapsulation in transient electronics [48].

It was noticed that SiO_2 dissolves to form silicic acid Si $(OH)_4$ with mechanism similar to that of silicon. The hydrolysis reaction can be described as $SiO_2 + 2H_2O \rightarrow Si(OH)_4$. The hydrolysis of Si_3N_4 on the other hand involves two steps, one of which is the oxidation to the SiO_2 ($Si_3N_4 + 6H_2O \rightarrow 3SiO_2 + 4NH_3$), followed by the hydrolysis of silicon oxide. The overall reaction can be described as $Si_3N_4 + 12H_2O \rightarrow 3Si(OH)_4 + 4NH_3$ [48].

Fabrication process can considerably affect the dissolution behavior of SiO_2 and Si_3N_4. To examine the effects, three different types of SiO_2 are prepared: (i) tg-oxide (thermally grown oxide using dry gas [O_2] and wet vapor [H_2O]), (ii) PECVD (plasma-enhanced chemical vapor deposited) oxide, and (iii) E-beam oxide (electron-beam evaporation). Figure 2.4a suggests that E-beam oxide possesses the fastest dissolution rate, while tg-oxide exhibits the lowest one. This stems from the difference in density and stoichiometry of the oxidation products [48].

Density can exert influence on diffusion rates and reactivity of dielectrics, consequently affecting the dissolution rates. This explains the low dissolution rate of tg-oxide, whose density is relatively high. Due to nanoscale fragmentation during evaporation, electron-beam oxide is oxygen richer than normal with low density, leading to fastest dissolution rate [48]. Consistent with silicon, the dissolution rate of SiO_2 increases with pH and temperature.

Three different types of Si_3N_4 were also examined: (i) LPCVD (low-pressure chemical vapor deposited) nitride, (ii) PECVD-HF (PECVD in high-frequency mode) nitride, and (iii) PECVD-LF (PECVD in low-frequency mode) nitride. The stoichiometry of the LPCVD nitride was found to be $Si_3N_{3.9}$, while that of the PECVD-LF nitride and PECVD-HF was $Si_3N_{4.3}$ and $Si_3N_{3.3}$, respectively [48]. As shown in Figure 2.4b, LPCVD nitride exhibited the lowest dissolution rate due to its relatively high density and favorable stoichiometry.

2.3 Fabrication Process for Nanomembrane Transient Devices

The manufacturing of transient electronics, owing to their unique material characters and applications, is more or less different from traditional electronic devices. The main issue is that many of the degradable materials, such as Mg and Zn for electrodes and those organic materials for substrates, cannot withstand high temperature and harsh chemical processing, which is common in state-of-the-art integrated circuit manufacturing. Various techniques have been developed to overcome the incompatibility of materials, and the presentative ones will be discussed in this section.

2.3.1 Solution-Based Processes

For organic transient electronics, a solution-based process is often preferred. It is a replacement for standard integrated circuit process, where spin-coating replaces vapor deposition and epitaxy to form multilayer structures and colloid mixture replaces doping and ion implantation to change the conductivity from intrinsic materials. The method is often associated with a shadow-mask-assisted pattering, for solvent in standard patterning process (i.e. photolithography) may harm the building blocks of devices. The first dissolvable electronic device, reported by Bao et al., is manufactured through this solution-based process and a shadow mask. Since then, a variety of organic transient electronics have arisen with the help of this cost-effective and versatile procedure.

Figure 2.5a illustrates the structure of a transient redox-based resistive switching memory (ReRAM), prepared by this method [49]. The device used an ITO (indium

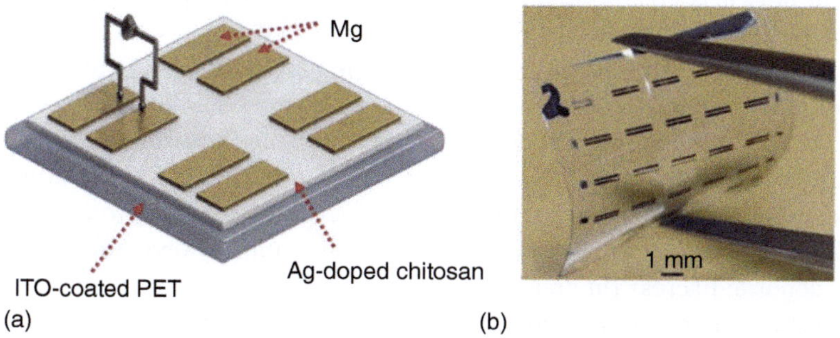

Figure 2.5 A transient organic ReRAM fabricated by solution-based process. (a) Schematic illustration of the layered structure of the device with Mg, Ag-doped chitosan, and ITO-coated PET and (b) optical photo of the resolvable memory devices. Source: Hosseini et al. [49]/with permission of John Wiley & Sons, Inc.

tin oxide)-coated PET (polyethylene terephthalate) as the substrate. Chitosan, made from crab shells, was dissolved in dilute acetic acid and then mixed with $AgNO_3$ powder. After spin coating of the mixture solution on the substrate and subsequent drying, Mg electrodes were deposited with thermal evaporation through utilizing a shadow mask for pattering. An optical image of the finished device is shown in Figure 2.5b. The ReRAM achieved typical memory characteristics, such as low power operation and long data retention, and quick water-resolving property is demonstrated due to the naturally abundant materials used to build the chip.

2.3.2 Transfer-Printing Method

Transfer-printing technique is widely used to turn rigid silicon devices into flexible counterparts. Researchers found that it can be useful to obtain transient devices with high electrical performance through this method by only changing the materials.

A special type of wafer named SOI (silicon-on-insulator), which has a $Si–SiO_2–Si$ structure, is frequently used in the transfer-printing method [21]. In transient electronics scenarios, the top silicon is often thinned to less than 100 nm to facilitate self-dissolution by repetitive oxidation and HF etching process [50]. Two different types of strategies could be applied to make the devices on the top silicon freestanding and ready to be transferred. Both of them use a high selectivity etching between Si and SiO_2 and use one of them as the sacrificial layer.

In the first approach, the buried oxide layer (BOX) in the middle of the SOI is functioned as the sacrificial layer [51, 52]. A schematic illustration of this type of transfer printing is presented in Figure 2.6. The top Si layer after thinning is doped first, which inevitably undergoes a high-temperature environment (whether in diffusion doping or in the annealing process of ion implantation). HF is then used to realize an undercut removal of the BOX leaving the upper layer freestanding. A polydimethylsiloxane (PDMS) stamp is used to stick the semi-finished device up and release it on a dissolvable substrate. Since HF also etches metals, the interconnection process is conducted after the transfer to finish the whole device.

The other approach, as shown in Figure 2.7, which allows the transfer of a complete electronic device, is a bit more complicated [51, 53, 54]. This approach depends on defined etching routes and an anisotropic wet chemical etching, with the BOX as a etch stop layer. A traditional integrated circuit process, including multi-steps of photolithography, dry and wet etching, and film growth, is then applied to the top silicon and sophisticated electronics are formed. After that, a SiN_x layer is first deposited by PECVD on the top of the wafer, which not only acts as a passivation layer but also compensates for the strain after the release. Then photolithography and dry etching define the trenches for anisotropic wet etching. Next, tethers, or anchors, are deposited across the trenches of the exposed silicon handle and join each corner as the support after the release. KOH or tetramethylammonium hydroxide (TMAH) is used as the etchant to undercut the silicon handler while leaving the devices between the BOX and SiN_x passivation layer unharmed and tethered on the wafer by the anchors at the corners. Finally, a PDMS stamp is then used to pick up the devices as the “ink” and release them on a new substrate.

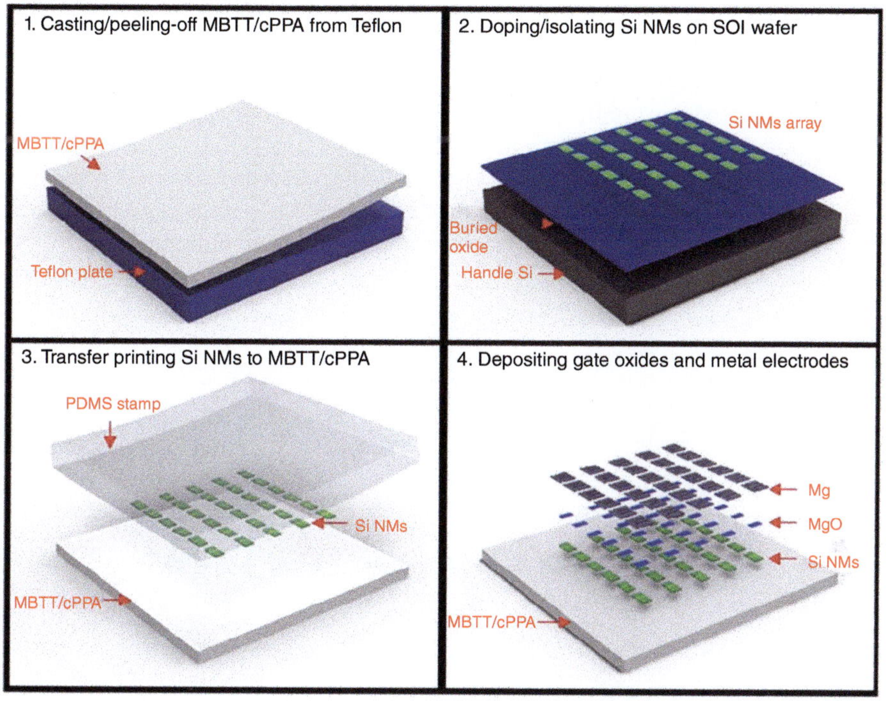

Figure 2.6 Schematics of the fabrication procedure for active electronics through transfer printing using the buried oxide as the sacrifice layer. Source: Reprinted from Hernandez et al. [51]. Copyright 2014, Wiley-VCH Verlag GmbH & Co. KGaA, Weinheim.

The transfer-printing method has been well developed and is a performance–cost balance and promising choice for the manufacturing of nanomembrane transient electronics [55]. Large-scale transfer printing on a 2-in. wafer has also been demonstrated, but compared to commercial size 8-in. or even 12-in. wafer production, the consistency and uniformity of the process still have a long way to go [56]. Furthermore, when the electronic device goes more complicated, the metal wiring goes inevitably multilayered and complex, whether the strain induced in the stick and release action would affect the interconnection layers is still unknown.

2.3.3 Back-Thinning Method

In back-thinning method (Figure 2.8), a SOI wafer is usually thinned by grinding or chemical–mechanical polishing to ~200 μm to facilitate further process [57]. A standard integrated circuit process is then applied to get functional electrical devices on SOI wafers. A substrate precursor is then spin-coated on the surface and developed as a handler for later procedure. After that, the wafer is flipped and various high-speed etching techniques, including deep reactive ion etching (DRIE), inductively coupled plasma reactive ion etching (ICP-RIE), and xenon difluoride (XeF_2) etching, could be applied to etch away the whole silicon handle layer thoroughly to the BOX leaving the nanomembrane devices fabricated at first [58]. Here, the BOX

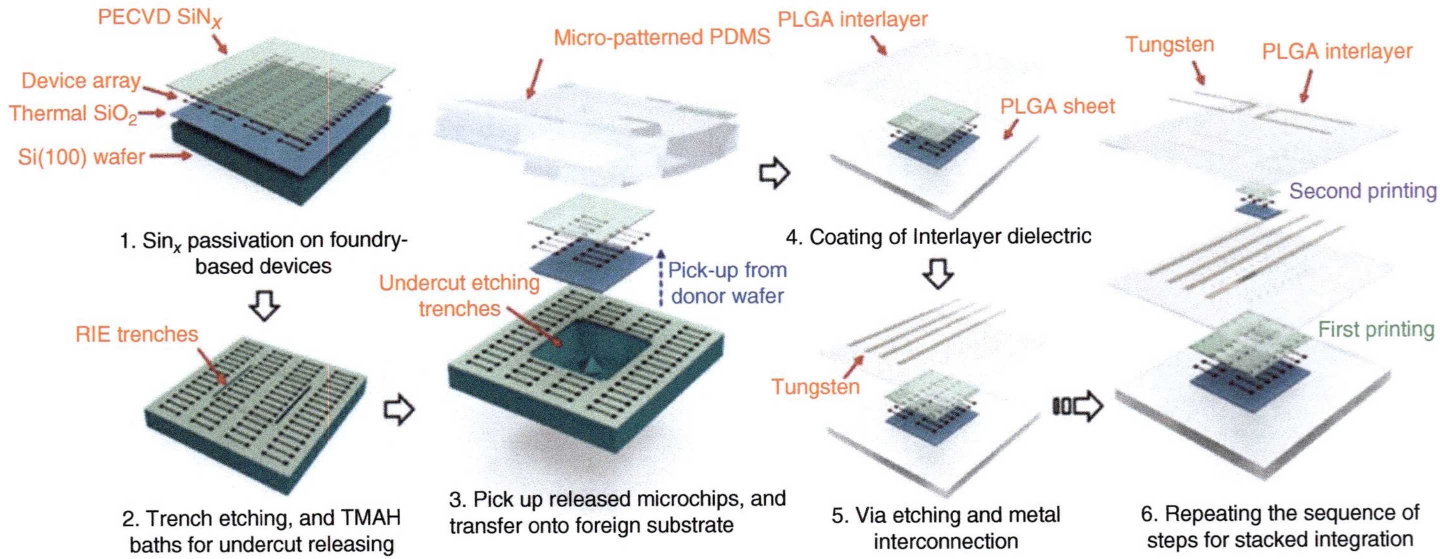

Figure 2.7 Schematics of the fabrication process of transient electronics through transfer printing using the silicon handle layer as the sacrifice layer. Source: Reprinted from Chang et al. [53]. Copyright 2018, Wiley-VCH Verlag GmbH & Co. KGaA, Weinheim.

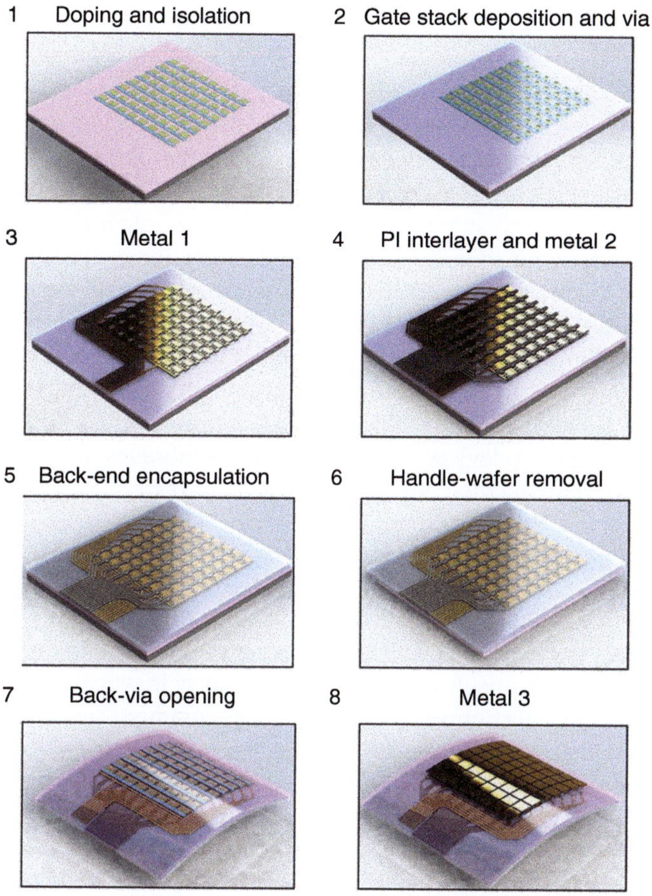

Figure 2.8 Schematics of the back-thinning method to fabricate transient electronics with state-of-the-art performance. Steps 1–8 show the detailed steps. Source: Adapted from Li et al. [52]. Copyright 2018, the National Academy of Sciences of the United States of America (NAS).

is also served as an etch stop layer; thus, a precise etching speed and high selectivity are essential in this approach.

Since all the apparatuses are ready in the integrated circuit industry and no theoretical problems lie in its generalization to large wafers, the future of industry-scale manufacture through back-thinning method is promising [59, 60]. Nonetheless, the excessive cost of long-time plasma or XeF_2 etching is indeed a problem [59, 61]. Thankfully, with the development of advanced packaging, which is striving for stacks of ultrathin polished chips in one die, the cost of a mechanical and physiochemical combined process could be lowered to an acceptable range [62].

2.4 Self-Destruction Mechanisms for Nanomembrane Devices

The core of transient electronics is that they can partially or completely degrade, dissolve, or disintegrate under certain conditions. The dysfunction process could be sorted into two categories, which are passive and active [1, 63]. The passive class refers to that the dysfunction process starts together with the use of the device and the lifetime of the device is determined once the design and fabrication are finished. While the active ones, also known as on-demand transient electronics, could work as a normal electronic device for a relatively long time until a trigger of self-dysfunction is activated. The strategies for designed triggering method of transient devices can be divided into three major groups by the triggering stimuli, which are thermal triggering [9, 64–67], optical triggering [51, 68, 69], and electrical triggering [70, 71]. Here, the representative triggering devices enabled by nanomembranes and their mechanisms are presented, showing their advantages/disadvantages.

2.4.1 Passive Dissolving Without Triggering

Passive dissolving in fluids is the most well-studied mechanism of the dysfunction of nanomembrane transient electronics. The ultrathin character of nanomembranes makes it possible for those usually unsolvable materials to dissolve in water or PBS. As mentioned in the material selection section, materials commonly used in the integrated circuit industry, such as Si, SiO_2, and W, can dissolve in days or even hours depending on the thickness without encapsulation [32]. By adding an encapsulation layer, the transience start point could be delayed. Figure 2.9 shows distinct dissolving start point of Mg when soaked in PBS at 70 °C with different encapsulation [72]. It could also be referred that the dysfunction/penetration mechanisms could vary depending on the materials.

Owing to the fixed lifetime, the most promising applications of this type of transient nanomembrane are disposable electronics and medical implantations. Since the therapy procedure and recovery time of wounding could vary a lot, it is necessary to modulate the lifetime of the device precisely. The key issue lies in that the encapsulation must be nontoxic for health concerns, low rigid for uniform and large-area coverage, and low ion/water permeability in the designed lifetime [73]. Traditional encapsulation materials, such as bulk metal/ceramic enclosures and multilayer stacks of organic/inorganic films, apparently do not meet the requirements.

Fang et al. first proposed an ultrathin and transferred SiO_2 that is thermally oxidized on a silicon wafer [72]. A device-first approach described hereinbefore is used to obtain the transient device, and an accelerated PBS immersion test was conducted. As shown in Figure 2.10, the scanning electron microscope (SEM) image shows a distinct decrease in the thickness of the encapsulation layer. In a case of 1000 nm thermal SiO_2 encapsulation, the devices could work properly for about 12 days in the PBS at 96 °C, which could be converted to a 70-year duration in normal implantation usage. This thermally grown SiO_2 is pinhole-free and uniform,

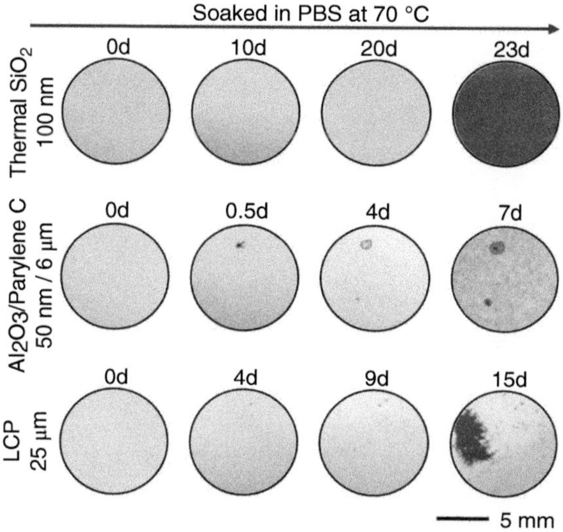

Figure 2.9 Sequential optical images of Mg with different encapsulations soaked in PBS solution at 70 °C. Source: Reprinted from Fang et al. [72]. Copyright 2016, the National Academy of Sciences of the United States of America (NAS).

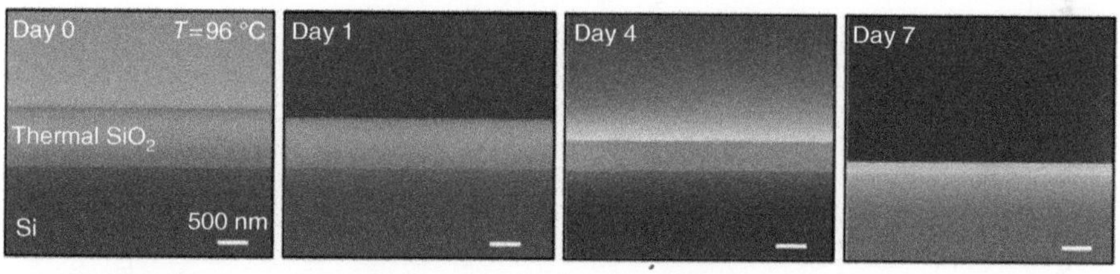

Figure 2.10 SEM images showing the dissolution of thermal SiO_2 soaking in PBS at 96 °C within seven days. Source: Reprinted from Fang et al. [72]. Copyright 2016, the National Academy of Sciences of the United States of America (NAS).

thus making it a suitable candidate for the encapsulation of transient electronics. However, to get a reliable water/ion penetration barrier, the SiO_2 could be thicker than hundreds of nanometers, which may affect the devices' capability of fitting curved surfaces in medical applications. In this regard, Song et al. carried out a series of experiments to achieve a reliable and ultrathin encapsulation [50, 73, 74]. A Parylene C/HfO_2/thermal SiO_2 tri-layer structure is selected. With a thickness of only 50/50/100 nm, the multilayer barrier achieved a comparative lifetime of a 1 μm thermal SiO_2.

By changing several parameters, the protection time of the water barrier could be tuned to be much diversified, suiting different application scenarios. A detailed parameter table containing various material selection and deposition methods is shown in Table 2.1 [72].

2.4.2 Thermally Triggered Devices

Thermosensitive materials are promising candidates as encapsulation or substrate due to their changes in physical and chemical properties (e.g. volume change [66], phase change [75], and chemical reaction [42]). Park et al. fabricated

Table 2.1 Properties of various barrier selections for passive transient electronics.

Category	Material	Deposition method	Thickness for Mg test	Barrier lifetime (at 37 °C)	Leakage mode
Organic	SU-8	Spin coating	1 μm	<1 day	Bulk
	Polyimide (PI)	Spin coating	1 μm	<1 day	Bulk
	Parylene C	CVD	1 μm	<1 day	Bulk
	PDMS	Spin coating	1 μm	<1 day	Bulk
	LCP	—	25 μm	147–309 days	Bulk
Inorganic	SiO_x	PECVD	1 μm	<1 day	Bulk
	SiN_x	PECVD	1 μm	<1 day	Bulk
	Al_2O_3	ALD	100 nm	<1 day	Pinhole
	HfO_2	ALD	100 nm	<1 day	Pinhole
Inorganic multilayer	$SiO_x/SiN_x \times 3$	PECVD	16.6/16.6 nm × 3	<1 day	Pinhole
	$Al_2O_3/HfO_2 \times 3$	ALD	16.6/16.6 nm × 3	<1 day	Pinhole
	Al_2O_3/Parylene C	ALD/CVD	50/950 nm	3–7 days	Pinhole
	Al_2O_3/Parylene C	ALD/CVD	50/6000 nm	5–11 days	Pinhole
	Al_2O_3/Parylene C × 3	ALD/CVD	50/283.3 nm	<1 day	Pinhole
Inorganic/organic multilayer	Al_2O_3/PI	ALD/spin coating	50/950 nm	<1 day	Pinhole
	Al_2O_3/PI × 3	ALD/spin coating	50/283.3 nm × 3	<1 day	Pinhole
	HfO_2/Parylene C	ALD/CVD	50/950 nm	102–214 days	Pinhole
	HfO_2/Parylene C × 3	ALD/CVD	50/283.3 nm × 3	<1 day	Pinhole
	HfO_2/PI	ALD/spin coating	50/950 nm	<1 day	Pinhole
	HfO_2/PI × 3	ALD/spin coating	50/283.3 nm × 3	<1 day	Pinhole
	SiN_x/Al_2O_3/Parylene C	PECVD/ALD/CVD	50/50/900 nm	2–3 days	Pinhole
Thermal oxide	SiO_2	Thermally grown	100 nm	6–7 years	Slow dissolution

Source: Adapted from Fang et al. [72]. Copyright 2016, the National Academy of Sciences of the United States of America (NAS).

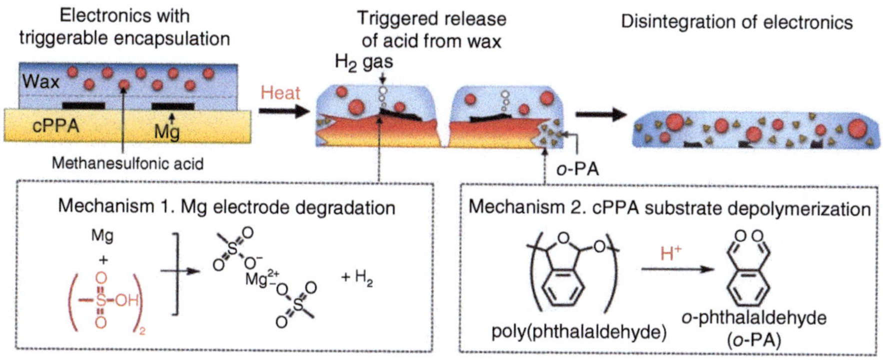

Figure 2.11 Mechanism of thermally triggered transient devices coated with wax containing MSA dispersions. The wax melts at 43 °C and results in the release of encapsulated MSA, leading to the degradation of Mg electrodes and depolymerization of cPPA. Source: Reprinted from Park et al. [65]. Copyright 2015 Wiley-VCH Verlag GmbH & Co. KGaA, Weinheim.

thermal-triggerable transient electronics with wax encapsulation [65]. As shown in Figure 2.11a, the example of thermal-triggerable transient devices was coated with wax/methane sulfonic acid (MSA), which act as sequestration and encapsulation [76]. Once the temperature reached the melting point, the encapsulated acid will release from the wax coating, leading to rapid destruction of the device, as the Mg electrodes were etched and the substrate severely wrecked. Compared with full solution triggers, acid-based trigger possesses higher efficiency for corrosion and destruction of electronic materials, such as metals and metal oxides. The degradation of Si NM-based PIN diode array on cyclic cPPA was achieved within 1 minute at a trigger temperature of 55 °C. This thermal-triggered system shows considerable potential for transient devices enabled by nanomembranes, and various heating methods can be applied to trigger the transience of electronics, such as near-infrared, magnetic field, and microwave radiation, providing more possibilities for applications. Moreover, Gao et al. designed a thermally triggered device that is dissolving-solution-free, including functional components and electronics based on an electrospun poly(ε-caprolactone) (PCL) nanofibrous polymer substrate [66]. The key to the transient process is the remarkable areal shrinkage of electrospun PCL polymer upon heating over melting temperature, arousing the irreversible deformation [77]. A 300-nm thick Si NM was transferred onto the electrospun PCL nanofibrous polymer film with Ag micro-heater below. Once the 100 V voltage was applied to the microheater, the whole electronic fragmentized within 2 seconds because of the transformation from meshwork to bulk solid caused by fast melting and abrupt shrinkage of PCL film, indicating promising on-demand electronics (Figure 2.12). The self-destructive strategy via mechanical destruction expands the application of transient devices in temporary functional components and reconfigurable electronics.

Precise and remote control of thermal-triggered nanomembrane transient devices has been proved to be feasible. The study by Lee et al. shows a wireless microfluidics

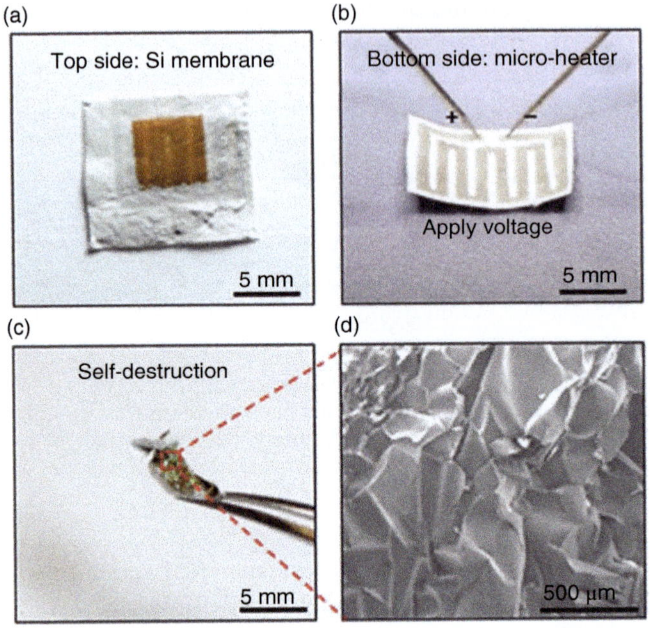

Figure 2.12 Mechanically destructed device with an electrospun PCL nanofibrous polymer film. (a, b) Optical images of the front and backside of the device before triggering. (c) Optical photo of the device after self-destruction. (d) SEM image of the fragmentized Si NM. Source: Copyright 2017, Springer Nature.

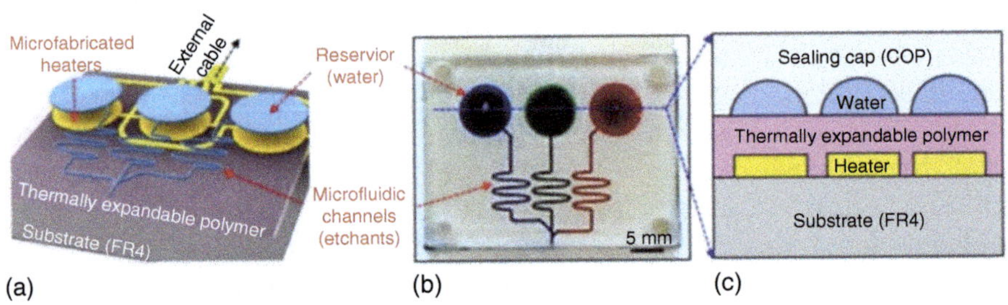

Figure 2.13 Schematics (a), optical image (b), and side view illustration (c) of a wireless transient system triggered by heating. Source: Adapted from Lee et al. [67]. Copyright 2015, Wiley-VCH Verlag GmbH & Co. KGaA, Weinheim.

system that consists of microheaters, thermally expandable polymers, fluid reservoirs, sealing caps (made of cyclic olefin polymer), and microfluidics channels on FR4 epoxy glass cloth (Figure 2.13) [67]. Polymers with a fast, severe, and irreversible volumetric change upon heating and the easily breakable water-containing reservoir were the key to the design. Once activating the self-destruction process, embedded microheater caused the volume of the thermally expansive polymers to expand 5 times within ~20 seconds, resulting in the controllable and precise release process of fluid reservoir. Water in the reservoir could serve as the etchant itself to start the transient process. Besides, with the combination of chemical powders, such as TMAH, in the microfluid channels, much stronger etchants could be formed to resolve various materials and accelerate the process.

Thermally triggered transience can also be achieved by fabricating structures that hold thermal-sensitive materials under the device layer. Li et al. demonstrated flexible transient phototransistors on a wafer-compatible transferred Si NM, which achieve thermal degradation by the thermal-sensitive poly-α-methyl styrene (PAMS) interlayer added below Si NM [9, 64, 78]. The PAMS started to produce volatile gases in the temperature range of ~250 to 300 °C, causing the distortion of Si NM devices. When heated over the decomposition temperature (~300 °C), nearly all PAMS decomposed into the gas phase, and therefore, a massive volume of volatile gas was generated due to the rapid decomposition of PAMS, resulting in increasing internal pressure and the total failure of SiNM devices eventually. The electrical performance of thermally triggered Si-based transistors and phototransistors failed completely after the thermal trigger due to the damage caused by the decomposition of PAMS. The surface morphologies of different stages of the process are shown in Figure 2.14. The degradation process of devices can be controlled by adjusting the heating temperature, which holds tremendous potential for smart on-demand transient electronics, such as privacy protection, circuit safety, and sensing devices.

Abovementioned thermally triggered strategies can be utilized in on-demand smart transient devices, where temperature can serve as sensitive trigger with remote and precise control. However, it should be noted that the storage and working condition of devices must be temperature stable to prevent unnecessary or ahead-of-time thermal trigger, which brings extra loss and low yield.

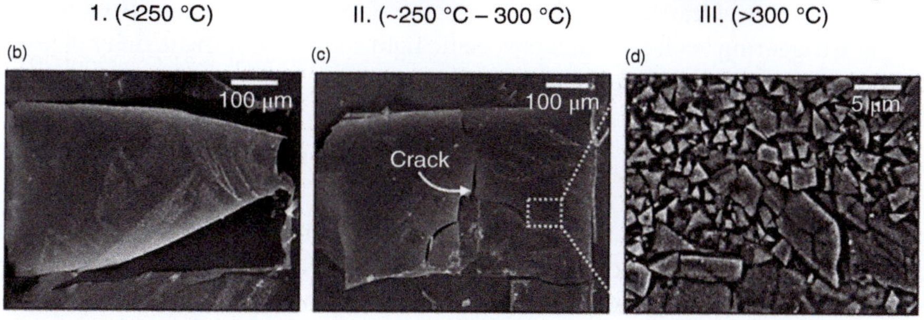

Figure 2.14 Surface morphologies of a thermally triggered transient device. (a) Schematics of the thermally triggered failure process. SEM images of different failure stages; (b) severe bending of the device, (c) generation of cracks, and (d) breaking into fragments. Source: Li et al. [9]/with permission of John Wiley & Sons, Inc.

2.4.3 Optically Triggered Devices

Optically triggered designs are potential and also suitable for the devices triggered in remote and noninvasive ways, such as sensors and monitors work in extreme conditions. As shown in Figure 2.15a [51], transient electronics consisting of a Si NM-based transistor array are built on a photoacid generator/cyclic poly(phthalaldehyde) (PAG/cPPA) substrate, which are representative optically triggered devices. The hydrogen chloride will be produced by the rapid depolymerization of degradable 2-(4-methoxystyryl)-4,6-*bis*(trichloromethyl)-1,3,5-triazine (MBTT)/cPPA substrate if the substrate exposed to the UV lights [79]. Eventually, the cPPA substrate began to decompose and the Mg electrodes were corroded by HCl, resulting in the failure of electronic components. After the 230-minutes irradiation of a 379-nm UV light, the substrate faced significant shrinkage, leading to the total failure of the electrical system (Figure 2.15b). It is worth noting that the transient time can be controlled by the adjustment of the PAG concentration and irradiation conditions of UV light [79]. The Mg resistance change under various exposure conditions on a substrate with 2.5% MBTT/cPPA is illustrated in Figure 2.15c. The transient process could be accelerated if a stronger exposure is selected. Figure 2.15d further depicts that a higher MBTT/cPPA concentration (5%) could distinctly make self-destruction faster. The degradation of metastable materials suggests a promising route for nanomembrane transient devices with faster degradation and more advance triggering modules. Recently, this novel strategy has been applied in dissolving gold in an avirulent aqueous solution by a photo-initiated cyanidation system [80], indicating potentially wider application of transient devices based on inert metal nanomembranes [69]. Photosensitive metastable polymers (e.g. poly(phthalaldehyde)) are able to depolymerize and generate etchants under visible light by incorporating photosensitive compounds, also paving ways for on-demand optically triggered smart transient devices, such as device based on transfer printed nanomembranes on the metastable polymer's substrates. It is believed that the alternative strategy in the application of transient devices on nanomembranes with quality assurance and high security has considerable potential for further study.

As an ideal type of transient strategy, optically triggered transient electronics excel at remote triggering via irradiation of specific light sources, but the toxicity of products is a key problem for practice applications.

2.4.4 Electrically Triggered Devices

Devices triggered by electrical stimuli usually result in multiple reactions, which contain electrochemical–mechanical distortions, electrochemical degradations, and electric sparks, having tremendous application value in data security since physical destruction of hardware is one of the most effective ways for resisting security risks.

To achieve the abovementioned function, Sim et al. [70] fabricated an electrically triggered module system, which could be triggered by DC voltage. The schematic illustration of the constructions of the system is illustrated in Figure 2.16a. The

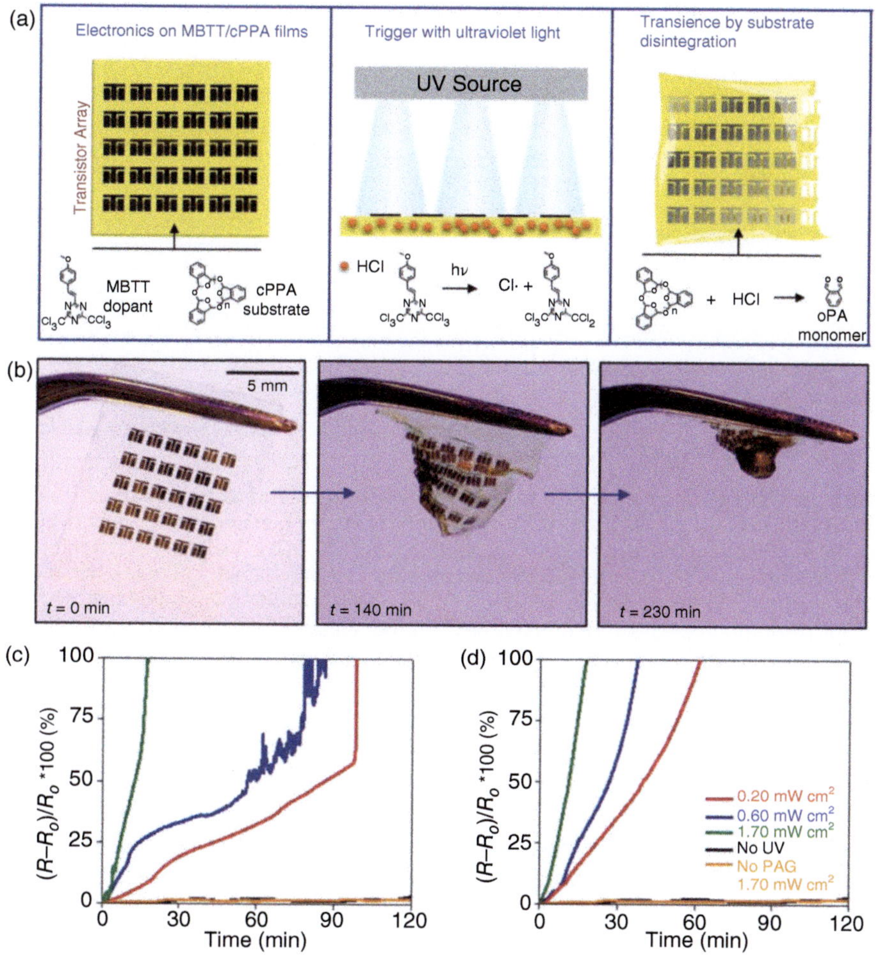

Figure 2.15 Optically triggered chemically destructive on-demand transient device. (a) Mechanism of optically triggered transient electronics on MBTT/cPPA films. The decomposition of acidic sensitive cPPA substrates caused by the HCl (middle) generated from the MBTT after UV irradiation (left). (b) Degradation process of a transistor exposed to 379 nm UV light. (c, d) The time-related change of resistance of a Mg resistor on a (c) 2.5% and (d) 5% MBTT/cPPA substrate with continuous exposure of 365 nm UV light. Source: Hernandez et al. [51]/with permission of John Wiley & Sons, Inc.

self-destruction module was composed of Au electrode, Si reservoir, corrosive solution, and a 400-nm thick Si_3N_4 nanomembrane. The corrosive solution was stored in the Si reservoir with a glass substrate that has gold electrodes attached to it. Once the DC voltage was applied, the electrolyzed etching solution in the reservoir generated gases, resulting in an increasing internal pressure. Eventually, the pressure reached a critical level, leading to the mechanical fracture of Si_3N_4 nanomembrane and the release of corrosive solution from Si reservoir, and therefore the functional devices were dissolved (Figure 2.16b). It is worth noting

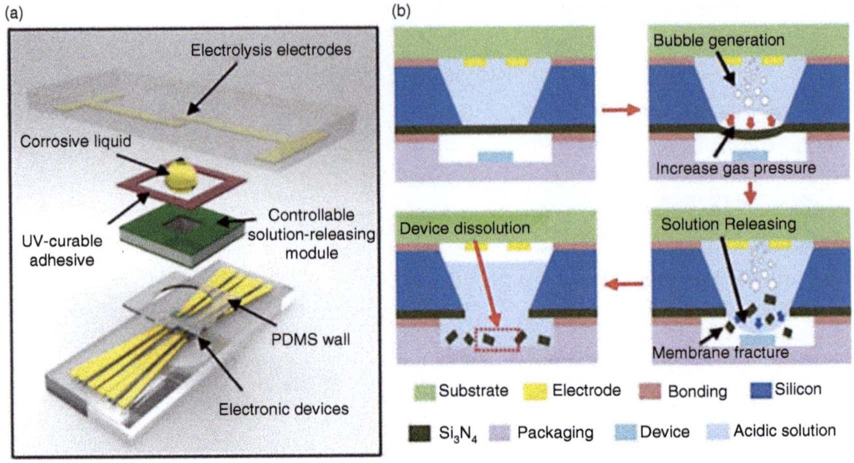

Figure 2.16 Design and mechanism of an electrically triggered transient device. (a) Schematic exploded view of electrically triggered transient electronics that consist of electrolysis electrodes, corrosive liquid, UV curable adhesive, PDMS wall, and electronic devices. (b) Key steps of the triggered transience process. When DC voltage is applied, the bubble generated by electrolyzation of acidic solution will increase the internal gas pressure, resulting in the fracture of Si_3N_4 NM and the release of the solution. The electronic devices are eventually dissolute by the acidic solution. Source: Adapted from Sim et al. [70]. Copyright 2014, IOP Publishing.

that the maximum strain, maximum displacement, and increasing rate of internal pressure depend on the nanomembrane size, indicating the potential of controllable transient time. The dissolution of Cu resistor and indium gallium zinc oxide (IGZO) field-effect transistors (FETs) demonstrate the complete destruction of electronics, ensuring the fast transience process. The ultrathin thickness of nanomembrane makes it prone to breakdown, and the novel strategy holds promise in various areas.

In addition, the study of Chen et al. [71] demonstrated a mechanism for electrochemically triggered transience of Si-based devices, which meets the demand of complete failure of transient electronics. The strategy inspired from the nanostructures (e.g. Si NMs) was designed to relax the lithiation-induced stress, which utilizes the lithiation to accelerate the degradation of Si-based devices through significant volume change [81, 82]. It is worth noting that the electronic devices will be destroyed more completely if combined with subsequent transience, such as reaction with moisture in the environment, providing promising routes for designing on-demand and smart transient electronics. Moreover, devices based on Si nanomembranes also have potential to be applied with the electrochemically triggered strategy with appropriate thickness.

2.5 Conclusion and Outlook

Since the introduction of nanomembrane-based transient electronics, the field has embraced great advances in scientific understanding and engineering realization.

Regarding the transient mechanism, extensive study has been carried out to explore the material behaviors and performances in the self-destruction scenario and determine their biocompatibility, especially for those passive transient electronics. The material systems, self-destruction methods, and preparation processes all have been expanded, thus suitable for more diversified applications. Coming down to applications, various devices with novel design and promising future applications have been demonstrated. The possibility of health monitoring, implantable disease treatment, and self-secured chips have been illuminated.

Despite all the advances, there are still a number of issues to be resolved, to meet the demand and standard for real commercial applications. Transience mechanisms triggered by electrical current, thermal effects, or optical effects are of interest partly because they can support fast responses. However, mainstream triggering methods developed yet, such as UV irradiation or high temperature, are not expectable to obtain in real scenarios, especially in medical applications. Furthermore, the additional components, e.g. microfluidic systems, wireless antennas, and heating coils, not only increase the manufacturing cost, but also raise stability issues during packaging as well as in usage. Last but not least, since all the transient electronic devices are still in their lab form, whether the behaviors still remain similar for very large-scale integrated circuit devicesis not clear. It is possible that the multilayer structure of passivation layers and metal interconnections could affect the transient process, e.g. mechanical fracture and chemical etching.

Still, great opportunities lie in the field of transient electronics. Ongoing emphasis on modified triggering approaches and lifetime modulation methods are likely to reshape our understanding of medical treatment, hardware-level security, and the life cycle of millions of electronic devices we made year by year.

References

1 Lee, G., Choi, Y.S., Yoon, H.-J. et al. (2020). Advances in physicochemically stimuli-responsive materials for on-demand transient electronic systems. *Matter* 3 (4): 1031–1052. https://doi.org/10.1016/j.matt.2020.08.021.

2 Han, W.B., Lee, J.H., Shin, J.W. et al. (2020). Advanced materials and systems for biodegradable, transient electronics. *Adv. Mater.* 32 (51): 2002211. https://doi.org/10.1002/adma.202002211.

3 Kang, S.K., Murphy, R.K., Hwang, S.W. et al. (2016). Bioresorbable silicon electronic sensors for the brain. *Nature* 530 (7588): 71–76. https://doi.org/10.1038/nature16492.

4 Hwang, S.W., Kim, D.-H., Tao, H. et al. (2013). Materials and fabrication processes for transient and bioresorbable high-performance electronics. *Adv. Funct. Mater.* 23 (33): 4087–4093. https://doi.org/10.1002/adfm.201300127.

5 Koo, J., Kim, S.B., Choi, Y.S. et al. (2020). Wirelessly controlled, bioresorbable drug delivery device with active valves that exploit electrochemically triggered crevice corrosion. *Sci. Adv.* 6 (35): eabb1093. https://doi.org/10.1126/sciadv.abb1093.

6 Shin, J., Yan, Y., Bai, W. et al. (2019). Bioresorbable pressure sensors protected with thermally grown silicon dioxide for the monitoring of chronic diseases and healing processes. *Natl. Biomed. Eng.* 3 (1): 37–46. https://doi.org/10.1038/s41551-018-0300-4.

7 Shin, J., Liu, Z., Bai, W. et al. (2019). Bioresorbable optical sensor systems for monitoring of intracranial pressure and temperature. *Sci. Adv.* 5 (7): eaaw1899. https://doi.org/10.1126/sciadv.aaw1899.

8 Hu, W., Jiang, J., Xie, D. et al. (2018). Transient security transistors self-supported on biodegradable natural-polymer membranes for brain-inspired neuromorphic applications. *Nanoscale* 10 (31): 14893–14901. https://doi.org/10.1039/c8nr04136a.

9 Li, G., Song, E., Huang, G. et al. (2018). High-temperature-triggered thermally degradable electronics based on flexible silicon nanomembranes. *Adv. Funct. Mater.* 28 (45): https://doi.org/10.1002/adfm.201801448.

10 Kwon, K.Y., Lee, J.S., Ko, G.J. et al. (2018). Biosafe, eco-friendly levan polysaccharide toward transient electronics. *Small* 14 (32): e1801332. https://doi.org/10.1002/smll.201801332.

11 Lee, B.H., Lee, D.I., Bae, H. et al. (2016). Foldable and disposable memory on paper. *Sci. Rep.* 6: 38389. https://doi.org/10.1038/srep38389.

12 Irimia-Vladu, M., Troshin, P.A., Reisinger, M. et al. (2010). Biocompatible and biodegradable materials for organic field-effect transistors. *Adv. Funct. Mater.* 20 (23): 4069–4076. https://doi.org/10.1002/adfm.201001031.

13 Jung, Y.H., Chang, T.H., Zhang, H. et al. (2015). High-performance green flexible electronics based on biodegradable cellulose nanofibril paper. *Nat. Commun.* 6: 7170. https://doi.org/10.1038/ncomms8170.

14 Rogers, J.A. and Ahn, J.-H. (2016). *Silicon Nanomembranes: Fundamental Science and Applications*. Wiley.

15 Sun, Y. and Rogers, J.A. (2007). Inorganic semiconductors for flexible electronics. *Adv. Mater.* 19 (15): 1897–1916. https://doi.org/10.1002/adma.200602223.

16 Kim, D.H., Ahn, J.H., Choi, W.M. et al. (2008). Stretchable and foldable silicon integrated circuits. *Science* 320 (5875): 507–511. https://doi.org/10.1126/science.1154367.

17 Kim, D.-H., Kim, Y.-S., Amsden, J. et al. (2009). Silicon electronics on silk as a path to bioresorbable, implantable devices. *Appl. Phys. Lett.* 95 (13): 133701. https://doi.org/10.1063/1.3238552.

18 Hwang, S.W., Tao, H., Kim, D.H. et al. (2012). A physically transient form of silicon electronics. *Science* 337 (6102): 1640–1644. https://doi.org/10.1126/science.1226325.

19 Hwang, S.W., Park, G., Cheng, H. et al. (2014). 25th anniversary article: materials for high-performance biodegradable semiconductor devices. *Adv. Mater.* 26 (13): 1992–2000. https://doi.org/10.1002/adma.201304821.

20 Katiyar, A.K., Davidson, A.A., Jang, H. et al. (2019). Ultrasoft silicon nanomembranes: thickness-dependent effective elastic modulus. *Nanoscale* 11 (32): 15184–15194. https://doi.org/10.1039/c9nr03995c.

21 Carlson, A., Bowen, A.M., Huang, Y. et al. (2012). Transfer printing techniques for materials assembly and micro/nanodevice fabrication. *Adv. Mater.* 24 (39): 5284–5318. https://doi.org/10.1002/adma.201201386.

22 Guo, Q., Zhang, M., Xue, Z. et al. (2015). Deterministic assembly of flexible Si/Ge nanoribbons via edge-cutting transfer and printing for van der Waals heterojunctions. *Small* 11 (33): 4140–4148. https://doi.org/10.1002/smll.201500505.

23 Shao, M., Ma, D.D.D., and Lee, S.-T. (2010). Silicon nanowires – synthesis, properties, and applications. *Eur. J. Inorg. Chem.* 27: 4264–4278. https://doi.org/10.1002/ejic.201000634.

24 Di, L., Yan, Y., Deng, Y. et al. (2020). Bioresorbable wireless sensors as temporary implants for in vivo measurements of pressure. *Adv. Funct. Mater.* 2003754: https://doi.org/10.1002/adfm.202003754.

25 Guo, Q., Koo, J., Xie, Z. et al. (2019). A bioresorbable magnetically coupled system for low-frequency wireless power transfer. *Adv. Funct. Mater.* https://doi.org/10.1002/adfm.201905451.

26 Koo, J., Macewan, M.R., Kang, S.-K. et al. (2018). Wireless bioresorbable electronic system enables sustained nonpharmacological neuroregenerative therapy. *Nat. Med.* 24 (12): 1830–1836. https://doi.org/10.1038/s41591-018-0196-2.

27 Boutry, C.M., Beker, L., Kaizawa, Y. et al. (2019). Biodegradable and flexible arterial-pulse sensor for the wireless monitoring of blood flow. *Natl. Biomed. Eng.* 3 (1): 47–57. https://doi.org/10.1038/s41551-018-0336-5.

28 Burton, A., Obaid, S.N., Vazquez-Guardado, A. et al. (2020). Wireless, battery-free subdermally implantable photometry systems for chronic recording of neural dynamics. *Proc. Natl. Acad. Sci. USA* 117 (6): 2835–2845. https://doi.org/10.1073/pnas.1920073117.

29 Cha, G.D., Kang, D., Lee, J. et al. (2019). Bioresorbable electronic implants: history, materials, fabrication, devices, and clinical applications. *Adv. Health Mater.* 8 (11): e1801660. https://doi.org/10.1002/adhm.201801660.

30 Bai, W., Yang, H., Ma, Y. et al. (2018). Flexible transient optical waveguides and surface-wave biosensors constructed from monocrystalline silicon. *Adv. Mater.* 30 (32): e1801584. https://doi.org/10.1002/adma.201801584.

31 Hwang, S.W., Kang, S.K., Huang, X. et al. (2015). Materials for programmed functional transformation in transient electronic systems. *Adv. Mater.* 27 (1): 47–52. https://doi.org/10.1002/adma.201403051.

32 Yin, L., Cheng, H., Mao, S. et al. (2014). Dissolvable metals for transient electronics. *Adv. Funct. Mater.* 24 (5): 645–658. https://doi.org/10.1002/adfm.201301847.

33 Zhang, X.G. (1996). *Corrosion and Electrochemistry of Zinc*. Plenum Press.

34 Song, G.L. and Atrens, A. (2003). Understanding magnesium corrosion – a framework for improved alloy performance. *Adv. Eng. Mater.* 5 (12): 837–858. https://doi.org/10.1002/adem.200310405.

35 Lang, W. and Zander, R. (1986). Salting-out of oxygen from aqueous-electrolyte solutions – prediction and measurement. *Ind. Eng. Chem. Fundam.* 25 (4): 775–782. https://doi.org/10.1021/i100024a050.

36 Anik, M. and Cansizoglu, T. (2006). Dissolution kinetics of WO_3 in acidic solutions. *J. Appl. Electrochem.* 36 (5): 603–608. https://doi.org/10.1007/s10800-006-9113-3.

37 Fu, K.K., Wang, Z., Dai, J. et al. (2016). Transient electronics: materials and devices. *Chem. Mater.* 28 (11): 3527–3539. https://doi.org/10.1021/acs.chemmater.5b04931.

38 Kang, S.K., Park, G., Kim, K. et al. (2015). Dissolution chemistry and biocompatibility of silicon- and germanium-based semiconductors for transient electronics. *ACS Appl. Mater. Interfaces* 7 (17): 9297–9305. https://doi.org/10.1021/acsami.5b02526.

39 Lee, Y.K., Yu, K.J., Song, E. et al. (2017). Dissolution of monocrystalline silicon nanomembranes and their use as encapsulation layers and electrical interfaces in water-soluble electronics. *ACS Nano* 11 (12): 12562–12572. https://doi.org/10.1021/acsnano.7b06697.

40 Rimstidt, J.D. and Barnes, H.L. (1980). The kinetics of silica-water reactions. *Geochim. Cosmochim. Acta* 44 (11): 1683–1699. https://doi.org/10.1016/0016-7037(80)90220-3.

41 Seidel, H., Csepregi, L., Heuberger, A. et al. (1990). Anisotropic etching of crystalline silicon in alkaline-solutions. 1. Orientation dependence and behavior of passivation layers. *J. Electrochem. Soc.* 137 (11): 3612–3626. https://doi.org/10.1149/1.2086277.

42 Yin, L., Farimani, A.B., Min, K. et al. (2015). Mechanisms for hydrolysis of silicon nanomembranes as used in bioresorbable electronics. *Adv. Mater.* 27 (11): 1857–1864. https://doi.org/10.1002/adma.201404579.

43 Hwang, S.W., Park, G., Edwards, C. et al. (2014). Dissolution chemistry and biocompatibility of single-crystalline silicon nanomembranes and associated materials for transient electronics. *ACS Nano* 8 (6): 5843–5851. https://doi.org/10.1021/nn500847g.

44 Seidel, H., Csepregi, L., Heuberger, A. et al. (1990). Anisotropic etching of crystalline silicon in alkaline-solutions. 2. Influence of dopants. *J. Electrochem. Soc.* 137 (11): 3626–3632. https://doi.org/10.1149/1.2086278.

45 Borenstein, J.T., Gerrish, N.D., Currie, M.T. et al. (1999). A new ultra-hard etch-stop layer for high precision micromachining. *Mems '99: 12 IEEE International Conference on Micro Electromechanical Systems, Technical Digest* 205–210. https://doi.org/10.1109/Memsys.1999.746810.

46 Lu, D., Liu, T.L., Chang, J.K. et al. (2019). Transient light-emitting diodes constructed from semiconductors and transparent conductors that biodegrade under physiological conditions. *Adv. Mater.* 31 (42): e1902739. https://doi.org/10.1002/adma.201902739.

47 Dagdeviren, C., Hwang, S.W., Su, Y. et al. (2013). Transient, biocompatible electronics and energy harvesters based on ZnO. *Small* 9 (20): 3398–3404. https://doi.org/10.1002/smll.201300146.

48 Kang, S.-K., Hwang, S.-W., Cheng, H. et al. (2014). Dissolution behaviors and applications of silicon oxides and nitrides in transient electronics. *Adv. Funct. Mater.* 24 (28): 4427–4434. https://doi.org/10.1002/adfm.201304293.

49 Hosseini, N.R. and Lee, J.-S. (2015). Biocompatible and flexible chitosan-based resistive switching memory with magnesium electrodes. *Adv. Funct. Mater.* 25 (35): 5586–5592. https://doi.org/10.1002/adfm.201502592.

50 Song, E., Fang, H., Jin, X. et al. (2017). Thin, transferred layers of silicon dioxide and silicon nitride as water and ion barriers for implantable flexible electronic systems. *Adv. Electron. Mater.* 3 (8): https://doi.org/10.1002/aelm.201700077.

51 Hernandez, H.L., Kang, S.K., Lee, O.P. et al. (2014). Triggered transience of metastable poly(phthalaldehyde) for transient electronics. *Adv. Mater.* 26 (45): 7637–7642. https://doi.org/10.1002/adma.201403045.

52 Li, J., Song, E., Chiang, C.H. et al. (2018). Conductively coupled flexible silicon electronic systems for chronic neural electrophysiology. *Proc. Natl. Acad. Sci. USA* 115 (41): E9542–E9549. https://doi.org/10.1073/pnas.1813187115.

53 Chang, J.K., Chang, H.P., Guo, Q. et al. (2018). Biodegradable electronic systems in 3D, heterogeneously integrated formats. *Adv. Mater.* 30 (11): https://doi.org/10.1002/adma.201704955.

54 Song, E., Chiang, C.H., Li, R. et al. (2019). Flexible electronic/optoelectronic microsystems with scalable designs for chronic biointegration. *Proc. Natl. Acad. Sci. USA* 116 (31): 15398–15406. https://doi.org/10.1073/pnas.1907697116.

55 Park, J., Lee, Y., Lee, H. et al. (2020). Transfer printing of electronic functions on arbitrary complex surfaces. *ACS Nano* https://doi.org/10.1021/acsnano.9b09846.

56 Wang, C., Linghu, C., Nie, S. et al. (2020). Programmable and scalable transfer printing with high reliability and efficiency for flexible inorganic electronics. *Sci. Adv.* 6 (25): eabb2393. https://doi.org/10.1126/sciadv.abb2393.

57 Fang, H., Yu, K.J., Gloschat, C. et al. (2017). Capacitively coupled arrays of multiplexed flexible silicon transistors for long-term cardiac electrophysiology. *Natl. Biomed. Eng.* 1: https://doi.org/10.1038/s41551-017-0038.

58 Hussain, A.M. and Hussain, M.M. (2016). CMOS-technology-enabled flexible and stretchable electronics for internet of everything applications. *Adv. Mater.* 28 (22): 4219–4249. https://doi.org/10.1002/adma.201504236.

59 Xu, D., Xiong, B., Wu, G. et al. (2012). Isotropic silicon etching with XeF_2 gas for wafer-level micromachining applications. *J. Microelectromech. Syst.* 21 (6): 1436–1444. https://doi.org/10.1109/jmems.2012.2209403.

60 Wang, S., Weil, B.D., Li, Y. et al. (2013). Large-area free-standing ultrathin single-crystal silicon as processable materials. *Nano Lett.* 13 (9): 4393–4398. https://doi.org/10.1021/nl402230v.

61 Winters, H.F. and Coburn, J.W. (1979). The etching of silicon with XeF_2 vapor. *Appl. Phys. Lett.* 34 (1): 70–73. https://doi.org/10.1063/1.90562.

62 Liebens, M., Jourdain, A., De Vos, J. et al. (2019). In-line metrology for characterization and control of extreme wafer thinning of bonded wafers. *IEEE Trans. Semicond. Manuf.* 32 (1): 54–61. https://doi.org/10.1109/tsm.2018.2885577.

63 You, C., Zhao, H., Guo, Q. et al. (2020). Material strategies for on-demand smart transient electronics. *MRS Bull.* 45 (2): 129–134. https://doi.org/10.1557/mrs.2020.20.

64 Li, G., Song, E., Huang, G. et al. (2018). Flexible transient phototransistors by use of wafer-compatible transferred silicon nanomembranes. *Small* 14 (47): 1802985. https://doi.org/10.1002/smll.201802985.

65 Park, C.W., Kang, S.K., Hernandez, H.L. et al. (2015). Thermally triggered degradation of transient electronic devices. *Adv. Mater.* 27 (25): 3783–3788.

66 Gao, Y., Sim, K., Yan, X. et al. (2017). Thermally triggered mechanically destructive electronics based on electrospun poly(ε-caprolactone) nanofibrous polymer films. *Sci. Rep.* 7 (1): 947. https://doi.org/10.1038/s41598-017-01026-6.

67 Lee, C.H., Jeong, J.-W., Liu, Y. et al. (2015). Materials and wireless microfluidic systems for electronics capable of chemical dissolution on demand. *Adv. Funct. Mater.* 25 (9): 1338–1343. https://doi.org/10.1002/adfm.201403573.

68 Gourdin, G., Phillips, O., Schwartz, J. et al. (2017). Phototriggerable, fully transient electronics: component and device fabrication. *IEEE 67th Electronic Components and Technology Conference (ECTC)*, Orlando, Florida, USA.

69 Chen, W.D., Kang, S.-K., Stark, W.J. et al. (2019). The light triggered dissolution of gold wires using potassium ferrocyanide solutions enables cumulative illumination sensing. *Sens. Actuators, B* 282: 52–59. https://doi.org/10.1016/j.snb.2018.10.094.

70 Sim, K., Wang, X., Li, Y. et al. (2017). Destructive electronics from electrochemical-mechanically triggered chemical dissolution. *J. Micromech. Microeng.* 27 (6): https://doi.org/10.1088/1361-6439/aa682f.

71 Chen, Y., Wang, H., Zhang, Y. et al. (2019). Electrochemically triggered degradation of silicon membranes for smart on-demand transient electronic devices. *Nanotechnology* https://doi.org/10.1088/1361-6528/ab2853.

72 Fang, H., Zhao, J., Yu, K.J. et al. (2016). Ultrathin, transferred layers of thermally grown silicon dioxide as biofluid barriers for biointegrated flexible electronic systems. *Proc. Natl. Acad. Sci. USA* 113 (42): 11682–11687. https://doi.org/10.1073/pnas.1605269113.

73 Song, E., Lee, Y.K., Li, R. et al. (2017). Transferred, ultrathin oxide bilayers as biofluid barriers for flexible electronic implants. *Adv. Funct. Mater.* 28 (12): https://doi.org/10.1002/adfm.201702284.

74 Song, E., Li, R., Jin, X. et al. (2018). Ultrathin trilayer assemblies as long-lived barriers against water and ion penetration in flexible bioelectronic systems. *ACS Nano* 12 (10): 10317–10326. https://doi.org/10.1021/acsnano.8b05552.

75 Kim, B.H., Kim, J.-H., Persano, L. et al. (2017). Dry transient electronic systems by use of materials that sublime. *Adv. Funct. Mater.* 27 (12): https://doi.org/10.1002/adfm.201606008.

76 Mellema, M., Van Benthum, W.A., Boer, B. et al. (2006). Wax encapsulation of water-soluble compounds for application in foods. *J. Microencapsulation* 23 (7): 729–740. https://doi.org/10.1080/02652040600787900.

77 Gernon, M.D., Wu, M., Buszta, T. et al. (1999). Environmental benefits of methanesulfonic acid. *Green Chem.* 1 (3): 127–140. https://doi.org/10.1039/a900157c.

78 McQuillan, B.W., Nikroo, A., Steinman, D.A. et al. (1997). The PAMS/GDP process for production of ICF target mandrels. *Fusion Technol.* 31 (4): 381–384. https://doi.org/10.13182/fst31-381.

79 Dilauro, A.M., Robbins, J.S., and Phillips, S.T. (2013). Reproducible and scalable synthesis of end-cap-functionalized depolymerizable poly(phthalaldehydes). *Macromolecules* 46 (8): 2963–2968. https://doi.org/10.1021/ma4001594.

80 Ašpergĕr, S. (1952). Kinetics of the decomposition of potassium ferrocyanide in ultra-violet light. *Trans. Faraday Soc.* 48: 617–624. https://doi.org/10.1039/tf9524800617.

81 Feng, K., Li, M., Liu, W. et al. (2018). Silicon-based anodes for lithium-ion batteries: from fundamentals to practical applications. *Small* 14 (8): 1702737. https://doi.org/10.1002/smll.201702737.

82 Yu, C., Li, X., Ma, T. et al. (2012). Silicon thin films as anodes for high-performance lithium-ion batteries with effective stress relaxation. *Adv. Energy Mater.* 2 (1): 68–73. https://doi.org/10.1002/aenm.201100634.

3

Diverse Polymer Nanomembranes Toward Task-Specific Applications

Hong Zhu, Feifei Wu, Jizhai Cui, Borui Xu, and Yongfeng Mei

Fudan University, Department of Materials Science, 220 Handan Road, Shanghai 200433, People's Republic of China

3.1 Introduction

A polymer is a large molecule composed of many repeating simple chemical units through the polymerization reaction of small molecules [1]. The small molecules, also known as monomers, can be synthesized as a polymer by combining with other molecules in the same or different types. And the number of repeating units in the polymer is defined as the degree of polymerization. After the polymerization, a series of chemical bonds and linkages are built between the repeating units, and the resulting molecule is often called a polymer chain similar to links in a chain [2]. Then these polymer chains entangle to linear, branched, ladder, and cross-linked structures.

Polymers can be classified in many different ways, such as by origin, polymer structure, polymerization mechanism, and thermal behavior [2]. According to the classification of the monomer materials, polymers range from natural biomaterials such as proteins, polysaccharides, and nucleic acids to synthetic materials such as plastics and rubbers. For example, DNA is a particularly long molecule constituted of up to 10^7 nucleotides, and polyethylene is a long, chain-like macromolecule made of ethylene molecules. For engineering purposes, polymers are classified as thermoplastics or thermosets, based on their thermomechanical properties. Thermoplastic polymers become softened and flow under heating and increasing pressure, and then reversibly recover to become hard and adapt to the shape of a container without structural breakdown after cooling. Thermoset polymers undergo a chemical process during heating, which is irreversible. The difference is generated from their basic structural differences. The thermoplastic polymers are mainly composed of linear and branched structures, whereas the thermosets are composed of cross-linked structures.

Polymers can be processed into different forms, such as rods, spheres, fibers, and thin films. Among all of these morphologies, self-supported quasi-2D structures

Nanomembranes: Materials, Properties, and Applications, First Edition.
Edited by Yongfeng Mei, Gaoshan Huang, and Xiuling Li.
© 2022 WILEY-VCH GmbH. Published 2022 by WILEY-VCH GmbH.

with a thickness of a few hundred nanometers, called polymer nanomembranes, contain quantities of nanometer-sized pores [3]. Polymer nanomembranes include pure polymer matrix and mixed polymer nanocomposites. Typical examples of pure polymer matrixes are cellulose acetate, polyvinylchloride, polyacrylonitrile, polysulfone, polycarbonate, etc. [4–7].

3.2 Fabrication and Functionalization of Polymer Nanomembranes

Polymer nanomembrane is a polymer membrane with a thickness of several hundred nanometers. In this part, we summarize various top-down and bottom-up approaches to fabricate or functionalize polymer nanomembranes. The top-down approaches, such as spin coating and dip coating, are the most investigated nanomembrane fabrication methods, which are suitable for most polymeric materials in a solvent or liquid phase. These top-down methods are usually perfect for the polymer nanomembrane with a thickness of 200–1000 nm. On the other hand, the bottom-up approaches are naturally superior to the top-down approaches in the fabrication of sub-100 nm, defect-free polymer nanomembranes.

3.2.1 Spin Coating and Dip Coating

Spin coating is an extremely simple, robust, and economical method to prepare a broad range of nanomembranes in polymeric materials. The quality and thickness of nanomembrane could be controlled through key parameters, including the speed, acceleration, and duration time of the spinning conditions and the viscosity and concentration of polymer solution [8]. Zheng et al. fabricated ultrathin hydrogel films with high toughness and fast response through a facile spin coating approach [9]. The obtained hydrogel nanomembranes possess excellent mechanical properties with Young's moduli ranging from 0.1 to 61.7 MPa. Topological features of substrates influence the film formation during the spin-coating process as well, due to the phase-separation process of polymeric surface on a topologically pre-patterned substrate [10]. Manipulating interfacial interactions via spin coating, selective deposition of patterned polymer nanomembranes on heterogeneous substrates could be achieved through a transition of glassy polymers from uniform wetting to complete dewetting [11]. Meanwhile, the flexible polymer layer fabricated through spin coating could also work as an encapsulation layer to modulate the bandgap of monolayer 2D materials [12]. The strong adhesion of spin-coated polymer nanomembrane and 2D materials and the high Young's modulus of polymer nanomembrane guarantee the mechanical strain transferred to the 2D functional materials. Figure 3.1 shows the detailed schematic illustration of the spin-coated encapsulation process.

Dip coating is another simple, fast, widely used, and facile one-step method to fabricate polymer nanomembranes. The process is just dipping the substrate into the polymer solution and then withdrawing it at a constant speed to form the

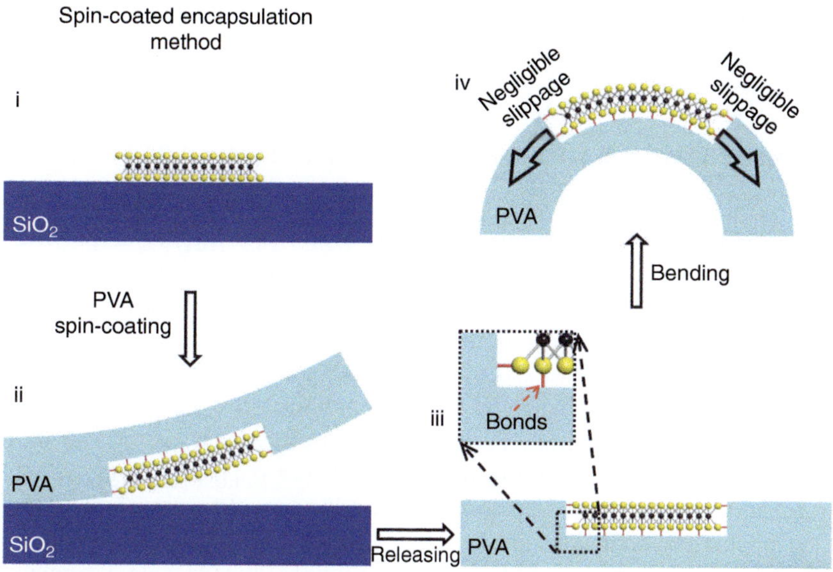

Figure 3.1 Schematic illustration of the spin-coated encapsulation method to efficiently modulate the bandgap of 2D materials. Source: Adapted from Li et al. [12]. Copyright 2020, Springer Nature.

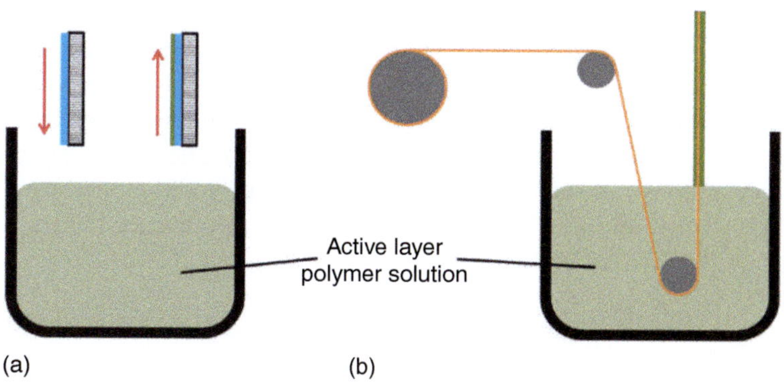

Figure 3.2 Schematic illustration of (a) simple dip coating and (b) continuous dip coating. Source: Reprinted from Xie et al. [13]. Copyright 2018, Elsevier.

nanomembrane. The schematic illustration of the procedure is shown in Figure 3.2. It could also be coupled with external controls, such as heat treatment or light illumination. Using a polystyrene-poly(4-vinyl pyridine) (PS-P4VP) block copolymer and two hydroxyl-functionalized small-molecule azo derivatives with different photochemical properties, Vapaavuori et al. controlled the surface pattern and the thickness of the polymer nanomembrane through in situ light illumination during the dip-coating process [14]. Using metal wire as a template and dipping into PDMS precursor solution, Xi et al. developed rapid prototyping of soft elastomeric tubular nanomembrane, which could work as the basic building blocks for microfluidic devices and assemble to 3D structures [15].

3.2.2 Layer-by-Layer (LbL) Assembly and Langmuir–Blodgett (LB) Films

Layer-by-layer (LbL) assembly is an approach to fabricate multilayer polymer nanomembrane based on the attractions of different layers. The attractions could be driven by a wide range of forces, such as electrostatic force, hydrophilic and hydrophobic interactions, hydrogen bonding, host–guest interactions, and covalent effect. Among all the driving forces, the electrostatic force between ions is the most common for LbL assembly, which deposits different polyelectrolytes with opposite charges. As illustrated in Figure 3.3a, a solid substrate with a charged surface is merged into a solution containing polyions with the opposite charge, which is absorbed on the substrate. When this substrate is exposed to another solution containing opposite-charge polyions, another layer of polyion is adsorbed onto the substrate. This procedure could be repeated multiple times to yield a stepwise growth of polymeric films. However, electrostatic LbL assembly is limited by the choice of template, which is expected to have charges or functional groups as the first layer. A typical LB film is formed by amphiphilic molecules with hydrophobic tails and hydrophilic heads. Due to the hydrophobic and hydrophilic interactions between the tail and head of the molecules, the monolayer of amphiphilic polymer could be transferred onto a substrate by passing the substrate through the interface several times, as shown in Figure 3.3b. However, fabricating a polymer nanomembrane without defect directly through this approach is difficult. Based on calix[n]arene (n = 4, 5, 6) surfactants and the poly(styrene sulfonate), poly(acrylic acid), or poly(acrylamide) "glues," the Regen group combined the electrostatic interactions with hydrophobic and hydrophilic interactions to fabricate LB films without defects in the thickness of 5–6 nm (Figure 3.3c) [16].

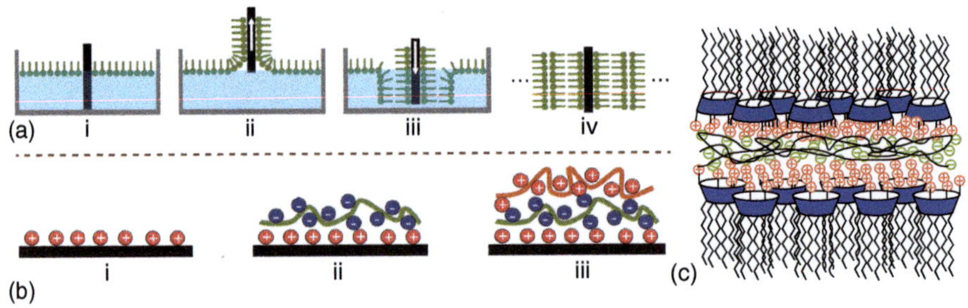

Figure 3.3 Schematic illustration of (a) layer-by-layer assembly and (b) Langmuir–Blodgett (LB) films. Source: Adapted from Xie et al. [13]. Copyright 2018, Elsevier. (c) A stylized illustration of an LB bilayer derived from a porous surfactant containing multiple cationic sites and glued together with polymeric anions. Source: Adapted from Wang et al. [16]. Copyright 2010, American Chemistry Society.

3.2.3 Chemical Vapor Deposition

Chemical vapor deposition (CVD) is a remarkable technique for fabricating high-quality thin films and coatings. Nowadays, CVD synthesis attracts significant interest and has been applied to the precise manufacturing of both inorganic nanomembranes of 2D materials and high-purity polymeric nanomembranes, which could be conformably deposited on various substrates.

Different from the previously described liquid-phase techniques for polymer nanomembrane synthesis, such as spin coating and dip coating, insoluble polymers deposited by the CVD process can avoid some solution effects or unwanted modification, creating a conformal coating [17]. Fluoropolymers, electrically conductive polymers, and cross-linked organic networks are typical CVD process-compatible polymeric materials. Regardless of the variations in different types, the fundamental CVD process consists of the following common elementary steps, as shown in Figure 3.4. The first step is to fill in the reactor with the reactant gases, which then either undergo gas-phase reactions to generate intermediate reactants and gaseous by-products via homogeneous reactions or diffuse directly through the boundary layer to the substrate. The reactant gases and the intermediate reactants adsorb onto the substrate and undergo the heterogeneous reactions at the gas–solid interface, which leads to continuous nanomembrane formation through the nucleation, growth, and coalescence process. Meanwhile, the gaseous products and unreacted species are carried away from the reaction zone through continuous gaseous flow. Polymeric nanomembranes grown by the CVD process follow the geometrical features of the underlying surface, yielding the same film thickness in all positions [19].

Although large amounts of CVD polymerization process utilize the same reaction mechanisms of step growth, chain growth, and condensation reactions, it is not suitable for all CVD syntheses, especially for the polymers that underlie the synthesis in solution. The adsorption of the CVD reactants prefers to grow at low surface temperature, typically in the range from −10 to 140 °C. Hence, rapid identification of the saturation pressure of a new vapor-phase reactant at the temperature of growth

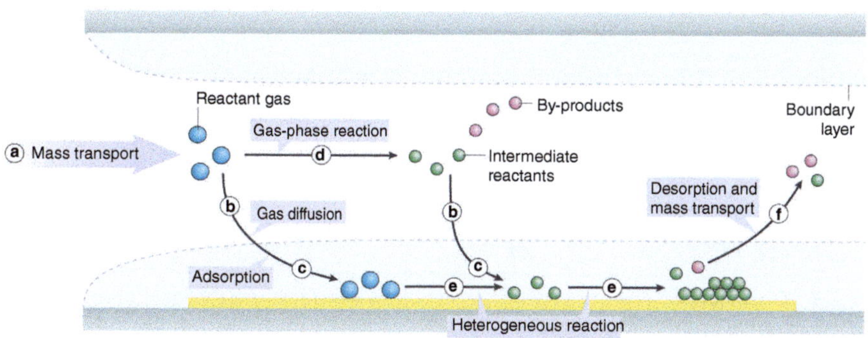

Figure 3.4 Schematic illustration of general elementary steps of a typical CVD process. Source: Reprinted from Sun et al. [18]. Copyright 2021, Springer Nature.

is quite necessary. This understanding of growth kinetics could be used to grow quantities of different homopolymers, copolymers, cross-linked organic networks, and organic/inorganic composite nanomembrane and adjust the condition to fabricate polymer nanomembrane above 10 μm thick [20].

3.2.4 Interfacial Polymerization

Interfacial polymerization is an important bottom-up method to fabricate polymeric nanomembranes. Many polymerization types are suitable for this interfacial polymerization, such as interfacial polycondensation, interfacial polyaddition, interfacial oxidative polymerization, interfacial polycoordination, interfacial supramolecular polymerization, and some others (shown in Figure 3.5a). In a typical interfacial polymerization process, two kinds of monomer species are dissolved in organic and aqueous (immiscible) solvents separately. At the interface of those solutions in two phases, the monomer species polymerized into a nanomembrane, which is governed by chemical kinetics and monomer diffusion through the polymerized nanomembrane. Thus, the reaction rate of the polymerized reaction quickly increases at first and reduces with the increase of the film thickness, and the polymerization finally stops due to the increased barriers of monomer diffusion (shown in Figure 3.5b) [21]. Quantities of factors such as temperature, solubility, the concentration of monomers, stoichiometry of the monomers [22], permeability of monomers in the polymer nanomembranes produced have a large effect on the thickness, morphology, and quality of the polymer nanomembranes produced. The high concentration of monomer induces to crumpled nanomembrane, and the nanomembrane is smooth and flat at the relatively low monomer concentration [23]. It is also possible to utilize the interfacial polymerization method to fabricate defect-free nanomembrane in difficult-to-machine materials,

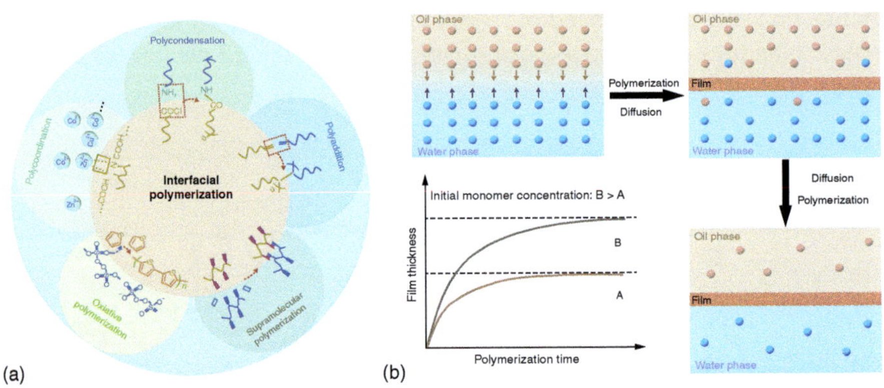

Figure 3.5 Classification and schematic illustration of interfacial polymerizations. (a) Classification of interfacial polymerizations. (b) Schematic illustration of the interfacial polymerization process and the relationship of film thickness and polymerization time at different monomer concentrations. Source: Adapted from Zhang et al. [21]. Copyright 2020, Wiley-VCH.

such as conductive polymers. However, the organic solvents for monomers are almost volatile organic compounds, which are a major source of pollution to the environment and limit the practical application in large areas. The separation of ionic liquids and water provides the possibility to work as "green" solvents for interfacial polymerization [24].

3.2.5 Release of Polymer Nanomembrane

The released polymeric nanomembrane is free-standing and exhibits fantastic properties. The typical methods for detaching polymeric nanomembranes from a solid substrate (e.g. glass, silicon wafers, plastic films) are mainly categorized into two steps: sacrificial layer and water-soluble supporting layer method (shown as Figure 3.6) [25]. In the initial state, the proper sacrificial layer is initially produced on the substrate, on which the polymeric nanomembrane is synthesized using the film-preparation methods mentioned above. Immersion into proper solutions dissolves the sacrificial layer to separate the perforated polymeric nanomembrane before the nanomembrane is submerged. Thus, the sacrificial layers are required to maintain undamaged against different solvents employed during the fabrication process and additionally able to dissolve in common organic solvents during the detachment and releasing process. Then the supporting layer (e.g. polyvinyl alcohol) is coated on the polymer nanomembrane and transferred from the substrate to the supporting layer. Finally, dissolution of the supporting layer releases the nanomembrane with the free-standing structure, and the polymeric nanomembrane is freely suspended on the surface of solutions. Kumar et al. fabricated a free-standing polysaccharide nanomembrane from chitosan using LbL assembly on a sacrificial layer [26]. The thickness of the nanomembrane is controlled through the number of polysaccharide layer pairs and the free-standing nanomembrane maintained the original shape and size of the substrate.

3.2.6 Incorporation of Nanomaterials

Incorporating inorganic or metallic nanoparticles into the polymeric nanomembranes matrix, the properties of the polymeric nanomembrane take advantage of both components and have been remarkably changed, including the mechanical

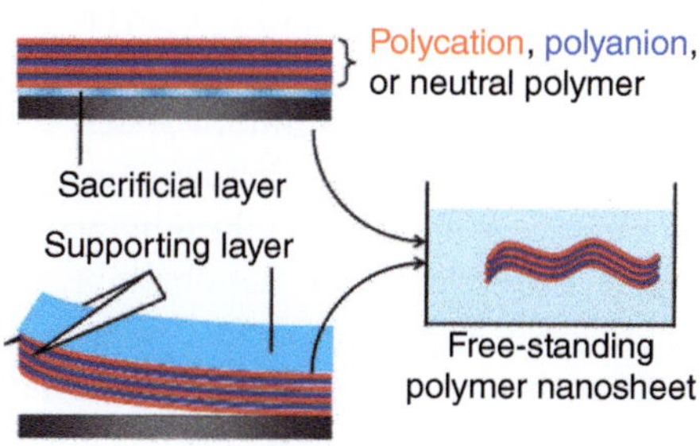

Figure 3.6 Schematic illustration of sacrificial and supporting layer methods for the fabrication of free-standing polymeric nanomembrane. Source: Reprinted from Fujie [25]. Copyright 2016, Springer Nature.

properties, selectivity, and permeability, which provide more opportunities for the researchers to develop more effective and suitable applications. For example, the incorporation of zinc oxide nanoparticles into chitosan film shows fantastic mechanical and antibacterial performance [27]. Generally, oxides, semiconductors, minerals, metals, and metal alloys, such as titanium dioxide (TiO_2), zinc oxide (ZnO), silver (Ag), and silica (Si) nanoparticles, are incorporated into the nanofiber membranes. Meanwhile, the most commonly used polymer matrixes contain amorphous thermoplastic (e.g. poly(methyl methacrylate) [PMMA], polycarbonate [PC], polystyrene [PS]) and thermosets (e.g. epoxy and silicone). However, incorporating nanoparticles in a polymer without any aggregation is a challenge, which might be caused by the aggregation state of the nanoparticles or poor affinity with a matrix. Thus, the global characteristics of polymer nanocomposites greatly depend on the dispersion of nanoparticles and their affinity with the matrix. Methods developed to control the polymerization process contain the adjustment of the polymerization process, the nanoparticles synthesis, and the polymer and the nanoparticle interface together with polymer casting or processing. The typical methodologies could be categorized into four parts: the blending of nanoparticles and polymer, the polymerization with the presence of preformed nanoparticles, the nanoparticle formation in a preformed polymer matrix, and the simultaneous polymerization and nanoparticle synthesis. The schematic illustrations of four methodologies are shown in Figure 3.7. Among all these four methodologies, direct mixing of nanoparticles with a polymer matrix (i.e. blending) is the simplest method, which mixes the nanoparticles into a melt polymer or a polymer solution. Through this method, massive nanocomposites could be manufactured in one step. Plasmonic nanoparticles incorporated in polymers at low volume ratios increase optical functionality or optical responsiveness to the material with distinct mechanical properties.

3.2.7 Metal/Polymer Nanomembranes

Deposition metal on polymer nanomembranes is another type of functionalization method for polymer nanomembranes. The simplest method is to spin-coat polymer nanomembrane and then deposit metal nanomembranes on it. Based on solid-state embossing of metal-coated polymer nanomembranes, a simple and versatile method is developed to create well-resolved metallic structures on polymer substrates [29]. These metal/polymer nanomembrane structures could also be patterned by microcutting, such as gold/FEP bilayer and gold/PVA. The extrusion-based additive manufacturing approach was applied to fabricate polymer–metal hybrids with polypropylene on aluminum sheets with electrochemically manufactured surface microstructures [30]. Contact temperature between the polymer and metal is controlled to adjust the adhesion incompatibility. A thermoresponsive and optically applicable copolymer nanomembrane of poly(*N*-isopropylacrylamide-R-glycidyl methacrylate) was produced to a responsive insulating layer, which was sandwiched between two porous silver nanomembranes to yield a thermoresponsive MIM cavity [31].

3.3 Properties of Polymer Nanomembranes

Based on various fabrication methods, polymer nanomembranes exhibit enhanced or new properties in mechanical properties, optical properties, permeability and stability, and so on. The enhancement might be induced by functional groups, cross-linking degree, the thickness of nanomembrane, percentage of incorporation, and so on. For each given property, we summarize factors that influence the properties of polymer nanomembranes.

3.3.1 Mechanical Properties

In the majority of the biomedical, pharmaceutical, robotic, and packaging applications, polymer nanomembranes are desired to be soft, flexible, and have a moderate toughness. Typical physical quantities to characterize the mechanical properties of polymer nanomembranes are tensile stress and Young's modulus. The composition of polymer is the key parameter to adjust the mechanical properties, such as incorporation of plasticizer or addition of cross-linker. The concentration of 3-glycidoxypropyltrimethoxysilane (GOPS) content in PEDOT:PSS has a significant effect to obtain mechanical integrity in aqueous media [32] (Figure 3.8a). Young's modulus of PEDOT:PSS nanomembranes increases from 90 to 150 MPa with the increasing amount of GOPS. Luqman et al. synthesized superabsorbent hydrogel

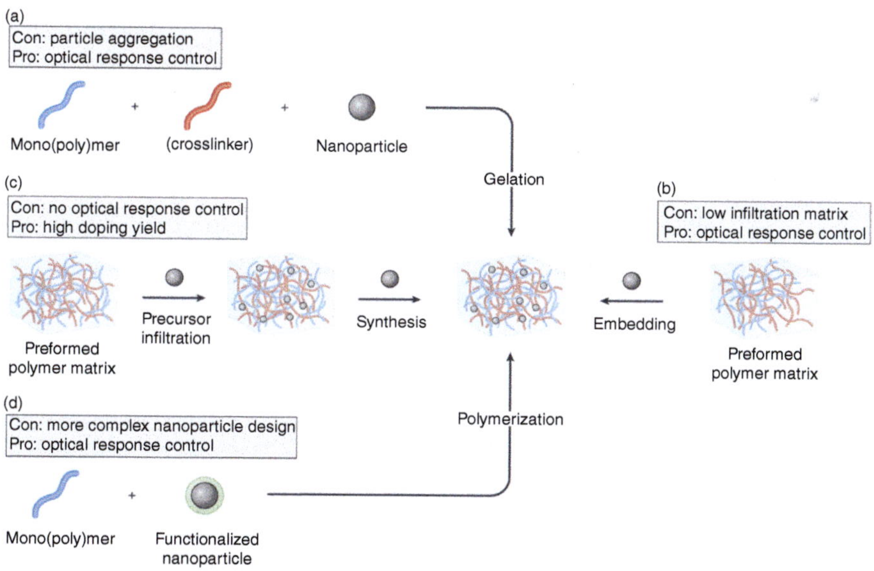

Figure 3.7 Schematic illustrations of fabrication methods of polymer nanocomposites. (a) The polymerization with the presence of pre-formed nanoparticles. (b) The blending of nanoparticles and polymer. (c) The nanoparticle formation in a preformed polymer matrix. (d) The simultaneous polymerization and nanoparticle synthesis. Source: Adapted from Pastoriza-Santos et al. [28]. Copyright 2018, Springer Nature.

with different ratios of acrylic acid to acrylamide [35]. The compressive hydrogel strength decreases with the increasing amount of acrylic acid, but the compressive hydrogel strength increases with the increasing amount of acrylamide. The mechanical robustness of polydopamine (PDA) nanomembranes might be raised by cross-linking of aggregated oligomers [36]. Meanwhile, the direction of fiber alignment within polymer nanomembrane determines the mechanical properties of polymer nanomembrane made by electrospun to a large degree [33] (Figure 3.8b). For aligned, random, and bilayered polycaprolactone, the tensile stresses are 1.3, 2.6, and 1.9 MPa correspondingly. Compared to bulk values at glassy state, the elastic moduli of polymer nanomembrane are gradually decreased when decreasing the thickness of the nanomembrane, especially for free-standing nanomembrane. A bilayer composite model is developed as well to predict the elastic moduli of free-standing nanomembrane [34]. The incorporation of inorganic nanoparticles into polymer nanomembrane could increase the mechanical properties as well.

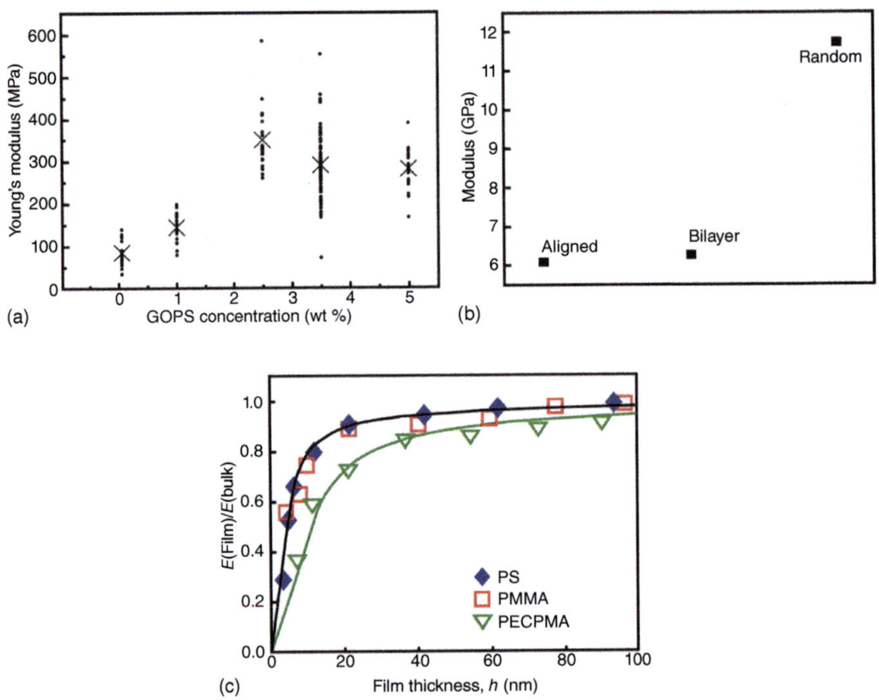

Figure 3.8 Mechanical properties of polymeric nanomembranes. (a) The change in Young's modulus of PEDOT:PSS as a function of GOPS content. Source: Adapted from ElMahmoudy et al. [32]. Copyright 2017, Wiley-VCH. (b) Young's modulus of electrospun polymeric nanomembranes in different alignment directions. Source: Adapted from Ghosal et al. [33]. Copyright 2018, Springer Nature. (c) Normalized elastic moduli of free-standing nanomembrane as a function of thickness. Source: Adapted from Xia and Lan [34]. Copyright 2019, American Chemical Society.

A significant increase of the tensile modulus was observed when 5 wt% ZnO nanoparticles were added into the PS matrix (Figure 3.8c).

3.3.2 Permeability

The permeation of molecules and ions through polymeric nanomembranes is an important requirement for the membrane separations of two phases. Rapid transportation needs using ultrathin layers of polymeric materials. At molecular level, the accurate design of film architecture endows selective and switchable permeation behaviors of polymer nanomembranes. The absorption and diffusion of permeant substances are typically increased with the thickness of the polymeric nanomembrane. External stimuli such as temperature, humidity, pH, and light affect the properties of polymeric nanomembranes as well [37]. On the other hand, functional groups in high densities can also lead to changes in film thickness changes. Quantities of organic functional groups, amine, carboxylic acid, epoxy, hydroxyl, perfluorocarbon, and zwitterionic moieties have been incorporated into polymeric nanomembranes to systematically tune the permeation property. Most of these moieties work as surface functionalization of polymer nanomembranes, which further allows the permeation of biomolecules and nanoparticles. Meanwhile, the functional groups of polymeric nanomembranes could also alter the surface energy from superhydrophobic to superhydrophilic, which alters the permeation of the polymer nanomembrane as well (Figure 3.9).

3.3.3 Stability

The long-term stability of polymer nanomembranes is strictly limited by the fact that polymers are materials generally out of equilibrium. Even without the influence of external conditions, the mechanical, thermal, and structural properties

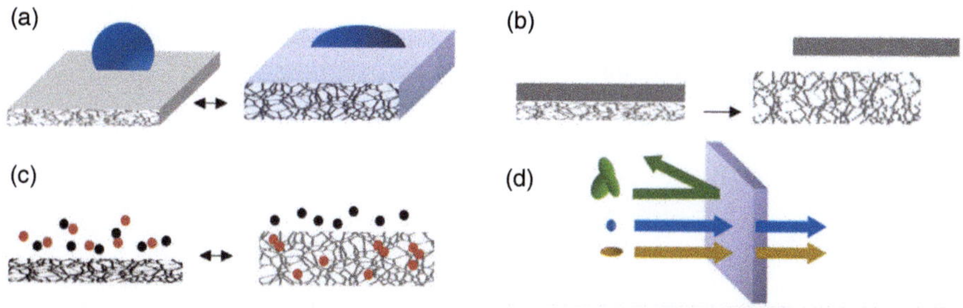

Figure 3.9 Permeation control of polymeric nanomembranes. (a) Collapsed and expanded states switch of hydrogel mesh upon the reversible water intake and expulsion. (b) Delamination and release of a free-standing hydrogel thin film induced by interfacial stress changes from the swelling. (c) Permeant species selectively pass through the polymeric nanomembranes. (d) Membrane separation relies on selective permeation across a layer. Source: Adapted from Gleason [37]. Copyright 2020, AVS.

of polymers might change with the changing storage time [38]. This effect is even enlarged when the polymer is nanoconstrained and its thickness is reduced to the nanometer scale. Therefore, the stability of polymer nanomembrane according to the properties for which they are fabricated is important to be explored. The abovementioned physical properties such as permeability, refractive index, and spectroscopic properties can be controlled. A simple theoretical model to set criteria of geometric features and analyze the mechanical stability for polymeric nanomembrane was developed [39].

3.3.4 Stimuli-Responsive Properties

A broad class of polymer nanomembranes possesses the stimuli responsiveness, which means they could be induced to undergo reversible or single-cycle phase transitions under different physical or chemical external stimuli [40]. These external stimuli include physical triggers containing thermal, magnetic, ultrasonic, electrochemical, or light stimuli and chemical triggers containing pH, redox reactions, supramolecular complexes, and bio-catalytically driven reactions [41]. Anionic hydrogel nanomembranes remain collapsed at low pH because of the existence of physical interactions that keep the network tightly complex. Once the pH increases above the pKa value of the polymer, the polymer nanomembrane swells because of the combination of electrostatic repulsion and water imbibition. Cationic hydrogel nanomembranes show an opposite swelling behavior to anionic hydrogel, which is shown in Figure 3.10.

The earliest work on electrical responsive polymers was published by Osada et al [43]. PAMPS hydrogel cross-linked from *N*,*N*-methylene bisacrylamide in a low cross-linking degree was immersed in a surfactant solution (*n*-dodecyl pyridinium chloride). The gel walked through a solution under the alternative and a reversible voltage of 20 V with an interval of two seconds. The gel is propelled by the reversible shrink of the polymer in an electric field, which is caused by the reversible and cooperative complexing of surfactant molecules onto the polymer. Through the self-imitated UV polymerization process, HEMA, $BMIMBF_4$, and TiO_2 are cross-linked to supramolecular tough nanocomposite [44]. This capacitor-like

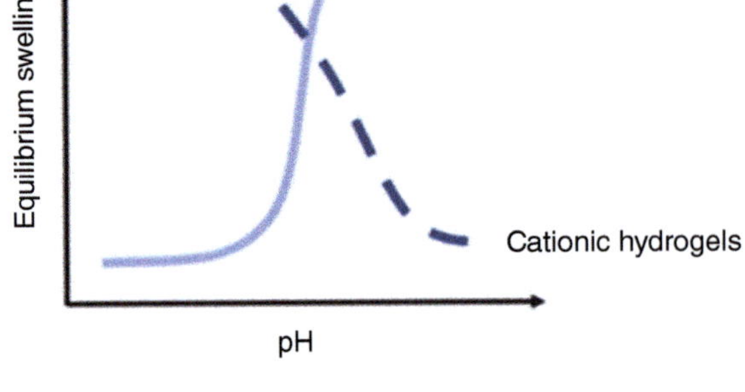

Figure 3.10 The equilibrium swelling behavior of anionic and cationic hydrogels. Source: Adapted from Koetting et al. [42]. Copyright 2015, Elsevier.

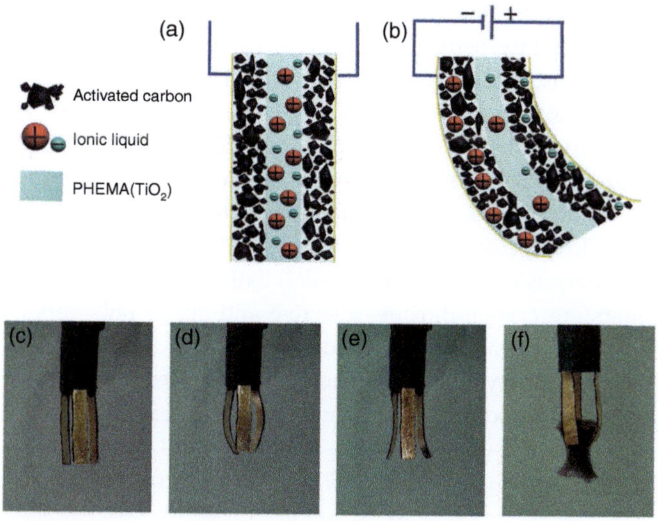

Figure 3.11 Schematic illustration of BMIMBF4 gel electrolyte-based actuator driven by an ion-transfer mechanism. (a) The electric double layer formed inside the porous media of activated carbon. (b) Displacement of the actuator under an applied voltage. (c–f) Photo of electrically driven polymer nanomembrane. Pristine state (c), under an applied voltage (d), under an inversely applied voltage (e), grabbing an object with the mass of 180 mg. Source: Liu et al. [44]. Reproduced with permission from Springer Nature.

soft actuator made of polymer nanomembrane showed a displacement response of electricity by electrically induced ion migration (shown in Figure 3.11).

3.4 Applications of Polymer Nanomembrane

Benefiting from the excellent properties of polymer nanomembrane in mechanical properties, optical properties, permeability, stability, and stimuli responsiveness, the polymer nanomembranes based on membrane technology show wide applications in various environmental applications, such as gas purification, wastewater treatment, and energy devices.

3.4.1 Gas Purification

Industrial manufacturing processes require separating a single gas constituent from a gas mixture. Among all gas-purification technologies, physical adsorption and membrane gas separation possess the flexibility to be applied to versatile gas-separation applications. For membrane gas separation, ideal materials are supposed to offer an optimal combination of permeability, selectivity, and sustainable performance over long-term operations. For economic viability, ideal materials are expected to be cost-effective and well balanced against production costs. Among the large family of materials, polymers with sub-nano-size cavities appear to be the appropriate choice for membrane gas separation.

Fundamental principles of gas sorption and transportation in polymer and material selection guidelines are discussed in detail by Ghosal et al. in 1993 [45]. The core of the gas-separation process by polymer nanomembrane is governed by the excellent gas selectivity and high permeability, which should be very close to or exceed the 2008 Robeson upper bound [46]. Polymeric nanomembrane applied in gas separation is limited by insufficient gas flow, which could be ideally overcome by simply reducing the thickness of membranes, which calls for nanomembrane. Generally, rubber polymers are carbon dioxide–selective and glassy polymers are hydrogen-selective, which meet the most important demands of CO_2 capture and H_2 purification in technological fields. Typical gas permeability and selectivity trade-off relationship with a large range of microporous polymers are shown in Figure 3.12.

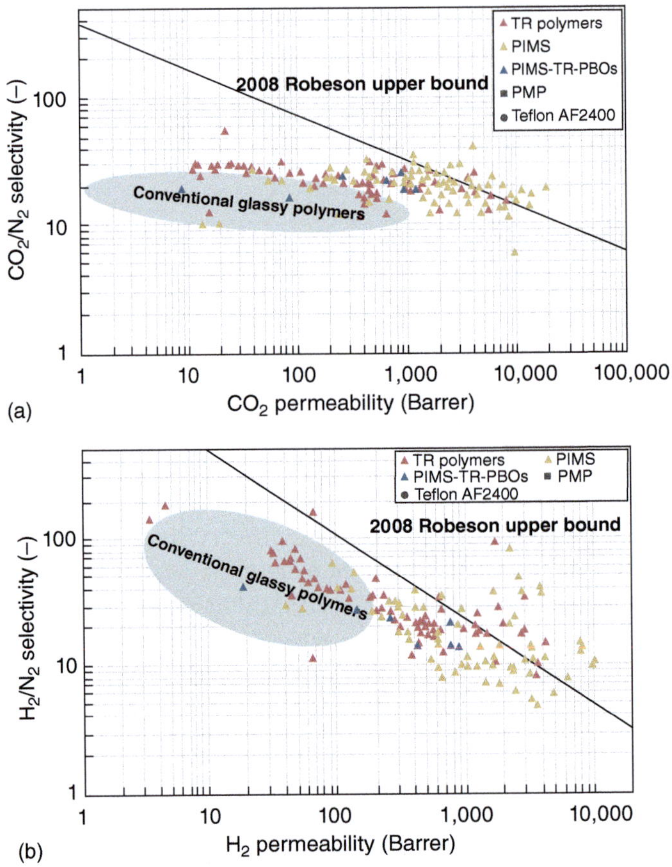

Figure 3.12 Permeability and selectivity trade-off relationship with a large range of microporous polymers. (a) Carbon dioxide/nitrogen and (b) hydrogen/nitrogen. The gray areas reveal the relative position of the conventional glassy polymers. Source: Adapted from Dong and Lee [47]. Copyright 2017, Royal Society of Chemistry.

Polyimides, substituted polyacetylenes, thermally rearranged polymers, and polymers with intrinsic microporosity are regarded as attractive materials for CO_2/N_2 separation membrane owing to the high carbon dioxide gas permeability and good selectivity between carbon dioxide and nitrogen. Incorporating inorganic fillers to form a composite and a mixed matrix membrane is a promising strategy to tackle CO_2 transportation efficiency and the filler facilitates the transportation of gas CO_2 but blocks the mass transfer pathway of gas N_2, which is shown in Figure 3.13 [48]. Multilayer poly(ethylene oxide) (PEO) and poly(methacrylic acid) (PMAA) deposited through LbL assembly show high gas-pair separation [49]. Eight- and ten-bilayered PEO/PMAA nanomembranes deposited on a 25 μm polystyrene show extremely high CO_2/N_2 selectivity of 142 and 136, respectively. Maximum CO_2 permeability is achieved by 100 000 g/mol PEO, but the highest CO_2/N_2 selectivity is achieved at the highest molecular weight of 4 000 000 g/mol. Carbon dioxide philic functional groups also provide excellent CO_2/light gases selectivity. For example, PIMS modified with triazole and methyl tetrazole proves this theory.

Hydrogen purification is another significant demand in several technological fields, such as fuel cell technology. Compared to other separation processes, membrane gas separation is an appropriate choice for producing highly pure hydrogen. Such polymeric nanomembranes are supposed to be selective for particular gas at a certain pressure and temperature range [50]. A nanoscopic thin selective layer of the block copolymer (PS-*b*-P4VP) with additive is fabricated on the polyethylene terephthalate (PET) substrate [51]. The porous PET nanomembranes with a block copolymer coating show a high selectivity of hydrogen against carbon dioxide. Similar results are reported by Kamakshi et al., and they coated carboxyl-functionalized PET nanomembranes with block copolymers [52]. Block copolymer also plays a crucial role in the improvement of hydrogen gas selectivity.

Other gas separation could be realized by the polymer nanomembrane as well. Amidoxime-functionalized polymer of intrinsic microporosity, ultrahigh hydrogen

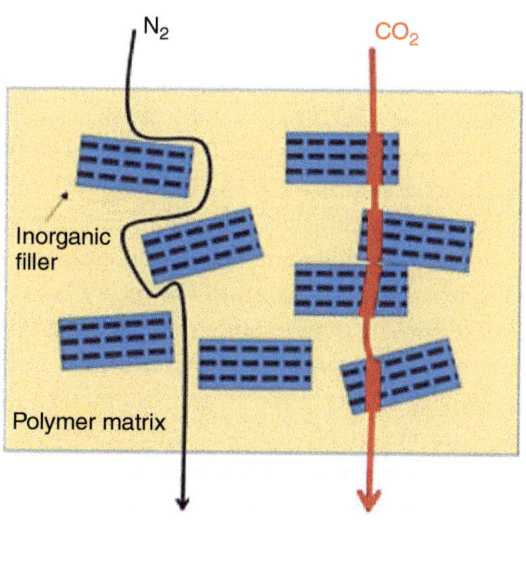

Figure 3.13 Function of fillers added to a mixed matrix membrane. The width of the line represents the mass transfer rate of gases across the nanomembrane. In this schematic illustration, the porous nanoparticles are beneficial to transport carbon dioxide gas but block the mass transfer pathway for nitrogen gas. Source: Adapted from Norahim et al. [48]. Copyright 2017, Wiley-VCH.

sulfide/methane selectivity up to 75, and hydrogen sulfide permeability above 4,000 is achieved under challenging feed pressure up to 77 bar of ternary feed mixture [53]. The extraordinary gas selectivity and permeability are two or three orders of magnitude higher than commercially available glassy polymeric membrane due to the unique design of surface modification.

3.4.2 Wastewater Treatment and Water Desalination

Due to the rapid growth of water demand, wastewater production, and reduction of available surface water reserves, effective wastewater treatment solutions and water recycling methods are required for the production of clean water. Detrimental contaminants, such as organic pollutants, chemicals, bacteria, particulates, minerals, and dissolved heavy metal ions, should be removed. Water purification technologies are mainly categorized into three types: physical separation (e.g. thermal distillation, filtration, and membrane separation), chemical processes (e.g. disinfection, coagulation, and ion exchange), and biological processes (e.g. slow sand filtration). Among all these methodologies, membrane separation occupies more than half of the current water purification market share.

The selective permittivity characteristic of polymer nanomembranes is exploited in a wide range of technological fields for water desalination technology, especially in reverse ultrafiltration (UF) and reverse osmosis (RO) processes. A synthetic polymeric nanomembrane is an interface acting as a selective barrier between two adjacent phases without additives to efficiently remove small organic molecules and inorganic salts. The pore size of polymer nanomembrane is typically 1 nm corresponding to 300–500 Da molecular weight cutoff [54]. For water filtration, polymer nanomembranes provide several advantages, including high flux, low maintenance, low operation cost, and low operating pressure. Polyvinylidene difluoride (PVDF) with extraordinary chemical and mechanical stability and good support porosity is an excellent material to work as a support polymer. Several typical monomers such as isophthaloyl chloride, tannic acid, and polyvinyl amine react with trimesoyl chloride (TMC), bisphenol A, *m*-phenylenediamine (MPD), diethylenetriamine, triethylenetetramine, tetraethylenepentamine, and piperazine to form active polymer nanomembrane for an efficient water treatment process. Working as UF film, Nanofiltration 270 (NF270), HODRA-CORE, ESNA-1-K1, and ESNA-1-LF were developed to remove the As(V) from wastewater [55]. Under optimal situations, ESNA-1-LF showed more than 94% rejection compared to other nanomembranes. Low ion concentrations and small amounts of addition of HA show high removal efficiency of As element. By irradiating polyethylene terephthalate Lumirror films with GeV heavy ions followed by ultraviolet exposure, the polymeric nanomembrane shows a high transport rate of potassium ions and a selectivity of alkali metal ions over heavy metal ions over 500 [56]. The selectivity is attributed to the interaction between the partially dehydrated ions and the negatively charged nanopore wall. Three representative ionic transport mechanisms with the pore size increasing negatively charged sidewalls (in blue) are shown in Figure 3.14. When the pore size is less than the hydration radius of

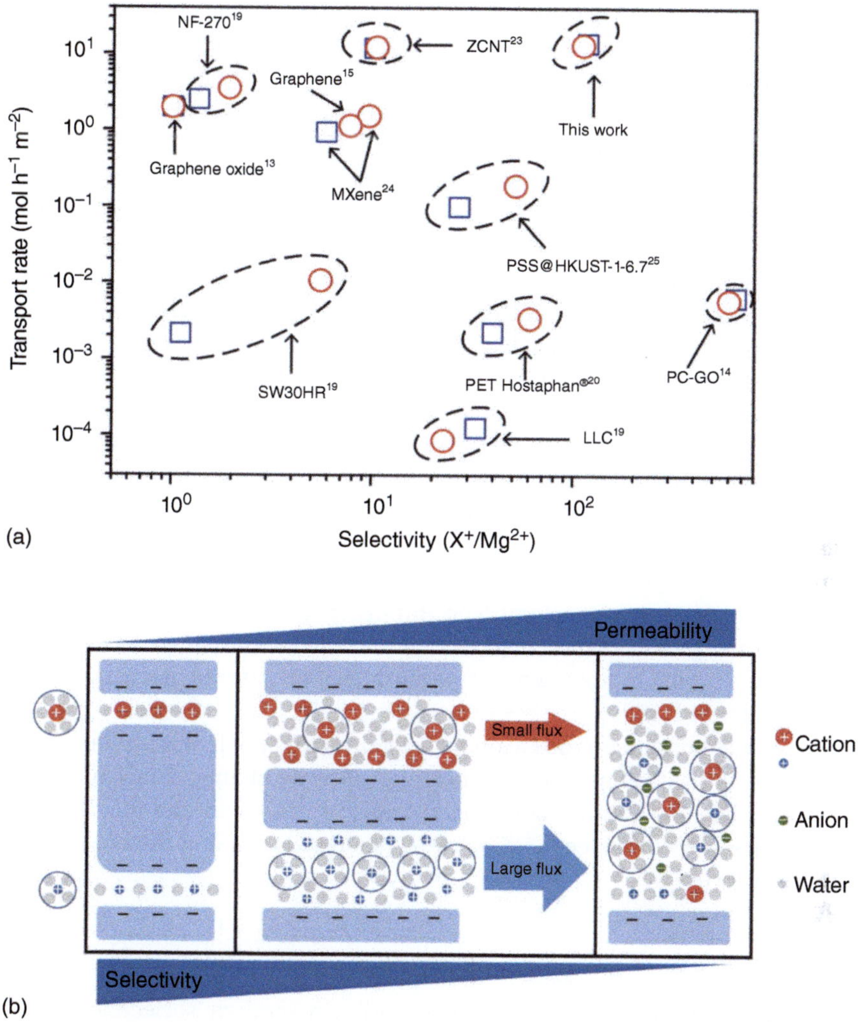

Figure 3.14 Ionic transport behaviors and representative mechanisms of polymer nanomembranes. (a) Transport rates versus selectivity of various nanoporous membranes. The selectivity of K^+/Mg^{2+} is in blue and the selectivity of Na^+/Mg^{2+} is in red. (b) Schematic illustration of three representative ionic transport mechanisms with different pore sizes. Source: Adapted from Wang et al. [56]. Copyright 2018, Springer Nature.

the cations, the dehydration energy decides the ionic selectivity. The left panel shows great selectivity but low permeability of polymeric nanomembrane with a pore size less than the hydration radius of cations. The right panel shows great permeability but no ionic selectivity of polymeric nanomembrane with a pore size much larger than the Debye screening length. The middle panel shows a high transport rate of one cation and high selectivity between the two cations with pore size in between.

Copolymerization of polymers is a method to combine two or more polymers to produce a new material with a target application. For wastewater treatment, the copolymerization technique provides an enhancement on the adsorption capacity of single polymers and improves the polymer aggregation in alkaline medium and polymer disintegration in aqueous solution as well [57]. A double polymer matrix of polystyrene sulfonate/3,4-ethlenedioxythiophene and lignin for the treatment of heavy metal ions was also reported by Checkol et al. [58]. The polystyrene moieties in the copolymer are regarded as a cation exchanger and the sulfonate groups and the lignin moieties increased the ability of toxic heavy metal ions' adsorption. The adsorption capacity reached to 245.5 mg/g by the double-polymer matrix and almost doubled (452.8 mg/g) when the lignin moieties were incorporated into the double-polymer matrix. Nano-enhanced and nano-engineered membranes are prepared to finely tune the selectivity, permeability, morphology, and physicochemical properties of nanomembrane to produce improved antifouling and antimicrobial properties and chlorine resistance. Superabsorbent polymer hydrogels (SPH) consisting of acrylic acid and acrylamide in different ratios have been synthesized, and the excellent swelling property is considered to be effective for removing the heavy metal ions in polluted water [33]. In other aspects, macromolecules are introduced into polymer nanomembrane through the interfacial polymerization system and the polyamide nanomembrane produced possesses spotted and striped Turing structures, which provide better performance of water treatment than the flat nanomembrane [59].

Besides the methods mentioned above, the incorporation of inorganic nanoparticles into the polymer nanomembrane matrix is an effective method to increase water purification efficiency. Kanjwal et al. incorporated $AgTiO_2$ particles into polymer nanomembrane, which induced a photocatalytic degradation and remove 75–95% of dairy effluents [60].

3.4.3 Polymer Electrolyte Nanomembrane

Energy-storage devices play vital important role in modern society and daily life. Organic electrodes, free of toxic metal species, show the possibility to lead to a new generation of high-efficient consumer energy-storage devices. Polymer nanomembranes, featuring open structures and tailorable chemical components, exhibit the possibility for new energy-storage devices. A systematic library of vinyl ether-based PEO side-chain copolymers was synthesized to control the crystallization of PEO [61]. The potential application of PEO side-chain copolymers to be polymer electrolytes was demonstrated. A homogeneous PEG network and $LiNTf_2$ were produced with excellent mechanical properties [62]. Polymer electrolyte membrane unitized regenerative fuel cells (PEM-URFCs) operate in two modes of hydrogen production (electrolysis cell mode) (EC mode) and power production (fuel cell mode) (FC mode) [63]. The hydrophobicity for water drainage in the FC mode and hydrophilicity for water supplement in EC mode, which are contradictory roles in the two modes. Thus, alternating hydrophobic and hydrophilic porous transportation layers were made to provide valuable insights into addressing

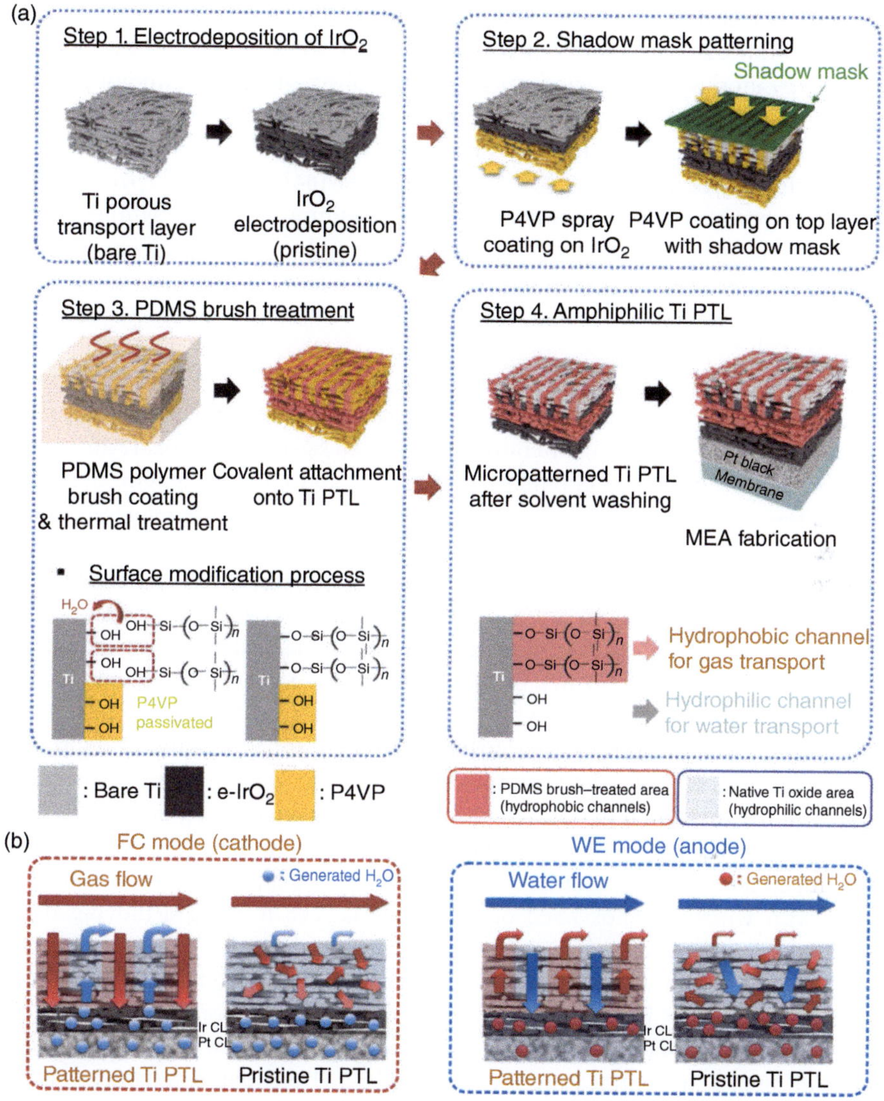

Figure 3.15 Fabrication and transport mechanisms of amphiphilic polymer nanomembranes utilized in regenerative fuel cells. (a) Schematic fabrication procedure of amphiphilic porous transportation layer. (b) Schematic illustration of mass transfer pathways in the patterned (left) and pristine (right) porous transportation layers under FC and EC modes. The red arrows represent the diffusion path of the oxygen, and the blue arrows indicate water entering the electrode. Source: Adapted from Lim et al. [63].

the contradictory and providing high efficiency (shown as Figure 3.15). Supercapacitors, featuring high power density and rapid charging, are another type of important energy-storage device, which could be categorized into electrical double-layer capacitors and pseudocapacitors. Polymeric nanomembranes are promising materials for capacitive energy storage because they provide high surface areas for increased double-layer capacitance, open porous structures for quick ion transport, and regulated redox-active moieties that allow faradaic (pseudocapacitive) energy storage. Conjugated microporous polyaminoanthraquinone polymer nanomembrane copolymerized from the Buchwald–Hartwig coupling reactions between 2,6-diaminoanthraquinone and aryl bromides was reported to demonstrate highly efficient supercapacitance properties [64]. Moreover, pyrene-based conjugated porous organic polymers with N-containing networks synthesized by Buchwald–Hartwig coupling reactions were also found to exhibit high capacitances with excellent long-term cyclic stability [65].

3.4.4 Actuators and Microrobots

An actuator is a mechanical device that converts the energy in other forms to forces and is used for moving or controlling a mechanism or system. Traditionally, actuators are propelled by pneumatic, hydraulic, and electromagnetic materials, which show good energy efficiency and low work since they are made with hard and heavy materials. Thus, soft materials made by polymer nanomembranes have attained active components that perform or transfer mechanical motions because of their responsiveness and softness, such as artificial muscles, soft robotics, programmable origami, and molecular motors. A thermal bimorph membrane structure of poly(3,4-ethylenedioxyhiophene) doped with *p*-toluenesulfonate (PEDOT-Tos) coated onto a polyethylene terephthalate (PET) is adopted to be a light-driven vibrotactile actuator [66]. Exposed to near-infrared light, the PEDOT-Tos layer converts the light energy into thermoelastic bending deformation because of the mismatch in coefficient of thermal expansion between PEDOT-Tos and PET. Liquid crystal elastomer doped with a near-infrared dye with one side cross-linked monodomain and the other side cross-linked polydomain is fabricated to flat strip with two ends fixed on substrate surface forming a moving bump [67]. Under laser scanning, the polymer nanomembrane strip shows the ability to move straight, climb an inclined surface, and undergo light-controlled arbitrary directional turning. Through the solvent molecule sorption mechanism, the porous polymer actuators possess the ability to bend in response to acetone vapor at a high speed and large-scale locomotion [68]. Dual photo-responsive liquid crystalline-based actuator containing a dye could locally convert to other forms with rewritable and reprogrammable ability [69]. Each dye absorbs at a different wavelength giving access to programmable actuators, and the folding process of dual-responsive actuators can be controlled by using different colors of light (Figure 3.16).

Hydrogels are three-dimensional network structures consisting of polymeric chains connected by tie points or joints by covalent bonds, ionic bonds, hydrogen bonds, physical entanglements, or dipolar interactions. They are insoluble and swell in water up to thermodynamic equilibrium. Some of the most commonly used hydrogels contain synthetics, such as polyethylene glycol, polyvinyl alcohol, or poly(2-hydroxyethyl methacrylate), and natural hydrogels, such as agarose, alginate, cellulose, chitosan, collagen, fibrin, and hyaluronan. Hydrogel nanomembranes,

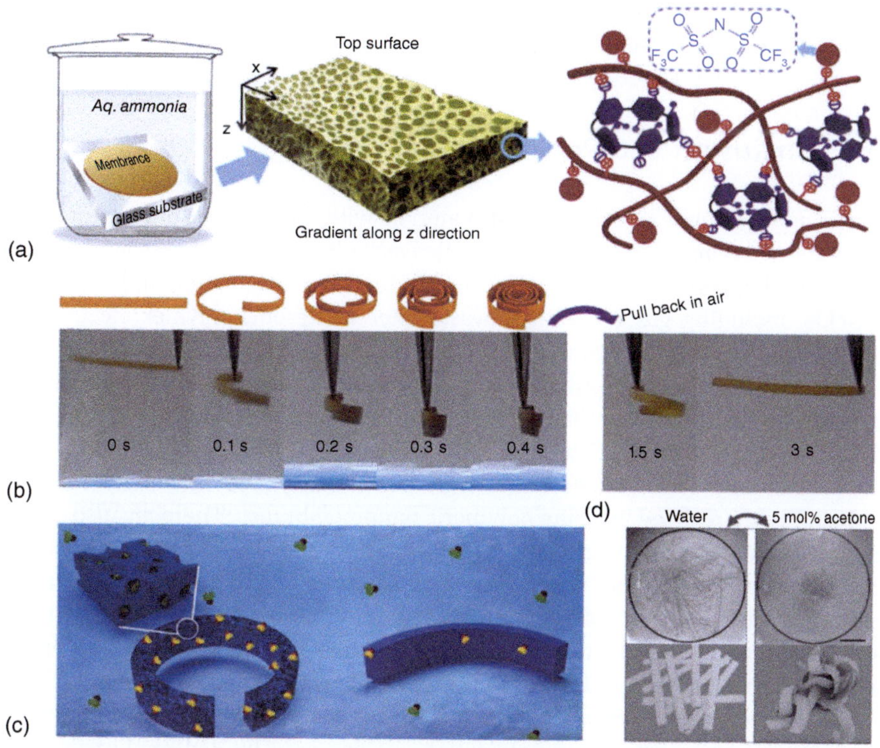

Figure 3.16 Schemes and photos of porous polymer nanomembranes working as soft actuators. (a) A schematic illustration of the fabrication routine (left), porous morphology (middle), and the chemical structure (right) of the PILTf$_2$/C-pillar[5]arene membrane actuator. The red line represents PILTf$_2$N polymer chains and the purple line represents the C-pillar[5]arene molecules. The schematic illustration shows the electrostatic complexation between the imidazolium cations on PILTf$_2$N and the carboxylate anions on C-pillar[5]arene molecules. (b) Adaptive movement of nanomembrane placed in acetone vapor (left) and then back in air (right). The acetone solvent is stained in blue. (c) A schematic mechanism of the sensitive actuation of a porous polymer nanomembrane actuator (left) compared to dense one (right). (d) Cooperative actuation of PCMVImTf$_2$N-PAA nanomembranes shuttled between water (left, top panel) and 5 mol% acetone and water mixture (right, top panel). Schematic illustration of the entangling and disentangling of nanomembrane stripes. The scale bar is 1 cm. Source: Wu et al. [70]. Reproduced with permission from Wiley-VCH.

featuring stimuli-responsive properties, are excellent actuators under various external stimuli, such as light stimulus, electrical stimulus, and chemical stimulus. A dual-layer, dual-composition polysiloxane-based liquid crystal soft actuator was synthesized as a plant tendril mimic material [71]. The actuator was able to perform two different three-dimensional reversible bendings versus chiral twisting under light stimuli of ultraviolet or near-infrared. The hydrogel walkers that were made of polyanionic poly(2-acrylamido-2-methylpropanesulfonic acid-*co*-acrylamide) were designed to an arc looper-like shape to show walk motion with two "legs" [72]. Such hydrogel systems create new opportunities for developing electro-controlled soft materials with simple fabrication strategies in the soft robotic field to control and manipulate the locomotion remotely.

3.5 Conclusion and Prospect

Polymer nanomembranes are low-cost materials with outstanding properties such as well-defined pore structures and controlled surface chemistry. These advantages provide more opportunities for applications in many environmental and engineering fields, including gas separation, wastewater treatment, polymer electrolyte, electrochemical energy storage, optoelectronics, actuators, and microrobots. Generally, the fabrication and modification processes of polymer nanomembranes are relatively simple, controllable, and scalable. Correspondingly, quantities of polymer nanomembranes with controlled compositions and functionalities have been successfully synthesized. Thus, high permeability and selectivity for gas purifications are realized by using polymeric nanomembranes. There are also three significant advantages of using polymer nanomembranes in microrobotics, compared to the use of inorganic materials, which are large propulsion force generated from large deformation of polymer nanomembrane, softness to adapt to a complex environment, and biocompatibility. The development of new synthetic strategies offers new opportunities for the applications of polymeric nanomembranes.

Although membrane technology is considered a promising technology for some environmental and engineering applications, it has only limited applications in industrial practice. The gap between laboratory research and the pilot plant still exists during this developing state. Taking gas purification as an example, flue gases contain minor components such as sulfide, oxynitride, carbon monoxide, and water vapor in various concentrations. Although they are not absorbed by polymer nanomembrane and the selectivity of target molecules is not reduced during the gas separation, these minor components largely reduce the lifetime of polymer nanomembrane [48]. Thus, in carbon dioxide industries, the polymer nanomembrane stability in long-term operation, membrane wetting, and membrane plasticization are still challenging problems. The wetting mechanism related to membrane properties and operating conditions is necessary to be established. More choices of polymer nanomembrane materials and synthesis mechanisms on molecular levels are needed to be explored. The contents covered in this chapter are just the tip of the iceberg in the development of polymer nanomembrane. With continuous contributions from interdisciplinary fields, practical applications of polymer nanomembrane with novel functionalities in diverse fields can be realized.

References

1 Doi, M. (1996). *Introduction to Polymer Physics*. Clarendon Press.
2 Ebewele, R.O. (2000). *Polymer Science and Technology*. CRC Press.

3 Jaber, J.A., Chase, P.B., and Schlenoff, J.B. (2003). Actomyosin-driven motility on patterned polyelectrolyte mono- and multilayers. *Nano Lett.* 3 (11): 1505–1509.

4 Hajicharalambous, C.S., Lichter, J., Hix, W.T. et al. (2009). Nano- and sub-micron porous polyelectrolyte multilayer assemblies: biomimetic surfaces for human corneal epithelial cells. *Biomaterials* 30 (23–24): 4029–4036.

5 Tai, Z., Ma, H., Liu, B. et al. (2012). Facile synthesis of Ag/GNS-g-PAA nanohybrids for antimicrobial applications. *Colloid. Surf. B Biointer.* 89: 147–151.

6 Wong, S.Y., Li, Q., Veselinovic, J. et al. (2010). Bactericidal and virucidal ultrathin films assembled layer by layer from polycationic N-alkylated polyethylenimines and polyanions. *Biomaterials* 31 (14): 4079–4087.

7 Berg, M.C., Zhai, L., Cohen, R.E., and Rubner, M.F. (2006). Controlled drug release from porous polyelectrolyte multilayers. *Biomacromolecules* 7 (1): 357–364.

8 Hall, D.B., Underhill, P., and Torkelson, J.M. (1998). Spin coating of thin and ultrathin polymer films. *Polym. Eng. Sci.* 38 (12): 2039–2045.

9 Zheng, S.Y., Tian, Y., Zhang, X.N. et al. (2018). Spin-coating-assisted fabrication of ultrathin physical hydrogel films with high toughness and fast response. *Soft Matter* 14 (28): 5888–5897.

10 Mondal, S. (2021). Impact of the process conditions on polymer pattern morphology during spin coating over topological surfaces. *Soft Matter* 17 (5): 1346–1358.

11 Zhang, Y., D'Ambra, C.A., Katsumata, R. et al. (2019). Rapid and selective deposition of patterned thin films on heterogeneous substrates via spin coating. *ACS Appl. Mater. Interfaces* 11 (23): 21177–21183.

12 Li, Z., Lv, Y., Ren, L. et al. (2020). Efficient strain modulation of 2D materials via polymer encapsulation. *Nat. Commun.* 11 (1): 1151.

13 Xie, K., Fu, Q., Qiao, G.G., and Webley, P.A. (2019). Recent progress on fabrication methods of polymeric thin film gas separation membranes for CO_2 capture. *J. Membr. Sci.* 572: 38–60.

14 Vapaavuori, J., Grosrenaud, J., Siiskonen, A. et al. (2019). Photocontrol of supramolecular azo-containing block copolymer thin films during dip-coating: toward nanoscale patterned coatings. *ACS Appl. Nano Mater.* 2 (6): 3526–3537.

15 Xi, W., Kong, F., Yeo, J.C. et al. (2017). Soft tubular microfluidics for 2D and 3D applications. *Proc. Natl. Acad. Sci.* 114 (40): 10590–10595.

16 Wang, M., Janout, V., and Regen, S.L. (2010). Glued Langmuir–Blodgett bilayers from Calix[*n*]arenes: influence of Calix[*n*]arene size on ionic cross-linking, film thickness, and permeation selectivity. *Langmuir* 26 (15): 12988–12993.

17 Yagüe, J.L., Coclite, A.M., Petruczok, C., and Gleason, K.K. (2013). Chemical vapor deposition for solvent-free polymerization at surfaces. *Macromol. Chem. Phys.* 214 (3): 302–312.

18 Sun, L., Yuan, G., Gao, L. et al. (2021). Chemical vapour deposition. *Nat. Rev. Methods Prim.* 1 (1): 5.

19 Moni, P., Al-Obeidi, A., and Gleason, K.K. (2017). Vapor deposition routes to conformal polymer thin films. *Beilstein J. Nanotechnol.* 8: 723–735.

20 Tao, R. and Anthamatten, M. (2012). Condensation and polymerization of supersaturated monomer vapor. *Langmuir* 28 (48): 16580–16587.

21 Zhang, F., Fan, J., and Wang, S. (2020). Interfacial polymerization: from chemistry to functional materials. *Angew. Chemie Int. Ed.* 59 (49): 21840–21856.

22 Zhang, P., Zhang, F., Zhao, C. et al. (2016). Superspreading on immersed gel surfaces for the confined synthesis of thin polymer films. *Angew. Chemie Int. Ed.* 55 (11): 3615–3619.

23 Karan, S., Jiang, Z., and Livingston, A.G. (2015). Sub-10 nm polyamide nanofilms with ultrafast solvent transport for molecular separation. *Science* 348 (6241): 1347–1351.

24 Zhang, J., Liu, H., and Jiang, L. (2017). Membrane-based strategy for efficient ionic liquids/water separation assisted by superwettability. *Adv. Funct. Mater.* 27 (20): 1606544.

25 Fujie, T. (2016). Development of free-standing polymer nanosheets for advanced medical and health-care applications. *Polym. J.* 48 (7): 773–780.

26 Kumar, M.N.V.R., Muzzarelli, R.A.A., Muzzarelli, C. et al. (2004). Chitosan chemistry and pharmaceutical perspectives. *Chem. Rev.* 104 (12): 6017–6084.

27 Shankar, S., Wang, L.-F., and Rhim, J.-W. (2018). Incorporation of zinc oxide nanoparticles improved the mechanical, water vapor barrier, UV-light barrier, and antibacterial properties of PLA-based nanocomposite films. *Mater. Sci. Eng., C* 93: 289–298.

28 Pastoriza-Santos, I., Kinnear, C., Pérez-Juste, J. et al. (2018). Plasmonic polymer nanocomposites. *Nat. Rev. Mater.* 3 (10): 375–391.

29 Stutzmann, N., Tervoort, T.A., Bastiaansen, K., and Smith, P. (2000). Patterning of polymer-supported metal films by microcutting. *Nature* 407 (6804): 613–616.

30 Hertle, S., Kleffel, T., Wörz, A., and Drummer, D. (2020). Production of polymer-metal hybrids using extrusion-based additive manufacturing and electrochemically treated aluminum. *Addit. Manuf.* 33: 101135.

31 Umar, M., Son, D., Arif, S. et al. (2020). Multistimuli-responsive optical hydrogel nanomembranes to construct planar color display boards for detecting local environmental changes. *ACS Appl. Mater. Interfaces* 12 (49): 55231–55242.

32 ElMahmoudy, M., Inal, S., Charrier, A. et al. (2017). Tailoring the electrochemical and mechanical properties of PEDOT:PSS films for bioelectronics. *Macromol. Mater. Eng.* 302 (5): 1600497.

33 Ghosal, K., Chandra, A., Praveen, G. et al. (2018). Electrospinning over solvent casting: tuning of mechanical properties of membranes. *Sci. Rep.* 8 (1): 5058.

34 Xia, W. and Lan, T. (2019). Interfacial dynamics governs the mechanical properties of glassy polymer thin films. *Macromolecules* 52 (17): 6547–6554.

35 Shah, L.A., Khan, M., Javed, R. et al. (2018). Superabsorbent polymer hydrogels with good thermal and mechanical properties for removal of selected heavy metal ions. *J. Cleaner Prod.* 201: 78–87.

36 Hong, S., Schaber, C.F., Dening, K. et al. (2014). Air/water interfacial formation of freestanding, stimuli-responsive, self-healing catecholamine Janus-faced microfilms. *Adv. Mater.* 26 (45): 7581–7587.

37 Gleason, K.K. (2020). Chemically vapor deposited polymer nanolayers for rapid and controlled permeation of molecules and ions. *J. Vac. Sci. Technol., A* 38 (2): 020801.

38 Cangialosi, D., Boucher, V.M., Alegría, A., and Colmenero, J. (2013). Physical aging in polymers and polymer nanocomposites: recent results and open questions. *Soft Matter* 9 (36): 8619.

39 Seol, C., Jang, S., Kim, J. et al. (2018). Fabrication and design of mechanically stable and free-standing polymeric membrane with two-level apertures. *Soft Matter* 14 (47): 9522–9527.

40 Willner, I. (2017). Stimuli-controlled hydrogels and their applications. *Acc. Chem. Res.* 50 (4): 657–658.

41 Kim, J., Kim, J.W., Kim, H.C. et al. (2019). Review of soft actuator materials. *Int. J. Precis. Eng. Manuf.* 20 (12): 2221–2241.

42 Koetting, M.C., Peters, J.T., Steichen, S.D., and Peppas, N.A. (2015). Stimulus-responsive hydrogels: theory, modern advances, and applications. *Mater. Sci. Eng. R Rep.* 93: 1–49.

43 Osada, Y., Okuzaki, H., and Hori, H. (1992). A polymer gel with electrically driven motility. *Nature* 355 (6357): 242–244.

44 Liu, X., He, B., Wang, Z. et al. (2015). Tough nanocomposite ionogel-based actuator exhibits robust performance. *Sci. Rep.* 4 (1): 6673.

45 Ghosal, K. and Freeman, B.D. (1994). Gas separation using polymer membranes: an overview. *Polym. Adv. Technol.* 5 (11): 673–697.

46 Robeson, L.M. (2008). The upper bound revisited. *J. Membr. Sci.* 320 (1–2): 390–400.

47 Dong, G. and Lee, Y.M. (2017). Microporous polymeric membranes inspired by adsorbent for gas separation. *J. Mater. Chem. A* 5 (26): 13294–13319.

48 Norahim, N., Yaisanga, P., Faungnawakij, K. et al. (2018). Recent membrane developments for CO_2 separation and capture. *Chem. Eng. Technol.* 41 (2): 211–223.

49 Song, Y., Lugo, E.L., Powell, S. et al. (2017). Highly selective multilayer polymer thin films for CO_2/N_2 separation. *J. Polym. Sci. Part B Polym. Phys.* 55 (23): 1730–1737.

50 Bakonyi, P., Kumar, G., Nemestóthy, N. et al. (2013). Biohydrogen purification using a commercial polyimide membrane module: studying the effects of some process variables. *Int. J. Hydrogen Energy* 38 (35): 15092–15099.

51 Kamakshi Kumar, R., Saraswat, V.K., Kumar, M. et al. (2017). Hydrogen gas separation with controlled selectivity via efficient and cost effective block copolymer coated PET membranes. *Int. J. Hydrogen Energy* 42 (31): 19977–19983.

52 Kamakshi Kumar, R., Saraswat, V.K., Kumar, M., and Awasthi, K. (2020). Active block copolymer layer on carboxyl-functionalized PET film for hydrogen separation. *Int. J. Hydrogen Energy* 45 (37): 18676–18684.

53 Yi, S., Ghanem, B., Liu, Y. et al. (2019). Ultraselective glassy polymer membranes with unprecedented performance for energy-efficient sour gas separation. *Sci. Adv.* 5 (5): eaaw5459.

54 Tul Muntha, S., Kausar, A., and Siddiq, M. (2017). Advances in polymeric nanofiltration membrane: a review. *Polym. Plast. Technol. Eng.* 56 (8): 841–856.

55 Yu, Y., Zhao, C., Wang, Y. et al. (2013). Effects of ion concentration and natural organic matter on arsenic(V) removal by nanofiltration under different transmembrane pressures. *J. Environ. Sci.* 25 (2): 302–307.

56 Wang, P., Wang, M., Liu, F. et al. (2018). Ultrafast ion sieving using nanoporous polymeric membranes. *Nat. Commun.* 9 (1): 569.

57 Berber, M.R. (2020). Current advances of polymer composites for water treatment and desalination. *J. Chem.* 2020: 1–19.

58 Checkol, F., Elfwing, A., Greczynski, G. et al. (2018). Highly stable and efficient lignin-PEDOT/PSS composites for removal of toxic metals. *Adv. Sustain. Syst.* 2 (1): 1700114.

59 Tan, Z., Chen, S., Peng, X. et al. (2018). Polyamide membranes with nanoscale turing structures for water purification. *Science* 360 (6388): 518–521.

60 Vecitis, C.D., Zodrow, K.R., Kang, S., and Elimelech, M. (2010). Electronic-structure-dependent bacterial cytotoxicity of single-walled carbon nanotubes. *ACS Nano* 4 (9): 5471–5479.

61 Butzelaar, A.J., Liu, K.L., Röring, P. et al. (2021). A systematic study of vinyl ether-based poly(ethylene oxide) side-chain polymer electrolytes. *ACS Appl. Polym. Mater.* 3 (3): 1573–1582.

62 Shioiri, R., Kokubo, H., Horii, T. et al. (2019). Polymer electrolytes based on a homogeneous poly(ethylene glycol) network and their application to polymer actuators. *Electrochim. Acta* 298: 866–873.

63 Lim, A., Jeong, H.-Y., Lim, Y. et al. (2021). Amphiphilic Ti porous transport layer for highly effective PEM unitized regenerative fuel cells. *Sci. Adv.* 7 (13): eabf7866.

64 Liao, Y., Wang, H., Zhu, M., and Thomas, A. (2018). Efficient supercapacitor energy storage using conjugated microporous polymer networks synthesized from Buchwald-Hartwig coupling. *Adv. Mater.* 30 (12): 1705710.

65 Bandyopadhyay, S., Singh, C., Jash, P. et al. (2018). Redox-active, pyrene-based pristine porous organic polymers for efficient energy storage with exceptional cyclic stability. *Chem. Commun.* 54 (50): 6796–6799.

66 Hwang, I., Kim, H.J., Mun, S. et al. (2021). A light-driven vibrotactile actuator with a polymer bimorph film for localized haptic rendering. *ACS Appl. Mater. Interfaces* 13 (5): 6597–6605.

67 Ge, F., Yang, R., Tong, X. et al. (2018). A multifunctional dye-doped liquid crystal polymer actuator: light-guided transportation, turning in locomotion, and autonomous motion. *Angew. Chemie Int. Ed.* 57 (36): 11758–11763.

68 Zhao, Q., Dunlop, J.W.C., Qiu, X. et al. (2014). An instant multi-responsive porous polymer actuator driven by solvent molecule sorption. *Nat. Commun.* 5 (1): 4293.

69 Gelebart, A.H., Mulder, D.J., Vantomme, G. et al. (2017). A rewritable, reprogrammable, dual light-responsive polymer actuator. *Angew. Chemie Int. Ed.* 56 (43): 13436–13439.

70 Wu, J., Xu, F., Li, S. et al. (2019). Porous polymers as multifunctional material platforms toward task-specific applications. *Adv. Mater.* 31 (4): 1802922.

71 Wang, M., Lin, B.-P., and Yang, H. (2016). A plant tendril mimic soft actuator with phototunable bending and chiral twisting motion modes. *Nat. Commun.* 7 (1): 13981.

72 Yang, C., Wang, W., Yao, C. et al. (2015). Hydrogel walkers with electro-driven motility for Cargo transport. *Sci. Rep.* 5 (1): 13622.

4

Inorganic Flexible Electronics: Materials, Strategies, and Applications

Guo Qinglei[1,2,3] and Di Zengfeng[2]

[1] *Shandong University, School of Microelectronics, 27 South Shanda Road, Jinan 250100, People's Republic of China*
[2] *Shanghai Institute of Microsystem and Information Technology, Chinese Academy of Sciences, State Key Laboratory of Functional Materials for Informatics, 865 Changning Road, Shanghai 200500, People's Republic of China*
[3] *Fudan University, State Key Laboratory of ASIC and Systems, 220 Handan Road, Shanghai 200433, People's Republic of China*

4.1 Introduction

Recent advances in materials science and fabrication techniques enable the design and build of advanced electronic components or systems with unusual formats that allow large mechanical deformations without significant changes in their performances [1]. These classes of electronics with features of lightweight, large area coverage, and mechanical flexibility are regarded as flexible electronics, which provide numerous opportunities for applications in consumer electronics [2, 3], soft robots or actuators [4, 5], biomedical (or transient) electronics [6–10], and many others.

In terms of active materials for fabricating flexible electronic devices, organic semiconductors [11, 12], for example polymers and small molecules, have unique advantages, because of their outstanding mechanical flexibility. Various types of organic flexible electronic devices/systems have been demonstrated, including organic light-emitting diodes [13, 14], organic photovoltaics [15], organic thin-film transistor or array [16, 17], sensors or monitors [18–20], and other related devices. To further boost the application of organic semiconductor in flexible electronics technology, continuous efforts should be devoted to developing new organic semiconductors with high or competitive carrier mobility [21] and robust physical properties [22] against external environments (such as humidity, temperature, or oxygen). Graphene and carbon nanotubes possess superior carrier mobility [23, 24]; however, the high-temperature growth condition [25–27] as well as the deficient ability of large-area synthesis may hinder their applications for designing and creating flexible electronic devices.

Inorganic semiconductors are of great interest in terms of flexible electronics because of their stable and superior physical properties [28–30]. Integrating amorphous silicon or polycrystalline silicon on a plastic substrate for subsequently

Nanomembranes: Materials, Properties, and Applications, First Edition.
Edited by Yongfeng Mei, Gaoshan Huang, and Xiuling Li.
© 2022 WILEY-VCH GmbH. Published 2022 by WILEY-VCH GmbH.

fabricating flexible electronic devices has been intensively explored [31, 32]. Despite the low carrier mobility of amorphous silicon and the high-temperature processing for gaining high-quality polycrystalline silicon, they are promising candidates for fabricating large-area solar cells. For metal oxide semiconductors, the capability of large-area and low-temperature growth (by solution-processed method or sputtering) [33], and the unique physical properties such as wide bandgap will provide strong competitiveness as the active materials for flexible electronics, especially transparent electronics. Current challenges are mainly concentrated upon developing p-type metal oxide semiconductors in a simple and convenient approach [34], although several works have demonstrated that SnO or SnO_2 is natural p-type materials [35]. III–V compound semiconductors have outstanding carrier mobility compared to other materials, thus offering significant opportunities for creating high-performance electronic devices [36]. Although the growth of III–V materials involves costly and complex equipment, their direct bandgap property draws significant attention for applications in high-performance optoelectronics, including on-chip laser [37], near-infrared photodetector [38], and light-emitting diode [39].

It is known that the first demonstration of both transistors and integrated circuits was conducted with germanium in 1947 and 1958, respectively, probably due to its highest hole mobility compared to any other known semiconductors. In terms of replacing silicon in advanced semiconductor technology [40], challenges still exist, mainly in the high-quality passivation of germanium [41]. Still, germanium is of wide interests in optoelectronic applications because of its small and tunable bandgap [42]. Silicon, with stable physical properties and competitive carrier mobility, is the most successful semiconductor for the application in microelectronics after the discovery of the convenient passivation technology (SiO_2) and the planar complementary metal–oxide–semiconductor (CMOS) technology. Although silicon is going to meet its physical limit for continuing Moore's law of traditional microelectronics, its potentials for fabricating emerging flexible electronics have attracted numerous attention "again." The following will mainly focus on inorganic materials and describe recent progress of inorganic flexible electronics, including design strategies, synthesis and assembly, and several functional examples and applications.

4.2 Strategies for Designing Flexible Electronics

Inorganic semiconductors gain great successes in traditional bulk microelectronics. However, bulk electronic device fabrication technique cannot be directly utilized to manufacture flexible electronics because of the large mechanical mismatch between bulk inorganic semiconductors and flexible substrates. Fortunately, anything thin enough is flexible. The layer thinning and transfer technique followed by the integration of thin film onto flexible substrates will provide possibilities for creating flexible electronics. To further enhance the flexibility, advanced designs and device fabrication technique are highly demanded. The following will briefly review the mechanical properties of thin films and the design principles for developing ultra-flexible electronics.

Bulk inorganic semiconductors are rigid and brittle, especially for silicon or germanium. Therefore, it is impossible to directly fabricate flexible electronic devices by utilizing bulk semiconductors. However, the mechanical properties of these bulk semiconductors, for example critical bending radius, flexural rigidity, and energy release rate, are strongly dependent on their thickness, thus providing possibilities to construct flexible electronic devices by integrating thin film materials onto flexible substrate. Taking silicon as an example, its critical bending radius (R_c) is defined as the bending radius at which it starts to fracture. When silicon is bended, the induced peak strain is expressed as $\varepsilon = t/(2^*R)$, where t is the thickness and R is the bending radius. Using 1% as the critical strain of silicon [43], the corresponding R_c at different thicknesses is calculated. As a result, R_c decreases linearly with the thickness of silicon [28]. For bulk silicon, the calculated bending radius approaches 0.1 m; therefore, the flexibility can be ignored. As the thickness decreases, for example 200 nm, the calculated critical bending radius is about 10 μm, creating possibilities for building flexible electronics. Similar to the critical bending radius, the flexural rigidity, which is defined as the resistance offered by a material when suffering from bending, also significantly decreases with the thinning of silicon [29], with a scaling law of $EI \sim Eh^3$, where E denotes Young's modulus, I is the second moment of area, and h is the thickness of membranes. Moreover, the energy-release rate, related to the separation energy for detaching a membrane from the supporting substrate, a process that dominates the conformal layer-stacking by transfer-printing technique [44], decreases linearly with the decrease of the silicon thickness. These thickness-dependent mechanical properties offer essential conditions (such as mechanical flexibility and the capability of being transferred and/or stacked on specified substrates) for building flexible electronic devices by utilizing silicon or other inorganic semiconductors with nanoscale thickness.

As mentioned in this section, transferring membranes with nanoscale thickness on flexible substrates will only provide bendability [28], as shown in Figure 4.1a. The bendability is strongly dependent on the thicknesses of functional membranes and flexible substrates. To further enhance the flexibility, researchers demonstrated that shaping functional membrane into wavy or buckled structures can accommodate the external strains by changing the amplitude or out-of-plane deformations like an accordion [43, 45, 48], as shown in Figure 4.1b. Two-dimensional wavy structure shaping of flexible devices will provide bi-axial stretchability [49]. Moreover, the concept of neutral mechanical plane [50], defined as the plane at which the strain or stress is zero when the device is bended, is proposed to further enhance the flexibility of flexible electronics devices. The functional silicon CMOS layer sandwiched by top and bottom protection layers can sustain either folding or twisting operation without significant change in the electrical performances [51].

Other strategies, such as serpentine interconnection or strain isolation [46, 47], can also provide opportunities to enhance the stretchability of flexible electronic devices, as shown in Figure 4.1c,d. More recently, the concept of "kirigami" that involves artificially patterning or cutting of the paper is also used to achieve super-stretchable electronics [52]. Followed by this kirigami concept, a stretchable graphene-based transistor is demonstrated. The conductance of the

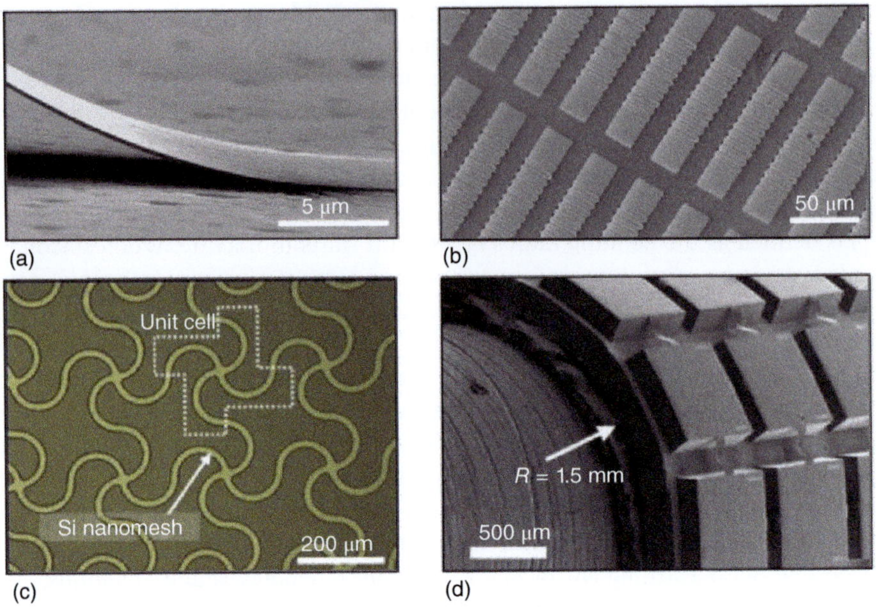

Figure 4.1 Mechanical designs for flexible electronics with enhanced flexibilities, from bendable to stretchable, foldable, or twistable. (a) A released Si nanomembrane with a thickness of 200 nm exhibiting superior bendability. Source: Guo et al. [28]. Reproduced with permission from Elsevier. (b) A stretchable Si film with a wavy structure design. Source: Guo et al. [45]. Reproduced with permission from IEEE. (c) A super stretchable silicon nanoribbon with the serpentine design, also revealing a good conformal contact on the unregular substrate. Source: Kim et al. [46]. Reproduced with permission from Wiley-VCH. (d) Stretchable GaAs photovoltaic modules wrapped on a cylinder with a radius of 1.5 mm. Source: Lee et al. [47]. Reproduced with permission from Wiley-VCH.

stretchable graphene kirigami transistor varying with the liquid gate is measured at unstretched (blue) and stretched (orange) states, respectively. Because the graphene has no lattice strain when the kirigami is extended, the conductance curves observed at unstretched and stretched conditions are expected to have no notable change, suggesting a promising candidate for stretchable electronics. Other similar strategies to achieve stretchable electronics have been demonstrated by direct nano/micro-pattern of the active materials, such as "serpentine" [53] or "nanomesh" [54] designs.

4.3 Strategies for Forming and Assembling Inorganic Nanomaterials

State-of-the-art strategies for the synthesis of inorganic nanomaterials can be divided into two categories: "bottom-up" technique and "top-down" technique. For the "bottom-up" technique, normally involving chemical reactions that occurred with selected precursors, two-dimensional nanomembranes [55], one-dimensional nanowires [56], and zero-dimensional quantum dots/nanoparticles [57] can be realized under controlled growth conditions. For example, utilizing the nanocluster-catalyzed vapor–liquid–solid method, silicon nanowires can be synthesized on target substrates [58]. The subsequent applications in devices or systems with various functionalities require the synthesized silicon nanowires to be physically integrated or transferred on desired substrates. For this purpose, a nanoscale-combing technique was proposed to realize the assembly of highly aligned nanowires over a large area [56]. In such an approach, a patterned target substrate that has anchoring regions with strong adhesion strength with nanowires and aligning regions with weak adhesion strength is utilized. Then, the nanowire donor substrate is flipped over and brought into contact with the target substrate. With controllable applied force and direction, the tip ends of nanowires will first fix to the anchoring region; then the remaining nanowires will be aligned onto the combing region during the transplant. The assembled nanowires exhibit a high degree of alignment in the predefined direction with the deviation smaller than 1°. Large areas of nanowire field-effect transistor array are demonstrated by this technique [56], with high uniformity, generality, and reproducibility, which can be extended for aligning other nanowires or for creating more complex structures. Similar schemes can be used to produce and assemble various types of semiconducting nanomaterials, such as nanomembranes and two-dimensional-layered materials as well. Capabilities of transferring and assembling nanowires on any desired substrate play vital roles for creating unusual functional devices, such as transparent electronics [59] and flexible electronics [60].

As a complementation to "bottom-up" technique, "top-down" technique, involving photolithography, patterning, and etching processes, with deterministic physical properties, is proposed to form nanomaterials released from their bulk forms. A sandwiched structure with the in-between silicon oxide as the sacrificial layer and the top nanomembrane as the functional layer is utilized as the donor [29]. Photolithography and patterning processes define dimensions and spatial orders of nanomembranes as well as the "etch-window," which allows the etchant pass through to remove the silicon oxide layer, leading to the release of the supporting nanomembrane. Followed by the transfer-printing technique [44, 61, 62], the released flexible nanomembrane can be integrated onto any desired substrate. Three key points should be taken into consideration when utilizing "top-down" technique, i.e. (i) the ability of selective removal of the sacrificial layer without any etch or degradation of the supporting nanomembrane, (ii) the ability of sustaining the spatial order of the patterned nanomembrane before and after the removal of sacrificial layers, and (iii) the ability of integrating or transferring the released nanomembrane onto any desired substrate, or with complex surfaces. The sacrificial layer can be heterogeneous, such as silicon dioxide for silicon or germanium releasing, AlGaSb for InAs nanomembrane releasing [36]; also can be homogeneous, for example, silicon nanoribbons can be released from their bulk wafer by anisotropic etch [63]. As for the deterministic control over the spatial assembly, special designs of anchors, consisting of photoresist or silicon dioxide etching ends [61, 64], can be utilized to fix the released nanomembranes without changing their spatial orders.

After the release of nanomembranes, various transfer-printing techniques have been demonstrated to assemble them onto desired flexible substrates, such as rolling transfer for large-area production of silicon nanomembranes [51], notch-assisted transfer for the control over the cracking behavior of brittle silicon nanomembranes [62], microstructured stamp with reversible adhesion for deterministic assembly [65], or conformal adhesive stamp for three-dimensional curvy electronics [66], providing numerous opportunities for fabricating various types of flexible electronic devices or systems, which will be discussed in Section 4.4.

4.4 Applications of Inorganic Flexible Electronics

With the ability of forming inorganic nanomaterials via either "bottom-up" or "top-down" technique, and the ability of assembling them onto desired substrates by the transfer-printing method, various types of flexible devices are created, as summarized in Figure 4.2. Figure 4.2a,b shows flexible transistors fabricated with silicon and germanium nanomembranes [67, 68], respectively, in which the device exhibits superior bendability. Figure 4.2c shows a thin-film transistor based on amorphous indium–gallium–zinc-oxide (IGZO) on flexible substrates, of which the transfer characteristics remain stable after the bending test at a radius of curvature of 10 mm over 10 000 cycles [69]. High-performance flexible optoelectronic devices are also available through the above method, although nanomaterials are normally regarded to have poor light absorption properties. By introducing a back-reflection gate, high-sensitivity silicon nanomembrane-based flexible phototransistors are demonstrated [70], as shown in Figure 4.2d. Other flexible optoelectronic devices, including light-emitting diode array for large-scale lighting [2], and solar cell array [71], are also demonstrated via "top-down" and transfer-printing techniques, as shown in Figure 4.2e,f. Flexible photonic devices, including photonic crystal cavity [72], silicon waveguide for bioresorbable biosensors [73], and silicon-based photonic circuit with tunable and reversible characteristics against the mechanical deformation [74], are summarized in Figure 4.2g–i. The demonstrations of flexible devices form the fundamentals for their following applications, including photo-imaging or sensing systems [75], epidermal electronics [76], bioimplantable electronics [77], power supply systems for bioimplants [9], and many others.

4.4.1 Flexible Photo-Imaging or Sensing Systems

One of the most promising applications for flexible optoelectronic devices or systems is flexible display, which involves light-imaging or -sensing optoelectronic components. High performances, including high-resolution imaging or directional sensitivity, are required in terms of fabricating photo-imaging or light-sensing systems. Flexible nanomembranes capable of reshaping into three-dimensional structures under mechanical deformations offer opportunities for creating novel flexible optoelectronic systems with high performances. As shown in Figure 4.3a, Zhang et al. utilized silicon nanomembranes and demonstrated an origami-inspired

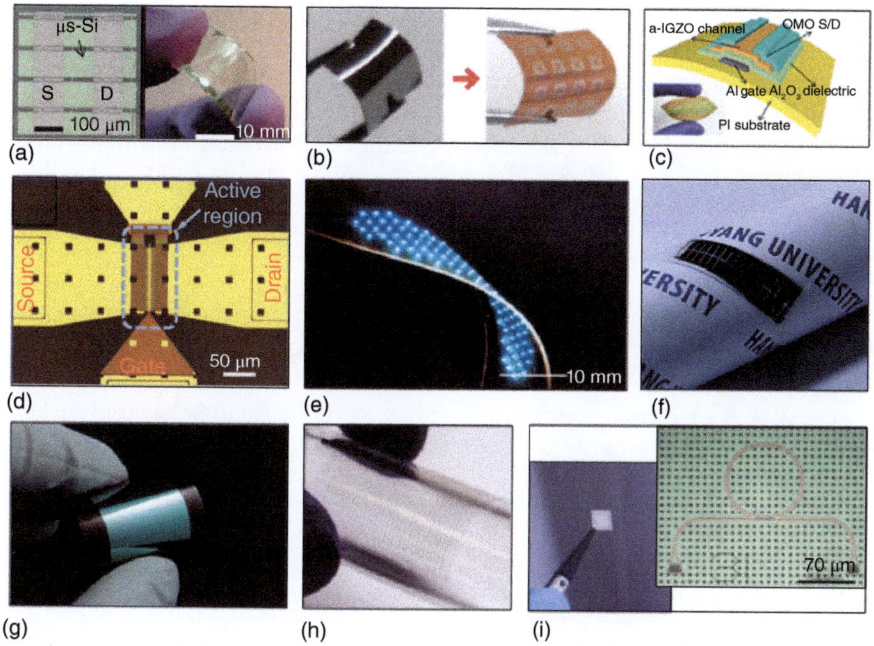

Figure 4.2 Various types of flexible electronic, optoelectronic, and photonic devices created with inorganic nanomaterials. (a) Optical images of silicon nanomembrane-based flexible transistors. Source: Menard et al. [67]. Reproduced with permission from AIP Publishing LLC. (b) Optical image of flexible germanium nanomembranes and flexible germanium-based transistors. Source: Higashi et al. [68]. Reproduced with permission from AIP Publishing LLC. (c) Optical image of IGZO-based thin-film transistors on flexible substrates. Source: Kim et al. [69]. Reproduced with permission from Elsevier. (d) An optical microscopy image of a silicon nanomembrane-based phototransistor. Source: Seo et al. [70]. Reproduced with permission from Wiley-VCH. (e) Flexible LED array embedded in luminescent waveguides. Source: Kim et al. [2]. Reproduced with permission from PNAS. (f) Flexible solar-cell array for sunlight harvesting. Source: Baek et al. [71]. Reproduced with permission from Royal Society of Chemistry. (g) Optical image of a silicon-based photonic crystal cavity. Source: Xu et al. [72]. Reproduced with permission from American Chemical Society. (h) Silicon waveguide for bioresorbable biosensors. Source: Bai et al. [73]. Reproduced with permission from Wiley-VCH. (i) Optical images of silicon-based flexible photonic circuits with tunable and reversible characteristics when applied with mechanical deformations. Source: Chen et al. [74]. Reproduced with permission from OSA Publishing.

hemispherical "electronic-eye" system [78]. Construction process of the system was finished sequentially by the fabrication of flexible silicon photodiode arrays, the fabrication of polymer microlens for compound eye, the origami process of silicon photodiode array mounted on either concave or convex fixtures, and the wire-bonding process for characterizations. The final three-dimensional system is shown in the left (sketched) and middle (optical image) panels of Figure 4.3a. The right panel of Figure 4.3a shows a high-resolution image of a laser point measured from the convex electronic eye camera, with the brighter regions indicating the photodiodes of the convex electronic eye camera that was illuminated by the laser point [78]. Applying this concept to state-of-the-art photo-imaging or light-sensing

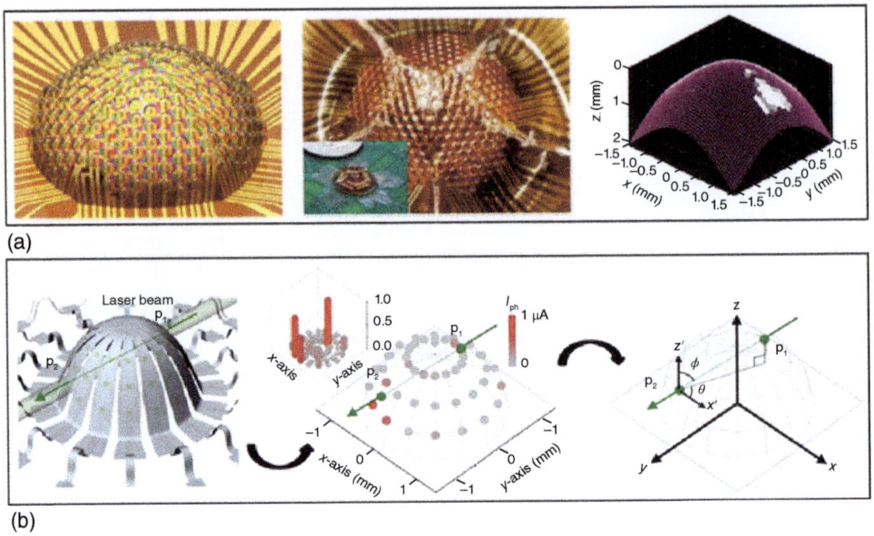

Figure 4.3 Origami-inspired photo-imaging and sensing systems. (a) Schematic diagram of the hemispherical "electronic eye" based on origami silicon photodiode array, where a polymer microlens is integrated with each photodiode to mimic the corneal lens in a compound eye (left), a photograph of a represented "electronic-eye" system (middle), high-resolution image of a laser point obtained from the built "electronic-eye" system (right). Source: Zhang et al. [78]. Reproduced with permission from Springer Nature. (b) Schematic illustration of a 3D hemispherical MoS_2 photodetection system with a laser beam penetrating at two points of the hemispherical surface (left), photocurrent distribution on the hemispherical surface (middle), and principle of determining the incident direction of the laser based on the 3D photocurrent distribution (right). Source: Adapted with permission from Lee et al. [79]. Copyright (2018) by Nature Publishing Group.

system will improve the performances, such as high-quality imaging, convenient focus tuning, or others of these systems. Besides, expanded capabilities that are required for some special uses are also in prospect. Further opportunities of this subject lie in scaling the density of photodiode pixel or developing hemispheres with dynamically tunable sizes and dimensions for capturing high-quality images.

In terms of light-sensing devices or systems, numerous efforts have been devoted to pursuing superior performances, including responsivity, detection limit, and response time. However, capabilities of three-dimensional photodetection or light imaging that can carry out the detection of direction and angular divergence properties of the incident light are rarely noticed. To address these issues, Lee et al. proposed an origami-inspired design that utilized two-dimensional materials (MoS_2 as channels and graphene as electrodes) to fabricate a functional three-dimensional photodetection system [79], which allows the tracking of both direction and intensity of the incident light. The working principle is schematically shown in the left panel of Figure 4.3b, where the incident laser can pass through the three-dimensional photodetection system. The locally illuminated two points: the entry P_1 and the exit P_2 can, therefore, determine the direction of the incident laser. The middle panel of Figure 4.3b presents a mapping result for a typical case, and the photoresponse of the device at P_1 is slightly higher than that of the device near P_2, which can be attributed to the light attenuation after reflection and absorption. Nevertheless, the direction of the incident laser exactly follows the determined $P_1 \rightarrow P_2$ vectors, as shown in the right panel of Figure 4.3b. The discussed work that uses flexible nanomaterials to create a functional system-level construction offers potentials for extending the capabilities of traditional sensing or actuating systems, including three-dimensional pressure sensing, three-dimensional vibrating, and other related.

4.4.2 Epidermal Electronics

As for human health care or monitoring, physiological or electrophysiological signals measured from the skin will offer straightforward information. However, nearly all related measurement systems at present still rely on small numbers of bulk electrodes that are mounted on the skin, conductive gels for good electrical interfaces between electrodes and the skin, conductive wires that connect to terminal instruments, and the terminal with a collection of power supply and data communication components. Although precise information can be provided, these systems are probably only suitable for lab- or clinical-level applications while unavailable for at-home health care. Moreover, the user experiences on these systems with bulk electrodes and complex wire connections are poor, especially for the requirement of none or cautious body movements to maintain good electrical connections. Flexible, stretchable, or soft electronics that are built onto ultrathin, light-weight, "skin-like" membranes can be conformally and invisibly laminated onto the skin surface, thus potentially providing complementation with traditional device technologies.

Figure 4.4a shows a demonstration of this type of platform, referred to as epidermal electronic system (EES) [76], with a collection of temperature sensor, electromyography (EMG) sensor, strain sensor, electrotactile stimulators, stimulation channels, and recording channels on a thin (about 60 μm), gas-permeable, elastic substrate, which can form robust adhesion to the skin via van der Waals forces. An exploded view of the multilayer construction, composed only of metals and dielectrics, is shown in Figure 4.4b. Moreover, the stretchable and compressible EES can be transferred and mounted on the skin of the forearm to form conformal and mechanically invisible connections, like a "tattoo," without any adhesive tapes or gels, as shown in Figure 4.4c. Advantages of the developed EES system mainly manifest in the simultaneous recordings of electrophysiological data and presentation of neural stimulatory inputs. The demonstrated EES platforms, capable of providing enough information for health monitor, are convenient for humans to get out-hospital health care, with comfortable experiences.

Extending the sizes of epidermal electronics allows their coverage over the entire or large region of the body, with large-area spatial mapping capabilities, which is of great interest. In addition, the ability of collecting multi-signal recordings offers complementary information associated with physical and psychological states. For example, Tian et al. demonstrated a class of large-area, multifunctional, and

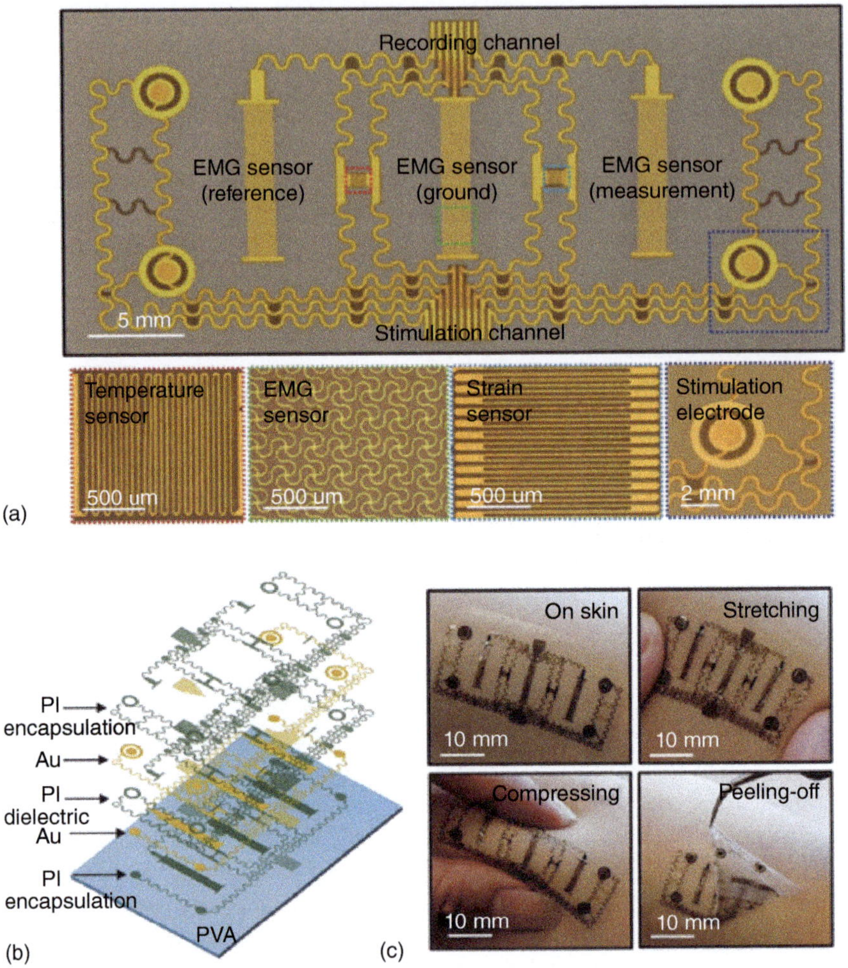

Figure 4.4 Epidermal electronics for biomedical applications. (a) Photograph of an epidermal electronic system (EES) with various devices and components, forming a multifunctional platform (top). The bottom displayed the magnified optical images of the multiple devices or sensors of the platform. (b) Exploded view of the schematic multifunctional EES, composed only of patterns of metals and dielectrics. (c) Optical images of the developed EES mounted on the skin of the forearm at various status of as-mounted, stretching, compressing, and peeling-off. Source: Xu et al. [76]. Reproduced with permission from Wiley-VCH.

magnetic resonance imaging (MRI)-compatible epidermal electronic devices, aiming for reliable and robust body-scale electrophysiological analyses [80]. Versatile applications of this unique system have been demonstrated, including robotic prosthesis control via EMG, full-scalp multichannel systems for EEG, and MRI compatibility of acquiring EMG and ECG. An example of these applications is the means for full-scalp EEG recordings. In Tian's work, four EEG arrays with 68 electrodes are mounted along the left–right direction over the scalp. The spatial map of the event-related response for standard and target stimuli exhibits significant potential differences, confirming the ability of acquiring high-quality, high-density EEG signals over the full scalp by utilizing the proposed large-area epidermal sensors. Moreover, the comfortable, mechanically stable interface between the epidermal EEG sensors and the skin allows chronic recordings (five days to two weeks) without interfering normal daily activities. Nevertheless, one of further efforts of epidermal electronics still involves material and device strategies to get perfect human–machine interfaces, aiming for more precise information, imperceptible attachments on the skin, reversible adhesion to the skin, and “skin-like” gas or sweat permeations.

4.4.3 Bioimplantable Electronics

Flexible electronics that can form a conformal contact with irregular surfaces of biological organs also draw significant attention for the applications in bioimplantable electronics. Various sensing or therapy platforms based on bioimplantable electronic devices or systems, capable of monitoring human health status or treating diseases, have been developed. Electric [81], thermal [82], mechanical [83], optoelectronic [84], and other related signals or properties of tissue organs can be monitored via these implanted electronic devices or systems, which give diagnostic and therapeutic guidance for relevant disease states.

For example, Yu et al. introduced a needle-shaped flexible piezoelectric microsystem, which can provide quantitative real-time measurements of variations in tissue modulus, to distinguish the abnormal tissue [83]. Key components of the microsystem are actuator (distant from the needle tip) and sensor (near the tip), both of which are constructed with the piezoelectric material lead zirconate titanate (PZT). With the specific design of substrate, the microsystem can be utilized as either a freestanding form (device 1, 75-μm-thick PI substrate) or a biopsy needle–supporting form (device 2, 3-μm-thick PI substrate). Device 1 has a sharp tip and narrow width, allowing penetration and injection into the targeted soft tissues. Device 2, using a conventional biopsy needle as the supporting substrate, can penetrate skin, fascia, and solid organs with minimized invasions. The working principle of the microsystem is described as follows. Once a given voltage is applied to the actuator, the induced sensing voltage will be strongly depended on the modules of the tissue target through mechanical coupling among actuator, sensor, substrate, and tissue target. In this work, both *in vivo* and *ex vivo* measurements on animal model tissues have been performed by the authors [83]. Explanted samples include lung, liver, spleen, kidney, and fat. Module values of these tissues or organs obtained from the microsystem are consistent with the reported values obtained by conventional indentation techniques [85, 86], indicating the superior reliability of the proposed microsystem. The authors also performed measurements on human organ tissues, including cancerous sites. Taking the thyroid as an example, the average module value measured from the normal areas is about 10 kPa; however, that of the thyroid tumor area is much higher, reaching about 30 kPa. This needle-shaped flexible piezoelectric microsystem demonstrated by Yu and collaborators can directly and precisely distinguish

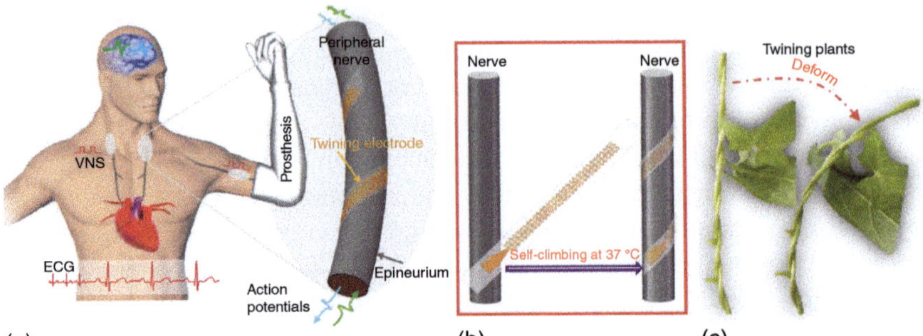

Figure 4.5 Bioimplantable electronics. (a) Schematic Illustrations of the concepts of neuromodulation (left) and the interface between self-climbing twining-electrode and nerve (right). (b) Schematic diagram of the self-climbing process of electrodes driven by the body temperature. (c) Images of twining plants duiring deformation. Source: Zhang et al. [87]. Reproduced with permission from AAAS.

normal from lesion areas of organ tissues, thus laying the foundation for subsequent diagnosis and treatment.

Electrode–nerve interface is considered as a new therapeutic strategy for treating various diseases related to the peripheral nervous system. However, the mechanical and geometrical mismatches between electrodes and nerves will induce irreversible nerve damage and poor long-term stability. To address these problems of current electrode–nerve interfaces, Feng's group demonstrated a twining electrode with climbing ability by integrating metal mesh serpentines onto a flexible shape memory substrate for peripheral nerve stimulation and recording [87]. The shape memory substrate, with permanent shape reconfigurability and distinct elastic modulus controllability, can be driven at body temperature. Figure 4.5 shows the schematic illustrations of the concept and twining electrode. Before the thermal stimulation, the twining electrode is temporarily flat, with one end adhering to the rounded nerve surface. Driven by 37 °C normal saline, the temporarily flattened twining electrode will self-climb along the nerve, forming a three-dimensional, flexible, and conformal electrode–nerve interface, as shown in the middle panel of Figure 4.5. This concept is similar to the twining plants (see the right panel of Figure 4.5), with super stable interface even under large deformations. By utilizing the proposed twining electrodes, *in vivo* experiments of vagus nerve stimulations and sciatic nerve action potential recordings were performed, suggesting the potential clinical utility.

Although bioimplantable electronics with minimized, flexible, and multifunctional (diagnostic or therapeutic) features have been rapidly developed, their potentials toward clinical applications are strongly challenged by the power supply devices or systems. Traditional route to address this problem is realized by bulky and rigid external power sources that form an electric connection with implanted devices or systems through skin puncture. However, the conductive-wire-connection-induced risk of local infection, as well as movement restriction (or possible) of the receptor, always accompanies this route. Recently, various types of power supply devices or systems for bioimplantable electronics have been demonstrated, such as electromagnetic energy harvesters [9, 88], batteries

[89], supercapacitors [90], piezoelectric harvesters [91], triboelectric devices [92], photovoltaic devices [93], and many others. Bioimplantable electronics for different application scenarios, such as biosensing, communicating, and stimulating, have certain requirements on output power and voltage of power supply systems, associated with other sophisticated functions. Therefore, based on the requirements of output properties for specific applications, one can choose appropriate device architectures as the power supply.

For the application of transient electronics [91], Dagdeviren et al. demonstrated a ZnO-based energy harvester through the combination of Mg as electrodes and interconnectors, MgO as dielectrics, and silk films as substrates and packages. The bending of ZnO energy harvester array (Figure 4.6a) can generate an output voltage of about 1.14 V and a current of 0.55 nA, as shown in Figure 4.6b,c. Aiming for the low-frequency power delivery of bioimplanted transient electronic devices, Guo et al. proposed a bioresorbable magnetically coupled system for wireless power transfer [88]. As shown in Figure 4.6d, the system consists of an Mg coil on PLGA substrate as the receiver and a magnet mounted on an electric motor as the transmitter. Once the magnet rotates, the magnetic flux through the coil will change as a rate

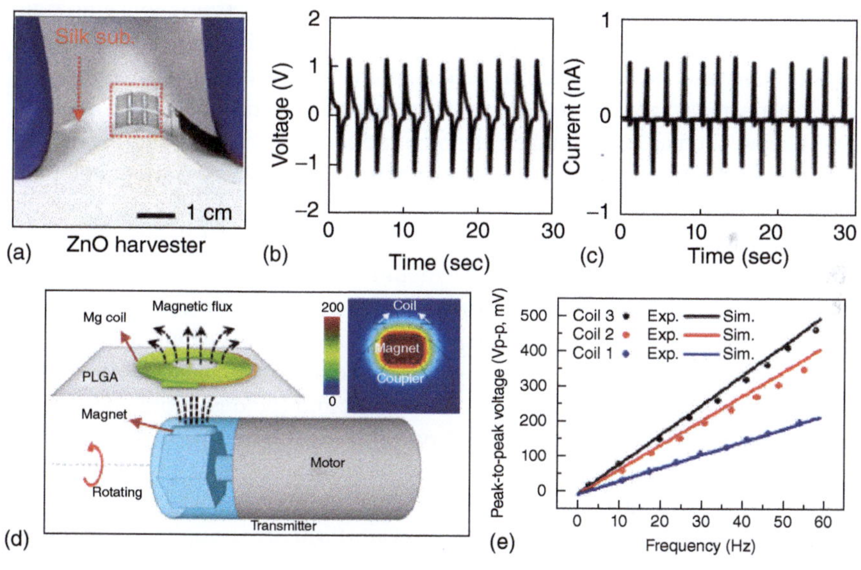

Figure 4.6 Power supply systems for bioimplants. (a) Optical image of a flexible ZnO-based energy harvester at the bending status. Source: Dagdeviren et al. [91]. Reproduced with permission from Wiley-VCH. (b) Open-circuit voltage varying with time during cycles of bending. (c) Short-circuit current of the harvester varying with time during bending cycles. Source: Adapted with permission from Dagdeviren et al. [91]. Copyright (2013) by Wiley-VCH Verlag GmbH & Co. KGaA, Weinheim. (d) Schematic illustration of the working principle of a bioresorbable magnetically coupled power supply system, including a Mg-based bioresorbable coil as the receiver and a rotating magnet as the transmitter. (e) Peak-to-peak voltage varying with the rotation frequency of the magnet collected from three coils with different resonant frequencies. Source: (d, e) Adapted with permission from Guo et al. [88]. Copyright (2019) by Wiley-VCH Verlag GmbH & Co. KGaA, Weinheim.

that depends on the rotation of the motor, thus producing an electromotive force (e.g. voltage) in the coil. By utilizing this scheme, power transfer from the rotating magnet to three individual coils with different resonant frequencies is demonstrated, and the output voltage (or delivered power) can be tuned by varying the rotation frequency of the motor, as shown in Figure 4.6e. The proposed magnetically coupled system can work as either a power source of implanted electronic devices or a bioresorbable electric stimulator for sciatic nerve in a Lewis rat model.

4.5 Conclusion

The results presented in this chapter illustrate the rapid growth of inorganic flexible electronics, with the combination of material strategies, mechanical designs, micro/nanofabrication techniques, device architectures, and system-level integrations. Inorganic nanomaterials with outstanding physical or chemical properties and mechanical flexibility can bypass the mismatch between conventional wafer-based electronics and emerging flexible electronics. With regard to the requirements on high flexibility or stretchability of these devices/systems, 3D micro/nano assembly technique, i.e. origami or kirigami architectures, may provide possibilities. With these schemes, high-performance flexible electronic, optoelectronic and photonic devices, system-level integrations are developed, suggesting huge application prospects of inorganic flexible electronics. One of the most promising opportunities lies in bioelectronics, capable of monitoring human health status and treating human diseases, such as epidermal electronics and bioimplantable electronics.

Future fundamental studies of this research area focus on: (i) developing cost-efficient and convenient approaches to fabricate flexible inorganic nanomaterials, and to assemble them or related devices on substrates with complex, curvy surfaces to match biological tissues; (ii) modulating the physical properties of inorganic nanomaterials via surface/interface modification, strain engineering, and/or heterointegrations for high-performance inorganic flexible electronics; (iii) continuing to improve the flexibility or stretchability of inorganic flexible electronics through mechanical designs or structure assemblies; and (iv) designing and exploring devices/systems with new functional or multifunction toward the practical applications for either consumer electronics, military electronics, or biomedical electronics. As for the engineering possibilities, continuous efforts may lie in developing integrated flexible electronic systems, capable of conformally wrapping onto biological tissues, dissolving in biofluids with a programmed rate, wirelessly transferring power and data, and monitoring health status and treating diseases, with practical utilities for biomedical electronics. These fundamental studies and engineering efforts will create significant potentials for inorganic flexible electronics.

Acknowledgments

This work is supported by the Qilu Young Scholar Program of Shandong University, Shandong University Multidisciplinary Research and Innovation Team of Young

Scholars (No. 2020QNQT015), the State Key Laboratory of ASIC & System (Grant No. 2020KF007), the State Key Laboratory of Functional Materials for Informatics (Grant No. SKL202101), Key Research Project of Frontier Science, Chinese Academy of Sciences (QYZDB-SSW-JSC021), and National Natural Science Foundation of China (Grant Nos. 61974157 and 51925208).

References

1 Rogers, J.A., Someya, T., and Huang, Y. (2010). Materials and mechanics for stretchable electronics. *Science* 327 (5973): 1603.

2 Kim, H.-s., Brueckner, E., Song, J. et al. (2011). Unusual strategies for using indium gallium nitride grown on silicon (111) for solid-state lighting. *Proc. Natl. Acad. Sci.* 108 (25): 10072.

3 Yu, X., Xie, Z., Yu, Y. et al. (2019). Skin-integrated wireless haptic interfaces for virtual and augmented reality. *Nature* 575 (7783): 473–479.

4 Xu, B., Tian, Z., Wang, J. et al. (2018). Stimuli-responsive and on-chip nanomembrane micro-rolls for enhanced macroscopic visual hydrogen detection. *Sci. Adv.* 4 (4): 8203.

5 Tian, Z., Xu, B., Hsu, B. et al. (2018). Reconfigurable vanadium dioxide nanomembranes and microtubes with controllable phase transition temperatures. *Nano Lett.* 18 (5): 3017–3023.

6 Chung, H.U., Kim, B.H., Lee, J.Y. et al. (2019). Binodal, wireless epidermal electronic systems with in-sensor analytics for neonatal intensive care. *Science* 363 (6430): 0780.

7 You, C., Zhao, H., Guo, Q. et al. (2020). Material strategies for on-demand smart transient electronics. *MRS Bull.* 45 (2): 129–134.

8 Boutry, C.M., Beker, L., Kaizawa, Y. et al. (2019). Biodegradable and flexible arterial-pulse sensor for the wireless monitoring of blood flow. *Nat. Biomed. Eng.* 3 (1): 47–57.

9 Koo, J., MacEwan, M.R., Kang, S.-K. et al. (2018). Wireless bioresorbable electronic system enables sustained nonpharmacological neuroregenerative therapy. *Nat. Med.* 24 (12): 1830–1836.

10 Chang, J.-K., Chang, H.-P., Guo, Q. et al. (2018). Biodegradable electronic systems in 3D, heterogeneously integrated formats. *Adv. Mater.* 30 (11): 1704955.

11 Someya, T., Bao, Z., and Malliaras, G.G. (2016). The rise of plastic bioelectronics. *Nature* 540 (7633): 379–385.

12 Ling, H., Liu, S., Zheng, Z. et al. (2018). Organic flexible electronics. *Small Method* 2 (10): 1800070.

13 Yin, D., Feng, J., Ma, R. et al. (2016). Efficient and mechanically robust stretchable organic light-emitting devices by a laser-programmable buckling process. *Nat. Commun.* 7 (1): 11573.

14 Bi, S., Li, Q., He, Z. et al. (2019). Highly enhanced performance of integrated piezo photo-transistor with dual inverted OLED gate and nanowire array channel. *Nano Energy* 66: 104101.

15 Gao, L., Zhang, Z.-G., Xue, L. et al. (2016). All-polymer solar cells based on absorption-complementary polymer donor and acceptor with high power conversion efficiency of 8.27%. *Adv. Mater.* 28 (9): 1884–1890.
16 Jiang, Y., Guo, Y.L., and Y. (2017). Engineering of amorphous polymeric insulators for organic field-effect transistors. *Adv. Elect. Mater.* 3 (11): 1700157.
17 Wang, S., Xu, J., Wang, W. et al. (2018). Skin electronics from scalable fabrication of an intrinsically stretchable transistor array. *Nature* 555 (7694): 83–88.
18 Khim, D., Ryu, G.-S., Park, W.-T. et al. (2016). Precisely controlled ultrathin conjugated polymer films for large area transparent transistors and highly sensitive chemical sensors. *Adv. Mater.* 28 (14): 2752–2759.
19 Boutry, C.M., Kaizawa, Y., Schroeder, B.C. et al. (2018). A stretchable and biodegradable strain and pressure sensor for orthopaedic application. *Nat. Electron.* 1 (5): 314–321.
20 Lee, H., Lee, S., Lee, W. et al. (2019). Ultrathin organic electrochemical transistor with nonvolatile and thin gel electrolyte for long-term electrophysiological monitoring. *Adv. Funct. Mater.* 29 (48): 1906982.
21 Kang, I., Yun, H.-J., Chung, D.S. et al. (2013). Record high hole mobility in polymer semiconductors via side-chain engineering. *J. Am. Chem. Soc.* 135 (40): 14896–14899.
22 Zhang, G., Lee, S., Gutiérrez-Meza, E. et al. (2019). Robust and stretchable polymer semiconducting networks: from film microstructure to macroscopic device performance. *Chem. Mater.* 31 (17): 6530–6539.
23 Dürkop, T., Getty, S.A., Cobas, E. et al. (2004). Extraordinary mobility in semiconducting carbon nanotubes. *Nano Lett.* 4 (1): 35–39.
24 Banszerus, L., Schmitz, M., Engels, S. et al. (2015). Ultrahigh-mobility graphene devices from chemical vapor deposition on reusable copper. *Sci. Adv.* 1 (6): 500222.
25 Wang, G., Zhang, M., Zhu, Y. et al. (2013). Direct growth of graphene film on germanium substrate. *Sci. Rep.* 3 (1): 2465.
26 Wang, G., Chen, D., Lu, Z. et al. (2014). Growth of homogeneous single-layer graphene on Ni-Ge binary substrate. *Appl. Phys. Lett.* 104 (6): 062103.
27 Li, J., Guo, Q., Zhang, N. et al. (2018). Direct integration of polycrystalline graphene on silicon as a photodetector via plasma-assisted chemical vapor deposition. *J. Mater. Chem. C* 6 (36): 9682–9690.
28 Guo, Q., Di, Z., Lagally, M.G. et al. (2018). Strain engineering and mechanical assembly of silicon/germanium nanomembranes. *Mater. Sci. Eng.: R: Rep.* 128: 1–31.
29 Rogers, J.A., Lagally, M.G., and Nuzzo, R.G. (2011). Synthesis, assembly and applications of semiconductor nanomembranes. *Nature* 477 (7362): 45–53.
30 Huang, G. and Mei, Y. (2012). Thinning and shaping solid films into functional and integrative nanomembranes. *Adv. Mater.* 24 (19): 2517–2546.
31 Kim, Y., Moon, J., Chung, C. et al. (2006). Performance improvement of ultralow temperature polycrystalline silicon TFT on plastic substrate by plasma oxidation of polycrystalline Si surface. *IEEE Electron Device Lett.* 27 (11): 896–898.

32 Long, K., Kattamis, A.Z., Cheng, I. et al. (2006). Stability of amorphous-silicon TFTs deposited on clear plastic substrates at 250 °C to 280 °C. *IEEE Electron Device Lett.* 27 (2): 111–113.

33 Guo, Q., Zhao, Y., Wang, G. et al. (2019). Sponge-templated production of ultra-thin ZnO nanosheets for printed ultraviolet photodetectors. *Appl. Phys. Lett.* 115 (12): 122106.

34 Yu, X., Marks, T.J., and Facchetti, A. (2016). Metal oxides for optoelectronic applications. *Nat. Mater.* 15 (4): 383–396.

35 Wang, Z., Nayak, P.K., Caraveo-Frescas, J.A. et al. (2016). Recent developments in p-type oxide semiconductor materials and devices. *Adv. Mater.* 28 (20): 3831–3892.

36 Ko, H., Takei, K., Kapadia, R. et al. (2010). Ultrathin compound semiconductor on insulator layers for high-performance nanoscale transistors. *Nature* 468 (7321): 286–289.

37 Yang, H., Zhao, D., Chuwongin, S. et al. (2012). Transfer-printed stacked nanomembrane lasers on silicon. *Nat. Photonics* 6 (9): 615–620.

38 Ren, A., Yuan, L., Xu, H. et al. (2019). Recent progress of III–V quantum dot infrared photodetectors on silicon. *J. Mater. Chem. C* 7 (46): 14441–14453.

39 Tian, Z., Li, Y., Su, X. et al. (2018). Super flexible GaN light emitting diodes using microscale pyramid arrays through laser lift-off and dual transfer. *Opt. Express* 26 (2): 1817–1824.

40 Pillarisetty, R. (2011). Academic and industry research progress in germanium nanodevices. *Nature* 479 (7373): 324–328.

41 Zheng, X., Zhang, M., Shi, X. et al. (2015). Fluorinated graphene in interface engineering of Ge-based nanoelectronics. *Adv. Funct. Mater.* 25 (12): 1805–1813.

42 Michel, J., Liu, J., and Kimerling, L.C. (2010). High-performance Ge-on-Si photodetectors. *Nat. Photonics* 4 (8): 527–534.

43 Khang, D.-Y., Jiang, H., Huang, Y. et al. (2006). A stretchable form of single-crystal silicon for high-performance electronics on rubber substrates. *Science* 311 (5758): 208.

44 Carlson, A., Bowen, A.M., Huang, Y. et al. (2012). Transfer printing techniques for materials assembly and micro/nanodevice fabrication. *Adv. Mater.* 24 (39): 5284–5318.

45 Guo, Q., Fang, Y., Zhang, M. et al. (2017). Wrinkled single-crystalline germanium nanomembranes for stretchable photodetectors. *IEEE Trans. Electron Devices* 64 (5): 1985–1990.

46 Kim, S., Li, Y., Song, J. et al. (2019). Biaxially stretchable ultrathin Si enabled by serpentine structures on prestrained elastomers. *Adv. Mater. Technol.* 4 (1): 1800489.

47 Lee, J., Wu, J., Shi, M. et al. (2011). Stretchable GaAs photovoltaics with designs that enable high areal coverage. *Adv. Mater.* 23 (8): 986–991.

48 Guo, Q., Zhang, M., Xue, Z. et al. (2013). Three dimensional strain distribution of wrinkled silicon nanomembranes fabricated by rolling-transfer technique. *Appl. Phys. Lett.* 103 (26): 264102.

49 Choi, W.M., Song, J., Khang, D.-Y. et al. (2007). Biaxially stretchable "wavy" silicon nanomembranes. *Nano Lett.* 7 (6): 1655–1663.

50 Kim, D.-H., Ahn, J.-H., Choi, W.M. et al. (2008). Stretchable and foldable silicon integrated circuits. *Science* 320 (5875): 507.

51 Ahn, J.-H., Kim, H.-S., Lee, K.J. et al. (2006). Heterogeneous three-dimensional electronics by use of printed semiconductor nanomaterials. *Science* 314 (5806): 1754.

52 Blees, M.K., Barnard, A.W., Rose, P.A. et al. (2015). Graphene kirigami. *Nature* 524 (7564): 204–207.

53 Kim, D.-H., Xiao, J., Song, J. et al. (2010). Stretchable, curvilinear electronics based on inorganic materials. *Adv. Mater.* 22 (19): 2108–2124.

54 Han, X., Seo, K.J., Qiang, Y. et al. (2019). Nanomeshed Si nanomembranes. *NPJ Flexible Electron.* 3 (1): 9.

55 Xu, B., Zhang, X., Tian, Z. et al. (2019). Microdroplet-guided intercalation and deterministic delamination towards intelligent rolling origami. *Nat. Commun.* 10 (1): 5019.

56 Yao, J., Yan, H., and Lieber, C.M. (2013). A nanoscale combing technique for the large-scale assembly of highly aligned nanowires. *Nat. Nanotechnol.* 8 (5): 329–335.

57 Wang, G., Guo, Q., Chen, D. et al. (2018). Facile and highly effective synthesis of controllable lattice sulfur-doped graphene quantum dots via hydrothermal treatment of durian. *ACS Appl. Mater. Interfaces* 10 (6): 5750–5759.

58 Cui, Y., Zhong, Z., Wang, D. et al. (2003). High performance silicon nanowire field effect transistors. *Nano Lett.* 3 (2): 149–152.

59 Li, W., Meredov, A., and Shamim, A. (2019). Coat-and-print patterning of silver nanowires for flexible and transparent electronics. *NPJ Flexible Electron.* 3 (1): 19.

60 Wang, J.-L., Hassan, M., Liu, J.-W. et al. (2018). Nanowire assemblies for flexible electronic devices: recent advances and perspectives. *Adv. Mater.* 30 (48): 1803430.

61 Guo, Q., Zhang, M., Xue, Z. et al. (2015). Deterministic assembly of flexible Si/Ge nanoribbons via edge-cutting transfer and printing for van der Waals heterojunctions. *Small* 11 (33): 4140–4148.

62 Chen, Y., Guo, Q., Huang, G. et al. (2018). Multifunctional nanocracks in silicon nanomembranes by notch-assisted transfer printing. *ACS Appl. Mater. Interfaces* 10 (30): 25644–25651.

63 Baca, A.J., Meitl, M.A., Ko, H.C. et al. (2007). Printable single-crystal silicon micro/nanoscale ribbons, platelets and bars generated from bulk wafers. *Adv. Funct. Mater.* 17 (16): 3051–3062.

64 Yang, Y., Hwang, Y., Cho, H.A. et al. (2011). Arrays of silicon micro/nanostructures formed in suspended configurations for deterministic assembly using flat and roller-type stamps. *Small* 7 (4): 484–491.

65 Kim, S., Wu, J., Carlson, A. et al. (2010). Microstructured elastomeric surfaces with reversible adhesion and examples of their use in deterministic assembly by transfer printing. *Proc. Natl. Acad. Sci.* 107 (40): 17095.

66 Sim, K., Chen, S., Li, Z. et al. (2019). Three-dimensional curvy electronics created using conformal additive stamp printing. *Nat. Electron.* 2 (10): 471–479.

67 Menard, E., Nuzzo, R.G., and Rogers, J.A. (2005). Bendable single crystal silicon thin film transistors formed by printing on plastic substrates. *Appl. Phys. Lett.* 86 (9): 093507.

68 Higashi, H., Nakano, M., Kudo, K. et al. (2017). A crystalline germanium flexible thin-film transistor. *Appl. Phys. Lett.* 111 (22): 222105.

69 Kim, Y.C., Lee, S.J., Oh, I.-K. et al. (2015). Bending stability of flexible amorphous IGZO thin film transistors with transparent IZO/Ag/IZO oxide–metal–oxide electrodes. *J. Alloys Compd.* 688 (15): 1108–1114.

70 Seo, J.-H., Zhang, K., Kim, M. et al. (2016). Flexible phototransistors based on single-crystalline silicon nanomembranes. *Adv. Opt. Mater.* 4 (1): 120–125.

71 Baek, S.-W., Shim, J.-H., Ko, Y.-H. et al. (2011). Low-cost and flexible ultra-thin silicon solar cell implemented with energy-down-shift via $Cd_{0.5}Zn_{0.5}S$/ZnS core/shell quantum dots. *J. Mater. Chem. A* 3 (2): 481–487.

72 Xu, X., Subbaraman, H., Chakravarty, S. et al. (2014). Flexible single-crystal silicon nanomembrane photonic crystal cavity. *ACS Nano* 8 (12): 12265–12271.

73 Bai, W., Yang, H., Ma, Y. et al. (2018). Flexible transient optical waveguides and surface-wave biosensors constructed from monocrystalline silicon. *Adv. Mater.* 30 (32): 1801584.

74 Chen, Y. and Li, M. (2014). Integrated silicon and silicon nitride photonic circuits on flexible substrates. *Opt. Lett.* 39 (12): 3449–3452.

75 Ran, W., Wang, L., Zhao, S. et al. (2020). An integrated flexible all-nanowire infrared sensing system with record photosensitivity. *Adv. Mater.* 32 (16): 1908419.

76 Xu, B., Akhtar, A., Liu, Y. et al. (2015). An epidermal stimulation and sensing platform for sensorimotor prosthetic control, management of lower back exertion, and electrical muscle activation, epidermal electronics. *Adv. Mater.* 28 (47): 4462–4471.

77 Shin, J., Yan, Y., Bai, W. et al. (2019). Bioresorbable pressure sensors protected with thermally grown silicon dioxide for the monitoring of chronic diseases and healing processes. *Nat. Biomed. Eng.* 3 (1): 37–46.

78 Zhang, K., Jung, Y.H., Mikael, S. et al. (2017). Origami silicon optoelectronics for hemispherical electronic eye systems. *Nat. Commun.* 8 (1): 1782.

79 Lee, W., Liu, Y., Lee, Y. et al. (2018). Two-dimensional materials in functional three-dimensional architectures with applications in photodetection and imaging. *Nat. Commun.* 9 (1): 1417.

80 Tian, L., Zimmerman, B., Akhtar, A. et al. (2019). Large-area MRI-compatible epidermal electronic interfaces for prosthetic control and cognitive monitoring. *Nat. Biomed. Eng.* 3 (3): 194–205.

81 Kang, S.-K., Murphy, R.K.J., Hwang, S.-W. et al. (2016). Bioresorbable silicon electronic sensors for the brain. *Nature* 530 (7588): 71–76.

82 Tian, L., Li, Y., Webb, R.C. et al. (2017). Flexible and stretchable 3ω sensors for thermal characterization of human skin. *Adv. Funct. Mater.* 27 (26): 1701282.

83 Yu, X., Wang, H., Ning, X. et al. (2018). Needle-shaped ultrathin piezoelectric microsystem for guided tissue targeting via mechanical sensing. *Nat. Biomed. Eng.* 2 (3): 165–172.

84 Zhang, H., Gutruf, P., Meacham, K. et al. (2019). Wireless, battery-free optoelectronic systems as subdermal implants for local tissue oximetry. *Sci. Adv.* 5 (3): 0873.

85 Rotsch, C., Braet, F., Wisse, E. et al. (1997). AFM imaging and elasticity measurements on living rat liver macrophages. *Cell Biol. Int.* 21 (11): 85–696.

86 Adler, A., Cowley, E.A., Bates, J.H.T. et al. (1998). Airway-parenchymal interdependence after airway contraction in rat lung explants. *J. Appl. Phys.* 85 (1): 231–237.

87 Zhang, Y., Zheng, N., Cao, Y. et al. (2019). Climbing-inspired twining electrodes using shape memory for peripheral nerve stimulation and recording. *Sci. Adv.* 5 (4): 1066.

88 Guo, Q., Koo, J., Xie, Z. et al. (2019). A bioresorbable magnetically coupled system for low-frequency wireless power transfer. *Adv. Funct. Mater.* 29 (46): 1905451.

89 Yin, L., Huang, X., Xu, H. et al. (2014). Materials, designs, and operational characteristics for fully biodegradable primary batteries. *Adv. Mater.* 26 (23): 3879–3884.

90 Lee, G., Kang, S.-K., Won, S.M. et al. (2017). Fully biodegradable microsupercapacitor for power storage in transient electronics. *Adv. Energy Mater.* 7 (18): 1700157.

91 Dagdeviren, C., Hwang, S.-W., Su, Y. et al. (2013). Transient, biocompatible electronics and energy harvesters based on ZnO. *Small* 9 (20): 3398–3404.

92 Zheng, Q., Zou, Y., Zhang, Y. et al. (2016). Biodegradable triboelectric nanogenerator as a life-time designed implantable power source. *Sci. Adv.* 2 (3): 1501478.

93 Lu, L., Yang, Z., Meacham, K. et al. (2018). Biodegradable monocrystalline silicon photovoltaic microcells as power supplies for transient biomedical implants. *Adv. Energy Mater.* 8 (16): 1703035.

5

Magnetic Nanomembranes

Guanghui Yan, Yi Ouyang, Anni Li, Yongfeng Mei, and Jizhai Cui

Fudan University, Department of Materials Science, No. 220 Handan Road, Shanghai 200433, China

5.1 Basic Theory of Magnetic Nanomembranes

5.1.1 Overview

Much endeavor has been made to take advantage of the magnetic nanomembranes since the ferromagnetic nanomembranes were successfully assembled into three-dimensional (3D) architecture. Fundamentals of magnetic materials and their corresponding magnetoelectric properties are discussed in this chapter to aid reader's understanding of magnetic nanomembranes.

Magnetic material comprises very tiny component, namely atom, which has magnetic moment. The magnetic moment of individual atom is primarily ascribed to circulating electrons. The alignment of magnetic moments of individual atom is called magnetic order. The magnetic order can be categorized into diamagnetism, paramagnetism, ferromagnetism, and antiferromagnetism. The corresponding magnetic materials with different magnetic orders are discussed in the very beginning of the section and followed by simple introduction to fabrication of the magnetic nanomembranes.

In magnetic nanomembranes, the magnetic anisotropies play important role in some functionalities of the nanomembranes. The magnetic anisotropies can be distinguished between magnetocrystalline anisotropy, which is associated with undisturbed crystal structures with anisotropy constant K_u, and induced anisotropy, which results from defects or partial atomic ordering. In Section 5.1.3, the magnetic anisotropies, including magnetocrystalline anisotropy, exchange-coupled anisotropy, stress-induced anisotropy, and shape anisotropy, are discussed in detail in the field of magnetic nanomembranes.

Besides magnetic order and magnetic anisotropy, the magnetoelectric transporting property is another important constituent of magnetic nanomembranes, which can be used in magnetic sensors. In recent years, magnetic nanomembranes with different magneto-transport effects, even spin-wave excitation, have been successfully fabricated. In Section 5.1.4, anisotropy magnetoresistance (AMR), giant

Nanomembranes: Materials, Properties, and Applications, First Edition.
Edited by Yongfeng Mei, Gaoshan Huang, and Xiuling Li.
© 2022 WILEY-VCH GmbH. Published 2022 by WILEY-VCH GmbH.

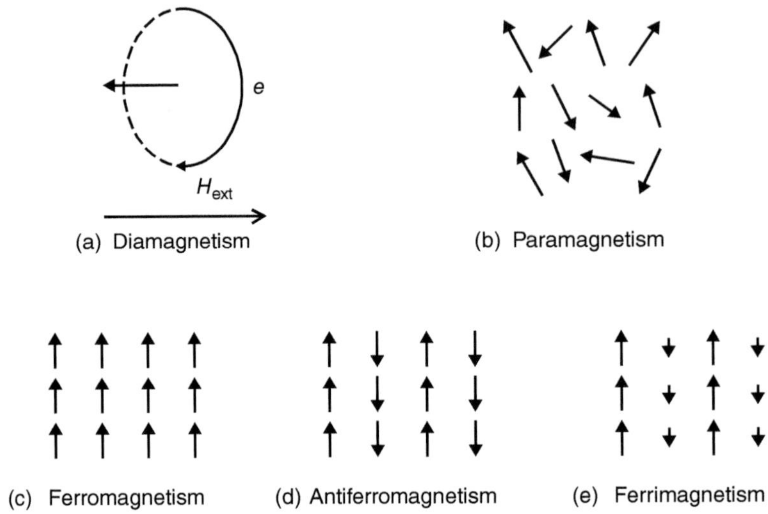

Figure 5.1 Schematic illustrating the different types of magnetic order based on the distribution of magnetization. The arrow signed by H_{ext} represents the direction of the external magnetic field H_{ext}, and the elliptic arrow denotes the arbitral direction of electron in (a). Other arrows represent the directions of spontaneous magnetization. (b–e) represent paramagnetic order, ferromagnetic order, antiferromagnetic order, and ferrimagnetic order, respectively.

magnetoresistance (GMR), tunneling magnetoresistance (TMR), giant magnetoimpedance (GMI), and planar Hall effect (PHE) are discussed in detail.

5.1.2 Magnetic Materials in Nanomembranes

With strain engineering, the membranes, deposited with magnetic materials with the thickness of several or dozens of nanometers via physical or chemical methods, can be assembled into magnetic nanomembranes with 3D architectures. Based on the magnetic order, magnetic materials can be categorized into diamagnetic, paramagnetic, ferromagnetic, and antiferromagnetic materials. Ferrimagnetic order, the net magnetization of which is not zero, can be treated as a special case of antiferromagnetic order. The ferromagnetic fine particles can be formed into superparamagnetic particles when the magnetization direction fluctuates under ambient temperatures. This section discusses abovementioned magnetic orders, as well as the fabrication of nanomembranes with different magnetic materials.

5.1.2.1 Types of Magnetic Materials

The circulating electron current under external magnetic fields will induce diamagnetism, as shown in Figure 5.1a. The induced diamagnetism is ubiquitous in the materials. For atoms whose electron orbits are fully occupied by electrons, the intrinsic magnetic moment is absent, resulting in diamagnetism, such as inert gas, Cu, Zn, and Au, to name a few.

In paramagnetic order, the magnetization is distributed disorderly (Figure 5.1b), and the net magnetization is zero, such as in Al, Li, Na, V, and air.

Figure 5.1c describes the ferromagnetic order, showing net magnetization in microscopic. Fe, Co, Ni, and their alloys are typical ferromagnetic materials. The magnetization of ferromagnetic materials decreases with the elevated temperature.

If the size of ferromagnetic particles is constrained to nanoscale in all three dimensions, this kind of ferromagnetic particles may reveal superparamagnetism. For a superparamagnetic particle with the volume (V), the magnetization direction of magnetic nanoparticles cannot stay with a stable direction when the thermal fluctuation ($k_B T$, k_B denoting the Boltzmann constant) exceeds the magnetocrystalline anisotropy energy ($K_u V$, K_u denoting the magnetocrystalline anisotropy constant) of particles. The superparamagnetism usually exists in the nanosized ferromagnetic particles, such as nanosized Fe_2O_3, CrO_2, and Fe_3O_4. γ-Fe_2O_3 nanoparticles are used as magnetic labels for biomolecules [1] and as an important component of ferrofluid in biomedical applications [2]. CrO_2 nanoparticles can also be encapsulated into polymeric hydrogel as micro-object in the application of microfluidic [3]. And nanosized Fe_3O_4 particles are extensively used in the fields of adaptive magnetic microrobots.

In ferromagnetic order, the directions of microscopic magnetization are parallel to each other, resulting in a common direction of all the magnetization. By contrast, the directions of magnetization in antiferromagnetic (AFM) order can be recognized as two sets of ferromagnetic orders with one ferromagnetic order being antiparallel to another (Figure 5.1d). Subsequently, the net magnetization in the antiferromagnetic order is zero. The typical antiferromagnetic materials are IrMn, FeMn, CoO, NiO, and MnO, to name a few.

Ferrimagnetism has special magnetic order, the microscopic magnetization of which resembles that in antiferromagnetic order, but exhibiting a nonzero net magnetization (Figure 5.1e). The typical ferrimagnetic material is ferrite, namely Fe_3O_4-type materials.

5.1.2.2 Fabrication of Magnetic Nanomembranes

The nanomembranes with different magnetic materials can be fabricated by physical vapor deposition (PVD), such as magnetron sputtering, electron beam evaporation, and thermal deposition. The chemical methods, such as microwave-assisted chemical process, electrodeposition, and atomic layer deposition (ALD), are also used in depositing magnetic materials.

Ferromagnetic (FM) materials, such as Ni, Fe, and Co, are widely used in the field of magnetic nanomembranes. Square-shaped ferromagnetic nanomembranes of Cr/Ni/Au were successfully prepared by e-beam evaporation and lifted off to be used as a "head" of magnetic helical microrobot [4]. Nickel nanomembranes were extensively used in the helical microrobots as magnetic nanomembranes, responding to the external magnetic fields [5–9].

Iron membranes can be rolled into three-dimensional architecture via strain engineering. For instance, Pd/Fe/Pd prepared by conventional e-beam evaporator was used to verify the rolling process of ferromagnetic membrane and investigate the rotation motion of microrobot [10]. Ti/Fe/Pt fabricated by e-beam evaporation for Ti/Fe membranes and magnetron sputtering for Pt membranes was finally rolled up into microtubes for microjet robots in hydrogen dioxide medium [11]. Via microwave-assisted chemical process [12] and electrodeposition [13], the iron membranes were successfully deposited on the surfaces of microrobots.

Cobalt nanomembranes with the easy magnetization axis perpendicular to the surface under some special process, such as $(Co/Pt)_N$ multilayers, can be rolled into microhelixes with radial magnetization for special locomotion mode [14, 15]. Meantime, the magnetic properties of ferromagnetic nanomembranes can be tuned via rolling process [16]. In 2019, Ti/Co/Al nanomembranes were thermally deposited for morphing nanomagnets and used in programmable microrobots [17].

Additionally, it was proved that the alloys of Fe, Co, and Ni can be used for various purposes in magnetic nanomembranes, such as NiFe [18, 19] nanomembrane and CoFe [20, 21] nanomembrane. Especially, Fe_3Si membrane prepared by co-evaporation of Fe and Si was rolled up, and the magnetoresistance was investigated for this kind of ferromagnetic materials [22].

Antiferromagnetic membranes play an important role in the magnetic nanomembranes. IrMn membrane prepared by magnetron sputtering method was used as exchange-biased layer for manipulating the superparamagnetic nanoparticles through the ferromagnetic microtube [20]. IrMn nanomembrane deposited with magnetron sputtering can also be utilized as pinning layer for the free magnetic layer due to the generated biased field between AFM and FM layers in spin valve (SV) to obtain giant magnetoresistance ratio [21]. The antiferromagnetic FeMn nanomembrane prepared by RF sputtering used in spin valves acted as the role of exchange-biased layer [23]. Additionally, the pinning behavior of antiferromagnetic materials IrMn was used in magnetic tunnel junctions (MTJs) as pinning layer to give rise to tunneling magnetoresistance effect [24].

5.1.3 Magnetic Anisotropy in Nanomembranes

The magnetic anisotropies can be distinguished between crystal anisotropies of undisturbed crystal structure and induced anisotropies describing the effects of deviations from ideal symmetry. The exchange-coupled anisotropy and stress-induced anisotropy are important in understanding the physical properties of magnetic nanomembranes. Theoretically, shape anisotropy results from the stray field effect, leading to preferred magnetization directions.

5.1.3.1 Magnetocrystalline Anisotropy

Magnetization in a microscopic magnetic domain is usually constrained to lie along one or more easy axes. Ferromagnetic monocrystals with the mentioned natural characteristics reveal magnetocrystalline anisotropy, which physically results from the spin–orbit interactions of atoms. A smaller external magnetic field can magnetize a ferromagnetic monocrystal to saturation state along the easy magnetization axis than the hard axis. For instance, iron crystal with easy axis [100] and hard axis [111], nickel crystal with easy axis [111] and hard axis [100], and cobalt crystal with easy axis [0001] and hard axis [1010]. Ferromagnetic crystal of hexagonal has only one easy magnetization axis, namely uniaxial anisotropy. The magnetization M under the uniaxial anisotropy can be denoted as anisotropy energy E_a, which is written as,

$$E_a = K_u \sin^2\theta \tag{5.1}$$

where θ is the angle between the magnetization direction and the easy magnetization axis. Here K_u, the anisotropy constant, is measured in J/m^3.

For multiaxial anisotropic ferromagnetic crystals, for instance, cubic crystals, the anisotropic energy of magnetization M can be represented by

$$E_a = K_1\left(\alpha_1^2\alpha_2^2 + \alpha_2^2\alpha_3^2 + \alpha_1^2\alpha_3^2\right) + K_2\alpha_1^2\alpha_2^2\alpha_3^2 \tag{5.2}$$

where K_1 and K_2 are anisotropy constants. And α_1, α_2, and α_3 represent the direction cosines with the angles between M with respect to the x, y, and z directions in Cartesian coordinates, respectively.

5.1.3.2 Shape Anisotropy

In the magnetic membranes with nanoscale thickness, the stray field effect is crucial to the formation of the directions of the microscopic ferromagnetic magnetization. The magnetization in magnetic layer essentially lies parallel to the plane of layers to avoid the formation of magnetic "charges" on the surfaces along the thickness direction. The in-plane magnetization can remarkably reduce the demagnetizing field in the thickness direction with demagnetizing factor N_d of almost 1, because of that, the magnetization states in magnetic nanomembranes will not be the lowest state of magnetic energy if larger demagnetizing field emerges on the surface.

$$H_d = -N_d \cdot M \tag{5.3}$$

where H_d is the vector demagnetizing field induced in the ferromagnetic body, and M denotes the magnetization of ferromagnetic body. This formula indicates that the demagnetizing field is antiparallel to the magnetization in ferromagnetic body. The demagnetizing energy E_d can be expressed as follows,

$$E_d = -\int_0^M \mu_0 H_d dM \tag{5.4}$$

where μ_0 is the vacuum permeability. When the demagnetizing field H_d is replaced by $-N_dM$, we get the following expression for the demagnetizing energy E_d,

$$E_d = \frac{1}{2}\mu_0 N_d M^2 \tag{5.5}$$

For easy understanding, the low demagnetizing energy suggests a small demagnetizing factor N_d, which is related to the shapes of ferromagnetic body. For instance, the N_d of sphere is 1/3 in all directions along diameter, tiny and long cylinder has $N_d \sim 0$ along the long axis direction, and thin circular specimen has the $N_d \sim 1$ in the direction of thickness. The perfect sphere eliminates the shape anisotropy, while the long nanowires exhibit great shape anisotropy in the direction of axis. Regarding the thin circular nanomembranes, in-plane shape anisotropy disappears due to the same demagnetizing factor in all directions in-plane.

The Ni nanomembrane with square shape has in-plane easy magnetization axis that can act as the "head" of magnetic microrobot to transform the rotation motion under rotating external magnetic fields to locomotion in low Reynolds number fluid [4, 25]. For planar magnetic nanomembranes, the magnetization usually tends to lie along the directions defined by its shape anisotropy. The morphologies of magnetic domains on the top surface of Py/Cu/Py multilayers [26] were investigated via longitudinal magneto-optical Kerr effect (MOKE), showing magnetic domains lying in-plane longitudinally for planar ferromagnetic nanomembranes.

5.1.3.3 Interface Anisotropy

The directions of magnetization usually lie parallel to the plane of thin magnetic membranes because the in-plane magnetization configuration is in the stable state. However, in perpendicular magnetic anisotropy (PMA), which is closely related to the effect of interfaces, the magnetization tends to lie along to the normal direction of the surface of magnetic membranes. PMA exists in special circumstances, such as in the microstructures of columnar crystals of polycrystal, fabrication process, and growth direction of polycrystals. Magnetic anisotropy of Au/Co/Au nanomembrane is strongly affected by the rolling process, which changes the direction of inclined deposition to longitudinal axis of microtube [16].

In a structurally isotropic medium, the surface energy density caused by magnetic surface can be expressed as [27],

$$e_s = K_s[1 - (m \cdot n)^2] \tag{5.6}$$

where n is the direction normal to the surface, and the unit of the coefficient K_s is J/m^2. For positive K_s, the surface energy density e_s is the smallest for perpendicular orientation of the magnetization at the surface ($m//n$). It is noted that the expression must be integrated over the surface.

In magnetic membranes, once the strength of interface anisotropy exceeds that of shape anisotropy, the ferromagnetic membranes show PMA, such as (Co/Pt) multilayers [14] and Co/Pd [28], in which the easy magnetization axis is perpendicular to the surface of planar nanomembranes, and they are distributed radically in rolled-up microtubes after rolling process.

5.1.3.4 Exchange-Coupled Anisotropy

Exchange-coupled anisotropy exists in ferromagnetic (FM) and AFM phases side by side, resulting in displaced asymmetrical magnetization curves along the field axis [29]. Actually, the exchange-coupled anisotropy is due to the interfacial effect between these two adjacent phases. The antiferromagnetic phase is given a specific magnetic order by cooling it below its Néel temperature under the influence of the coupled saturated ferromagnet. Subsequently, the magnetization of AFM phase side close to the interface tends to trap the magnetization in FM phase side, resulting in a displaced hysteresis. The M–H curve obtained from the adjacent FM and AFM phases when cooling through Néel temperature under absence of external magnetic field, shows symmetrical hysteresis curve, while the ordinary exchange-coupled M–H curve demonstrates displaced hysteresis when cooling through Néel temperature under external magnetic field.

Experimentally, Ueltzhoffer et al. designed and prepared the magnetic microtubes with exchange-bias field, with which the magnetic micro-objects were demonstrated to be manipulated through the inner tube and surface of the tubes [20]. Besides, the exchange bias has also been used in the giant magnetoresistance effect as a pinning layer to trap the magnetization of adjacent ferromagnetic layers with relatively low coercivity to form spin valves.

5.1.3.5 Stress-Induced Anisotropy

Under the influence of external or built-in stress, the magnetization configurations inside the ferromagnets can be changed due to the magnetoelastic energy coming from the stress. This phenomenon is termed as Villari effect, which is also called as inverse magnetostrictive effect. In contrast, magnetostrictive effect describes the induced strain occurring in the ferromagnets under the influence of external magnetic fields. The magnetostrictive coefficient (λ) is used to describe the capability of magnetostrictive effect in the ferromagnet. This coefficient also affects the value of magnetoelastic energy and subsequently the spontaneous magnetization distribution. Ni–Fe alloy is termed as permalloy or Py, and is typical soft magnetic material with high magnetic permeability at low magnetic fields. For varied compositions, it has different signs of magnetostrictive coefficients: positive, negative, and even near-zero magnetostrictive coefficient [30, 31]. The external or built-in stress may influence the direction of spontaneous magnetization inside the ferromagnets. Under stress on the ferromagnet, the magnetoelastic energy E_{me} resulting from the stress can be written as,

$$E_{\mathrm{me}} = -\frac{3}{2}\lambda\sigma \cdot \cos^2\theta \tag{5.7}$$

where λ denotes the magnetostrictive coefficient, σ denotes the stress applying on the material, and θ is the angle between the magnetization M and the stress σ. For the ferromagnets with different magnetostrictive coefficients, the stress will result in varied magnetization configurations in the ferromagnets, as shown in Figure 5.2. The compressive stress ($\sigma < 0$) and negative magnetostrictive constant ($\lambda < 0$) render spontaneous magnetization lie parallel ($\theta = 0°$ or $180°$) to the stress since the orientations between magnetization and stress result in lowest magnetoelastic energy

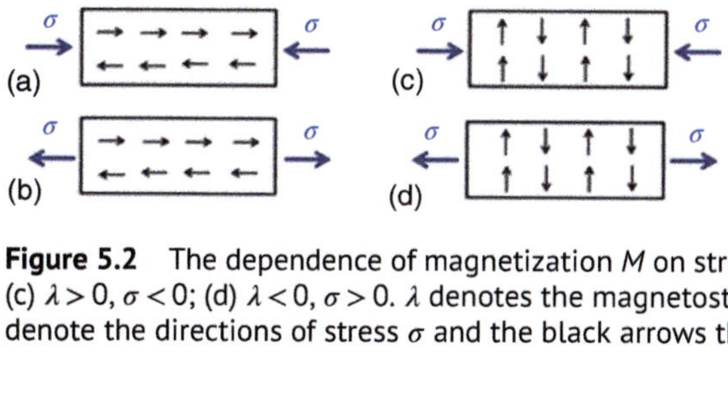

Figure 5.2 The dependence of magnetization M on stress σ. (a) $\lambda < 0, \sigma < 0$; (b) $\lambda > 0, \sigma > 0$; (c) $\lambda > 0, \sigma < 0$; (d) $\lambda < 0, \sigma > 0$. λ denotes the magnetostrictive coefficient, the blue arrows denote the directions of stress σ and the black arrows the direction of magnetization M.

(Figure 5.2a). The same orientation relationship occurs at tensile stress $\sigma > 0$ and positive magnetostrictive coefficient $\lambda > 0$, as indicated in Figure 5.2b. For negative sign of $\lambda\sigma$ (<0), the lowest magnetoelastic energy is in the situation of θ being 90° or 270°, with spontaneous magnetization perpendicular to stress direction, as schematically illustrated in Figure 5.2c,d. Along the directions of stable spontaneous magnetization or easy axis, the ferromagnet can reach the saturated magnetization state easily. As a result, the magnetoelastic energy exhibits anisotropy.

In the above discussion, the magnetic anisotropies, including magnetocrystalline anisotropy, shape anisotropy, interface anisotropy, exchange-coupled anisotropy, and stress-induced magnetic anisotropy, are briefly reviewed. As for magnetic membranes, the magnetic configurations play a crucial role in *M–H* curves, giant magnetoimpedance effect, and the like. The perspective of micromagnetism is usually used to investigate the magnetic configurations of magnetic membranes. From the perspective of micromagnetism, the magnetic Gibbs free energy of magnetic membranes mainly consists of five parts: exchange energy, anisotropy energy, demagnetizing energy, external field (Zeeman) energy, and magnetoelastic energy, which eventually determine the microscopic magnetic configurations of magnetic membranes and magnetic microtubes as well. The geometry and complicated energies in the rolled-up or curved membranes give rise to complex magnetic domain configurations, resulting in many difficulties in analyzing and visualizing the magnetic domain configurations. However, some progresses in characterization of the microscopic magnetic configurations of microtubes have been made. The following main techniques are used to successfully characterize the microscopic magnetic configurations of rolled-up magnetic membranes: hard or soft X-ray vector tomography [32–34], MOKE microscopy [26, 35], and magnetic force microscopy (MFM) [28]. With MOKE technique, the longitudinal, spiral, and azimuthal magnetic domain configurations were found in rolled-up permalloy microtubes with different compositions and sizes (Figure 5.3) [35].

5.1.4 Magnetoelectronic Properties in Magnetic Nanomembranes

5.1.4.1 Anisotropic Magnetoresistance (AMR) Effect

AMR effect is typically found in 3d ferromagnetic metals, such as Fe, Co, Ni, and their alloys [36]. This effect indicates that the electrical resistance of a material differs from the relative orientation between the supplied current and the magnetization direction of the material. For easy understanding, the external magnetic fields will generate Lorentz force on the conducting electrons, which induces different transport routes in the material, consequently causing varied electrical resistance. The physical origin is anisotropic scattering cross section of orbitals anisotropy at different orientations between supplied current and the applied magnetic field (Figure 5.4) [37]. Typically, the magnetoresistance ratios are usually small, c. 1–3% in fields of a few Gauss (Gs). The electrical resistances of the rolled-up Py nanomembrane with 1.5 windings exhibit different MR ratios, and this can be explained from the magnetic domain patterns formed before and after rolling process [38].

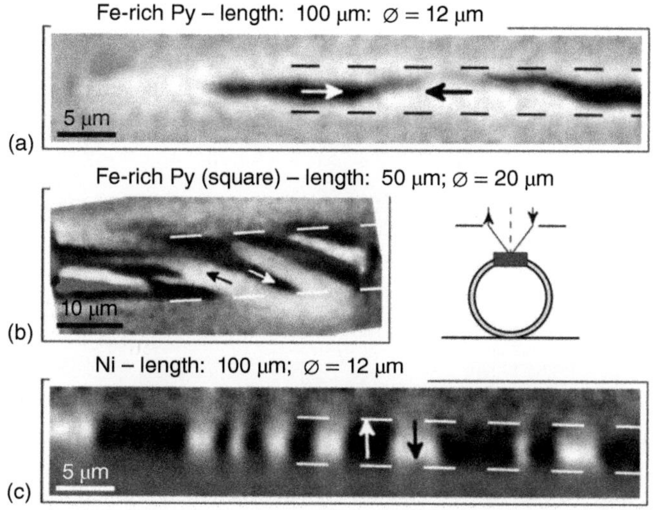

Figure 5.3 Magnetic domains structure on the surfaces of rolled-up permalloy and Ni nanomembranes characterized by magneto-optical Kerr effect (MOKE) microscopy. The dark and bright contrasts in microtubes denote the magnetic domain, with white and dark arrows denoting the directions of spontaneous magnetization inside the magnetic domains. (a) Longitudinal magnetic domain structures in Fe-rich Py microtube with length of 100 μm and diameter of 12 μm. (b) Spiral magnetic domain structures in Fe-rich Py microtube with length of 50 μm and diameter of 20 μm. (c) Circular magnetic domain structures in Ni microtube with length of 100 μm and diameter of 12 μm. Source: Streubel et al. [35]. Reproduced with permission from Wiley-VCH.

Figure 5.4 Schematic sketch illustrating the origin of anisotropic magnetoresistance. AMR depends on the scattering cross section, which is determined by the orientations of current (*I*) and magnetization (*M*).

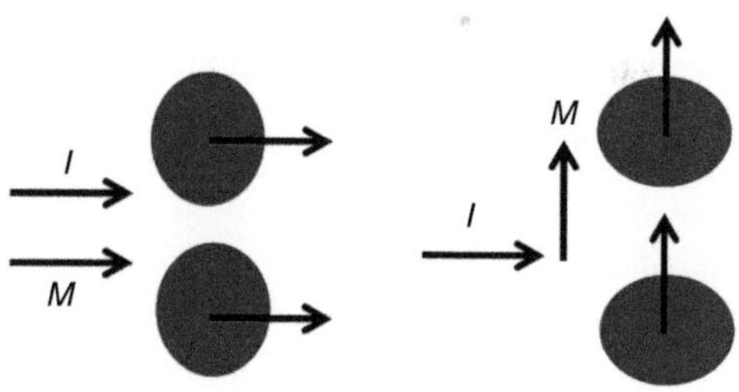

5.1.4.2 Giant Magnetoresistance (GMR) Effect

The GMR effect can date back to the discovery of exchange coupling effect in the Fe/Cr multilayer stack by Albert Fert and Peter Grünberg (2007 Nobel Prize) in the late of 1980s [39, 40]. Different magnetization orientations of the successive ferromagnetic layers result in different magnetoresistance. An antiparallel magnetization configuration induces larger magnetoresistance in ferromagnetic layers than parallel magnetization configuration does. The magnetoresistance (MR) ratio can be calculated based on the following formula,

$$\mathrm{MR} = \frac{R(H) - R_{\mathrm{Sat}}}{R_{\mathrm{Sat}}} \times 100\% \tag{5.8}$$

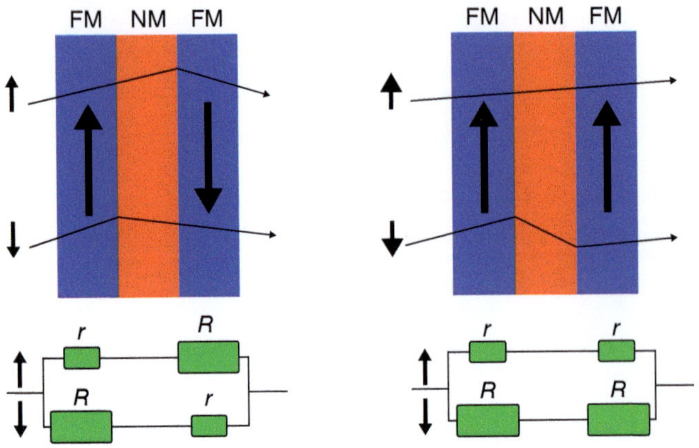

Figure 5.5 Schematic sketch illustrating the physical origin of giant magnetoresistance from multilayer stack consisting of ferromagnetic and nonmagnetic layers. The spin-up and spin-down electrons are scattered differently in the interfaces between the ferromagnetic layers with different magnetizations and nonmagnetic layers. R and r ($R > r$), denote the electrical resistances under spin orientation of electrons antiparallel and parallel to the magnetization of ferromagnetic layers, respectively.

where $R(H)$ denotes the electric resistance at the magnetic field of H, and R_{Sat} is a resistance when the multilayer stack is magnetized to saturation state. For example, in the exchange-coupled $[Co(1.0\,nm)/Cu(1.2\,nm)]_{50}$ multilayer stacks at the first AFM coupling maximum, the typical value of MR ratio can reach 50% [41].

Physically, the GMR effect originates from the asymmetry spin-dependent scattering at the interfaces of the multilayer stack (Figure 5.5). From the perspective of dual-current model proposed by Mott, the conduction electrons in ferromagnetic layers transport via spin-up and spin-down channels [43], as shown in Figure 5.5. The electrons with spin directions parallel and antiparallel to the magnetization directions of ferromagnetic layer are termed as "majority" and "minority" electrons, respectively. The conductions of electrons in the two spin channels are closely associated with the density of empty states where they can be scattered to. And the empty states can be different in the 3d ferromagnetic transition metals. As a result, the two spin channels have different scattering probabilities and the conductivities of the electrons. The total resistance is deemed to be the sum of the resistance of each spin conduction channel in different layers. As for multilayer stack GMR sensor, the MR properties can be tuned dramatically by changing the thickness of the nonmagnetic (NM) layer due to the Ruderman–Kittel–Kasuya–Yosida (RKKY) oscillatory effect of the exchange coupling strength [44, 45].

In the field of nanomembranes, GMR multilayer stacks were extensively studied in recent years. Chen, Mei and coworkers investigated the MR properties of the exchange-coupled (Co/Cu) multilayer stack sensors on different substrates [42]. The GMR sensor on flexible substrates, such as polyester and Kapton, exhibits comparable MR ratio compared with the sensors on inorganic silicon substrate (Figure 5.6). It is reported that rolled-up (Co/Cu) multilayer stacks can exceed their planar counterparts, reaching MR ratio of c. 20% [46]. The dependence of GMR

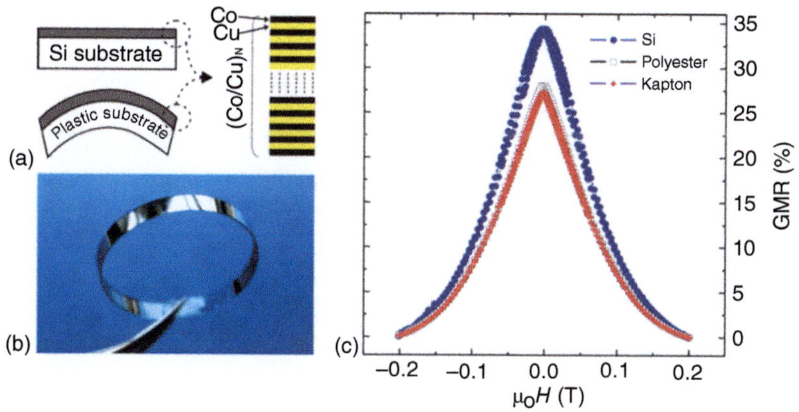

Figure 5.6 Comparison of magnetoresistance ratio of $(Co/Cu)_N$ multilayer stacks on plastic substrates and silicon substrate. (a) Schematic sketch of $(Co/Cu)_N$ multilayers stack, and the stacks on Si substrate and plastic substrate. (b) Optical image of bendable GMR sensor on plastic substrate. (c) The GMR ratio curves with GMR sensorics on different substrates. Source: Chen et al. [42]. Reproduced with permission from Wiley-VCH.

properties on the numbers of N in $(Co/Cu)_N$ multilayer stacks was also studied in the rolled-up sensors [47].

Consumer electronics and medical applications require the feasibility of re-shapeable sensors. The GMR sensors were proved to be fabricated successfully into flexible, stretchable, printable, and imperceptible shapes. (Co/Cu) [48] and (Py/Cu) [49] multilayer stacks were fabricated into flexible GMR sensors. And rolled-up (Py/Cu) GMR multilayer stacks successfully demonstrated in-flow detection of magnetic micro-objects [3]. The tensile strain test was applied on the stretchable (Co/Cu) GMR multilayer stacks, showing almost tensile-invariant properties in the regime of strain of 25% [50]. Printable magnetoresistive devices were reported in 2012 [51]. Later, GMR flakes with size of c. 40–150 μm consisting of $(Co/Cu)_{50}$ multilayer stacks with first AFM coupling maximum were fabricated and successfully printed via brushing technique [52]. Additionally, the lightweight, imperceptible GMR sensors with $(Py/Cu)_N$ multilayer stacks were demonstrated [53].

Except exchange-coupled multilayer stacks for GMR sensors, spin-valve sensors, which were invented in the beginning of the 1990s [54, 55], are another type of GMR sensors. Normally, a spin-valve structure can be seen as sandwiched structures consisting of free layer/nonmagnetic layer/pinning layer, namely FM/NM/FM/AFM. The pinning layer is composed of FM/AFM, in which one ferromagnetic layer is pinned by an adjacent antiferromagnetic layer through unidirectional interface coupling effect. However, the magnetization of the free layer, another ferromagnetic layer, is free to be switched by an external magnetic field. The stacks of free layer and pinned layer have displaced magnetization hysteresis. In this respect, different magnetic fields are required to reverse the magnetizations of the free layer and the pinned layer to give rise to a high resistance when the magnetizations of neighboring ferromagnetic layers are antiparallel to each other in between the switching fields.

Spin-valve GMR sensors also have great potential for shapeable magnetoelectronics [56]. In recent years, a spin-valve sensor with (Py/CoFe/Cu/CoFe/Py/IrMn) was demonstrated to have potential to act as a supplement for augmented reality [21].

The successive ferromagnetic layers with different coercive fields (H_c) can also result in the functions of spin valves, with the structure of FM (low-H_c)/NM/FM (high-H_c) besides antiferromagnetic coupling in ferromagnetic layers. It is worth noting that the curves of MR ratio of the abovementioned spin valves are normally asymmetric with magnetic field, and the applied magnetic fields at maximum MR ratios differ between increasing and decreasing the magnetic fields, due to the different magnetization configurations in the two ferromagnetic layers.

5.1.4.3 Tunneling Magnetoresistance (TMR) Effect

Normally, the structure of spin valves is composed of FM/Conductor/FM, in which the layer of conductor is nonmagnetic. If the conducting layer is replaced by a thin insulating layer, such as oxide, the structure of FM/Insulator/FM makes up MTJs, in which the TMR effect is found. The band structure of TMR effect is schematically illustrated in Figure 5.7. The figure indicates that different orientations of magnetizations result in varied tunneling currents through MTJs. For the ferromagnetic layers with parallel magnetizations, the tunnel current through the insulating barrier is higher than the case in antiparallel magnetizations. In other words, antiparallel magnetizations in the ferromagnetic layers result in larger electric resistance than in the case of parallel magnetizations. The spin polarization of the tunneling electrons, P, is determined by the spin-dependent density of states (DOS) of the majority and minority electrons near the Fermi surface of the ferromagnetic layers. Based on the assumption that the tunneling probability of electrons is simply proportional to corresponding DOS, the TMR ratio is given by the following formula according to Julière's model,

$$\mathrm{TMR} = \frac{2P_1P_2}{P_1 - P_2} \tag{5.9}$$

where P_1 and P_2 are the polarization of two ferromagnetic layers of MTJs [57].

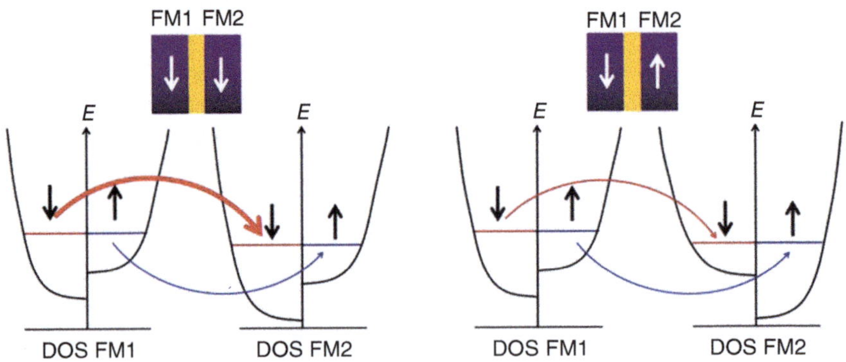

Figure 5.7 Schematic band structure illustrating the spin-polarized tunneling current through the insulating barrier of magnetic tunnel junctions with parallel and antiparallel magnetization configurations.

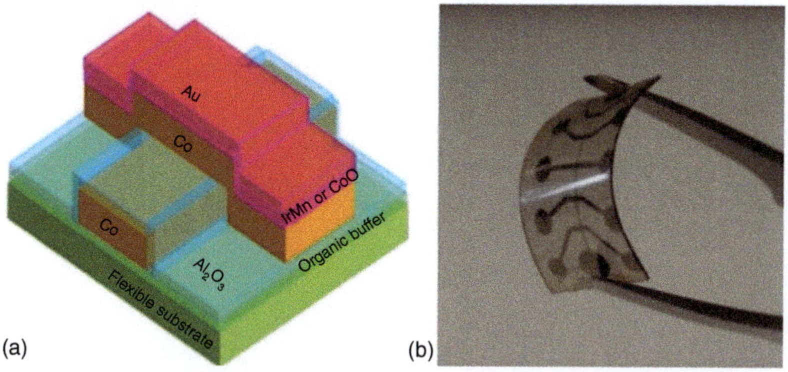

Figure 5.8 Flexible TMR sensor on organic substrate. (a) Schematic sketch illustrating the structures of TMR sensor on flexible substrate. (b) Optical microscopy image of bendable TMR sensor on flexible substrate. Source: Barraud et al. [24]. Reproduced with permission from AIP Publishing LLC.

It is challenging to make free-standing nanomembranes with TMR sensor due to very smooth interface that is mandatory to ensure homogeneity of the transport properties in MTJs. With the advanced technology, a breakthrough of flexible TMR sensors has been made (Figure 5.8). TMR sensor composed of IrMn/Co/Al_2O_3/Co MTJs was situated on flexible organic substrate gel film via photolithographic patterning and membrane deposition. The spin-dependent tunneling properties were preserved after twisting and bending the MTJs on flexible substrate [24]. Other bendable MTJs (Co/Al_2O_3/NiFe) on polyimide substrate, due to the varied coercive fields for NiFe and Co in nature, were reported a TMR ratio of 12% at room temperature [58].

Besides the amorphous AlO_x insulating barrier in MTJs, crystalline MgO is an important insulating barrier material, which can give rise to large MR ratio in TMR effect. Compared to incoherent tunnel current with aluminum oxide barrier, coherent tunnel current is generated through crystalline MgO. MR ratio of 100% at room temperature was reported in MTJs with MgO(100) insulating barrier [59]. To acquire a large TMR ratio in MgO-based MTJs, high-temperature (>300 °C) heat treatment procedure is required to crystallize MgO barrier. In 2016, MgO-based MTJs (CoFeB/MgO/CoFeB) were grown on thermally oxidized silicon substrates with post-annealing, followed by being transferred onto flexible polyethylene terephthalate (PET) substrate after releasing from silicon substrate, obtaining an increased TMR ratio, which was attributed to the effect of strain on coherent tunnel current [60]. It is noted that the coercivities of magnetic layers in CoFeB/MgO/CoFeB MTJs were tuned by the strain employed on different thickness CoFeB layers. To simplify the fabricating process of MgO-based MTJs, CoFeB/MgO/CoFeB MTJs were fabricated directly on polyimide substrate via photolithographic patterning and deposition, followed by high temperature (623 K) post-annealing heat treatment [61].

5.1.4.4 Giant Magnetoimpedance (GMI) Effect

When exposed to a magnetic field, soft ferromagnetic conductor driven by an alternating current (AC) will get a large variation of the real and imaginary part of the

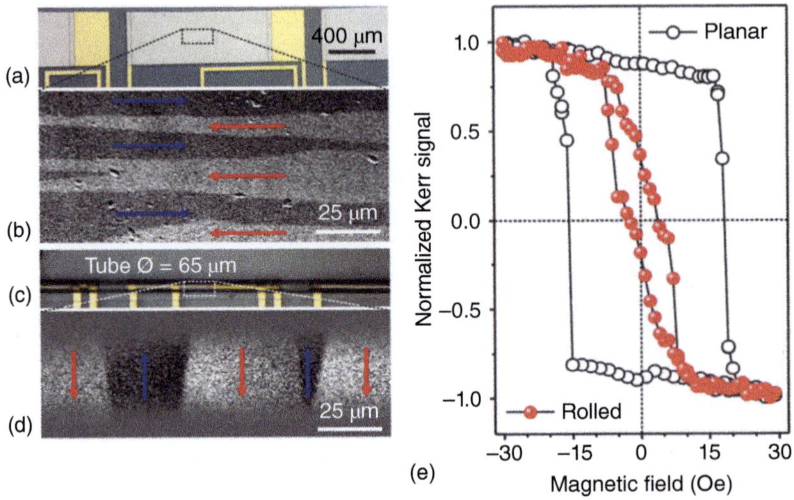

Figure 5.9 Magnetic domain configurations and magnetic properties of planar $Ni_{81}Fe_{19}/Cu/Ni_{81}Fe_{19}$ nanomembrane and rolled-up microtube. (a) and (c), the optical microscopy image of planar $Ni_{81}Fe_{19}/Cu/Ni_{81}Fe_{19}$ nanomembrane and microtube, respectively. (b) and (d), magnetic domain structures of the areas circled in (a) and (c), respectively, and the bright and dark contrasts representing the magnetic domains with different magnetization directions indicated the blue and red arrows. (e) The normalized magnetic hysteresis (*M–H*) curve of planar and rolled-up $Ni_{81}Fe_{19}/Cu/Ni_{81}Fe_{19}$ nanomembranes. Source: Karnaushenko et al. [26]. Reproduced with permission from Wiley-VCH.

complex impedance. This is known as GMI effect. The GMI ratio can be calculated as [26, 62],

$$\frac{\Delta Z}{Z_0} = \frac{(|Z(H)| - |Z_{\max}|)}{|Z_{\max}|} \times 100\% \tag{5.10}$$

where ΔZ is the difference between the absolute values of the impedance $Z(H)$ under an applied field (H) and the maximum absolute impedance $Z_{\max}$. Occasionally, the impedance Z_0 at zero magnetic field is used to estimate the $Z_{\max}$ [26].

This GMI phenomenon was first discovered in soft ferromagnetic amorphous materials [63–65]. The circumferential magnetic permeability plays an important role in GMI effect of soft ferromagnetic microwires. In other words, the circular magnetic domain configurations are required to give rise to GMI. For amorphous Co–Fe–B wires [65] after quenching process, the negative magnetostriction and quenched-in stresses may render surface anisotropy generated in the shell of wire circularly, leading to circumferential magnetic domain structures. It is feasible to get circumferential magnetic domain configurations in rolled-up ferromagnetic microtubes via rolling process due to the comprehensive magnetic interactions in the ferromagnetic nanomembranes. A flexible GMI sensor on Kapton consisting of $Ni_{81}Fe_{19}/Cu/Ni_{81}Fe_{19}$ tri-layer was successfully fabricated [66]. And rolled-up $Ni_{81}Fe_{19}/Cu/Ni_{81}Fe_{19}$ nanomembranes exhibited GMI effect because of a transition from the planar magnetic domain structure in planar membrane to the circular magnetic domain structure on the surface of rolled-up microstructure (Figure 5.9) [26]. The prepared GMI sensor was demonstrated to have strong responsiveness to small magnetic fields at ambient conditions.

5.1.4.5 Planar Hall Effect (PHE)

Conventional Hall effect physically comes from the Lorentz force applied on the conduction electrons by the magnetic fields. On the contrary, PHE in anisotropic ferromagnetic membranes originates from the scattering of electrons caused by ferromagnets. As for conventional Hall effect, the induced transverse Hall voltage is perpendicular to the applied current and external magnetic fields. However, the external magnetic field, applied current, and induced voltage all lie in the plane of ferromagnetic membranes in PHE. The structure of PHE sensor is usually formed into cross-shaped patterning, while the short side detects induced voltage and the long side applies current. This effect can be used in detection of the strengths and angles of weak magnetic fields. For highly compliant PHE sensors, elliptically shaped patterning of Py nanomembrane crosses was reported to detect the angle of applied magnetic field [67].

5.1.5 Magnetic Torque and Force

The magnetic force (F_m) and torque (T_m) exerted on a magnetic microbot can be evaluated by using the magnetic volume (V), magnetization (M), and magnetic field (B) as follows,

$$F_m = V(M \cdot \nabla) B \tag{5.11}$$

$$T_m = VM \times B \tag{5.12}$$

The magnetic force could be provided by either permanent magnet or electric magnet; such a force can be used to directly attract micro magnetic objects toward a target place [68], or to steer the direction of a moving magnetic machine in liquid [69]. The main problem of the use of magnetic force is the significant decay of field strength with increasing distance. On the other hand, magnetic torque, which could be provided by the uniform magnetic field generated by one or several pairs of Helmholtz coil, is used more often for precise remote control of magnetic microbots [70].

As a magnetic field (B) is applied at an angle α to the net magnetization vector M of an object with magnetic volume V, the field exerts a torque T_m to the net magnetization vector M [71].

For polycrystalline soft magnetic nanomembranes, the easy axis usually lies in the in-plane directions because of the large shape anisotropy [72]. As the magnetic torque rotates M at angle θ away from its equilibrium direction, the magnetic anisotropy energy E_a increases according to the following equation:

$$E_a = K_a \sin^2_\theta \tag{5.13}$$

where K_a is the magnetic anisotropy constant [73], given by,

$$K_a = \frac{1}{2\mu_0}(N_a - N_c)M^2 \tag{5.14}$$

where N_a and N_c are the length and thickness shape-anisotropy constants [72], and μ_0 is the permeability of free space.

Magnetic actuation of nanomembranes enables mechanical flexures of individually addressable, magnetically activated devices [73, 74].

Large deflections in magnetic cantilevers, springs, helix, and rolled-up microbots can be achieved remotely by using magnetic forces. On the other hand, folding of nanopatterned membranes has been proposed as an on-chip assembling method for three-dimensional nanostructures.

5.1.6 Curie Temperature/Metamagnetism

When heated above a certain temperature, magnetic materials lose their permanent magnetic properties, and that temperature is called the Curie point. For permanent magnetic or ferromagnetic materials, their ordered magnetic moments are aligned in the same direction, thus creating a strong response to magnetic sources. Above the Curie temperature (T_c), the magnetic moments are disordered. Correspondingly, the material loses its spontaneous magnetism and turns into paramagnetic state. Magnetic susceptibility above the T_c can be determined from the Curie–Weiss law:

$$\chi = \frac{C}{T - T_C} \tag{5.15}$$

where C is the Curie constant.

The particle size of a magnetic material changes the Curie temperature. When decreasing the particle size in nanomembranes, the fluctuation of electron spins becomes prominent and causes disorder of magnetic moments, which results in the decrease of T_c [75].

Metamagnetism is not a permanent magnetic property of magnetic materials. It refers to a sudden change of magnetic property (e.g. from paramagnetic to ferromagnetic) caused by magnetic phase change of a magnetic material upon modulation of applied magnetic field. The metamagnetic behavior may have quite different physical mechanisms for different types of material systems. Here we list some typical mechanisms of metamagnetic behavior as follows:

(1) *Itinerant metamagnetism* – In a paramagnetic system, the exchange splitting of the Fermi surface of itinerant electrons causes an energetically favorable transition to magnetic ordering, such as ferromagnetism [76].
(2) *Antiferromagnetic transition* – Field-induced spin flips in antiferromagnets cascade at critical energy determined by the applied magnetic field [77].

5.2 Applications of Magnetic Nanomembranes

5.2.1 Overview

The electronics of tomorrow will be soft and compliant to form a link between the materials and the virtual world. Magnetic sensorics is one important member of the family of electronics. With appropriate procedures, the magnetic nanomembranes can be assembled into compact 3D architectures and fabricated on flexible substrates to form magnetic sensorics. And the magnetic field sensor elements readily conform to any objects, even human skin, due to their compliant designs and mechanical properties. They are believed to have great potential in magnetoelectronics and micro/nanorobotics applications.

For the purpose of the compliant designs of magnetoelectronics, the magnetoresistive sensorics have advanced into re-shapeable forms, including flexible, printable, stretchable, and even imperceptible magnetoelectronics in recent years. The main functionalities of the magnetoelectronics can be grouped into two categories: acquisition of displacement and detecting weak magnetic fields. Displacement-sensing capability of magnetic sensorics is used in electronic skin. The non-touch capability of magnetic sensing elements would provide an ideal way to touchless human–machine interaction. And if the magnetic sensing elements are integrated into functional electronics and implanted into organism, they can be used to monitor the components of the heart in real time for the purpose of diagnostic at the very beginning of dysfunctions. The magnetic nanomembranes with magnetic sensorics on organic substrates also play an important role in the consumer electronics and e-skin. Such magnetoelectronics can be used in tracking and navigation. In the field of advanced fluidics, high-sensitive magnetoelectronic systems are used in the detection of magnetically labeled analytes for bioassays. Recently, the magnetic nanomembranes are demonstrated to have potentials in some other applications, such as spin-wave instruments and strain detection.

Additionally, the magnetic nanomembranes can be assembled into micro/nanorobots which are able to swim in the fluids with low Reynolds numbers in which the inertial effect can be neglected. And the micro/nanorobots have great potentials in the biomedical applications.

5.2.2 Magnetic Sensorics

5.2.2.1 Detection of Magnetic Field Strength

In the field of biology and medicine, magnetic particles are usually used for the sake of diagnosis or treatment. An important objective is to detect these magnetic particles with weak field strength, which is the main functionality of the magnetic sensing elements. For this purpose, one can use a dipole field model to approximately identify the static magnetic field of magnetic particle, the stray field $B(\mathbf{m}, \mathbf{r})$ generated from the dipole is given by [78]:

$$B(\boldsymbol{m}, \boldsymbol{r}) = \frac{\mu_0}{4\pi}\left[\frac{3\boldsymbol{r}(\boldsymbol{m}\cdot\boldsymbol{r})}{\boldsymbol{r}^5} - \frac{\boldsymbol{m}}{\boldsymbol{r}^3}\right] \tag{5.16}$$

where $\boldsymbol{m}$ is a vector magnetic moment induced by the magnetic particle, $\boldsymbol{r}$ denotes a vector distance from the particle, and μ_0 is the magnetic permeability. In some cases, the detection of magnetic fields is usually dynamic, such as in-flow detections of magnetic microdroplets. Since the magnetic stray field of a dynamic particle can act on a certain volume of sensor, the magnetic stray field of the particle should be an average value over the active sensing volume (v). At this circumstance, the effective magnetic stray field (B_{eff}) acting on the sensor can be written as [78]:

$$B_{\text{eff}} = \iiint \frac{\mu_0}{4\pi wlt} \left[\frac{3\boldsymbol{r}(\boldsymbol{m} \cdot \boldsymbol{r})}{\boldsymbol{r}^5} - \frac{\boldsymbol{m}}{\boldsymbol{r}^3} \right] dv \tag{5.17}$$

where w, l, and t are the width, length, and thickness of free layer in the MR sensor, respectively.

The effective stray fields of a particle acting on the free layer of MR sensor would result in different resistance. Under the measuring current (I), the voltage change ΔV is obtained from the discrepancy of resistance. At the assumption of linear response of a MR sensor, the voltage change is then given by [79]:

$$\Delta V = \frac{R_{\text{AP}} - R_P}{R_P} \times R_{\text{sq}} \times \frac{l}{w} \times I \times \frac{B_{\text{eff}}}{cB_k} \tag{5.18}$$

where R_{AP} is the resistance at antiparallel magnetic states, R_P is the resistance at parallel magnetic states, R_{sq} is the square resistance under parallel magnetic state, B_k is the anisotropy field of free layer, and c is a correcting factor for MR sensor (if sensor is spin valve, $c = 2$; if sensor is GMR multilayer stack, $c = 1$) [79]. This formula describes the simplest model for estimating the sensor response to the dipole field of a magnetic particle on top of the free layer of MR sensor.

Magnetic sensing elements consisting of Py(1.5 nm)/[Py(1.5 nm)/Cu(2 nm)]$_{30}$ multilayer stacks and Wheatstone bridge were reported to be rolled up into microtube, through which the magnetic ferrofluidic flow consisting of CrO_2 and polymeric hydrogel can move, and the magnetic field generated by CrO_2 will be probed (Figure 5.10a) [3]. Besides, the flexible GMR stacks consisting of [Py(1.5 nm)/Cu(2.3 nm)]$_{30}$ multilayers can be bent to a radius of 2 mm to probe the magnetic fields induced by a magnetic microdroplet (Figure 5.10b) [2]. The limit detection for this type of GMR element in an emulsion droplet of 1 nl is about 4 mg/ml. GMI effect can also be used in detection of the magnetic fields. Magnetoelectronics for in-flow detection of magnetic particles [3] or magnetically labeled analytes [2] may boost the health monitoring, point-of-care (PoC) diagnostics, and environmental sensing capabilities, in advanced fluidics and lab-on-a-chip platforms.

Besides the detection of dynamic magnetic fields generated by in-flow magnetic particles, magnetic sensors can realize the detections of static magnetic fields as well. GMI effect is also used to detect the weak magnetic fields, and it can put the detection limitation to pico-Tesla [80], which is more sensitive than that of GMR and TMR sensors with the regime of mT or below. Conventional GMI elements use the amorphous quenched CoFeSiB microwires [62, 65]. Instead of amorphous CoFeSiB microwires, rolled-up microtubes consisting of NiFe/Cu/NiFe multilayers were prepared to give rise to GMI due to the resultant circular magnetic permeability coming from the circumferential magnetic domain configurations around the surface of the microtubes. The platform of rolled-up magnetic microtubes exhibits high detection sensitivity with 45 μV/Oe at excitation current of 1 mA, which may have potential in magnetoencephalography [26].

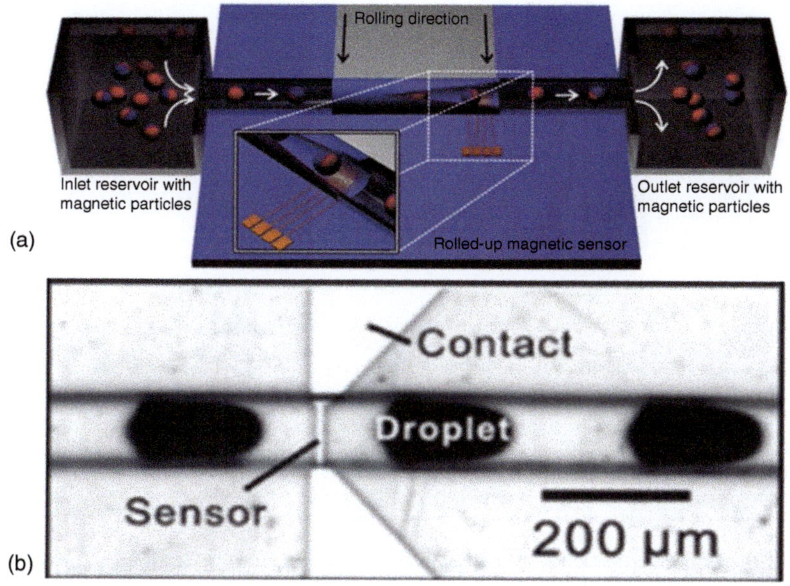

Figure 5.10 Applications of GMR devices in detection of magnetic hydrogel and microdroplet. (a) Schematic sketch illustrating rolled-up GMR sensor for detection of magnetic hydrogel. Source: Mönch et al. [3]. Reproduced with permission from American Chemical Society. (b) Optical image of bendable GMR multilayer stacks for detection of ferrofluid-Fe_2O_3 microdroplet. Source: Lin et al. [2]. Reproduced with permission from Royal Society of Chemistry.

5.2.2.2 Acquisition of Motion and Displacement

Another main functionality of the magnetic sensorics is acquisition of motion and displacement. With the magnetic sensing capabilities, GMR sensor elements equipped with soft electronics can act as motion tracking and touchless control. It was demonstrated that spin-valve sensors consisting Py(4 nm)/CoFe(1 nm)/Cu(2.4 nm)/CoFe(1 nm)/Py(4 nm)/IrMn(8 nm) multilayers conformed to a flexible thin substrate with a thickness of 1.7 μm were utilized as e-skin to probe the position of hand, leading motion tracking of hand in an artificial magnetic environment for the purpose of augmented reality [21]. Similarly, GMR sensing elements consisting Py(1.5 nm)/$[Py(1.5\,nm)/Cu(2.3\,nm)]_{30}$ multilayer stacks fabricated on the ultrathin polymeric substrate (<2 μm), show extreme bendability (bending radii <3 μm) and lightweight (~3 g/m^2), which are termed as imperceptible magnetoelectronics were successfully prepared by film deposition and photographic patterning and demonstrated to be wearable on human fingers as e-skin [53]. With the merit of imperceptible magnetoelectronics, the magnetic sensing elements were reported to be used in bio-integrated electronics and functional implants to monitor the heart for diagnosing early stages of dysfunctions [81].

With the displacement sensing capability of magnetic sensorics, magnetoresistive sensors have been successfully demonstrated in magnetic microelectromechanical system (m-MEMS) platform in recent years [82]. The typical structure

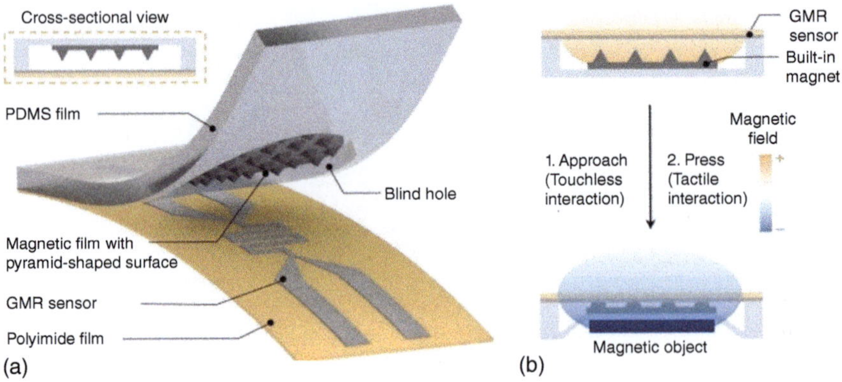

Figure 5.11 Schematic sketch illustrating combination of magnetic microelectromechanical system (m-MEMS) and GMR sensor to realize tactile and touchless purposes. (a) Schematic structure of the designed m-MEMS. (b) Mechanism of m-MEMS sensing mode. Source: Adapted from Ge et al. [82]. Copyright 2019, Nature.

of m-MEMS includes pyramids with a height of 28 μm consisting of NdFeB microparticles, polydimethylsiloxane (PDMS), and GMR multilayer stacks comprise $[Py(1.5\,nm)/Cu(2.3\,nm)]_{30}$ multilayers situated on polyimide membrane (Figure 5.11). With the capability of acquisition of magnetic fields, tactile and even touchless interactions were demonstrated to be realized in an artificial magnetic environment.

Except the GMR sensorics, the PHE sensor also has the capability of acquisition of magnetic fields to implement tracking, navigation, and orientation. PHE sensors with elliptically shaped patterning of Py crosses situated on polymeric foil with the thickness of 6 μm, bended to a radius of c. 0.5 mm, was proved to act as angle detection, which can be used in the fields of navigation and orientation (Figure 5.12) [67]. Basically, referring to the structure of PHE sensor lying in *xy*-plane, a uniform current I_x applied through the ferromagnetic thin membrane of thickness t_{FM} will generate a transverse voltage V_y given by [83]:

$$V_y = \frac{I_x \Delta\rho \cdot \sin(2\theta)}{2t_{FM}} \tag{5.19}$$

with $\Delta\rho = \rho_{\parallel} - \rho_{\perp}$, where $\rho_{\parallel}$ and $\rho_{\perp}$ are the resistivities parallel and perpendicular to the magnetization direction, respectively. A ferromagnetic thin membrane in a single domain state has in-plane magnetization. θ is the angle between the magnetization $\boldsymbol{M}$ and applied current $\boldsymbol{I}_x$. This expression is valid for a Hall bar with infinitesimal voltage probes but is usually used for describing the response of PHE sensor cross.

5.2.3 Magnetic Microbot

In the classic Sci-Fi film *Fantastic Voyage (1966)*, artists imagined a miniaturized submarine carrying its crew into human body to perform microsurgery. It has been a long-standing goal to create untethered microscale intelligent robots that can

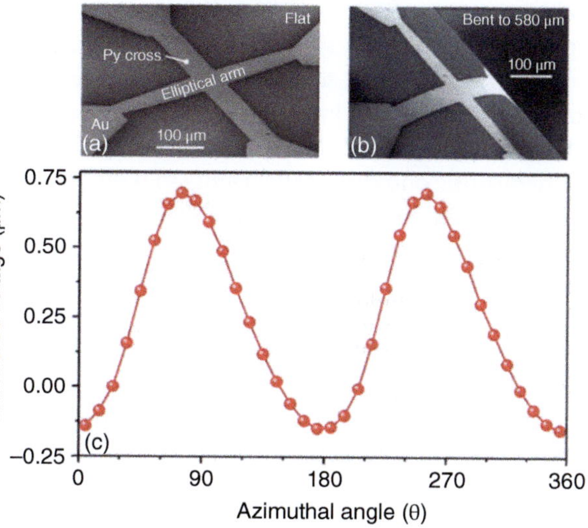

Figure 5.12 Compliant PHE sensor for detection of angles. (a) and (b) SEM images of planar and bendable elliptically shaped PHE sensor. (c) The response curve of transverse voltage induced by the azimuthal angle with the prepared bendable PHE sensor. Source: Granell et al. [67]. Reproduced with permission from Springer Nature.

operate inside human body. Robots have been an indispensable enhancement to the daily life, industry, and medical treatment of humans. Traditional surgical instruments can be driven by electromagnetic motors at macroscales, for example, the Da Vinci surgical system has been applied to perform minimally invasive procedures. But it is still a major challenge to power micromachines in tether-free manner [84]. The traditional view of intelligent robots relies on the collective functionality by the combination of computational units and software. While a manmade microbot is too small to host electronic circuit, the microorganisms produced by nature, however, are equipped with onboard sensing, motion, and deformation tools to dynamically adapt to the changing environment. Inspired by microorganisms, the challenge of development of propulsion, programmability, and multifunctionality in cell-size microbots could be addressed alternatively by embedding intelligence through reconfigurable geometry and smart materials [85]. Considering a microrobot with dimension α swims in a biological fluid with velocity ν, the Reynolds number is determined as: $Re = \alpha\nu\rho/\eta$, where ρ and η are the density and viscosity of the fluid, respectively. For microorganisms in nature, their typical dimension and velocity are on the order of 10^{-6} to 10^{-4}, such that the Reynolds number $Re \ll 1$. The low *Re* number means that viscous forces dominate over inertial forces in a fluid. As Purcell pointed out in his 1977 paper *Life at low Reynolds number*, a swimmer cannot achieve net propulsion with reciprocal motions at $Re \ll 1$ because it is exposed in an extremely viscous environment where friction dominates over inertia of mass, known as his "scallop theorem." It tells us that a micro-swimmer must break the symmetrical motion to achieve net propulsion in liquid [86].

Through the research endeavors in the past years, solutions for the power of micro and nanoscale machines have been well researched. The tremendous innovation and refinement of this field have pushed the advance of micromachines toward potential applications. Typical strategy for the energy conversion at small scale could be achieved through chemical reaction [87, 88], acoustic field [89, 90], light field [91, 92], electric field [93, 94], or magnetic field [4, 95, 96]. Among these approaches, magnetic actuation is the most applicable method for *in vivo* tasks because it is free from toxic chemical fuels, penetrable for most of human tissues, and can be operated in confined and enclosed spaces with excellent durability. Microscale propulsion in an untethered manner may help to achieve the delivery of therapeutic and diagnostic agents to previously hard-to-reach areas in the body. However, it is challenging for the scale-down fabrication and design of magnetic microbots that are capable of directional motion at small scales. With the advances in micro and nanotechnology, the self-folding approach, namely micro-origami, has been widely applied for producing 3D microbots from 2D nanomembranes. A common approach is the rolled-up technology through the release of gradient strain induced by the lattice mismatch, surface tension [84, 97], etc. The rolled-up process can be used to produce either helical microbots [98] or rolled-up microjets [69]. The former chiral structures transform the axial rotation around their helical axis into a directional propulsion [98], and the later tubular structure is basically fueled by chemical energy and can be remotely steered by magnetic field when embedded with magnetic layer in the microbots body [69]. Similar multilayer structure can also be found in the thermo-magnetically responsive micro grippers that pick and place soft cargo when exposed to temperature change [99]. As an alternative to the complex top-down fabrication process, the directed self-assembly of magnetic particles provides a pathway to fabricate reconfigurable colloidal micromachines [100]. This approach takes advantage of the encoded interaction mechanism of colloidal building blocks and utilizes magnetic field to switch the configuration of the assembled micromachines [101].

Here we discuss the main mechanisms for self-propulsion at small sizes and elucidate the design principles and propelling strategies of 3D magnetic microbots assembled from 2D magnetic nanomembranes.

5.2.3.1 Helical Microbot

Microorganisms in nature, such as *Escherichia coli bacteria*, swim in viscous liquids with the rotation of helical flagella driven by molecular motors at the end of their body [102]. Such chiral motion is efficient for propulsion as compared to reciprocal configurations. Inspired by nature, nanoscale robots with helix structures were found to be a very promising solution for efficient propulsion at small scales [102]. The first magnetically driven helical prototype was made by Honda et al. from Tohoku University in 1996, which can swim in a viscous silicon oil [103]. In 2007, the first microscale prototype of helical swimming robot was demonstrated by a group from ETH Zurich, which achieved the highest axial velocity of 3.9 μm/s [98]. Theoretically, the maximum velocity of helical microbots driven by magnetic torque outperforms those of magnetic force-driven microbots at small scales [104]. Specifically, they have a body shape that transforms axial rotation into directional propulsion along the helical axis [7]. Magnetically propelled helical microbots hold potential in a variety of biomedical applications, including drug delivery [105], cell manipulation [7], and minimally invasive surgery [84]. These helical micromachines are capable of navigation through various types of biological fluids,

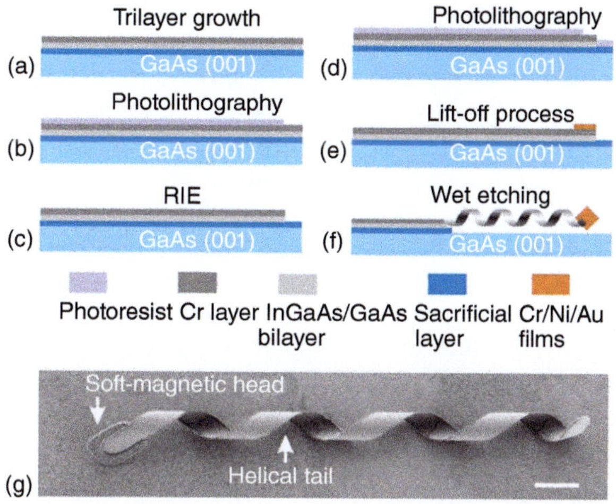

Figure 5.13 Rolled-up helical microbot: (a–f) Fabrication procedure of the helical microbot with InGaAs/GaAs/Cr helical tail. (g) SEM image showing the structure of a released microbot (scale bar: 4 μm). Source: Zhang et al. [4]. Reproduced with permission from AIP Publishing LLC.

ranging from blood, urine, to more complex environments, such as cerebrospinal fluids, gastrointestinal tract, or brain matter [96, 98, 106].

For the rolled-up technology, the thin-film deposition is the first step to fabricate helical microbots. Figure 5.13 shows a typical fabrication procedure of a helical microbot with InGaAs/GaAs/Cr helical tail [4]. The rolled-up procedures usually include film growth with molecular beam epitaxy and e-beam evaporation, patterning procedure by reactive ion etching, lift-off process, etching of the sacrificial layer, and finally, the tail will scroll into helical structure due to the local strain distribution [98]. The rolled-up direction and diameter are tunable by applying different parameters, e.g. the trigger direction of liquid, thickness, or lattice mismatch between layers [107]. In 2009, Ghosh and Fischer proposed a batch-fabrication method for helical microbots. First, a helical SiO_2 template is deposited on Si substrate by glancing angle deposition, then a ferromagnetic layer is grown onto the surface by thermal evaporation, and finally, the helical robots coated with magnetic layer are released from the substrate [108]. The helix template can also be fabricated through 3D direct laser writing, which allows one to construct more complex structures, such as fixing a microholder at the end of the helix [7]. From a biocompatible perspective, Gao et al. developed micro magnetic helix from the template of spiral vessels of plants [106]. This simple fabrication process was cost effective for mass production.

Rotation of magnetic field provides an attractive way to control the swimming of helical microrobot. The most used magnetic source for microscale manipulation is electromagnetic coils [109]. By placing a pair of coils at a distance equal to the coil radius from each other, one can obtain a magnetic-field generation device called Helmholtz coil (Figure 5.14). In the center of the coil, uniform fields can be achieved, which enable remote control of helical microrobots.

A magnetic microbot can be directly manipulated by pure magnetic torque generated by a uniform magnetic field. When the applied magnetic field is not in line with the net magnetization vector of the microbot, it causes rotation of the microbot. As the easy axis of the microrobot tends to align with the field direction, continuous

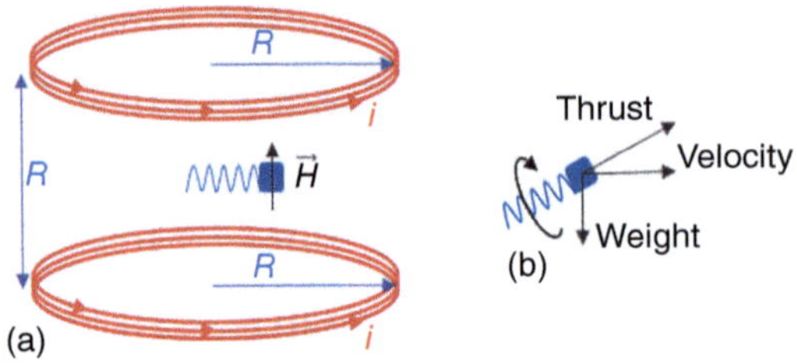

Figure 5.14 The actuation mechanism of helical microbots: (a) The schematic diagram of the Helmholtz coil setup. (b) The motion of a helical microbot.

rotation of magnetic field leads to the conversion into the propulsion of a magnetic object [4]. The translational movement generated by helix rotation makes the helical microbots swim forward [70]. The frequency-dependent swimming behavior of a helical microbot is different in three different types, including the wobbling, corkscrew, and step-out regions, corresponding to low to high field frequencies. The wobbling phenomenon is usually observed in frequency <10 Hz. In this region, the rotation of the microbot follows the rotation of magnetic field, but it wobbles around the helical axis and moves forward at the same time. The wobbling disappears when the frequency increases and is replaced by a stable corkscrew-type motion. Beyond a critical frequency Ω_c, a retarding action is observed in the rotational dynamics. The critical frequency can be described by the equation $\Omega_c = mB/\gamma$, where Ω_c is the critical frequency, m is the magnetization, B is the external magnetic field strength, and γ is the rotational drag coefficient on a liquid surface [70]. The effect of gravity on the drifting of helical microbots also plays a significant role in their swimming behavior. The density of the microbots is usually larger than the surrounding liquid. Therefore, the compensation of gravity must be considered when determining the rotation direction, as shown in Figure 5.14b.

5.2.3.2 Tubular Microbots

The significantly reduced size makes the viscous forces dominate the inertial effects of microbots. Consequently, nonreciprocal motion is essential to overcome propulsion challenges of viscous environment and Brownian effects. While the size limit does not allow for on-board batteries or other forms of power sources, a "microengine" that generates propulsion in microscale must be encoded with energy conversion mechanisms. Stimuli-responsive microbots allow remote steering by different stimuli, such as light, ultrasound, magnetic field, and chemical gradient. Biochemical liquids exist in many organs in human body, such as stomach, intestinal canal, or cholecyst. Therefore, great research endeavors have been made in the use of chemistry to propel microengines in a similar way that biological motors power themselves with natural sources with all kinds of biochemical reactions. It is advantageous that microengine directly takes use of energy from its surroundings; thus, it can operate enduringly in positions where light or ultrasound can hardly penetrate. Based on this concept, a large group of self-powered microbots, which harvest chemical energy from surroundings by the catalytic engines, has been proposed. The functionality and sophistication of these microbots has been increased with the better use of smart materials, enabling their potential applications in a wide range of biomedical tasks, including targeted drug delivery, tissue- and artery-cleaning/removal schemes, precision nanosurgery, and cancer therapeutics.

The microengines are developed by using some typical chemically active materials, including noble metals (platinum, gold, and silver) [10, 113], transient metals (iron [13], zinc [111, 112], and magnesium [113]), reactive minerals (calcium carbonate) [114], and semiconductors [115]. The integration of different catalytic materials into featured structural designs, coupled with different kinds of environmental fuels, leads to a variety of propulsion mechanisms. Whitesides et al. first put this idea into practice with a millimeter-scale object: hydrophobic plates were fixed with a Pt-covered porous glass [116]. When placed into the hydrogen peroxide (H_2O_2) fuel solution, the Pt layer decomposes the fuel hydrogen peroxide into water (H_2O) and oxygen gas (O_2) and propelled the plates to move forward. The microscale catalytic engines were fabricated by Walter et al. in 2004, composed of 1 μm long Pt and Au segments where the platinum end functions with the same catalytic mechanism in H_2O_2 fuel. Subsequently, the reaction generated a self-electrophoretic gradient over their surface and resulted in directional motion [117]. However, these early prototypes suffered from low propulsion speed and lack of efficient steering methods. To generate stronger propulsive force and improve navigation capabilities of bubble-propelled microbot, an efficient way is to fabricate 3D tubular structure, which propels itself like the chemical-fueled macroscale rocket. Tubular structures are suitable for biomedical applications because they provide a large surface area for chemical reactions and can be easily coated with magnetic layer for remote magnetic manipulation.

By exploiting the release of prestressed nanomembranes to obtain rolled-up microstructures, bubble-propelled tubular microbots that accumulate gas by the catalytic reaction of chemical fuel have been developed [87, 118, 119]. In 2008 Mei et al. demonstrated the self-propelled tubular microbot assembled by rolled-up Ti/Fe/Au/Ag multilayer nanomembrane [10]. The deposition of ferromagnetic layer (Fe/Co/Ni) onto the tube wall enables maneuver of tubular microbots remotely by an external magnetic field [11]. By adjusting the bubble radius and ejection frequency, such catalytic microjet engine can reach a speed of 50 body lengths per second [69]. Due to the excellent performance in propulsion and control, the functionality of rolled-up microbots in performing useful tasks was subsequently studied. In 2010, the manipulation of nanoplate was achieved by rolled-up microbots (Ti/Fe/Pt) by the research group of Schmidt [11]. Then in 2011, their group demonstrated the first attempt to transport cells using artificial microbots made by Ti/Fe/Pt rolled-up tubes [120]. Shankar et al. demonstrated a monoclonal antibody (mAb)-coated rolled-up Ti/Fe/Au/Pt microbots, which reached a relatively high speed of about 85 μm/s. The microbot was able to transport circulating tumor cells in diluted serum medium [121]. One defect of microbot is its poor penetration power when encountering biological tissues; this limits its delivery application for therapeutic cargo. Solovev et al. demonstrated that the asymmetric tip of the rolled-up tubes leads to different trajectories, such as circular or corkscrew, and the latter one causes rotation around the tube's axial direction, which allows the tubes to drill and penetrate biomaterials [122]. Wei et al. reported biocompatible polyaniline/Zn tubular microengine that is propelled by a spontaneous redox reaction, which generates hydrogen bubbles on Zn surface, and enables them to navigate in raw biological media and toward new extreme environments (e.g. human digestive system) [112]. The further advance of practical biomedical applications needs the improvement in propulsion efficiency and power of the microbot. Further functionalization and sophistication could be achieved by loading functional materials onto the rolled-up walls of the microbot.

The general fabrication approach for rolled-up microbot is to first deposit metallic multilayer (e.g. Fe, Ti, Ni, and Pt) on sacrificial photoresist layer by electron-beam or magnetron sputtering [108, 123]. Thereafter, the thin films were released and rolled up from the photoresist sacrificial layer with acetone, as shown in Figure 5.15a. A supercritical point drier helps to avoid collapsing of the tube structure due to the high surface tension of the etchant. Internal stress could be introduced by carefully choosing bilayers or trilayers of material. The stresses in the structure result in deformation of the thin films, which allows one to fabricate well-positioned tubes in predefined areas of a substrate and with good tube-to-tube uniformity [16, 124]. The increased surface area on tube walls can be further functionalized to satisfy different needs, which helps them outperform their counterparts in bulky or planar structures [11]. The unique tubular structures of rolled-up micromotors provide a high shape anisotropy and different distribution of drag forces along different directions [120]. Solovev et al. take advantage of the large strain in deposited nanomembranes and fabricated tubular microbots 20 times smaller than the usually reported rolled-up catalytic counterparts [122]. They grew InGaAs/GaAs thin films on sacrificial AlAs substrate using molecular beam epitaxy followed by magnetron sputtering of Pt catalytic layer. After selective etching of sacrificial layer, the rolled-up nanotubes with diameter ranging from 280 to 600 nm could be fabricated. The mechanical rigidity of the tubes could be improved by employing a thin adhesion layer, such as Cr, between the InGaAs and Pt layers. Down to this microscale, it has been shown that catalytic nanotubes can overcome Brownian diffusion and swim through high viscous fluid.

After microtube is released from substrate, it can be driven by a recoiling mechanism produced by ejecting microbubbles. By immersing a catalytic microtube into a H_2O_2 solution, microbubbles are dissolved from inside of the tube. They are ejected from one of the tube openings with a certain frequency. The increase of H_2O_2 concentration reduces the bubble radius (R), while it increases the generation frequency (f), as shown in Figure 5.15b. The bubble radius mainly depends on the surface tension (γ) of the fluid and the pressure difference (ΔP) between the generated gas and the surrounding fluid. The bubble radius is calculated by the equation $R = 2\gamma/\Delta P$. The bubbles were generated inside the tube, expanded, and then moved to the larger opening of the tube because of the difference in Laplace pressure at the liquid/gas interfaces with different diameters [125]. Finally, these mechanisms allow bubble ejection on the larger end of the tube.

5.2.3.3 Reconfigurable Micromachines

The development of miniaturized microbots furnished with multifunctionalities in self-propulsion, diagnosis, and transportation has the application of miniaturized

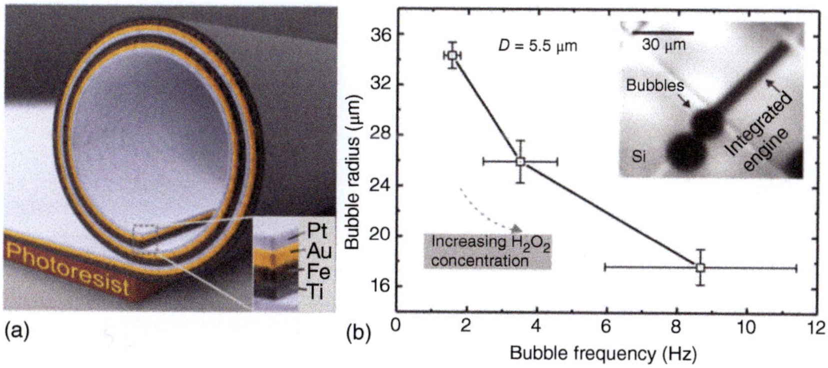

Figure 5.15 Structure and propulsion mechanism of tubular microbots: (a) Schematic diagram of a rolled-up microtube consisting of Pt/Au/Fe/Ti multilayer on a photoresist sacrificial layer. (b) Bubble radius as a function of the bubble frequency with increasing H_2O_2 concentration. The inset shows an optical image of an integrated rolled-up microjet engine generating bubbles. Source: Solovev et al. [69]. Reproduced with permission from Wiley-VCH.

microbots toward cell-level detection, drug delivery, and minimally invasive surgery. The next generation of intelligent microbots requires on-demand reconfiguration, and shape-morphing ability with the integration of artificial muscle and other biomimetic devices. A macroscale robot could be encoded with collective intelligence by integrating processing units and algorithms; however, it is not a viable path to translate the same technology down to microscale robot, because the small size of microbot limited the fabrication of sophisticated electronic structure as well as the integration of on-board power sources. As an alternative, the bottom-up colloidal-scale assembly provides a feasible way to fabricate reconfigurable micromachines. Self-assembly is a feasible approach to fabricate complex 3D architecture and electronic devices at small scale [126]. Self-assembly of colloidal matter helps to fabricate diverse target functional structures. In principle, elementary building blocks can be assembled into complex superstructures via a cascade of sequential interactions [100]. The major challenge for reconfigurable assembly is to establish encoded local interaction between the neighboring particles. Han et al. [95] demonstrated colloidal origami and chain-like microstructure from the assembly of patchy magnetic cubes (Figure 5.16a). The magnetic membrane coated on the cube surface leads to local magnetic dipole interactions between cubes [127]. Furthermore, switching the external field (on/off) allows the assembled structure to switch between different configurations dominated by either directional field–dipole or dipole–dipole interactions.

Self-propelled microbots swimming in biological liquids is significantly affected by viscous forces at low Reynolds numbers. Although reciprocal motion cannot enable net propulsion in low Reynolds numbers in Newtonian fluids (Purcell's scallop theorem), it should be noted that most of biological fluids are non-Newtonian (blood, saliva, and synovial fluid). When a microbot swims using time-asymmetric

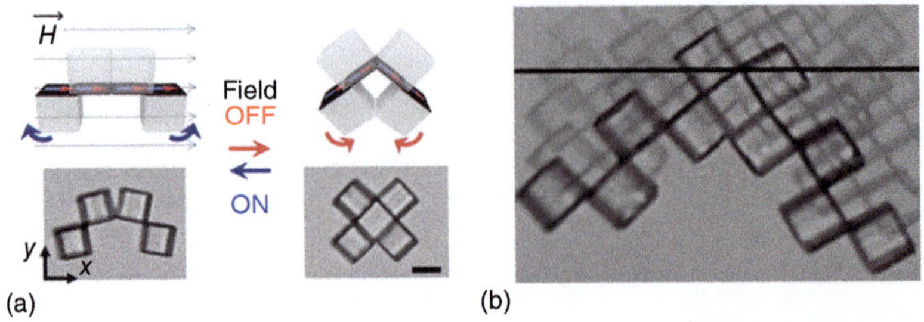

Figure 5.16 Structure and propulsion mechanism of assembled microbots: (a) Schematics of assembly pathways for two basic assembly configurations of a couple of patchy microcubes coated with a magnetic membrane. (b) Temporal asymmetric motion of assembled microscallop in a shear-thinning fluid. Source: Han et al. [95]. Reproduced with permission from American Chemical Society.

stroke in the shear-thinning fluid, even reciprocal motion can achieve directional movement (Figure 5.16b).

In 2014, Fischer et al. first demonstrated a magnetic submillimeter single-hinged scallop, which swam in non-Newtonian fluids by time-asymmetric flapping motions [128]. However, it has remained a challenging task to decrease the size of such single-hinged swimmers to the microscale by traditional top-down fabrication processes. Recently, Han and his coworkers demonstrated the magnetic assembly of cobalt-coated patchy microcubes into "microscallops" in non-Newtonian fluids further [95]. In their experiment, a foldable scallop-like structure was formed by assembling colloidal particles. Because of the magnetic shape anisotropy of the cobalt membrane coated on the surface of the particles, the assembled chains were aligned with the field upon applying external magnetic field. After removing the field, the assembled scallop closed to minimize the magnetic interaction energy. Although the assembled structure can only perform reciprocal motion in magnetic field, Han and his coworkers placed their assembled scallop into a shear-thinning xanthan gum solution to investigate its motion capabilities. Interestingly, it was demonstrated that net propulsion can be achieved by time-asymmetric stroke (slow opening followed by rapid closing of the scallop).

Miniaturized soft robots made by stimuli-responsive smart materials can be fabricated into compact and flexible structures that are much smaller than the organs and arteries inside human body. Such soft robots have been widely studied recently since their soft body can interact with human body more safely than rigid machines. Shape-morphing systems, which can achieve morphological transformations upon exertion of external stimuli, are of great interest for robotic applications. However, current shape-morphing configurations are predetermined by structural design, e.g. spatial distribution of heterogeneous materials [129, 130]. They cannot be reprogrammed because the system is restricted to a single type of transformation that is determined by the structure design. The capability to reconfigure in response to external stimuli could be a key step toward future intelligent microbot systems because it enables the microbots to step over obstacles, different textures, and materials in unstructured environments [131]. In magnetic shape-morphing systems, it

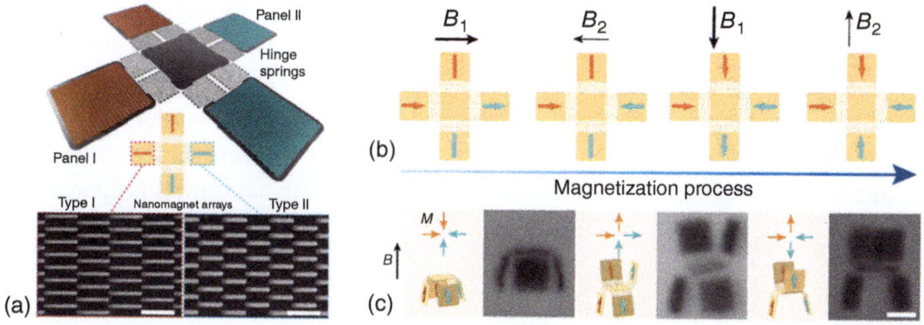

Figure 5.17 Structure and shape-morphing mechanism of a reconfigurable microbot: (a) Schematics show the structure of a four-panel shape-morphing micromachine; each panel was deposited with an array of nanomagnets featured with different size and shape ratio, as shown in the inserted SEM images of the nanomagnet arrays. (b) Schematic of the encoding of the micromachine using different magnetic fields. (c) Schematics of the magnetic configurations and micromachine folding behavior on application of the controlling magnetic field, with optical microscope images showing the three different conformations of the fabricated devices. Source: Cui et al. [17]. Reproduced with permission from Springer Nature.

utilizes magnetic torque to deform the robot body. Based on this mechanism, magnets of varying sizes have been integrated into flexible structures or soft materials to fabricate soft magnetic robots with time-varying shapes from millimeter to micrometer scale [132–134]. In 2019, Cui et al. developed a strategy to integrate multiple shape-morphing patterns into a microbot by programming the single-domain nanomagnetic arrays on connected panels (Figure 5.17a) [17]. As shown in the inserted SEM image, arrays of nanoscale cobalt magnets of 100–500 nm can be grown on flexible silicon nitride (Si_3N_4) panels by using an electron-beam lithography technique. A microscale reconfigurable robot could be fabricated by assembling several panels. The nanomagnets are single domains with stable magnetic properties, which have a tunable magnetic shape anisotropy at room temperature. In this size, the stadium-shaped single-domain nanomagnets can be encoded with shape-morphing information by magnetic fields that exceeds the nanomagnets' coercivity (less than a hundred microtesla due to size effect). This unique feature makes it possible to assemble nanomagnet arrays into micromachines with sophisticated morphing capabilities by finely tuning the robots' structures, magnetization patterns, and strengths. Figure 5.17b shows the schematic of the programming of the microbot using different magnetic fields. The tendency of the nanomagnets to align with externally applied magnetic fields makes it possible to quickly move or change the shape of these untethered robots remotely (Figure 5.17c). Since the forces and torques generated by external magnetic fields can be accurately calculated, strategies can be developed to precisely control the shape-morphing of the specific robot designs.

5.2.4 Other Functionalities

Magnetoelectronic sensors, such as GMR sensors, can be used in other applications. It was reported that the spin-valve sensor consisting of Co/Cu/NiFe/FeMn GMR stacks on polyimide can detect the strain direction applied on the substrate because of the strain-sensitive Co layer and strain-insensitive NiFe layer, which is pinned by antiferromagnetic FeMn layer [23].

Besides, the ferromagnetic nanomembranes, such as rolled-up Py microtube, also have potential applications in spin-wave instruments, which were demonstrated to have the capability of detecting spin-wave interference with azimuthal propagation [19, 135].

References

1 Baselt, D.R., Lee, G.U., Natesan, M. et al. (1998). A biosensor based on magnetoresistance technology. *Biosens. Bioelectron.* 13: 731–739.

2 Lin, G., Makarov, D., Melzer, M. et al. (2014). A highly flexible and compact magnetoresistive analytic device. *Lab Chip* 14 (20): 4050–4058.

3 Mönch, I., Makarov, D., Koseva, R. et al. (2011). Rolled-up magnetic sensor: nanomembrane architecture for in-flow detection of magnetic objects. *ACS Nano* 5 (9): 7436–7442.

4 Zhang, L., Abbott, J.J., Dong, L. et al. (2009). Artificial bacterial flagella: fabrication and magnetic control. *Appl. Phys. Lett.* 94 (6): 064107.

5 Walker, D., Käsdorf, B.T., Jeong, H.-H. et al. (2015). Enzymatically active biomimetic micropropellers for the penetration of mucin gels. *Sci. Adv.* 1: e1500501.

6 Lin, X., Zhu, H., Zhao, Z. et al. (2020). Hydrogel-based Janus micromotors capped with functional nanoparticles for environmental applications. *Adv. Mater. Technol.* 5: 2000279.

7 Tottori, S., Zhang, L., Qiu, F. et al. (2012). Magnetic helical micromachines: fabrication, controlled swimming, and cargo transport. *Adv. Mater.* 24 (6): 811–816.

8 Servant, A., Qiu, F., Mazza, M. et al. (2015). Controlled in vivo swimming of a swarm of bacteria-like microrobotic flagella. *Adv. Mater.* 27 (19): 2981–2988.

9 Medina-Sanchez, M., Schwarz, L., Meyer, A.K. et al. (2016). Cellular cargo delivery: toward assisted fertilization by sperm-carrying micromotors. *Nano Lett.* 16 (1): 555–561.

10 Mei, Y.F., Huang, G.S., Solovev, A.A. et al. (2008). Versatile approach for integrative and functionalized tubes by strain engineering of nanomembranes on polymers. *Adv. Mater.* 20: 4085–4090.

11 Solovev, A.A., Sanchez, S., Pumera, M. et al. (2010). Magnetic control of tubular catalytic microbots for the transport, assembly, and delivery of micro-objects. *Adv. Funct. Mater.* 20: 2430–2435.

12 Venugopalan, P.L., Sai, R., Chandorkar, Y. et al. (2014). Conformal cytocompatible ferrite coatings facilitate the realization of a nanovoyager in human blood. *Nano Lett.* 14 (4): 1968–1975.

13 Alcântara, C.C.J., Kim, S., Lee, S. et al. (2019). 3D fabrication of fully iron magnetic microrobots. *Small* 15: 1805006.

14 Smith, E.J., Makarov, D., Sanchez, S. et al. (2011). Magnetic microhelix coil structures. *Phys. Rev. Lett.* 107 (9): 097204.

15 Smith, E.J., Makarov, D., and Schmidt, O.G. (2011). Polymer delamination: towards unique three-dimensional microstructures. *Soft Matter* 7 (24): 11309.

16 Müller, C., Khatri, M.S., Deneke, C. et al. (2009). Tuning magnetic properties by roll-up of Au/Co/Au films into microtubes. *Appl. Phys. Lett.* 94: 102510.

17 Cui, J., Huang, T.Y., Luo, Z. et al. (2019). Nanomagnetic encoding of shape-morphing micromachines. *Nature* 575 (7781): 164–168.

18 Ureña, E.B., Mei, Y., Coric, E. et al. (2009). Fabrication of ferromagnetic rolled-up microtubes for magnetic sensors on fluids. *J. Phys. D: Appl. Phys.* 42: 055001.

19 Balhorn, F., Mansfeld, S., Krohn, A. et al. (2010). Spin-wave interference in three-dimensional rolled-up ferromagnetic microtubes. *Phys. Rev. Lett.* 104: 037205.

20 Ueltzhoffer, T., Streubel, R., Koch, I. et al. (2016). Magnetically patterned rolled-up exchange bias tubes: a paternoster for superparamagnetic beads. *ACS Nano* 10 (9): 8491–8498.

21 Bermúdez, G.S.C., Karnaushenko, D.D., Karnaushenko, D. et al. (2018). Magnetosensitive e-skins with directional perception for augmented reality. *Sci. Adv.* 4: eaao2623.

22 Schumann, J., Lisunov, K.G., Escoffier, W. et al. (2012). Magnetoresistance of rolled-up Fe_3Si nanomembranes. *Nanotechnology* 23 (25): 255701.

23 Matsumoto, H., Ota, S., Ando, A., and Chiba, D. (2019). A flexible exchange-biased spin valve for sensing strian direction. *Appl. Phys. Lett.* 114: 132401.

24 Barraud, C., Deranlot, C., Seneor, P. et al. (2010). Magnetoresistance in magnetic tunnel junctions grown on flexible organic substrates. *Appl. Phys. Lett.* 96 (7): 072502.

25 Zhang, L., Abbott, J.J., Dong, L. et al. (2009). Characterizing the swimming properties of artificial bacterial flagella. *Nano Lett.* 9 (10): 3663–3667.

26 Karnaushenko, D., Karnaushenko, D.D., Makarov, D. et al. (2015). Self-assembled on-chip-integrated giant magneto-impedance sensorics. *Adv. Mater.* 27 (42): 6582–6589.

27 Hubert, A. and Schäfer, R. (1998). *Magnetic Domains*. Berlin: Springer.

28 Puwenberg, N., Reiche, C.F., Streubel, R. et al. (2019). Magnetization reversal and local switching fields of ferromagnetic Co/Pd microtubes with radial magnetization. *Phys. Rev. B* 99: 094438.

29 Nogues, J. and Schuller, I.K. (1999). Exchange bias. *J. Magn. Magn. Mater.* 192: 203–232.

30 McKeehan, L.W. and Cioffi, P.P. (1926). Magnetostriction in permalloy. *Phys. Rev.* 28 (1): 146–157.

31 McKeehan, L.W. (1926). The significance of magnetostriction in permalloy. *Phys. Rev.* 28 (1): 158–166.

32 Donnelly, C., Guizar-Sicairos, M., Scagnoli, V. et al. (2017). Three-dimensional magnetization structures revealed with X-ray vector nanotomography. *Nature* 547 (7663): 328–331.

33 Streubel, R., Kronast, F., Fischer, P. et al. (2015). Retrieving spin textures on curved magnetic thin films with full-field soft X-ray microscopies. *Nat. Commun.* 6: 7612.

34 Streubel, R., Han, L., Kronast, F. et al. (2014). Imaging of buried 3D magnetic rolled-up nanomembranes. *Nano Lett.* 14 (7): 3981–3986.

35 Streubel, R., Lee, J., Makarov, D. et al. (2014). Magnetic microstructure of rolled-up single-layer ferromagnetic nanomembranes. *Adv. Mater.* 26 (2): 316–323.

36 McGuire, T.R. and Potter, R.I. (1975). Anisotropic magnetoresistance in ferromagnetic 3d alloys. *IEEE Trans. Magn.* 11 (4): 1018–1038.

37 Tsymbal, E.Y. and Pettifor, D.G. (2001). Perspectives of giant magnetoresistance. *Solid State Phys.* 56: 113–237.

38 Streubel, R., Makarov, D., Lee, J. et al. (2013). Rolled-up permalloy nanomembranes with multiple windings. *Spin* 03 (03): 1340001.

39 Binasch, G., Grunberg, P., Saurenbach, F., and Zinn, W. (1989). Enhanced magnetoresistance in layered magnetic structures with antiferromagnetic interlayer exchange. *Phys. Rev. B* 39 (7): 4828–4830.

40 Baibich, M.N., Broto, J.M., Fert, A. et al. (1988). Giant magnetoresistance of (001)Fe/(001)Cr magnetic superlattices. *Phys. Rev. Lett.* 61 (21): 2472–2475.

41 Parkin, S.S.P., Li, Z.G., and Smith, D.J. (1991). Giant magnetoresistance in antiferromagnetic Co/Cu multilayers. *Appl. Phys. Lett.* 58 (23): 2710–2712.

42 Chen, Y.-f., Mei, Y., Kaltofen, R. et al. (2008). Towards flexible magnetoelectronics: buffer-enhanced and mechanically tunable GMR of Co/Cu multilayers on plastic substrates. *Adv. Mater.* 20: 3224–3228.

43 Mott, N.F. (1936). The resistance and thermoelectric properties of the transition metals. *Proc. Royal Soc. London. Series A – Math. Phys. Sci.* 156 (888): 368–382.

44 Bruno, P. and Chappert, C. (1991). Oscillatory coupling between ferromagnetic layers separated by a nonmagnetic metal spacer. *Phys. Rev. Lett.* 67 (18): 2592–2592.

45 Parkin, S.S., More, N., and Roche, K.P. (1990). Oscillations in exchange coupling and magnetoresistance in metallic superlattice structures: Co/Ru, Co/Cr, and Fe/Cr. *Phys. Rev. Lett.* 64 (19): 2304–2307.

46 Müller, C., Bufon, C.C.B., Fuentes, M.E.N. et al. (2012). Towards compact three-dimensional magnetoelectronics – magnetoresistance in rolled-up Co/Cu nanomembranes. *Appl. Phys. Lett.* 100 (2): 022409.

47 Muller, C., Bof Bufon, C.C., Makarov, D. et al. (2012). Tuning giant magnetoresistance in rolled-up Co–Cu nanomembranes by strain engineering. *Nanoscale* 4 (22): 7155–7160.

48 Münzenrieder, N., Karnaushenko, D., Petti, L. et al. (2016). Entirely flexible on-site conditioned magnetic sensorics. *Adv. Electron. Mater.* 2 (8): 1600188.

49 Pérez, N., Melzer, M., Makarov, D. et al. (2015). High-performance giant magnetoresistive sensorics on flexible Si membranes. *Appl. Phys. Lett.* 106 (15): 153501.

50 Melzer, M., Kopylov, A., Makarov, D., and Schmidt, O.G. (2013). Stretchability and self-healing of wrinkled gmr multilayers on elastomeric membranes. *Spin* 03 (03): 1340005.

51 Karnaushenko, D., Makarov, D., Yan, C. et al. (2012). Printable giant magnetoresistive devices. *Adv. Mater.* 24 (33): 4518–4522.

52 Karnaushenko, D., Makarov, D., Stober, M. et al. (2015). High-performance magnetic sensorics for printable and flexible electronics. *Adv. Mater.* 27 (5): 880–885.

53 Melzer, M., Kaltenbrunner, M., Makarov, D. et al. (2015). Imperceptible magnetoelectronics. *Nat. Commun.* 6: 6080.

54 Dieny, B. (1994). Giant magnetoresistance in spin-valve multilayers. *J. Magn. Magn. Mater.* 136: 335–359.

55 Dieny, B., Speriosu, V.S., Parkin, S.S.P. et al. (1991). Giant magnetoresistive in soft ferromagnetic multilayers. *Phys. Rev. B* 43 (1): 1297–1300.

56 Melzer, M., Lin, G., Makarov, D., and Schmidt, O.G. (2012). Stretchable spin valves on elastomer membranes by predetermined periodic fracture and random wrinkling. *Adv. Mater.* 24 (48): 6468–6472.

57 Julière, M. (1975). Tunneling between ferromagnetic films. *Phys. Lett. A* 54A (17): 225–226.

58 Bedoya-Pinto, A., Donolato, M., Gobbi, M. et al. (2014). Flexible spintronic devices on Kapton. *Appl. Phys. Lett.* 104 (6): 062412.

59 Parkin, S.S., Kaiser, C., Panchula, A. et al. (2004). Giant tunnelling magnetoresistance at room temperature with MgO(100) tunnel barriers. *Nat. Mater.* 3 (12): 862–867.

60 Loong, L.M., Lee, W., Qiu, X. et al. (2016). Flexible MgO barrier magnetic tunnel junctions. *Adv. Mater.* 28 (25): 4983–4990.

61 Ota, S., Ono, M., Matsumoto, H. et al. (2019). CoFeB/MgO-based magnetic tunnel junction directly formed on a flexible substrate. *Appl. Phys Express* 12 (5): 053001.

62 Phan, M.-H. and Peng, H.-X. (2008). Giant magnetoimpedance materials: fundamentals and applications. *Prog. Mater. Sci.* 53 (2): 323–420.

63 Machado, F.L.A. and Rezende, S.M. (1996). A theoretical model for the giant magnetoimpedance in ribbons of amorphous soft-ferromagnetic alloys. *J. Appl. Phys.* 79: 6558.

64 Panina, L.V., Mohn, K., Uchyama, T. et al. (1995). Giant magneto-impedance in Co-rich amorphous wires and films. *IEEE Trans. Magn.* 31 (2): 1249–1260.

65 Panina, L.V. and Mohri, K. (1994). Magneto-impedance effect in amorphous wires. *Appl. Phys. Lett.* 65 (9): 1189–1191.

66 Li, B., Kavaldzhiev, M.N., and Kosel, J. (2015). Flexible magnetoimpedance sensor. *J. Magn. Magn. Mater.* 378: 499–505.

67 Granell, P.N., Wang, G., Cañon Bermudez, G.S. et al. (2019). Highly compliant planar Hall effect sensor with sub 200 nT sensitivity. *npj Flex. Electron.* 3: 1–6.

68 Leong, T.G., Randall, C.L., Benson, B.R. et al. (2009). Tetherless thermobiochemically actuated microgrippers. *Proc. Natl. Acad. Sci. USA* 106 (3): 703–708.

69 Solovev, A.A., Mei, Y., Bermudez Urena, E. et al. (2009). Catalytic microtubular jet engines self-propelled by accumulated gas bubbles. *Small* 5 (14): 1688–1692.

70 McNaughton, B.H., Agayan, R.R., Clarke, R. et al. (2007). Single bacterial cell detection with nonlinear rotational frequency shifts of driven magnetic microspheres. *Appl. Phys. Lett.* 91 (22): 224105.

71 Kim, S., Qiu, F., Kim, S. et al. (2013). Fabrication and characterization of magnetic microrobots for three-dimensional cell culture and targeted transportation. *Adv. Mater.* 25 (41): 5863–5868.

72 Osborn, J.A. (1945). Demagnetizing factors of the general ellipsoid. *Phys. Rev.* 67 (11–12): 351.

73 Judy, J.W. and Muller, R.S. (1997). Magnetically actuated, addressable microstructures. *J. Microelectromech. Syst.* 6 (3): 249–256.

74 Judy, J.W. and Muller, R.S. (1996). Magnetic microactuation of torsional polysilicon structures. *Sensors Actuat. A Phys.* 53 (1–3): 392–397.

75 Michels, D., Krill Iii, C., and Birringer, R. (2002). Grain-size-dependent Curie transition in nanocrystalline Gd: the influence of interface stress. *J. Magn. Magn. Mater.* 250: 203–211.

76 Fujita, A., Fujieda, S., Hasegawa, Y., and Fukamichi, K. (2003). Itinerant-electron metamagnetic transition and large magnetocaloric effects in $La(Fe_xSi_{1-x})_{13}$ compounds and their hydrides. *Phys. Rev. B* 67 (10): 104416.

77 Stryjewski, E. and Giordano, N. (1977). Metamagnetism. *Adv. Phys.* 26 (5): 487–650.

78 Jackson, J.D. (1998). *Classical Electrodynamics*. Wiley.

79 Loureiro, J., Andrade, P.Z., Cardoso, S. et al. (2011). Magnetoresistive chip cytometer. *Lab Chip* 11 (13): 2255–2261.

80 Uchiyama, T., Mohri, K., Honkura, Y., and Panina, L.V. (2012). Recent advances of pico-tesla resolution magneto-impedance sensor based on amorphous wire CMOS IC MI sensor. *IEEE Trans. Magn.* 48 (11): 3833–3839.

81 Kim, D.H., Lu, N., Ghaffari, R. et al. (2011). Materials for multifunctional balloon catheters with capabilities in cardiac electrophysiological mapping and ablation therapy. *Nat. Mater.* 10 (4): 316–323.

82 Ge, J., Wang, X., Drack, M. et al. (2019). A bimodal soft electronic skin for tactile and touchless interaction in real time. *Nat. Commun.* 10: 4405.

83 Henriksen, A.D., Dalslet, B.T., Skieller, D.H. et al. (2010). Planar Hall effect bridge magnetic field sensors. *Appl. Phys. Lett.* 97: 013057.

84 Xi, W., Solovev, A.A., Ananth, A.N. et al. (2013). Rolled-up magnetic microdrillers: towards remotely controlled minimally invasive surgery. *Nanoscale* 5 (4): 1294–1297.

85 Soto, F., Karshalev, E., Zhang, F. et al. (2021). Smart materials for microrobots. *Chem. Rev.* 122 (5): 5365–5403.

86 Purcell, E.M. (1977). Life at low Reynolds number. *Am. J. Phys.* 45 (1): 3–11.

87 Huang, G., Wang, J., Liu, Z. et al. (2017). Rocket-inspired tubular catalytic microjets with grating-structured walls as guiding empennages. *Nanoscale* 9 (47): 18590–18596.

88 Pumera, M.J.N. (2010). Electrochemically powered self-propelled electrophoretic nanosubmarines. *Nanoscale* 2 (9): 1643–1649.

89 Aghakhani, A., Yasa, O., Wrede, P., and Sitti, M. (2020). Acoustically powered surface-slipping mobile microrobots. *Proc. Natl. Acad. Sci. USA* 117 (7): 3469–3477.

90 Ahmed, D., Baasch, T., Jang, B. et al. (2016). Artificial swimmers propelled by acoustically activated flagella. *Nano Lett.* 16 (8): 4968–4974.

91 Palagi, S., Mark, A.G., Reigh, S.Y. et al. (2016). Structured light enables biomimetic swimming and versatile locomotion of photoresponsive soft microrobots. *Nat. Mater.* 15 (6): 647–653.

92 Xu, L., Mou, F., Gong, H. et al. (2017). Light-driven micro/nanomotors: from fundamentals to applications. *Chem. Soc. Rev.* 46 (22): 6905–6926.

93 Yan, J., Han, M., Zhang, J. et al. (2016). Reconfiguring active particles by electrostatic imbalance. *Nat. Mater.* 15 (10): 1095–1099.

94 Alapan, Y., Yigit, B., Beker, O. et al. (2019). Shape-encoded dynamic assembly of mobile micromachines. *Nat. Mater.* 18 (11): 1244–1251.

95 Han, K., Shields, C.W., Bharti, B. et al. (2020). Active reversible swimming of magnetically assembled "Microscallops" in non-Newtonian fluids. *Langmuir* 36 (25): 7148–7154.

96 Peyer, K.E., Tottori, S., Qiu, F. et al. (2013). Magnetic helical micromachines. *Chem. – A Eur. J.* 19 (1): 28–38.

97 Mei, Y., Solovev, A.A., Sanchez, S., and Schmidt, O.G. (2011). Rolled-up nanotech on polymers: from basic perception to self-propelled catalytic microengines. *Chem. Soc. Rev.* 40 (5): 2109–2119.

98 Bell, D.J., Leutenegger, S., Hammar, K. et al. (2007). Flagella-like propulsion for microrobots using a nanocoil and a rotating electromagnetic field. *Proc. 2007 IEEE Int. Conf. Robot. Autom.* pp. 1128–1133.

99 Kobayashi, K., Yoon, C., Oh, S.H. et al. (2019). Biodegradable thermomagnetically responsive soft untethered grippers. *ACS Appl. Mater. Interfaces* 11 (1): 151–159.

100 Sacanna, S., Rossi, L., and Pine, D.J. (2012). Magnetic click colloidal assembly. *J. Am. Chem. Soc.* 134 (14): 6112–6115.

101 Gębara, P., Marcin, J., and Škorvánek, I. (2017). Effect of partial substitution of La by Ce on the nature of phase transition in magnetocaloric $La_{1-x}Ce_xFe_{11.2}Co_{0.7}Si_{1.1}$ alloys. *J. Electron. Mater.* 46 (11): 6518–6522.

102 Qiu, F. and Nelson, B.J. (2015). Magnetic helical micro- and nanorobots: toward their biomedical applications. *Therm. Eng.* 1 (1): 021–026.

103 Honda, T., Arai, K., and Ishiyama, K. (1996). Micro swimming mechanisms propelled by external magnetic fields. *IEEE Trans. Magn.* 32 (5): 5085–5087.

104 Abbott, J.J., Peyer, K.E., Lagomarsino, M.C. et al. (2009). How should microrobots swim? *Int. J. Robot. Res.* 28 (11–12): 1434–1447.

105 Qiu, F., Mhanna, R., Zhang, L. et al. (2014). Artificial bacterial flagella functionalized with temperature-sensitive liposomes for controlled release. *Sens. Actuators B* 196: 676–681.

106 Gao, W., Feng, X., Pei, A. et al. (2014). Bioinspired helical microswimmers based on vascular plants. *Nano Lett.* 14 (1): 305–310.

107 Xu, B., Zhang, X., Tian, Z. et al. (2019). Microdroplet-guided intercalation and deterministic delamination towards intelligent rolling origami. *Nat. Commun.* 10 (1): 5019.

108 Ghosh, A. and Fischer, P. (2009). Controlled propulsion of artificial magnetic nanostructured propellers. *Nano Lett.* 9 (6): 2243–2245.

109 Fischer, P. and Ghosh, A. (2011). Magnetically actuated propulsion at low Reynolds numbers: towards nanoscale control. *Nanoscale* 3 (2): 557–563.

110 Soler, L., Magdanz, V., Fomin, V.M. et al. (2013). Self-propelled micromotors for cleaning polluted water. *ACS Nano* 7 (11): 9611–9620.

111 Sattayasamitsathit, S., Kou, H., Gao, W. et al. (2014). Fully loaded micromotors for combinatorial delivery and autonomous release of cargoes. *Small* 10 (14): 2830–2833.

112 Gao, W., Uygun, A., and Wang, J. (2012). Hydrogen-bubble-propelled zinc-based microrockets in strongly acidic media. *J. Am. Chem. Soc.* 134 (2): 897–900.

113 Nourhani, A., Karshalev, E., Soto, F., and Wang, J. (2020). Multigear bubble propulsion of transient micromotors. *Research* 2020: 1–9.

114 Guix, M., Meyer, A.K., Koch, B., and Schmidt, O.G. (2016). Carbonate-based Janus micromotors moving in ultra-light acidic environment generated by HeLa cells in situ. *Sci. Rep.* 6 (1): 1–7.

115 Dai, B., Wang, J., Xiong, Z. et al. (2016). Programmable artificial phototactic microswimmer. *Nat. Nanotechnol.* 11 (12): 1087–1092.

116 Ismagilov, R.F., Schwartz, A., Bowden, N., and Whitesides, G.M. (2002). Autonomous movement and self-assembly. *Angew. Chem. Int. Ed.* 41 (4): 652–654.

117 Paxton, W.F., Kistler, K.C., Olmeda, C.C. et al. (2004). Catalytic nanomotors: autonomous movement of striped nanorods. *J. Am. Chem. Soc.* 126 (41): 13424–13431.

118 Li, J., Esteban-Fernandez de Avila, B., Gao, W. et al. (2017). Micro/nanorobots for biomedicine: delivery, surgery, sensing, and detoxification. *Sci. Robot.* 2: eaam6431.

119 Bandari, V.K., Nan, Y., Karnaushenko, D. et al. (2020). A flexible microsystem capable of controlled motion and actuation by wireless power transfer. *Nat. Electron.* 3 (3): 172–180.

120 Sanchez, S., Solovev, A.A., Schulze, S., and Schmidt, O.G. (2011). Controlled manipulation of multiple cells using catalytic microbots. *Chem. Commun.* 47 (2): 698–700.

121 Balasubramanian, S., Kagan, D., Jack Hu, C.-M. et al. (2011). Micromachine-enabled capture and isolation of cancer cells in complex media. *Angew. Chem.* 123 (18): 4247–4250.

122 Solovev, A.A., Xi, W., Gracias, D.H. et al. (2012). Self-propelled nanotools. *ACS Nano* 6 (2): 1751–1756.

123 Li, J.-X., Lu, B.-R., Shen, Z. et al. (2011). Magnetic and meniscus-effect control of catalytic rolled-up micromotors. *Microelectron. Eng.* 88 (8): 1792–1794.

124 Gabler, F., Karnaushenko, D.D., Karnaushenko, D., and Schmidt, O.G. (2019). Magnetic origami creates high performance micro devices. *Nat. Commun.* 10 (1): 3013.

125 Bico, J. and Quéré, D. (2002). Self-propelling slugs. *J. Fluid Mech.* 467: 101.

126 Philp, D. and Stoddart, J.F. (1996). Self-assembly in natural and unnatural systems. *Angew. Chem. Int. Ed.* 35 (11): 1154–1196.

127 Han, K., Shields, C.W., Diwakar, N.M. et al. (2017). Sequence-encoded colloidal origami and microbot assemblies from patchy magnetic cubes. *Sci. Adv.* 3 (8): e1701108.

128 Qiu, T., Lee, T.C., Mark, A.G. et al. (2014). Swimming by reciprocal motion at low Reynolds number. *Nat. Commun.* 5: 5119.

129 Tan, L., Davis, A.C., and Cappelleri, D.J. (2020). Smart polymers for microscale machines. *Adv. Funct. Mater.* 31: 2007125.

130 Rus, D. and Tolley, M.T. (2015). Design, fabrication and control of soft robots. *Nature* 521 (7553): 467–475.

131 Zhao, X. and Kim, Y. (2019). Soft microbots programmed by nanomagnets. *Nature* 575 (7781): 58–59.

132 Huang, H.W., Sakar, M.S., Petruska, A.J. et al. (2016). Soft micromachines with programmable motility and morphology. *Nat. Commun.* 7: 12263.

133 Hu, W., Lum, G.Z., Mastrangeli, M., and Sitti, M. (2018). Small-scale soft-bodied robot with multimodal locomotion. *Nature* 554: 81–85.

134 Palagi, S. and Fischer, P. (2018). Bioinspired microrobots. *Nat. Rev. Mater.* 3 (6): 113–124.

135 Balhorn, F., Jeni, S., Hansen, W. et al. (2012). Axial and azimuthal spin-wave eigenmodes in rolled-up permalloy stripes. *Appl. Phys. Lett.* 100 (22): 222402.

6

Mechanics of Spontaneous Deformation in Nanomembranes: Theory, Simulations, and Experiments

Haning Xiu[1], *Guangchao Wan*[2], *Ziao Tian*[2], *Shicheng Huang*[2], *Gaoshan Huang*[3], *Ian Trase*[2], *Yongfeng Mei*[3], *and Zi Chen*[1]

[1]*Brigham and Women's Hospital/Harvard Medical School, Department of Surgery, 75 Francis Street, Boston, MA 02115, USA*
[2]*Dartmouth College, Thayer School of Engineering, 14 Engineering Dr., Hanover, NH 03755, USA*
[3]*Fudan University, Department of Materials Science, 220 Handan Rd, Shanghai 200433, People's Republic of China*

6.1 Introduction

The deformation and morphology of thin objects such as plates, ribbons, rods, and membranes have been studied over the past centuries. While the fundamental theories have been well established, recent development of new technology in micro- and nano-fabrication has brought forth new results that have garnered increasing interest in addressing some emerging albeit highly nontrivial problems.

The mechanics of thin plates, ribbons, and rods are highly nontrivial both in terms of their equilibrium morphologies and instability [1]. Mechanical stresses are considered to play a key role in morphogenesis during plant and animal development. Moreover, there is increasing attention on elastic thin films and ribbons undergoing large deformations, largely owing to the advancements of novel micro-fabrication techniques to produce ultrathin materials with exceptional properties, such as carbon nanotubes [2] and graphene [3]. In particular, helical structures have attracted scientific attention for their potential role in nanoelectromechanical systems (NEMS) [4], advanced sensing techniques, drug delivery [5–7], magnetic field detection [8], optoelectronics [9], and microrobotics [10]. Moreover, multistable structures appear in a variety of natural and engineered systems, with promising applications ranging from bioinspired robotics to deployable, morphing structures in aerospace engineering. Further quantitative understanding of the nonlinear mechanics of thin objects will meet the emerging need of the development of novel devices, such as stretchable electronics, artificial muscles, and drug delivery systems. The large deformation behavior and instability of rods have also received ample attention because they are physically ubiquitous and mathematically nontrivial.

Nanomembranes: Materials, Properties, and Applications, First Edition.
Edited by Yongfeng Mei, Gaoshan Huang, and Xiuling Li.
© 2022 WILEY-VCH GmbH. Published 2022 by WILEY-VCH GmbH.

Many topics in mechanical instability involve solving a complex system of equations, such as ordinary differential equations (ODEs) [11]. One case is the buckling of rods. As the internal moment and shear force on the cross-section area of rods can be expressed as the second and third derivatives of the deflection, the force balance equation of the rod is usually expressed in a fourth-order ODE [12]. Analytical solutions are only available in very limited situations, but many numerical studies have examined the buckling of rods. Examples include variational iteration method [12], double-scale perturbation method [13], approximate analytic technique [14], and exponential matrix approach [15].

Helical structures are commonly observed in natural and engineered systems, such as tendrils [16], cardiac growth [17], sensors [18], and active materials [19]. The formation of chiral characteristics can be related to the asymmetry of the growth rate of the organism [20], swelling [17], surface stress [21], or material anisotropy [22]. It has been shown [23] that hemihelices with alternating chirality can be fabricated by using a bilayer strip with constant intrinsic curvature. The multistability of helical ribbons is also an intriguing phenomenon. One previous paper from our research group [24] examined the multistability of helical ribbons due to the nonlinear geometric effect.

Bistable structures are structures that have two stable configurations. They abound in natural and artificial systems. Examples include the lobes of Venus flytraps [25, 26], the trap doors of bladderworts [27], slap bracelets [28], poppers [29], multilayer cross-ply composites [30], smart airfoils [31], and micro-claws [32]. The emergence of bistability or multistability unveils the nonlinear mechanical system and in most cases the nonlinearity originates from the interplay between the geometry and mechanics when the slender structure undergoes large deformation [24]. Typically, the bistability is determined through some control parameters (e.g. boundary conditions and external stimuli). With changing these control parameters, there exist bifurcation points when a stable state is destroyed, or a new stable equilibrium is created. If the current state loses its stability, the bistable structure will jump to the remote stable equilibrium. This process leads to a dramatic shape transition caused by the release of elastic energy [33]. This snap-through phenomenon has been used for application for actuators [34], energy harvesting [35, 36], and mechanical metamaterials [37].

Many distinguished characteristics of the bistable structures can be utilized to enrich the design space. To name a few, one bistable structure can fulfill two different tasks if each stable shape has its own functionality. In addition, it can switch from one stable shape to the other and can stay in the new shape without continuing energy input since the new stable shape can maintain itself, which can reduce the energy cost in reconfiguration. More importantly, if snap-through can be induced on demand, we can achieve fast motion induced by the prompt release of the stored elastic energy that can improve the performance of actuators. All these unprecedented properties motivate us to examine how to generate and control the bistable behaviors.

A review of linear elasticity theory of thin structures, including beams, plates, shells, rods, and ribbons, is presented, followed by applications of these thin structures.

6.2 Linear Elasticity Theory

Here, a review of classical linear elasticity theory is given. It provides the necessary theoretical foundation for later derivations of (i) thin beams using linearized Kirchhoff theory; (ii) thin plates using Föppl–von Kármán plate theory; and (iii) continuum elasticity theory for helicity of ribbons.

6.2.1 Stresses and Strains

Solids undergo deformations under applied stresses. The deformation of a solid body can be mathematically described as follows. Let $\boldsymbol{r}$ denote the position of a point in the body in some coordinate system (such as a three-dimensional Cartesian frame $\{\boldsymbol{E}_x, \boldsymbol{E}_y, \boldsymbol{E}_z\}$). After deformation, its position vector becomes $\boldsymbol{r}'$, and the displacement vector $\boldsymbol{u}$ (with components u_i, $i = 1, 2, 3$) is defined as

$$\boldsymbol{u} = \boldsymbol{r}' - \boldsymbol{r} \tag{6.1}$$

The strain tensor that describes the extent to which the body is deformed can be defined as

$$\gamma_{ij} = \frac{1}{2}\left(\frac{\partial u_i}{\partial x_j} + \frac{\partial u_j}{\partial x_i} + \frac{\partial u_i}{\partial x_j}\frac{\partial u_j}{\partial x_i}\right) \tag{6.2}$$

For large deformations, one should use Eq. (6.2) for nonlinear analysis. But for small deformations, higher-order terms in Eq. (6.2) can be neglected and the analysis can then be simplified such that the strain tensor becomes

$$\gamma_{ij} = \frac{1}{2}\left(\frac{\partial u_i}{\partial x_j} + \frac{\partial u_j}{\partial x_i}\right) \tag{6.3}$$

In an undeformed body, the resultant of the forces on a given portion of the body is zero since all parts are in mechanical equilibrium. However, when a body deforms, the rearrangement of the molecules gives rise to internal forces, called internal stresses, that tend to help the body establish a new equilibrium state. The internal stresses arise from the displacements of the molecules from the original unstressed equilibrium positions. As a convention, σ_{ij} denotes the stress tensor. In linear elasticity theory, the stresses and strains satisfy Hooke's law

$$\sigma_{ij} = c_{ijkl}\gamma_{kl} \tag{6.4}$$

The force balance on an infinitesimal volume element gives

$$\frac{\partial \sigma_{ij}}{\partial x_j} + P_i = 0 \tag{6.5}$$

where P_i denotes the component of the external force along x_i axis and the summation over repeated index j has been assumed as in the Einstein convention.

6.2.2 Linearized Kirchhoff Theory for Thin Beams

The 2D Kirchhoff equation, $\mathrm{d}^2M/\mathrm{d}s^2 + F_y \cos\theta - F_x \sin\theta = 0$, describes the deformation of an inextensible rod with a free boundary condition, and it assumes that the beam's deflection is purely two dimensional. M represents the internal moment at the cross section at the arc length of s, which is the distance from the origin O, θ is the rotation angle at the position s, F_x and F_y are the inter-force at the position of s over the cross section projected on the x and y axes, respectively. The moment differential equation with respect to s, $M = EI\, \mathrm{d}\theta/\mathrm{d}s$ (EI is the flexural rigidity), can be solved using finite element method (FEM) by given boundary conditions.

$$EI\frac{\mathrm{d}^2\theta}{\mathrm{d}s^2} - F\sin(\theta - \theta_0) = 0 \tag{6.6}$$

where $F = \pm\sqrt{F_x{}^2 + F_y{}^2}$ is the resultant force (positive or negative sign indicates the tensile or compressive force) and $\theta_0 = \arctan(F_y/F_x)$ is a constant.

Assuming small transverse displacement, w, of the beam, Eq. (6.6) can be linearized and simplified to $EI\,\mathrm{d}^4w/\mathrm{d}s^4 - F\,\mathrm{d}^2w/\mathrm{d}s^2 = 0$. Besides, to ensure the linearized equation solvable, the assumption of the inextensibility of the beam requires $\int_0^L (\mathrm{d}w/\mathrm{d}s)^2\mathrm{d}s = 2(L-d)$ and the boundary conditions should satisfy $w(s=0)=0$, $w(s=L)=0$, $\mathrm{d}w/\mathrm{d}s(s=0) = \sin\theta_1$ and $\mathrm{d}w/\mathrm{d}s(s=L) = \sin\theta_2$. Following the previous study [34], we rescale the parameters as $w = \sqrt{L(L-d)}\widetilde{w}$, $s = Lt$, $p^2 = -FL^2/EI$, $\beta_1 = \sqrt{L/(L-d)}\sin\theta_1$, $\beta_2 = \sqrt{L/(L-d)}\sin\theta_2$ and obtain the nondimensional, governing equations as

$$\frac{\mathrm{d}^4\widetilde{w}}{\mathrm{d}t^4} + p^2\frac{\mathrm{d}^2\widetilde{w}}{\mathrm{d}t^2} = 0 \quad \int_0^1 \left(\frac{\mathrm{d}\widetilde{w}}{\mathrm{d}t}\right)^2 \mathrm{d}t = 2 \tag{6.7}$$

The only control parameters, two nondimensional parameters β_1 and β_2, are substituted into the boundary conditions as $\widetilde{w}(t=0)=0$, $\widetilde{w}(t=1)=0$, $\mathrm{d}\widetilde{w}/\mathrm{d}t(t=0) = \beta_1$, $\mathrm{d}\widetilde{w}/\mathrm{d}t(t=1) = \beta_2$. Therefore, the number of the control parameters of the clamped beam is reduced from three (i.e. θ_1, θ_2, d/L) to two (i.e. β_1, β_2) under this scaling treatment. Parameters β_1 and β_2 are determined by the beam's length L, the distance d, and the rotation angles θ_1 and θ_2. This characteristic determines that only the beam's geometry influences the bistable behavior and there is no relation to its constituent materials. The equilibria of the clamped beam are solved using Eq. (6.7).

The solution to Eq. (6.7) takes the form as

$$\begin{aligned}\widetilde{w} = {}& \frac{\beta_1(p\cos(p) - \sin(p)) - \beta_2(p - \sin(p))}{2p(\cos(p)-1) + p^2\sin(p)}[(1-\cos(p))t + \cos(pt) - 1] \\ &+ \frac{\beta_1(\cos(p) + p\sin(p) - 1) + \beta_2(1-\cos(p))}{2p(\cos(p)-1) + p^2\sin(p)}[\sin(pt) - t\sin(p)]\end{aligned} \tag{6.8}$$

where the unknown parameter p is obtained by putting Eq. (6.8) into the constraint in Eq. (6.7) and solving the nonlinear equation

$$\Phi_1(p)\left(\beta_1^2 + \beta_2^2\right) + \Phi_2(p)\beta_1\beta_2 = 2p[p\sin(p) + 2(\cos(p)-1)]^2$$

$$\Phi_1(p) = \frac{1}{2}p^3 - \left(\frac{1}{4}\sin(2p) + \sin(p)\right)p^2 + (\cos(p) - \cos(2p))p + \frac{1}{2}\sin(2p) - \sin(p)$$

$$\Phi_2(p) = -\cos(p)p^3 + 3\sin(p)p^2 + 6(\cos(p) - 1)p + (2\sin(p) - \sin(2p)) \qquad (6.9)$$

Equation (6.9) contains multiple roots of the parameter p, corresponding to the multiple equilibrium shapes of the clamped strip in Eq. (6.8).

Although Eqs. (6.8) and (6.9) yield the expression of the equilibrium shape, we still need to determine whether the equilibrium shape is stable or not. To achieve this goal, the stability of the beam can be determined by giving a small-amplitude oscillation about the equilibrium state [38, 39]. Here, alternatively, we rely on the variational principle to explore the stability criterion. To employ the variational principle, it is essential to write down the total potential energy of the mechanical system incorporating all the elastic energy, constraints, and work from the boundaries. In our case, boundaries perform no external work because the clamped ends prohibit free displacement and rotation. The total elastic energy of the system is mainly contributed from the bending energy U_b, which is expressed as

$$U_b = \frac{1}{2}EI\int_0^L \left(\frac{d^2w}{ds^2}\right)^2 ds = EI\frac{L-d}{L^2}\int_0^1 \frac{1}{2}\left(\frac{d^2\widetilde{w}}{dt^2}\right)^2 dt = EI\frac{L-d}{L^2}\widetilde{U_b} \qquad (6.10)$$

where $\widetilde{U_b}$ is the nondimensional bending energy. Apart from the elastic energy, the total potential energy should also include the geometrical constraint through the introduction of Lagrangian multiplier λ. Therefore, the nondimensional potential energy $\widetilde{U}$ is written as

$$\widetilde{U} = \frac{1}{2}\int_0^1 \left(\frac{d^2\widetilde{w}}{dt^2}\right)^2 dt - \lambda\left(\int_0^1 \left(\frac{d\widetilde{w}}{dt}\right)^2 dt - 2\right) \qquad (6.11)$$

The stable equilibrium shape will emerge when the first-order variation of the total potential energy is zero and the second-order variation is positive $\delta\widetilde{U} = 0$, $\delta^2\widetilde{U} > 0$.

6.2.3 Föppl–von Kármán Plate Theory

The configuration of a deformed, thin plate can be well described by the displacement field of its middle plane since the thickness is relatively small and the reference configuration is chosen as the initial flat shape. The displacement field has three components, $u(x, y)$, $v(x, y)$, and $w(x, y)$, which represent the displacement of a point with coordinate (x, y) on the middle plane along the x, y, and z directions, respectively. In the meantime, the deformation of the middle plane is depicted by two tensor fields, which are the in-plane strain tensor $\overline{\overline{\varepsilon}} = [\varepsilon_x, \varepsilon_y, 2\varepsilon_{xy}]^T$ and curvature tensor $\overline{\overline{\kappa}} = [\kappa_x, \kappa_y, 2\kappa_{xy}]^T$, respectively. The in-plane strain tensor measures the planar stretching of the middle plane, whereas the curvature tensor denotes the plate's bending extent. The expressions of these two tensors follow Voigt notation and it is hypothesized that displacement can be large while the strain components remain small. The FvK kinematics connect the components of these two tensor fields with the displacement components as

$$\varepsilon_x = \frac{\partial u}{\partial x} + \frac{1}{2}\left(\frac{\partial w}{\partial x}\right)^2$$

$$\varepsilon_y = \frac{\partial v}{\partial y} + \frac{1}{2}\left(\frac{\partial w}{\partial y}\right)^2$$
$$2\varepsilon_{xy} = \frac{\partial u}{\partial y} + \frac{\partial v}{\partial x} + \frac{\partial w}{\partial x}\frac{\partial w}{\partial y}$$
$$\kappa_x = \frac{\partial^2 w}{\partial x^2}$$
$$\kappa_y = \frac{\partial^2 w}{\partial y^2}$$
$$\kappa_{xy} = \frac{\partial^2 w}{\partial x \partial y} \tag{6.12}$$

The FvK kinematics keep the nonlinear coupling between the planar stretching and bending through the second-order terms of the vertical displacement w in the expressions of the in-plane strain components. Therefore, the components in the in-plane strain tensor and curvature tensor are not independent, and they need to satisfy the compatibility condition as

$$\frac{\partial^2 \varepsilon_x}{\partial y^2} + \frac{\partial^2 \varepsilon_y}{\partial x^2} - 2\frac{\partial^2 \varepsilon_{xy}}{\partial x \partial y} = -\Phi(\overline{\overline{\varepsilon}}) = (\kappa_{xy})^2 - \kappa_x \kappa_y = -K \tag{6.13}$$

where Φ represents the differential operator and K denotes the Gaussian curvature. This differential equation can be interpreted as the simplified form of the famous *Gauss Theorema Egregium*, which implicates that any variation of the Gaussian curvature will inevitably lead to the stretching of a surface.

Accordingly, another two tensor fields are required to represent the intrinsic configuration as $\overline{\overline{\varepsilon^{\mathrm{i}}}} = \left[\varepsilon_x^{\mathrm{i}}, \varepsilon_y^{\mathrm{i}}, 2\varepsilon_{xy}^{\mathrm{i}}\right]$ and $\overline{\overline{\kappa^{\mathrm{i}}}} = \left[\kappa_x^{\mathrm{i}}, \kappa_y^{\mathrm{i}}, 2\kappa_{xy}^{\mathrm{i}}\right]$. Here, superscript i represents the components concerning the intrinsic configuration. Any deviation from the intrinsic configuration will result in elastic deformation as $\overline{\overline{\varepsilon^{\mathrm{E}}}} = \overline{\overline{\varepsilon}} - \overline{\overline{\varepsilon^{\mathrm{i}}}}$ and $\overline{\overline{\kappa^{\mathrm{E}}}} = \overline{\overline{\kappa}} - \overline{\overline{\kappa^{\mathrm{i}}}}$, where the superscript E represents the components related to the elastic deformation.

Unlike the tensor fields of actual configuration $\overline{\overline{\varepsilon}}$ and $\overline{\overline{\kappa}}$, the tensor fields of intrinsic shape $\overline{\overline{\varepsilon^{\mathrm{i}}}}$ and $\overline{\overline{\kappa^{\mathrm{i}}}}$ do not need to satisfy the compatibility constraint equation (6.13). If $\Phi(\overline{\overline{\varepsilon^{\mathrm{i}}}}) \neq K^{\mathrm{i}}$, this type of intrinsic shape is sometimes referred to as a non-Euclidean configuration since it cannot be isometrically embedded into a 3D Euclidean space [40]. The non-Euclidean intrinsic configuration indicates that $\overline{\overline{\varepsilon}} = \overline{\overline{\varepsilon^{\mathrm{i}}}}$ and $\overline{\overline{\kappa}} = \overline{\overline{\kappa^{\mathrm{i}}}}$ cannot be satisfied simultaneously; otherwise, the compatibility criterion will not be satisfied. Consequently, residue stress must exist and drive the thin plate into complex 3D morphology.

The equilibrium configuration of the plate can be sought as the stationary value of the total elastic energy. Assuming the material is isotropic and linear elastic, the bending energy density u_{b} can be rewritten in terms of the combination of tensor $\overline{\overline{\kappa^{\mathrm{E}}}}$ as

$$u_{\mathrm{b}} = \frac{Eh^3}{24(1-\nu^2)}\left[\left(\kappa_x - \kappa_x^{\mathrm{i}}\right)^2 + \left(\kappa_y - \kappa_y^{\mathrm{i}}\right)^2 + 2\nu\left(\kappa_x - \kappa_x^{\mathrm{i}}\right)\left(\kappa_y - \kappa_y^{\mathrm{i}}\right)\right] + \frac{Eh^3}{12(1+\nu)}\left(\kappa_{xy} - \kappa_{xy}^{\mathrm{i}}\right)^2 \tag{6.14}$$

where E and ν are Young's modulus and Poisson's ratio, respectively. u_b is constant if the curvature tensor and intrinsic curvature tensor are both uniform. Thus, the bending energy U_b can be directly integrated over the whole 2D domain as $U_\mathrm{b} = \iint u_\mathrm{b}\,\mathrm{d}S = u_\mathrm{b}S$, where S is the area of the middle plane.

Stretching energy density u_s can be expressed based on the planar stress components:

$$u_\mathrm{s} = \frac{h}{2E}\left[\sigma_x^2 + \sigma_y^2 - 2\nu\sigma_x\sigma_y + 2(1+\nu)\tau_{xy}^2\right] \tag{6.15}$$

where σ_x, σ_y, and τ_{xy} represent the normal and shear stress components, respectively. Based on the constitutive relationship, the compatibility condition of the in-plane strain components can also be written in terms of in-plane stress components as

$$\nu\left(\frac{\partial^2\sigma_y}{\partial y^2} + \frac{\partial^2\sigma_x}{\partial x^2}\right) - \left(\frac{\partial^2\sigma_x}{\partial y^2} + \frac{\partial^2\sigma_y}{\partial x^2}\right) + 2(1+\nu)\frac{\partial^2\tau_{xy}}{\partial x\partial y} = E\left[K - \Phi\left(\overline{\overline{\varepsilon^\mathrm{i}}}\right)\right] \tag{6.16}$$

For the plane elastic problem without body force, the stress components can be expressed through Airy stress function $\phi(x, y)$ as $\sigma_x = \partial^2\phi/\partial y^2$, $\sigma_y = \partial^2\phi/\partial x^2$, and $\tau_{xy} = -\partial^2\phi/\partial x\partial y$. Inserting these expressions into Eq. (6.16), one can yield the fourth-order differential equation regarding Airy stress function ϕ as

$$\Delta\Delta\phi = -E\left(K - \Phi\left(\overline{\overline{\varepsilon^\mathrm{i}}}\right)\right) \tag{6.17}$$

where Δ denotes Laplace operator and the boundary condition is set as there is no normal traction or shear stress on the edges. In a bilayer structure with uniform misfit strain, $\Phi\left(\overline{\overline{\varepsilon^\mathrm{i}}}\right) = 0$, which means the plate contains no stretching only when the Gaussian curvature is zero.

Equation (6.17) is linear, indicating that $u_\mathrm{s}{\sim}K^2$ because the Gaussian curvature is assumed constant. By integrating the stretching energy density over the middle plane, one can obtain the stretching energy as $U_\mathrm{s} = EhS^3K^2/24\Lambda$, where the coefficient Λ depends on the shape of plate's platform. Analytical solution exists for the coefficient Λ when platform is elliptical having its semi-major/minor axis a and b [41] as $\Lambda = \pi^2(6a^2/b^2 + 6b^2/a^2 + 4)$. For other shapes, a numerical scheme is needed to solve Eq. (6.17) to acquire the stress distribution first, and then calculate Λ through integration. For a square platform, Λ is estimated as $16\pi^2/0.738$ for good accuracy [42].

The total elastic energy will be sum of bending energy and stretching energy as $U = U_\mathrm{b} + U_\mathrm{s}$. It can take the nondimensional form by utilizing the relationship $U = \overline{U}ESh/24(1-\nu^2)$ and $\kappa = \overline{\kappa}/h$ as

$$\begin{aligned}\overline{U} = {} & \left(\overline{\kappa_x} - \overline{\kappa_x^\mathrm{i}}\right)^2 + \left(\overline{\kappa_y} - \overline{\kappa_y^\mathrm{i}}\right)^2 + 2\nu\left(\overline{\kappa_x} - \overline{\kappa_x^\mathrm{i}}\right)\left(\overline{\kappa_y} - \overline{\kappa_y^\mathrm{i}}\right) \\ & + 2(1-\nu)\left(\overline{\kappa_{xy}} - \overline{\kappa_{xy}^\mathrm{i}}\right)^2 + \frac{(1-\nu^2)S^2}{\Lambda h^4}\left(\overline{\kappa_x}\,\overline{\kappa_y} - \overline{\kappa_{xy}}^2\right)^2\end{aligned} \tag{6.18}$$

The assumption of uniform curvature makes the reduced model contain only three degrees of freedom, namely $\overline{\kappa_x}$, $\overline{\kappa_y}$, and $\overline{\kappa_{xy}}$, in the expression of total elastic energy. As a result, the partial differentiation of the total elastic energy with $\partial\overline{U}/\partial\overline{\kappa_x} = 0$,

$\partial\overline{U}/\partial\overline{\kappa_y} = 0$, and $\partial\overline{U}/\partial\overline{\kappa_{xy}} = 0$ will yield the equilibrium solutions, while the Hessian matrix will determine whether the solution is stable or not. Previously research has proven that twisting is absent when the material property is isotropic elastic [43], indicating that $\overline{\kappa_{xy}} = 0$. Therefore, the governing equations can be expressed as

$$\left(\overline{\kappa_x} - \overline{\kappa_x^{\mathrm{i}}}\right) + \nu\left(\overline{\kappa_y} - \overline{\kappa_y^{\mathrm{i}}}\right) + (1-\nu^2)\phi\overline{\kappa_x}\,\overline{\kappa_y}^2 = 0$$
$$\left(\overline{\kappa_y} - \overline{\kappa_y^{\mathrm{i}}}\right) + \nu\left(\overline{\kappa_x} - \overline{\kappa_x^{\mathrm{i}}}\right) + (1-\nu^2)\phi\overline{\kappa_y}\,\overline{\kappa_x}^2 = 0 \tag{6.19}$$

where $\phi hS^2/\Lambda h^4$ is a nondimensional parameter determined by the plate's geometry. The stability is determined by two criteria as

$$\left[1 + (1-\nu^2)\phi\overline{\kappa_y}^2\right]\left[1 + (1-\nu^2)\phi\overline{\kappa_x}^2\right]$$
$$-\left[\nu + 2(1-\nu^2)\phi\overline{\kappa_x\kappa_y}\right]^2 > 0, \quad (1+\nu)\phi\overline{\kappa_x\kappa_y} < 1 \tag{6.20}$$

6.2.4 Continuum Elasticity Theory for Helicity of Ribbons

Helical structures arise in a variety of natural and engineered systems [44] and play a significant role in applications in NEMS [4], drug delivery and biochemical/biomedical sensing [5–7], magnetic field detection [8], optoelectronics [6], and micro-robotics [10]. The mechanics of helices of many micro- and nanoscale systems is driven by material surface stress and other residual stresses. These stresses are similar to the elastic restoring forces of bulk deformation [45–49]. Generally, ribbon helicity is caused by the following anisotropic mechanical properties: surface stress [21], external stress [50], residual strain, and elastic modulus [51], and meanwhile, there exists nonaxiality between the principal axes of curvature and the principal geometric axes of the ribbon. Here, differential geometry is used to establish a mathematical model of a ribbon deformed with given magnitudes and orientations of principal curvatures. The model shows that the centerline develops to a circular helix that has constant radius and pitch.

A ribbon is simplified to an elastic strip with length of L, width of $w \ll L$, and thickness of $H \ll w$ [52]. The ribbon has a rectangular cross section and the principal geometric axes are oriented along the ribbon's length ($\boldsymbol{d}_x$), width ($\boldsymbol{d}_y$), and thickness ($\boldsymbol{d}_z$), which gives an orthonormal triad of vectors $\{\boldsymbol{d}_x, \boldsymbol{d}_y, \boldsymbol{d}_z\}$ rotating along with the ribbon [53, 54]. In planar bending [55], curvature is restricted to the $\boldsymbol{d}_x$ (lengthwise) direction; when the ribbon helically deforms, the curvature will occur along with principal directions

$$\boldsymbol{r}_1 = \boldsymbol{d}_x\cos\phi - \boldsymbol{d}_y\sin\phi \quad \text{and} \quad \boldsymbol{r}_2 = \boldsymbol{d}_x\sin\phi - \boldsymbol{d}_y\cos\phi \tag{6.21}$$

where ϕ notifies the direction of the principal curvatures according to the principal geometric axes and κ_1 and κ_2 denote the $\boldsymbol{r}_1$'s and $\boldsymbol{r}_2$'s directed principal curvatures, respectively. Any points along the centerline follow a space curve

$$\boldsymbol{P} = \boldsymbol{P}(x) = X(s)\boldsymbol{E}_x + Y(s)\boldsymbol{E}_y + Z(s)\boldsymbol{E}_z \tag{6.22}$$

where $s \in \{0, L\}$ is the convective arc length. We then analyze a relationship between the ribbon curvatures and orientation $\{\kappa_1, \kappa_2, \phi\}$ and corresponding parameters,

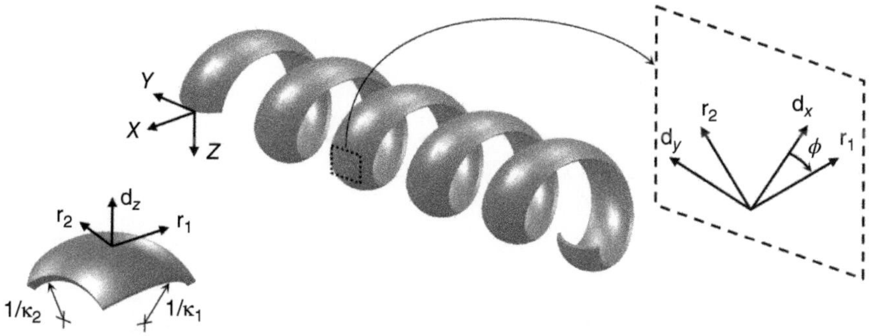

Figure 6.1 Schematic of a helical ribbon. The vectors $\mathbf{d}_x$ and $\mathbf{d}_y$ are directions of the lengthwise and widthwise of the ribbon, respectively. The bases $\mathbf{r}_1$ and $\mathbf{r}_2$ represent the principal directions of the curvature.

such as the pitch, radius, axis, and chirality ($\boldsymbol{P} = \boldsymbol{P}(s)$). For convenience, the unit normal $\boldsymbol{N} = \boldsymbol{d}_z \equiv \boldsymbol{d}_x \times \boldsymbol{d}_y = \boldsymbol{r}_1 \times \boldsymbol{r}_2$ is defined and results in

$$\frac{\mathrm{d}\boldsymbol{r}_1}{\mathrm{d}s} = \boldsymbol{N}\kappa_1 \cos\phi \quad \text{and} \quad \frac{\mathrm{d}\boldsymbol{r}_2}{\mathrm{d}s} = -\boldsymbol{N}\kappa_2 \sin\phi \tag{6.23}$$

Lastly, the identities $\mathrm{d}\boldsymbol{P}/\mathrm{d}s = \boldsymbol{d}_x$ and $\boldsymbol{N} = \boldsymbol{r}_1 \times \boldsymbol{r}_2$ together imply (Figure 6.1)

$$\frac{\mathrm{d}\boldsymbol{P}}{\mathrm{d}s} = \cos\phi \boldsymbol{r}_1 + \sin\phi \boldsymbol{r}_2 \quad \text{and} \quad \frac{\mathrm{d}\boldsymbol{N}}{\mathrm{d}s} = \kappa_1 \cos\phi \boldsymbol{r}_1 + \kappa_2 \sin\phi \boldsymbol{r}_2 \tag{6.24}$$

6.2.4.1 Derivation of the Helix Centerline Coordinates

Next, a differential equation for $N(s)$ is defined and solved explicitly. Derived from Eqs. (6.23) and (6.24), it can be rewritten as

$$\frac{\mathrm{d}^2\boldsymbol{N}}{\mathrm{d}s^2} = -\left(\kappa_1^2 \cos\phi + \kappa_2^2 \sin\phi\right)\boldsymbol{N} \tag{6.25}$$

Equation (6.25) is subject to the initial conditions $\boldsymbol{N}(0) = \boldsymbol{E}_z$ and

$$\frac{\mathrm{d}\boldsymbol{N}(0)}{\mathrm{d}s} = \kappa_1 \cos\phi \boldsymbol{r}_1(0) + \kappa_2 \sin\phi \boldsymbol{r}_2(0) \tag{6.26}$$

where

$$\boldsymbol{r}_1(0) = \cos\phi \boldsymbol{E}_x - \sin\phi \boldsymbol{E}_y \quad \text{and} \quad \boldsymbol{r}_2(0) = \sin\phi \boldsymbol{E}_x + \cos\phi \boldsymbol{E}_y \tag{6.27}$$

Solving for $\boldsymbol{N}(s)$ yields

$$\boldsymbol{N}(0) = \left(\frac{\beta}{\alpha}\sin\alpha s\right)\boldsymbol{E}_x - \left(\frac{\tau}{\alpha}\sin\alpha s\right)\boldsymbol{E}_y + \cos\alpha s \boldsymbol{E}_z \tag{6.28}$$

where

$$\alpha = \sqrt{\kappa_1^2\cos^2\phi + \kappa_2^2\sin^2\phi}, \beta = \kappa_1\cos^2\phi + \kappa_2\sin^2\phi, \text{and } \tau = (\kappa_1 - \kappa_2)\sin\phi\cos\phi \tag{6.29}$$

Integrating Eq. (6.23) with the initial conditions from Eq. (6.27) and integrating d$\boldsymbol{P}$/ds in Eq. (6.24) yield

$$X = s - \frac{\beta^2}{\alpha^3}(\alpha s - \sin \alpha s)$$

$$Y = \frac{\beta}{\alpha^3}(\alpha s - \sin \alpha s)(\kappa_1 - \kappa_2)\sin\phi\cos\phi$$

$$Z = \frac{\beta}{\alpha^2}(\cos \alpha s - 1) \tag{6.30}$$

Geometric properties of the helical space curve P are as follows.

The space curve $\boldsymbol{P}$ expressed above represents a helix with curvature β, torsion τ, axis $\boldsymbol{M}$, helix angle Φ (the angle between the helix tangent and the plane normal to the helix axis), radius R, and pitch D. The corresponding Frenet–Serret frame $\{\boldsymbol{T}, \boldsymbol{N}, \boldsymbol{B}\}$ [56] satisfies the following differential equations:

$$\frac{\mathrm{d}\boldsymbol{T}}{\mathrm{d}s} = \beta\boldsymbol{N}(s),\quad \frac{\mathrm{d}\boldsymbol{N}}{\mathrm{d}s} = \beta\boldsymbol{T}(s) + \tau\boldsymbol{B}(s),\text{ and } \frac{\mathrm{d}\boldsymbol{B}}{\mathrm{d}s} = -\tau\boldsymbol{N}(s) \tag{6.31}$$

Here, $\boldsymbol{T} = \boldsymbol{d}_x$ is the tangential component,

$$\boldsymbol{T}(s) = \left[1 - \frac{\beta^2}{\alpha^2}(1 - \cos\alpha s)\right]\boldsymbol{E}_x + \frac{\beta\tau}{\alpha^2}(1 - \cos\alpha s)\boldsymbol{E}_y - \frac{\beta}{\alpha}\sin\alpha s\boldsymbol{E}_z \tag{6.32}$$

and $\boldsymbol{B} = \boldsymbol{T} \times \boldsymbol{N}$ [57] is a bi-normal vector:

$$\boldsymbol{B} = \frac{\beta\tau}{\alpha^2}(1 - \cos\alpha s)\boldsymbol{E}_x + \left[\frac{\beta^2}{\alpha^2}(1 - \cos\alpha s) + \cos\alpha s\right]\boldsymbol{E}_y - \frac{\tau}{\alpha}\sin\alpha s\boldsymbol{E}_z \tag{6.33}$$

The axis $\boldsymbol{M}$ is parallel to the vector generated by an entire recurrent unit:

$$\boldsymbol{P}(2\pi/\alpha) - \boldsymbol{P}(0) = 2\frac{\pi\tau}{\alpha^2}\left(\frac{\tau}{\alpha}\boldsymbol{E}_x + \frac{\beta}{\alpha}\boldsymbol{E}_y\right) \tag{6.34}$$

Normalization yields

$$\boldsymbol{M} = \frac{\tau}{\alpha}\boldsymbol{E}_x + \frac{\beta}{\alpha}\boldsymbol{E}_y = \frac{(\kappa_1 - \kappa_2)\sin\phi\cos\phi}{\sqrt{\kappa_1^2\cos^2\phi + \kappa_2^2\sin^2\phi}}\boldsymbol{E}_x + \frac{\kappa_1\cos^2\phi + \kappa_2\sin^2\phi}{\sqrt{\kappa_1^2\cos^2\phi + \kappa_2^2\sin^2\phi}}\boldsymbol{E}_y \tag{6.35}$$

As expected for a helix, $\boldsymbol{T}\cdot\boldsymbol{M} = \tau/\alpha$ is a constant corresponding to the helix angle

$$\Phi = \sin^{-1}\frac{\tau}{\alpha} = \sin^{-1}\frac{(\kappa_1 - \kappa_2)\sin\phi\cos\phi}{\kappa_1\cos^2\phi + \kappa_2\sin^2\phi} \tag{6.36}$$

Thus, helical chirality is determined by the sign of the helix angle with a positive sign indicating right-handed helices and negative sign representing left-handed helices. Equation (6.36) implies that the chirality is subject to the signs of both $(\kappa_1 - \kappa_2)$ and $\sin\phi\cos\phi$. It is expected that by switching the values of κ_1 and κ_2, the helix shape transitions to its mirror image about the plane $y = z$ with reversed handedness. Lastly, the helix has a radius

$$R = \frac{\beta}{\alpha^2}\frac{\kappa_1\cos^2\phi + \kappa_2\sin^2\phi}{\kappa_1^2\cos^2\phi + \kappa_2^2\sin^2\phi} \tag{6.37}$$

$$D = \|\boldsymbol{P}(2\pi/\alpha) - \boldsymbol{P}(0)\| = \frac{2\pi\tau}{\alpha^2} \tag{6.38}$$

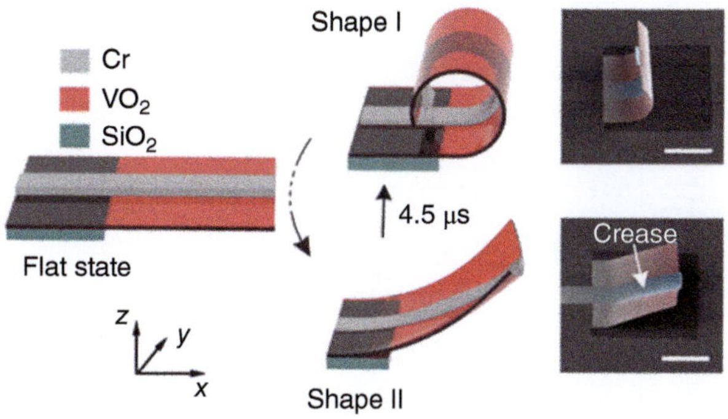

Figure 6.2 Schematic illustration and colorized scanning electron microscope (SEM) images of the bistable Cr/VO_2 nanomembranes. The scale bar is 40 μm.

6.3 Bistability of Thin Structures: Bistable and Reconfigurable Nanomembrane

We create a microscale reconfigurable structure with bistability by releasing the bilayer Cr/VO_2 nanomembrane from the 2D precursor. This bilayer structure can be stable either in a rolled shape (shape I) or a "folded" configuration with crease-like feature (shape II), as shown in Figure 6.2. Such bistability stems from the misfit strain between Cr and VO_2 layers and the clamped boundary condition on one end. The Cr/VO_2 nanomembrane can switch from shape II to shape I when pushed by a mechanical probe and switch back to shape II under a heating–cooling process, based on changing temperature to induce the metal-insulator phase transition of VO_2. Owing to the "snap-through" instability of a bistable structure and the rapid phase transition of VO_2, the typical timescale for this shape transition is observed from 4.5 μs to 1 ms. This shape-changing time is by far the shortest that has been recorded among microscale mechanical actuators (Figure 6.3). Furthermore, we perform FEM to predict the shapes of the bistable structure and plot the stability diagram versus geometric parameters, which are in good agreement compared with experimental observations. It is found that bistability is primarily determined by the geometry of Cr/VO_2 nanomembrane, suggesting that the bistable behavior of such 3D structure can be precisely controlled by designing the geometry of the 2D precursor. Moreover, we show that the bistable nanomembranes can be appropriately assembled into various mesostructures, such as a mechanical gripper, which can grab and release microscale items. Therefore, our results not only demonstrate the feasibility of tailoring bistability in microscale, rigid structures, but also provide an alternative prototype for the future design of fast switches and actuators across multiple length scales.

6.3.1 Experiments

To illustrate the concept, we first experimentally demonstrate the bistability of the proposed prototype due to a given boundary condition. Initially, a narrow strip of Cr with thickness of 30 nm is deposited on the VO_2 layer and VO_2 is attached to the SiO_2 (silicon dioxide) substrate, forming a tri-layered structure as shown in Figure 6.2.

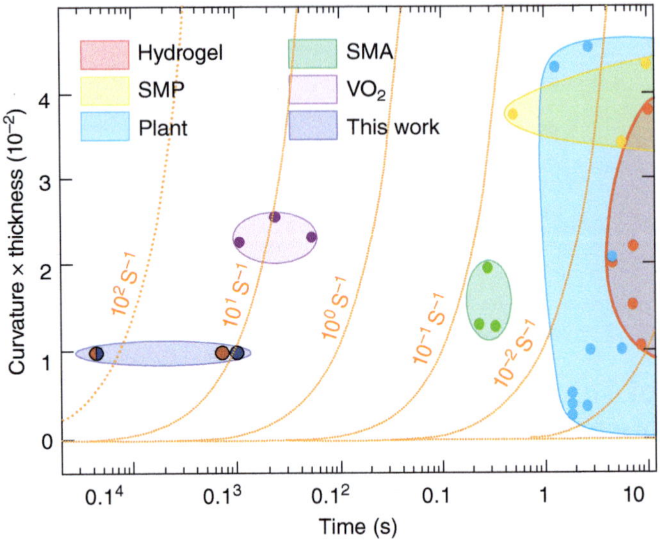

Figure 6.3 Summary of "curvature × thickness" against time for membrane actuators from this work and literature, including the membranes made of hydrogel, shape memory polymer (SMP), VO_2, natural plant, and shape memory alloy (SMA). Yellow dashed lines denote constant actuation rates.

Here, the thickness of VO_2 is chosen as 20–28 nm to optimize its phase transition property. A misfit strain exists between the Cr and VO_2 layer because of the Cr layer's thermal contraction when it is cooled down. To release the Cr/VO_2 thin film, we etch away the SiO_2 substrate while keeping one end of Cr/VO_2 nanomembrane attached to the SiO_2 substrate. Driven by the misfit strain, the Cr/VO_2 thin film bends along the longitudinal direction (x direction) into a rolled shape (shape I) after being released. Then we gradually increase the temperature from 321 to 339 K. During this heating process, the curvature of the rolled shape gradually decreases (Figure 6.4) because VO_2 can generate a compressive strain when it undergoes the metal-insulator phase transition around 338–345 K [58]. When the Cr/VO_2 bilayer becomes flat, we decrease the temperature back to 321 K. However, instead of returning to the rolled shape (shape I), this bilayer structure exhibits an irreversible shape transition – the bilayer region in the middle bends along the transverse direction (y direction) and two crease-like deformations are generated near the clamped end, folding the entire structure upwards with a tilting angle θ with respect to the substrate (shape II). This new shape suggests that the clamped nanomembrane can adopt two stable shapes without changing the misfit strain or boundary conditions. By using a mechanical probe, we can force the structure to snap from shape II back to shape I like a microscale "catapult" (Figure 6.5).

To illustrate the aforementioned process, by applying FEM, spatially averaged curvatures of the shapes I and II in x and y directions (κ_x and κ_y, red and blue lines, respectively) are calculated by varying the temperature from 321 to 339 K, which is presented in Figure 6.4. Corresponding simulation results of the morphable nanomembrane's configurations are also shown as the insets. Experimental data

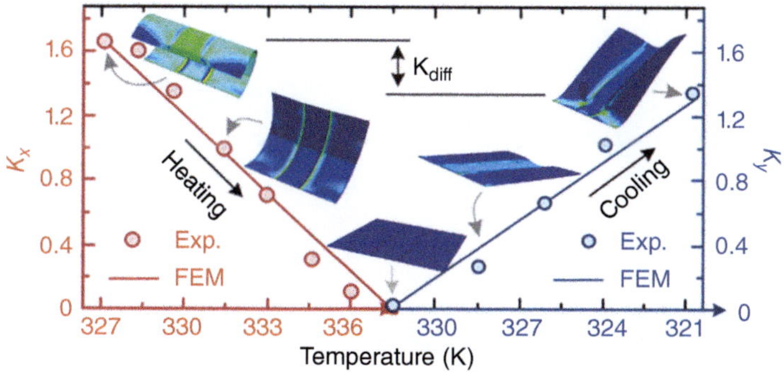

Figure 6.4 Experimental and FEM results of the spatially average curvature in *x* and *y* direction versus temperature (K). Insets show FEM predictions of the nanomembrane's shape during the heating and cooling process. The coordinates are illustrated in Figure 6.2. Red line and dots refer to heating process from shape I to flat state. Blue line and dots refer to cooling process from flat state to shape II. $h\kappa_{diff}$ is the difference between *h* in shape I and *h* in shape II.

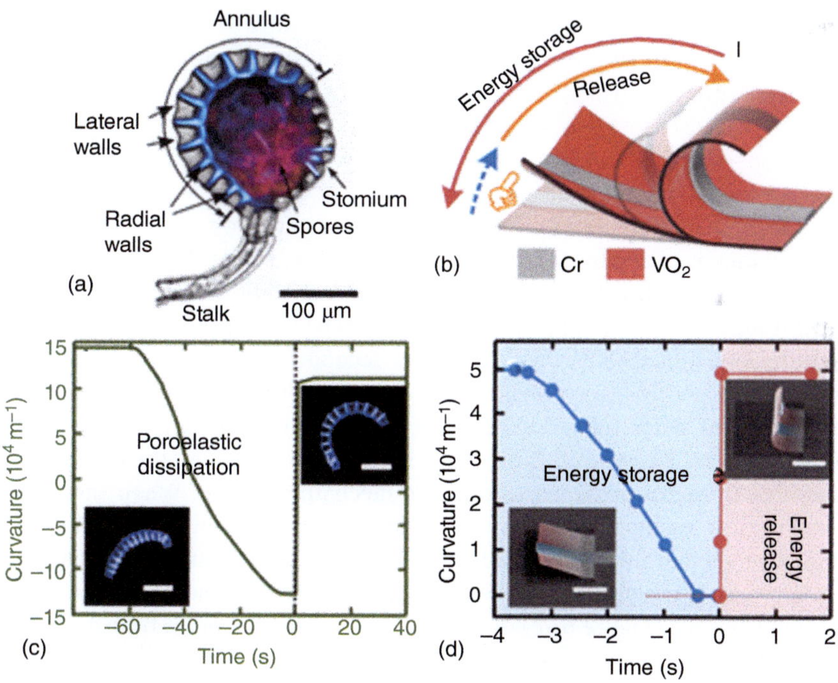

Figure 6.5 Micro-catapult compared with the natural sporangium. (a) The sporangium of *Polypodium aureum*. The scale bar is 100 μm [59]. (b) Annulus curvature during opening and closing procedure. Insets show annulus geometry during sporangium opening and closing. The scale bar is 100 μm [59]. Source: Reprinted with permission from Noblin et al. [59]. (c) The structural and actuating scheme of micro-catapult. The red and blue arrows represent the energy storage process via heating and cooling cycle. Orange arrow represents the release process from shape II to shape I. (d) Curvature change of micro-catapult as a function of time. The catapult was stimulated at $t = 0$ seconds. The closure dynamics are characterized by two phases: a slow phase presenting energy storing, a rapid phase presenting energy releasing. Insets show open and closed catapult. The scale bar is 30 μm.

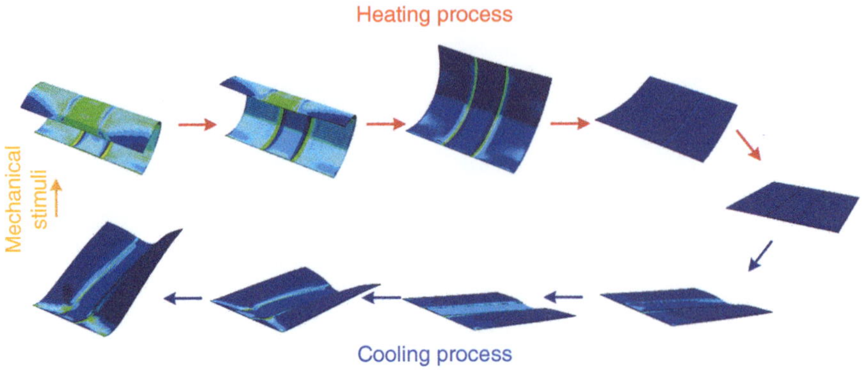

Figure 6.6 FEM results of the whole heating and cooling cycle.

are plotted with red and blue circle markers for the heating and cooling processes, respectively. In the heating process, the nondimensional curvature of shape I in x direction starts from $h\kappa_x = 1.65$ (h is the nanomembrane's thickness) and decreases with the temperature increasing. At the critical temperature 337.5 K, the curvature could decrease to zero ($h\kappa_x = 0$, flat state). Upon cooling, the nondimensional curvature in y direction ($h\kappa_y$) increases and forms the folded structure. When temperature decreases to 321 K, $h\kappa_y$ rises up to 1.35, which is slightly smaller than the nondimensional curvature $h\kappa_x$ in shape I. The whole actuation cycle from FEM simulation is shown in Figure 6.6.

The two steps of shape transitions in a complete actuation cycle are actually based on distinct mechanisms (Figure 6.7). The first step, where shape I switches to shape II, utilizes the ultrafast metal-insulator phase transition of VO_2 in response to temperature change [60, 61] (about 40 fs), and hence the speed mainly depends on the diffusion of heat that can take several seconds. With an efficient heating method, the time of this shape transition can be reduced to below 1 ms. Meanwhile, the second step of shape transition adopts a different mechanism by virtue of the "snap-through" instability [62]. During this step, the structural elastic energy first increases due to the external work from the mechanical probe. When the probe reaches a critical position, the thin film suddenly loses its stability, resulting in jumping from the current stable configuration to the other stable one (shape I). During this snap-through process, the stored elastic energy typically fast converts to the kinetic energy (Figure 6.7c) only with a very short period of time, which is as short as 4.5 ms in our case, much shorter than the first step. The timescale of the elastic snap-through can be approximated as $t \sim L\sqrt{\rho/E} = 0.06$ ms (the characteristic length of $L \approx 100$ mm, Young's modulus $E \approx 25$ GPa, and the mass density of $\rho \approx 7.19$ g/cm^3) [33, 63]. Because of the rigid material-based actuator having high Young's modulus, the estimated snapping time can be significantly shorter than that of polymeric actuators. Notably, this combination of two dramatically different timescales in our experiments also exists in nature, such as the snapping of Venus flytrap [25, 26], the sucking of carnivorous bladderwort [27], and the sporangium snapping of *Polypodium aureum* [59]. Inspired by the sporangium's

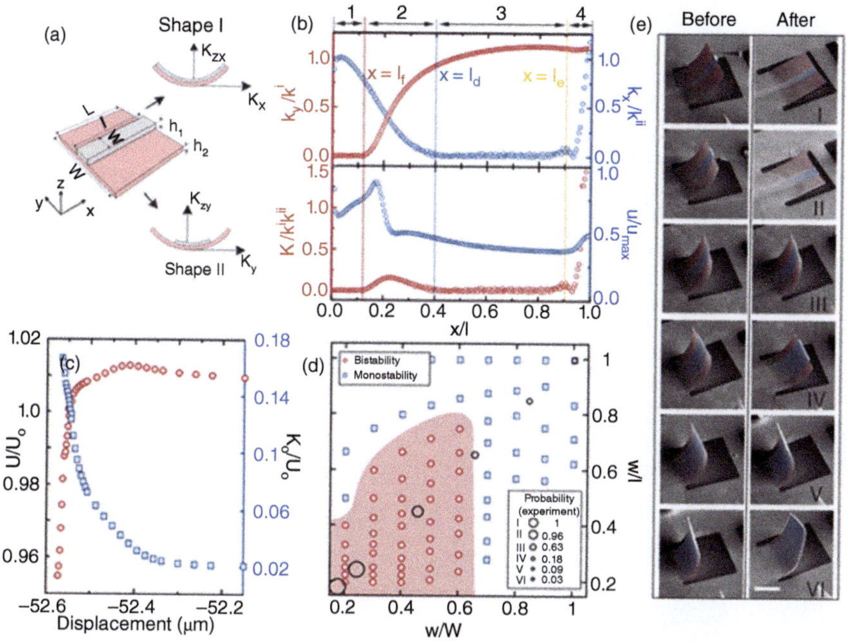

Figure 6.7 Finite element analysis of the bistable behavior. (a) Schematic of the geometry model used in FEM. (b) The variations of κ_y/κ^{i} (top panel, left axis), κ_x/κ^{ii} (top panel, right axis), $K/\kappa^{i}\kappa^{ii}$ (bottom panel, left axis), and u/u_{max} (bottom panel, right axis, u_{max} is the maximum value of u) versus the x/l in shape II from FEM. Three dashed lines divide the bilayer into four regions: Region 1 (folding region), Region 2 (doubly curved region), Region 3 (far-field region), and Region 4 (edge region). (c) The variations of the elastic energy U and the kinetic energy K_e of the Cr/VO_2 bilayer during snap-through from FEM. U_0 is the elastic energy of shape II and d (μm) is the displacement of the probe. (d) The stability diagram of the Cr/VO_2 nanomembrane in terms of w/W and w/l ($W = 100\,\mu m$, $L = l$). The red (bistable) and blue (monostable) unfilled dots are results from FEM. The black unfilled circles are results from experiments. (e) SEM images of two stable shapes with Cr width in 15 (I), 30 (II), 45 (III), 70 (IV), 85 (V), and 100 nm (VI) before and after energy storage. The scale bar is 20 μm.

ejection, an artificial micro-catapult was developed for energy harvesting and releasing (Figure 6.5).

6.3.2 Assembly of Bistable Nanomembranes into Mesostructures

Each Cr/VO_2 unit has two stable equilibria. If we combine N bistable units together, in theory, the assembled polymer can have 2^N multistable configurations at most. Based on this idea, various 3D mesostructures can be obtained by appropriately assembling several Cr/VO_2 bistable or monostable units together. In addition, the constructed stability diagram implies that the bistability of a single Cr/VO_2 nanomembrane is mainly determined by its geometry. Therefore, the multistable states of the mesostructures can also be programmed by changing the geometry of their 2D precursors. To demonstrate this, we first clamp two bistable units to the

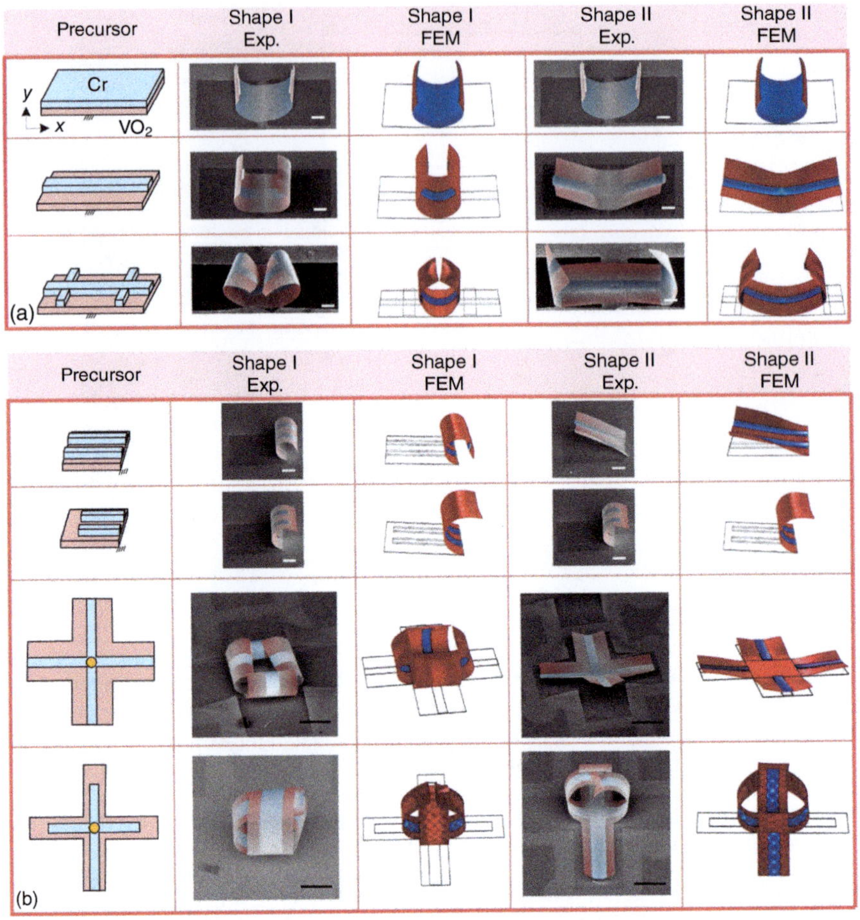

Figure 6.8 Experimental and FEM studies of different reconfigurable 3D mesostructures. (a) SEM images and FEM simulations on Cr/VO_2 bimorphs with different Cr ribbon precursors in shapes I and II. Oblique lines indicate the bonding position with substrate. The scale bar is 10 μm. (b) SEM images and FEM simulations on Cr/VO_2 bimorphs with symmetric and asymmetric precursors. First two rows are the design of single strip with different width of Cr ribbons, which show tuning bistable states. Third and fourth rows are four-arm capturing actuators with symmetric and asymmetric precursors. The yellow circles in precursors indicate the bonding position with the substrate. The scale bar is 40 μm.

same edge. When both the Cr and VO_2 layers have the same width, the unit has only one stable configuration and can only bend along the longitudinal direction (Figure 6.8a, first row). When the width of Cr layer w decreases, each unit starts to two stable shapes and the assembled structure can form shapes resembling a "two-armed gripper" or a "flapping bird" (Figure 6.8a, second row). A new "clamped edge" inside each unit is then formed by attaching another narrow Cr strip in the transverse direction near the end of the tip. As such, two stable configurations can only be found in the region between the clamped edge and the transverse Cr strip and the strip can only roll up along the longitudinal direction near the tip of nanomembrane (Figure 6.8a, third row). Moreover, two units are bonded side by side along the transverse direction rather than along the longitudinal direction, the idea of which can be realized by bonding two parallel Cr strips on one VO_2 layer. It is found that this assembled new material exhibits similar bistability to the unit – the folded shape will lose stability when the length of Cr layer decreases (Figure 6.8b, first and second row).

These ideas can be further extended and explored for more multistable behaviors by assembling more units together. Here, four Cr/VO_2 units are bonded to four sides of squared SiO_2 substrate. If four units are all bistable, then the assembled structure can become a "four-armed gripper" (Figure 6.8b, third row) to grab (left) and release (right) items under thermal stimuli. Such design can also be modified to promote its maneuverability. For instance, three bistable units are replaced by three monostable units (decrease the lengths of their Cr layers), only the last bistable unit can switch between shapes I and II and become flat when releasing items, with application for achieving a directional release of cargo (Figure 6.8b, fourth row).

6.4 Wrinkling, Rolling, and Twisting of Micro-/Nanostructures

6.4.1 Wrinkling in Thin Flexible Layers and Nanomembrane

A multiple structure within which one layer was pre-strained can spontaneously deform into a 3D structure to minimize strain energy [64, 65]. Examples of such spontaneous deformations include wrinkling [57, 66], bending [64, 67], folding [68–70], and ridging [71], controlled by the strain gradient and magnitude. Typically, wrinkles are generated by a small-strain gradient albeit a bent or roll-up structure will form under a large-strain gradient [72]. Due to the competition between bending and stretching energy, in real materials transitions between wrinkling and rolling states can occur after the pre-strained layer is released in one end with the other end being fixed.

An isotropically strained structure with two bonded layers is used for a simple study. As shown in Figure 6.9 [72], the bilayer structure, consisting of two thicknesses d_1 and d_2, is subjected to biaxial strains ε_1 and ε_2, respectively. The structure is clamped over its half of widthwise edge with a distance h and is initially pre-strained over the length L. The released portion can be elastically relaxed with no constraints except by the fixed boundary, shown with dashed line in Figure 6.9a [72]. The bilayer structural average strain and strain gradient are given by $\overline{\varepsilon} = (\varepsilon_1 d_1 + \varepsilon_2 d_2)/(d_1 + d_2)$ and $\Delta\varepsilon = \varepsilon_1 - \varepsilon_2$, respectively, and $\Delta\varepsilon > 0$. The elastic energy per unit area of the bilayer is initially $E_0 = Y\left(d_1\varepsilon_1^2 + d_2\varepsilon_2^2\right)/(1 - \nu)$, where Y and ν denote Young's modulus and Poisson's ratio, respectively. In the situation of bending, the fixed boundary limits relaxation in the x direction and the strain is relaxed via bending in the y direction to form a curved structure with inner radius R (Figure 6.9b). Since the layers are thin, the stress component in the radial direction (through the thickness direction) must be zero at equilibrium [11]. The total elastic energy E_{bent} is calculated by integrating the elastic strain energy density throughout the film [72]. The

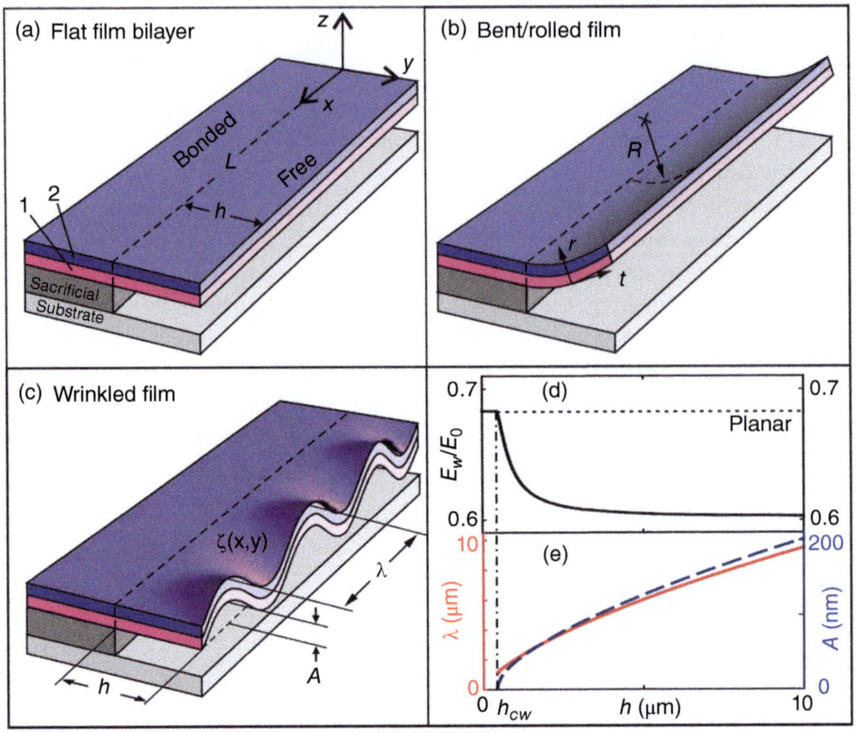

Figure 6.9 Schematics of (a) strain released bilayer, (b) bilayer bent with an inner radius R, and (c) wrinkled structure profiled as a function of $\zeta(x, y)$ with amplitude A and wavelength λ. (d) Ratio of E_w and energy of planar relaxation versus the length, h. (e) Wavelength λ (solid line) and amplitude A (dashed line) versus the length h for the wrinkled structure. The critical length of h_{cw} is indicated by the vertical dotted-dashed line. Source: Reprinted with permission from Čendula et al. [72].

equilibrium elastic energy of the roll-up structure, E_{bent}, can be obtained by energy minimization and normalized to E_0 to be compared with the normalized energy E_w of the wrinkled structure. The energy of the wrinkled structure as a function of the wavelength is shown in Figure 6.9d, and the minimum critical wrinkle length h_{cw} is identified to be $h_{cw} \approx 2.57d_2\sqrt{-\bar{\varepsilon}}$. For $h < h_{cw}$, energy minimization cannot provide enough energy for wrinkling and $A = 0$ and $\lambda \to \infty$ [73], transitioning back to a planar relaxation in the y direction (dashed line in Figure 6.9d). For $h > h_{cw}$, wrinkling happens, and both λ and A increase as h increases, demonstrated in Figure 6.9e. The preferred equilibrium shape of a cantilevered film forms based on these two normalized energies, E_{bent} and E_w. However, $h > h_{cw}$ does not guarantee the generation of wrinkling. Large h and large strain gradient (or $\Delta\varepsilon$) allow the bilayer to roll, while only small strain gradient can allow the film to wrinkle [32, 72]. For example, a bi-layered film consisting of 10 nm $In_{0.1}Ga_{0.9}As$ and 10 nm GaAs ($Y = 80$ GPa and $\nu = 0.31$) was calculated to obtain the favorable shape as a function of h and $\Delta\varepsilon$ by systematically changing ε_1 and ε_2, and the phase diagram is shown in Figure 6.10. Bent structure is favored when $h < 700$ nm, and when h becomes larger, the wrinkled

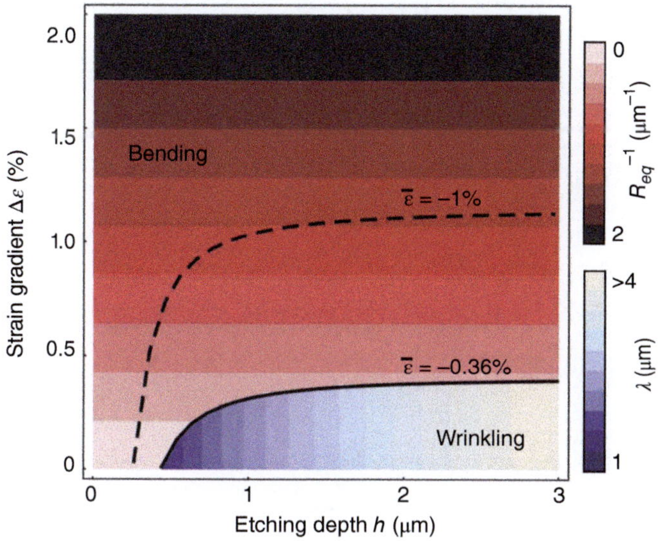

Figure 6.10 Phase diagram of preferred equilibrium shapes of pre-strained bilayer structures based on the ratio of E_w and E_{bent}. Solid curve indicates the boundary separating the bent and wrinkled domains. Source: Reprinted with permission from Cendula et al. [72].

structure is dominant with lower energy [72]. The boundary between the two equilibrium shapes is indicated by a solid line in Figure 6.10 [72]. Figure 6.10 shows that for wrinkled structure, the wavelength λ increases as h increases, while the inner radius R_{eq} of the bending structure decreases with increasing $\Delta\varepsilon$. The continuum theory may not always model precisely for small-length-scale structures, resulting from increasing dislocations in the boundary, surface properties, and size effects.

In reality, two typical methods are used to generate and control wrinkling structures. The first one, the internal control method, introduces a pre-strain into the active layer, which is consistent with model above. This method is applied for generating SiGe wrinkle arrays in 2007 from Mei et al. group [74]. A thin pre-strained active layer (e.g. SiGe layer), bonded on a sacrificial buffer layer, is partially released on the substrate by etching off away some parts of the sacrificial layer. Once the strained layer completes its partial free process, the strain relaxes at etching-off positions and forms wrinkles perpendicular to the etching front [74]. The channels generated from wrinkling films play an important role in applications for micro-/nano-fluidics and biology. Even the wrinkles presented in Figure 6.11 are not sinusoidal shapes resulting from the bon-back influences [74]. Complicated channel arrays to satisfy specific functional demands can be produced because of wrinkles normal to the etching front that can be made using traditional photolithography [76]. The self-assembly of the channels is dependent on the size, shape, and spacing of the etching start windows, as well as layer thickness [76]. Micro-/nano-channel arrays with specific orders can be produced in optimal conditions.

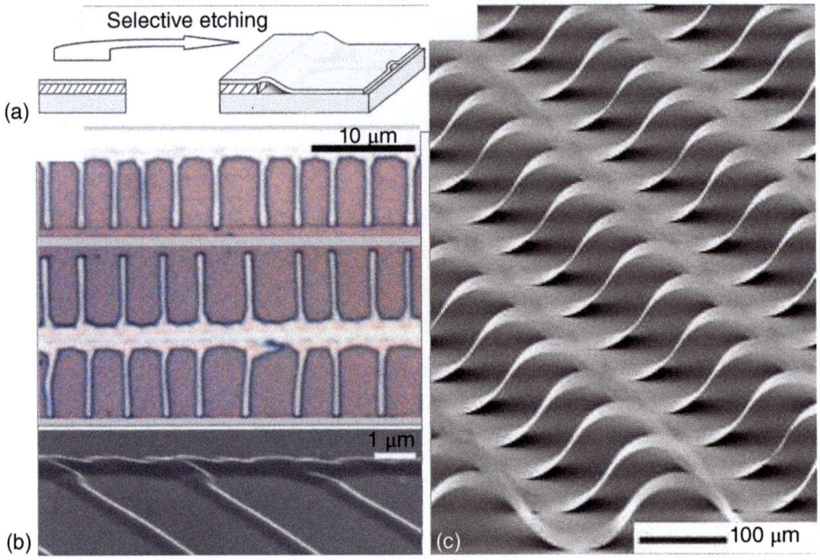

Figure 6.11 (a) A schematic describes how the wrinkles form perpendicular to the etching front. Source: Adapted with permission from Mei et al. [74]. (b) Optical microscopy images illustrate a linear nanochannel network with single-sided (upper) (corresponding SEM image shown below) and double-sided branch channels (middle). Source: Mei et al. [74] Reproduced with permission from Wiley-VCH. (c) Si ribbon structures bonded on a PDMS substrate are pre-strained to 50%. Source: Sun et al. [75] Reproduced with permission from Springer Nature.

On the other hand, the second approach, called external control method, does not need the release of the active layer from a pre-strained situation. In this approach, the substrate is compressed while the surface functional layer is under stretching related to the substrate, leading to generate wrinkles [75, 77]. Figure 6.11b,c shows a typical wrinkle structure that is created by this approach. Ordered wrinkle arrays can be produced by [75] illuminating UV light via a photomask to pattern chemical adhesion sites on pre-strained polydimethylsiloxane (PDMS) substrate. Ozone proximal areas are created on the surface of the PDMS under UV light exposure [78]. These ozone proximal areas transition the unmodified hydrophobic surface (inorganic surfaces) to a polarized surface (strong chemical bonds [79]). As a result, the unexposed areas have weak interaction with other surfaces [80]. The geometry of the wrinkles is able to reconfigure by changing the size and distribution of the ozone proximal areas. Inorganic semiconductor wrinkles are commonly used for flexible electronics and stretchable electronics, which are completely described from the Rogers' group at University of Illinois Urbana-Champaign (UIUC) [81–83].

Note that the external control approach not only can release pre-strained substrates [84, 85], but also can be used to produce substrate deformation. In Bowden et al. work [86], thermal contraction of a substrate can produce wrinkles in thin metal films. In fact, wrinkles can be simply formed on a treated PDMS substrate. Chung et al. [87] reported plasma-assisted wrinkle formation, which involves the steps of (i) heating the PDMS; (ii) exposing the substrate to UV light to generate a thin film of silica-like materials bonded on the substrate; and (iii) cooling down the structure to generate wrinkles. The wavelength and amplitude of wrinkles are controlled by changing the heating temperature and the exposure time. A higher temperature can induce larger strain and produce larger wrinkle amplitude accordingly. Increased the time of oxygen plasma can thicken the silica-like layers influence both the wavelength and amplitude of the wrinkle [87].

6.4.2 Mechanical Self-Assembly of Helical Ribbons

Generally speaking, a helical ribbon starts bending around its two principal bending axes, $\mathbf{r}_1$ and $\mathbf{r}_2$, with principal curvatures, κ_1 and κ_2, shown in Figure 6.12a. A misorientation angle ϕ is defined based on the principal axes and the geometric axes $\mathbf{d}_x$ and $\mathbf{d}_y$. In Chen et al. study, the coordinates of the centerline are defined by the arc length s, and the geometric properties, especially the pitch p, the helix angle θ, and the helix radius R, can then be expressed by three independent geometric parameters, κ_1, κ_2, and ϕ:

$$p = 2\pi\tau/\alpha^2, \quad R = \beta/\alpha^2, \quad \theta = \sin^{-1}(\tau/\alpha) \tag{6.39}$$

where $\alpha = \sqrt{\kappa_1^2 \cos^2\phi + \kappa_2^2 \sin^2\phi}$, $\beta = \kappa_1 \cos^2\phi + \kappa_2 \sin^2\phi$, and $\tau = (\kappa_1 - \kappa_2) \sin\phi \cos\phi$.

Additionally, the deformation of a ribbon is determined by minimizing the total energy using the surface stress (f^+ and f^-) from the top and bottom surfaces,

$$\Pi = f^- : \gamma|_{z=-H/2} + f^+ : \gamma|_{z=H/2} + \int_{-H/2}^{H/2} \frac{1}{2}\gamma : C : \gamma \, dz \tag{6.40}$$

where γ is the strain tensor, C is the tensor of elastic stiffness, and the z is defined to be the distance from the midplane toward the thickness direction. Analytical

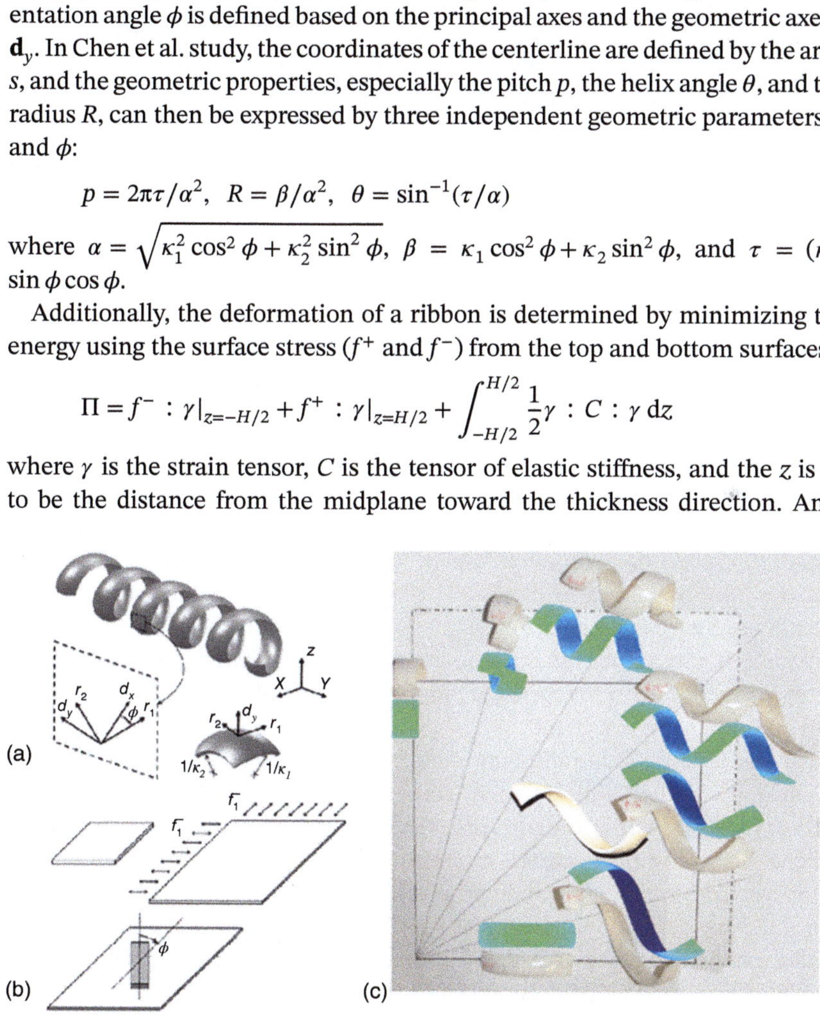

Figure 6.12 Tunable helical ribbons. (a) Illustration of a helical ribbon. (b) Schematic of a latex rubber bonded with an elastic strip adhesive. (c) A piece of latex rubber sheet is pre-stretched with different misorientation angles ($\phi = 0°, 15°, 30°, 45°, 60°, 75°$, and $90°$). Source: (c) Chen et al. [52] Reproduced with permission from AIP Publishing LLC.

solutions are then achieved when f^+ is set to be zero. The theoretical results predict accurately compared with the experiments (experiment setup is shown in Figure 6.12b). Helical ribbons are created by bonding an elastic strip adhesive to a pre-stretched latex rubber sheet. Different shapes are formed based on different misorientation angles ranging from 0 to 90° (Figure 6.12c).

A similar theoretical model was proposed by Armon et al. [20] to describe the formation of helical ribbons as a result of incompatible target metrics (Figure 6.13). The model can investigate the chiral opening of Bauhinia seed pods (Figure 6.13a). Pre-stretched latex rubber sheets are bonded to form different helical ribbon shapes. The theoretical pitches and radii agreed well with the experimental ribbon shapes. The transitions between cylindrical helical ribbons and helicoids also show a good agreement. Bauhinia seed pods apply this mechanism to twist open into two pieces by pushing handedness in the opposite direction. Forterre and Dumais [88] further developed paper models to demonstrate the "phase space" of helical ribbons. The fibers in their papers are generally aligned parallel to one side, and as a result, an anisotropic bilayer is created by bonding two pieces of papers that have perpendicularly aligned fibers. When the structure is submerged in water, this bilayer attempts to bend along both directions, and then form saddle-shaped structures, which have negative Gauss curvature. Furthermore, helical ribbons of different shapes are formed by cutting the strips along different directions (Figure 6.13h).

Since the theoretical approaches mentioned above do not need to consider any microscopic interactions, the models can predict helical shapes from any initially flat elastic sheet caused by residual strains, surface stresses, swelling/shrinkage, and/or differential growth [88].

A class of stretchable nanoribbons was made by Pham et al. [89], which has capacity of transitioning into helices when immersed in water. Metal, polymer, and ceramic materials are used to produce ribbons via evaporative deposition. The cross section of the nanoribbons is an isosceles triangle, and the ribbons will spontaneously curl to reduce energy due to their geometric asymmetry (Figure 6.14a). The helices are formed only at the angles where they prevent the ribbon self-intersecting, resulting in tight coils with low pitch (Figure 6.14b). Handedness is governed by the asymmetry of the cross section.

Helices are commonly found in plant tendrils and tend to reverse their chirality midway along their life. Godinho et al. studied this reversal and created cellulose liquid crystal fibers that imitate this effect depending on temperatures [92]. Active micro- and nano-fibers were formed through acetylizing, crystalizing, and subsequent electrospinning of hydroxypropylcellulose. These fibers can curl upon increasing the temperature while decreasing their tension in the meantime. The authors also explored how to fabricate the intrinsic curvature of the electrospun cellulose fibers, and why the twist is caused by off-core defects [93].

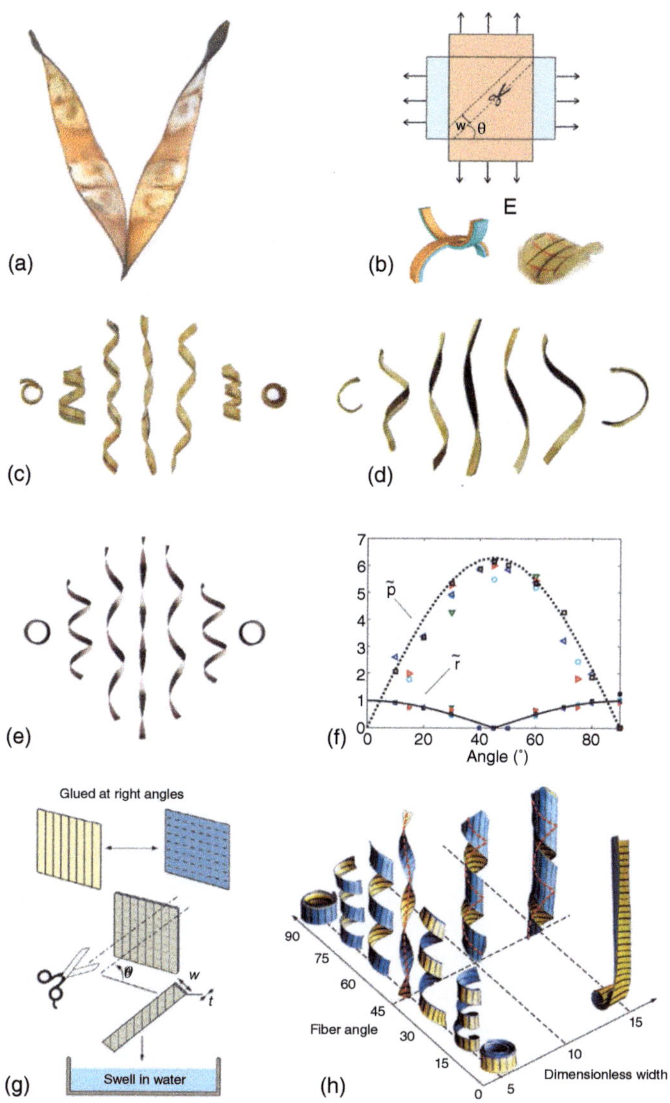

Figure 6.13 (a) Open Bauhinia pods. (b) Fabrication of a mechanical analog of Bauhinia pod. (c–e) Narrow helical strips cut along different directions with angle $\theta = 0°, 15°, 30°, 45°, 60°, 75°$, and 90° from a latex sheet, Bauhinia pods and analytical results, respectively. (f) Radius and pitch versus angle θ. Source: (a–f) Adapted from Armon et al. [20] with permission. (g) Schematics for the method used to create paper models to make coiling behavior analog to plant structures. (h) Helical shapes of paper models for different parameters: the nondimensional width and the misorientation angle. Source: (g, h) Adapted from Forterre and Dumais [88] with permission.

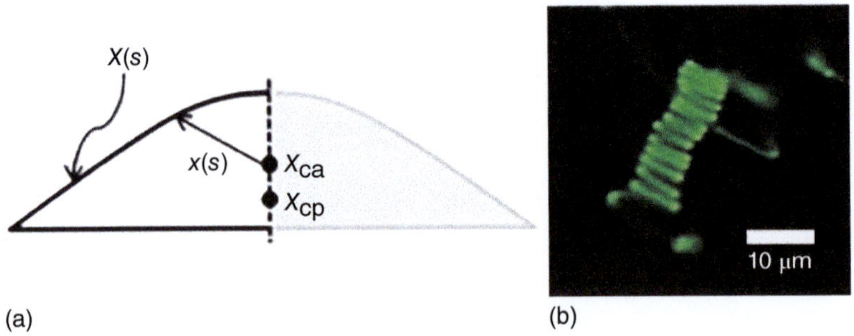

Figure 6.14 (a) Cross section of isosceles nanoribbons, and $X(s)$ shows the perimeter path. Source: Adapted from Pham et al. [89] with permission. (b) Coiled nanoribbons immersed in water. Source: Pham et al. [89] Reproduced with permission from Wiley-VCH.

Interfacial energy can also stimulate helical and/or multiple-stranded helical shapes. Ji et al. [94] investigated that interfacial adhesion is influenced by creating some double helices at the microscopic scale. Such double helices have been widely observed in a variety of biological and high polymer systems, including DNA [95, 96], carbon nanofibers [97], and carbon nanotubes [98, 99].

6.4.3 Shape Instability and Transitions in Helical Ribbons

Shape transitions between twisted ribbons, helicoids, spiral ribbons, and tubules have been demonstrated by controlling the relevant geometric parameters in recent studies. These parameters involve the magnitudes and signs of the principal curvatures, the geometric dimensions, and the misorientation angle [20, 24, 100]; and these parameters are often changed by external stimuli, including changes in temperature [101, 102], pH, or swelling/deswelling [20, 100].

Transitions between helicoids and spiral ribbons were investigated by Oda et al. [103] where the transition occurred when the length of the molecular chain changed (Figure 6.15a). Twist-nematic elastomers are also able to achieve these shapes [101] (Figure 6.15b–g). Sawa et al. [101] found that thin ribbons can switch between different twisting shapes when the width of the ribbons is changed. They then proposed a theoretical model accounting for this twist by introducing a term in the total potential energy. The transitions between purely twisted cylindrical helical shapes of the twist-nematic-elastomer ribbons are dependent on the temperature and ribbon's width. This shape preference is determined by the competition between bending and in-plane stretching energy. Additionally, they also revealed that with changing the temperature, some liquid crystal elastomer ribbons are capable of switching their chirality when undergoing transitions.

Both Armon et al. [20, 100] and Guo et al. [104] succeeded to examine shape transitions of strain-engineered elastic ribbons by their own approaches, while they proposed similar criteria: when the nondimensional width $W/\sqrt{\kappa/H}$ (where W is the width of the ribbon, H is the thickness, and κ is the principal curvature) exceeds a critical value, a transition of elastic ribbon will occur. The shape transition

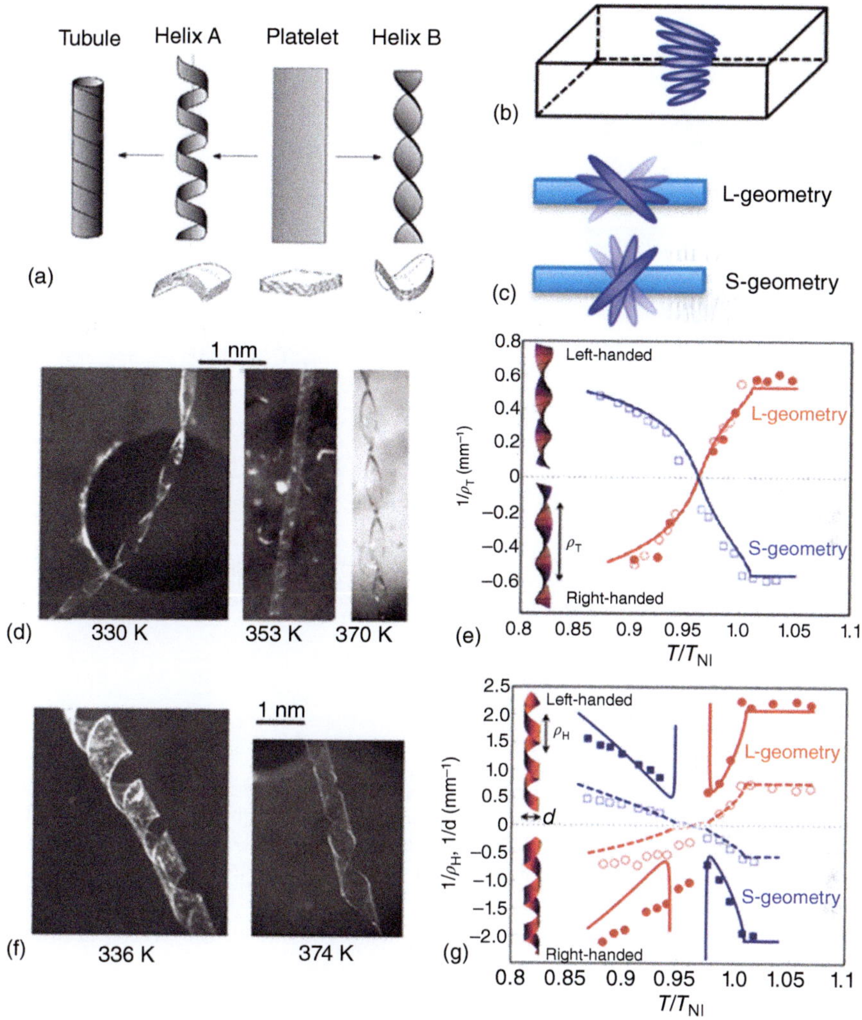

Figure 6.15 (a) Schematic of cylindrical helix and purely twisted ribbons. Source: Adapted from Oda et al. [103] with permission. (b) Side view the direction of twisted-nematic-elastomer (TNE) ribbons. (c) Top view of L- and S- geometry. (d, e) The inverses of pitch and diameter of the helices versus the temperature. (f, g) Cylindrical helical ribbons are generated by the wide twisted-nematic-elastomer (TNE). Source: (b, c, e, g) Adapted from Sawa et al. [101] with permission. (f) Sawa et al. [101] Reproduced with permission from PNAS.

(the ribbon exists multistability) is actually determined by these three geometric parameters. Guo et al. [104] further developed theoretical analysis to match experiments (Figure 6.16) to deeply understand this shape transitions corresponding multistability of the ribbon. In his theory, the deformed shape of ribbons is controlled by the following key parameters, involving the misfit strains, the width and thickness of the ribbon, the misorientation angle, and the elastic layer properties.

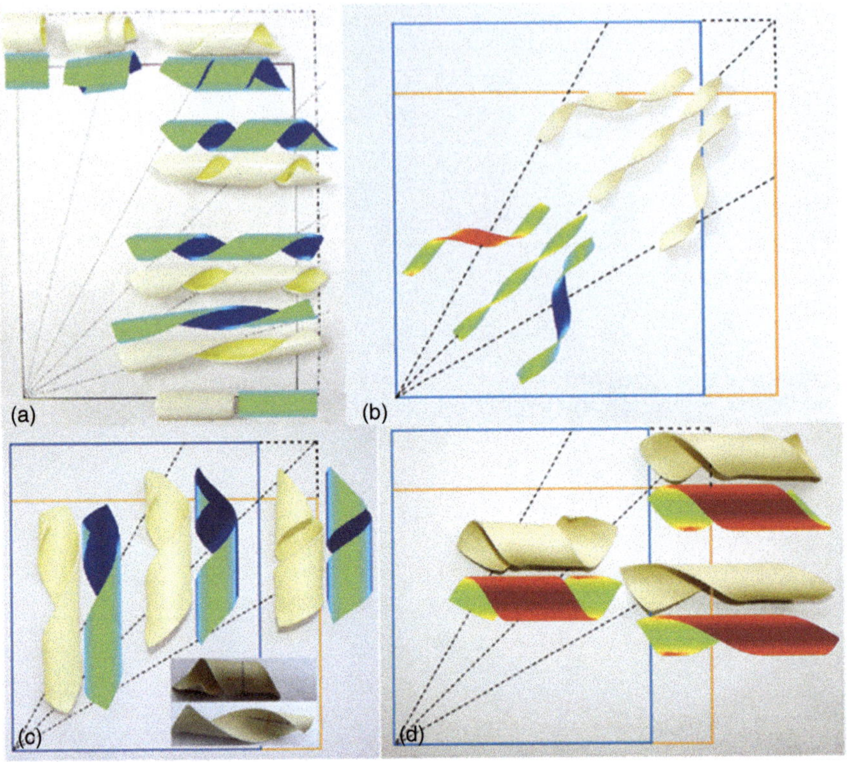

Figure 6.16 (a) Mono-stable helical ribbons. The pre-stretches are set to be 0.12 and 0.24 parallel and perpendicular to the width of the ribbon, respectively. (b) Mono-stable helical ribbons. (c) Bistable-helical ribbons. (d) Other achievable multistable shapes of the same ribbons in (c). Source: Guo et al. [104]. Reproduced with permission from AIP Publishing LLC.

A purely twisted ribbon can be generated (Figure 6.16b) when the principal misfit strains obey

$$\varepsilon_1 \cos^2 \phi + \varepsilon_2 \sin^2 \phi = 0 \tag{6.41}$$

This result shows a good agreement with the one derived by Chen et al. [52] ($\kappa_1 \cos^2 \phi + \kappa_2 \sin^2 \phi = 0$).

What calls for special attention is that the differences between a helicoid and a purely twisted ribbon can be confusing sometimes [105]. In addition to a plane, a helicoid becomes the only ruled minimal surface. In fact, the centerline of a helicoid is unnecessary to be straight, so, in some cases, it can also be a helix. For example, a purely twisted ribbon has a straight centerline with the radius of the bounding cylinder being zero. A helicoid has principal curvatures κ_1 and κ_2, and satisfies $\kappa_1 + \kappa_2 = 0$. Based on this rule, when $\kappa_1 = -\kappa_2$ and $\phi = \pi/4$, a purely twisted ribbon can generate a helicoid.

In Cranford and Buehler's work [106], a multiscale dynamics model was developed to investigate motion of molecules based on a two-dimensional coarse-grained model. The proposed model depicted the mechanical self-assembly of graphenes made by single and multilayers. A spontaneous transition of graphene ribbons was observed from a purely twisted ribbon to a coiled one, leading to a strain occupied with more homogeneity. Their findings have a good agreement with the experiments and theoretical predictions done by Armon et al. [20, 100] and Guo et al. [104]. The results are also similar to those studied by Lee et al. [107], Kit et al. [108], and Chen et al. [109], and additionally, the transition to the coiled shape is dominated by a mechanical instability determined by the relation of the imposed strain and the bending stiffness of the twisted graphene [106]. Consequently, this transition would be highly possible to happen in stiffer or thicker graphene ribbons.

The mechanical–chemical coupling influence in these (theoretical) models can enhance the understanding of the mechanical deformation and actuation of fibrous living organs [90]. In particular, Iamsaard et al. [110] illustrated that responses of plant tendrils unfolding can be replicated by helical ribbons which are created from liquid crystal polymer networks (Figure 6.17a–f). The authors created a kind of helical coil connected with a right-hand chiral coil and a left-hand chiral coil on each side. This behavior induced to move away from both ends of the synthetic tendril and become off-axis from the other end.

Liquid crystal polymer networks can also be used to the change humidity. De Haan et al. [111] implemented experiments on an Levi Civita (LC) polymer sheet to achieve undergoing bending and twisting induced by uniform humidity (Figure 6.17g,h). The sheet started bending toward the untreated side when one side was dipped in KOH and rinsed in water, but it bent toward the other side when drying it (Figure 6.17f). The resulting shape transition of the sheets was observed by changing the humidity when the sheets were treated with basic solution in some specific patterns.

Most physical or engineered systems discussed above have been inspired by biological phenomena. In turn, the study of chemical–mechanical coupling effects boosts development of how deformation and actuation of fibrous living organs work in reality [90].

6.4.4 Mechanical Self-Assembly of Rolled-Up Tubes

The rolling of flexible materials is a type of mechanical self-assembly. To design and fabricate a complex 3D structure, some characteristics need to be carefully considered, which includes the rolling direction and its misalignment with the crystal orientation. In case the strain gradient is determined, the rolling direction is significantly influenced by the mechanical properties of the flexible layer. The anisotropic properties are decided by the structural anisotropy in a single-crystal epitaxial layer [112]. To experimentally validate this phenomenon, Chun and Li [113] devised a wheel pattern that has eight rectangle stripes symmetrically distributed along the <100> and <110> directions (Figure 6.18a). Figure 6.18b shows the release of the $In_{0.3}Ga_{0.7}As/GaAs$ bilayers on the (001) GaAs substrate. The rolling occurred in four longer stripes oriented along the <100> direction and stopped at the foot anchors, and hence formed tubes. The rolling still occurred in the <100> direction, but "turn-over" triangular patterns were generated in this

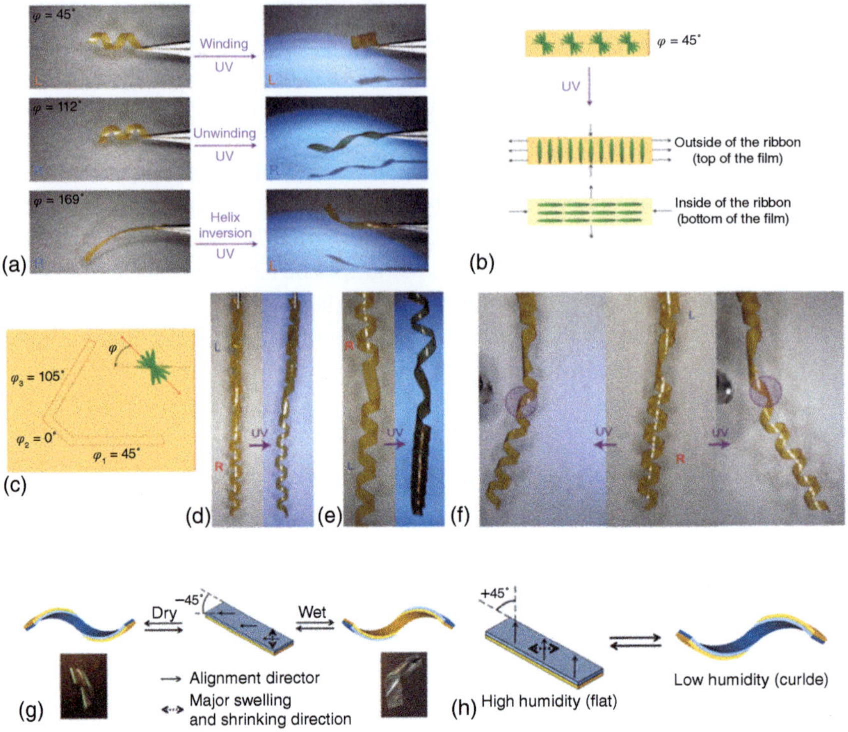

Figure 6.17 (a, b) Three phases of LC helix under UV light exposure: curling, uncurling, and inversion. Helices curls where the handedness is strengthened by the alignment director but uncurls and ultimately inverts where handedness is opposite to the alignment director. (c–f) When selectively applied the UV radiation close to the kink, the end of the helix shifts away and off axis from the opposite end. Source: Adapted from Iamsaard et al. [110] with permission. (g, h) Schematic of how the helical handedness is informed by the alignment director constructed from a LC network. Source: (g, h) Adapted from De Haan et al. [111] with permission.

direction. The persistence of the rolling along the <100> direction demonstrates the stiffness anisotropy of cubic GaAs crystals, regardless of the orientations of rectangle stripes [113].

The rolling behavior of rectangular layer with both sides oriented in the softest directions was studied by Chun et al. [114]. The rolling direction is plotted in Figure 6.19 versus two nondimensional parameters, b/a, c/a, and the length of the membrane, a, where b and c are the width of the membrane and the circumference of the tube, respectively. Three types of rolling behavior are observed in the experiments plotted in different colored dots. The blue and red dots indicate long- and short-side rolling, respectively, the green ones are complicated cases where some tubes rolled up from the long side while others from the short side. It is implied that the rolling direction is related not only to the dimension of the membrane (a and b)

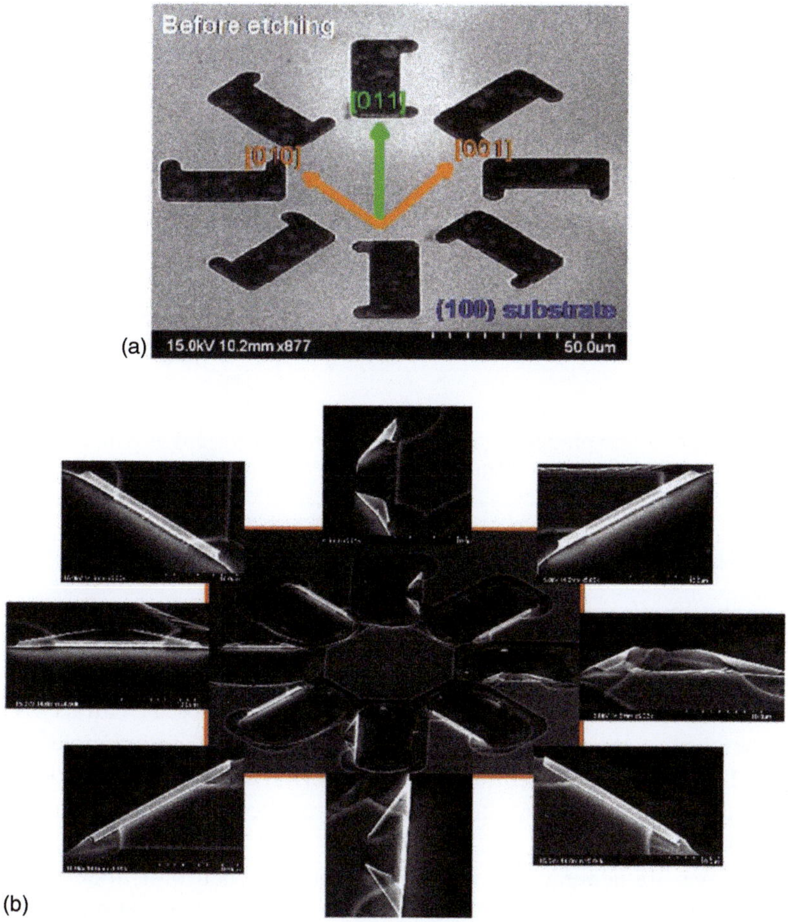

Figure 6.18 SEM images of a wheel pattern of anchored rectangular pads (a) before and (b) after etching. $In_{0.3}Ga_{0.7}As/GaAs$ film was grown on (100) GaAs. Source: Chun and Li [113] Reproduced with permission from IEEE.

but also to the tube circumference (c). Rolling can only occur on the long side when $c/a > 2$ and $b/a > 9$, but can be observed on both long- and short sides when c/a and b/a are at the very low level of values [114].

Chemical etching anisotropy potentially influences the rolling behavior through another experimental study [112]. The final rolling direction has multiple dependent aspects, including the length and width of the flexible layer, the final shape energy, the time evolution of the rolling process, the control of the etching isotropy, and tube circumference [114]. Poisson's ratio may also be changed if the material is anisotropic [115]. The release of the strain along the tube axis also takes effects on the rolling process and tube's geometry, which has been experimentally detected by X-ray diffraction [116] and optical characterization [117].

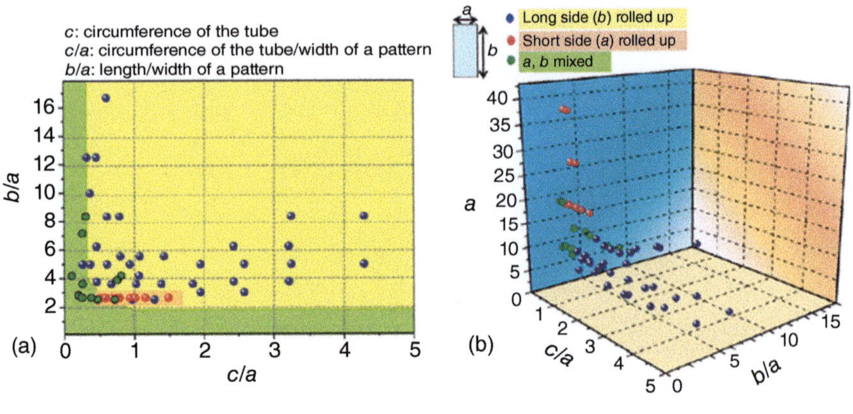

Figure 6.19 Rolling direction plotted in (a) two and (b) three dimensions as a function of tube circumference *c*, width *a*, and rectangular layer length *b*. Source: Reprinted with permission from Chun et al. [114].

6.4.5 Chiral Nanoarchitectonics

Chiral nanoarchitectures are widely explored by many researchers since they are attractive and elegant structures with excellent and unique functions. Several worthwhile creations and functions of chiral nanoarchitectures, such as precise fabrication of chiral nanoarchitectures, are summarized in this section. Nanoscale chiral twists and helices are the most common chiral nanoarchitectures found in supramolecular chirality [118–121] and nanoscale chirality [122]. Here, the design and fabrication of the fundamentals of chiral nanoarchitectures are introduced based on hierarchical self-assembly.

Helical structures are generated at intermediate stages when they undergo the self-assembly dynamic process. To understand the formation of various chiral nanoarchitectures, it is necessary to completely clarify this self-assembly transformation. Several transitions from twists to helices have been observed, for example, the evolution of a twist to a helix of a nanoribbon or a nanofiber, and the transition from a chiral twist to a helix with barber-spot. The self-assembly behavior of a peptide amphiphile (PA) was studied by Stupp and coworker, and they directly observed a self-assembly transition was induced over time [123]. The behavior changed from a short-twisted ribbon to a long twisted one and then transformed into a helical ribbon. In summary, this structural transition from twisted to helical ribbons was strongly determined by increasingly stable molecular packing during the assemblies. Based on the findings from the previous study, the transformation of a ribbon from a twisted shape to a helical shape most likely occurs by a uniform twisting of β-sheets to maximize molecular packing. The understanding of the transitions of PA assemblies will enhance the exploration of amyloid fibrils and contribute to understanding the chiral nanoarchitectures.

Nanotubular structures generally form from helical ribbons [123]. They were found through cryo transmission electron microscopy (cryo-TEM) during assemblies over time, from micelles of $C_{12} - \beta_{12}$ to closed nanotubes. During the transition, nanotubes underwent a couple of intermediate stages, including elongated fibrils, twisted ribbons, and coiled helical ribbons in sequence (Figure 6.20A). The fundamental unit was made by a monolayer lamella of *N*-α-lauryl-lysyl-amino and lauryl-lysyl-amide (abbreviated as $C_{12}-\beta_{12}$). The $C_{12}-\beta_{12}$ monomers are induced to assemble to form ribbons due to the combined effects of hydrophobic interactions and hydrogen bonding. The twisted ribbons were first created as the precursors of coiled ribbons, and then the coiled ribbons transitioned to nanotubes. After the investigation, the authors concluded that the transition from monomers to nanotubes cannot be precisely modeled by the growing width model and the closing pitch model, since both the width and pitch of ribbons grew with maturation. Furthermore, they validated that the width of the ribbon controlled its shape. Narrow chiral ribbons prefer to twist with a Gaussian saddle-like curvature, but wider ribbons are more likely to transform to a helix with a cylindrical curvature.

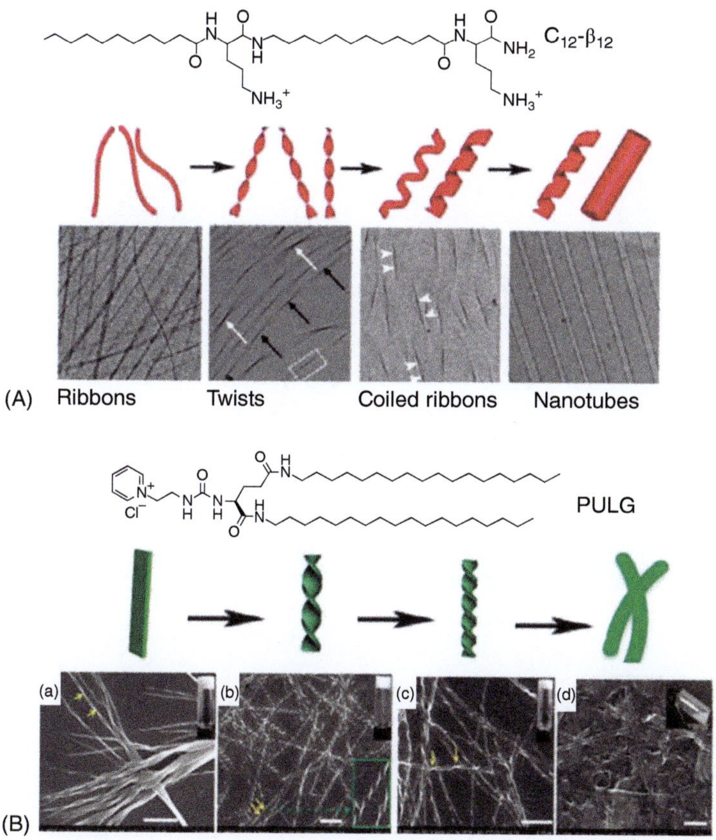

Figure 6.20 (A) Molecular structure of $C_{12}-\beta_{12}$, schematic of chiral self-assembly, and cryo-TEM images of corresponding structures. Source: Ziserman et al. [124]. Reproduced with permission American Chemical Society. (B) Molecular structure of PULG, schematic of helical patch modified by water, and SEM image of the PULG assembly at H_2O/ethanol ratios (vol%) of: (a) 5%, (b) 10%, (c) 20%, and (d) 30%. Source: Liu et al. [125]. Reproduced with permission Royal Society of Chemistry.

These results provide a comprehensive understanding to distinguish chiral twists and chiral nanotubes, as well as the formation of different chiral nanoarchitectures.

Morphology of chiral amphiphiles formed during the self-assembly of ribbons is determined by elastic forces and the chiral orientation of the constituent molecules [126]. Selinger et al. [127] tried to implement a new mesoscale simulation method, called Monte Carlo simulations, to investigate the influences of elasticity, chirality, and molecular tilt on chiral self-assembly. It is reported that due to elastic properties of the membrane, the curvature of the structure transitions from cylindrical to saddle-like shapes. Meanwhile, the simulation also found that the left- or right-handedness is related to the molecular tilting directions chiral handedness interactions, and further explored that single-handedness molecules can be generated by either right- or left-handed structures.

Other approaches have also been used to form diverse chiral nanostructures, for example, a delicate molecular design. Barclay et al. [128] systematically investigated

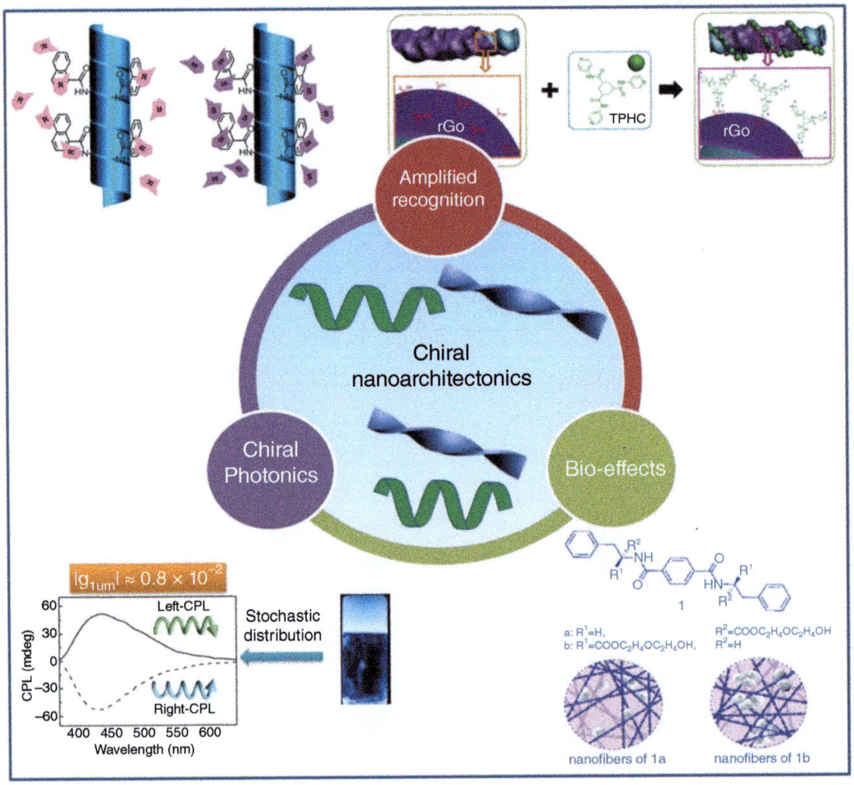

Figure 6.21 Functions of chiral nanoarchitectures. Source: Reprinted with permission Zhang et al. [133]. Top left panel: Reproduced with permission Wang et al. [134]. Copyright 2014, Wiley-VCH; top right panel: Reproduced with permission Jung et al. [135]. Copyright 2013, American Chemical Society; bottom left panel: Reproduced with permission Shen et al. [136]. Copyright 2015, Royal Society of Chemistry; bottom right panel: Reproduced with permission Liu et al. [137]. Copyright 2014, Wiley-VCH.

how molecular structures, temperature, and solvents exert influences on the chirality of assemblies. Briefly, their study revealed that the chiral morphologies can be achieved in the case of balancing the self-assembly of the hydrophilic headgroup and the hydrophobic tail that the choice of solvent and temperature also take effects on formation of the chiral order of aggregate morphologies [129–132]. For example, the morphology of the gel of a pyridine-linked glutamide derivative gelator (PULG) was determined by the solvent. In organic solvents, PULG molecules can only form nanofiber structures without chirality, when adding a small amount of water, chiral structure formation can be activated, and the pitch of the helical ribbon is probable to be varied by the amount of water (Figure 6.20B).

In biological systems, double-helical DNA and proteins can supply accurate and efficient functions to satisfy certain needs, but it is unclear what and how new functions chiral nanoarchitectures can. Assembled twisted and helical structures are potentially used to amplify the molecular properties and develop new/unique functions that single chiral molecules or nonchiral structures cannot realize, shown in Figure 6.21.

6.5 Multistability of Multilayer Structures

6.5.1 Tristability of Bilayer Composite

More and more researchers have devoted to the research of programmable shape transition induced by external stimuli. Multistable structures can present multiple stable equilibrium shapes and transition from one to another when they respond to external excitation, with applications of energy harvesting, sensors, and actuators [138–140] being examples. Among those applications, a tunable helical ribbon can experience transformation from a 2D to a 3D shape by changing the strain of the structure [52]. Such strain-engineering approach can be used to fabricate and analyze bistable helical ribbons [24].

Bistable structures have been investigated for several decades, while tristability of structures has not been deeply explored and understood. In this section, a partially bonded bilayer structure was designed and fabricated using a shape memory polymer and a rubber strip. This structure can appear tri-stable shapes, hemihelical, left-handed, and right-handed helical shapes, and is capable of transitioning between them via changing temperatures. Finite element simulations are implemented to validate this tristability property and to predict the various configurations given geometric and material parameters. Our work provides a concise description of how to produce smart reconfigurable structures for intelligent materials.

6.5.1.1 Theoretical Representation of the Partially Bonded Strip

One way to study such multistability is to model the tri-stabile helices as rods. Using the elasticity theory of rod, the moment on the cross-section area is

$$\boldsymbol{M} = EI_1(\kappa_1 - K)\boldsymbol{d}_1 + EI_2\kappa_2\boldsymbol{d}_2 + GJ\kappa_3\boldsymbol{d}_3 \tag{6.42}$$

where K is the internal bending moment induced by the bonding of two layers. The bending in both directions, $\mathbf{d}_1$ and $\mathbf{d}_2$, is superposed and represented by a single bending component $EI\boldsymbol{t} \times \frac{d\boldsymbol{t}}{ds}$. $\boldsymbol{t}$ and $\mathbf{d}_3$ are synonymous. If we assume that the internal bending moment is in the direction of $\boldsymbol{t} \times \frac{d\boldsymbol{t}}{ds}$, which is a reasonable assumption when the internal bending moment is the only bending moment, or is large enough to cause the rod to bend primarily along its direction, then the total moment in tri-stable helices is:

$$\boldsymbol{M} = EI\left(1 - K/\left|\frac{d\boldsymbol{t}}{ds}\right|\right)\boldsymbol{t} \times \frac{d\boldsymbol{t}}{ds} + M_z\boldsymbol{t} \tag{6.43}$$

The BEHM model can thus be modified to investigate the shape of tri-stable helices.

6.5.1.2 Numerical Simulation of the Partially Bonded Strip

To show the tristability of the partially bonded helices, we conducted a series of simulations using the software ABAQUS. We modeled the behavior of a double-layered strip with the top layer attached to only half of the upper surface of the bottom layer, as shown in Figure 6.22. Each layer is 1 mm wide, 0.1 mm tall, and 10 mm long. The strip is clamped on one end. We used a linear elastic model with Young's modulus set to be 200 MPa and Poisson's ratio set to be 0.47. One layer is a thermo-deformable material with a thermal expansion coefficient of 0.1. The element used is C3D8R.

We use the dynamic explicit feature in ABAQUS to construct two steps in the simulation. During the first step, a torque is applied on the free end as a perturbation to trigger different stable states by applying two-concentrated loads of the same magnitude (0.001 N) but with a different direction on two nodes. In the second step, the temperature of the structure is increased gradually from 0 to 3°C. The strip bends due to the differential thermal expansion between the two layers. Meanwhile, the perturbation is removed. So different stable states can be observed at the end of the second step due to different paths of removal of the perturbation. The torque can be set to decrease linearly to zero throughout the whole step or within the first half of

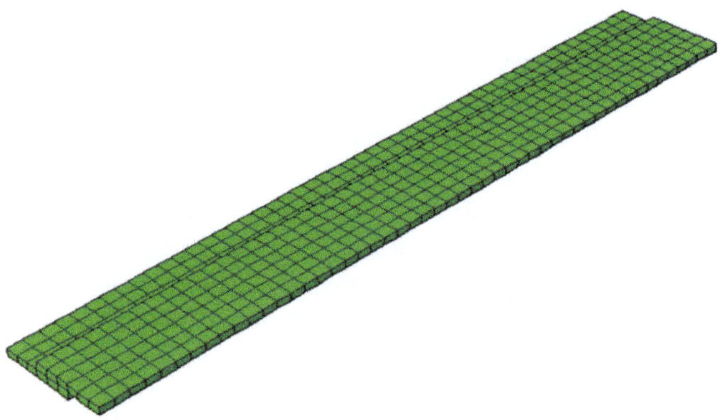

Figure 6.22 Configuration of the bilayer strip.

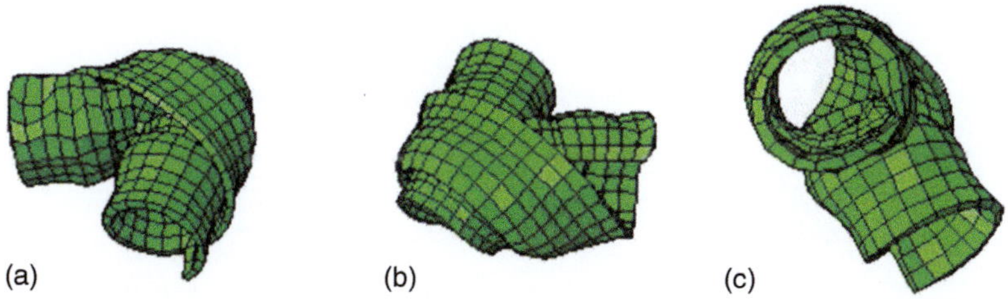

Figure 6.23 Different views of the stable states (a) x–y plane, (b) x–z plane, and (c) y–z plane.

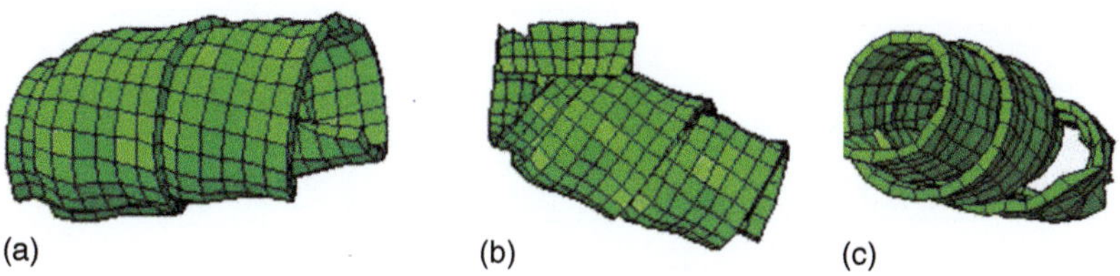

Figure 6.24 Different views of helical stable shapes.

the second step. The strain energy, kinetic energy, and viscous dissipation energy are compared with each other to demonstrate the stability of the ribbon configuration.

Figure 6.23 shows multiple hemihelices stable shapes resulting from the temperature difference and perturbation. A chirality transformation appears in Figure 6.23 – the strip started from one handedness, and due to the self-contact, nevertheless, half of the strip has reversed handedness.

Figure 6.24 shows the helical stable state of the strip. This is the one without the chirality perversion. The different configurations are induced by different triggering torque as the perturbation. Self-contact also plays a role in forming different shapes.

6.5.2 Bistable and Neutral Stable Shells

A theoretical model and proving experiments are developed to study bistability of strained multilayer systems by Chen et al. [24]. Monostable and bistable shapes were fabricated by two pieces of rubber sheets with prestrains along perpendicular directions. These two sheets were then bonded with an adhesive elastic acrylic sheet (Figure 6.25a,c–f). The composite layers can either exhibit mono- or bi-stable equilibria determined by the geometry. Monostability appears when the sheet has smaller width or thicker thickness but bistability vice versa. The formation of bistability was modeled by the deformation of an initially flat elastic strip onto the surface of a torus (Figure 6.25b) with two principal curvatures (κ_1 and κ_2). κ_1 and κ_2 were identified to determine bistability by comparing the relation of the bending and in-plane stretching energy. In this model, the regions of monostability and bistablity

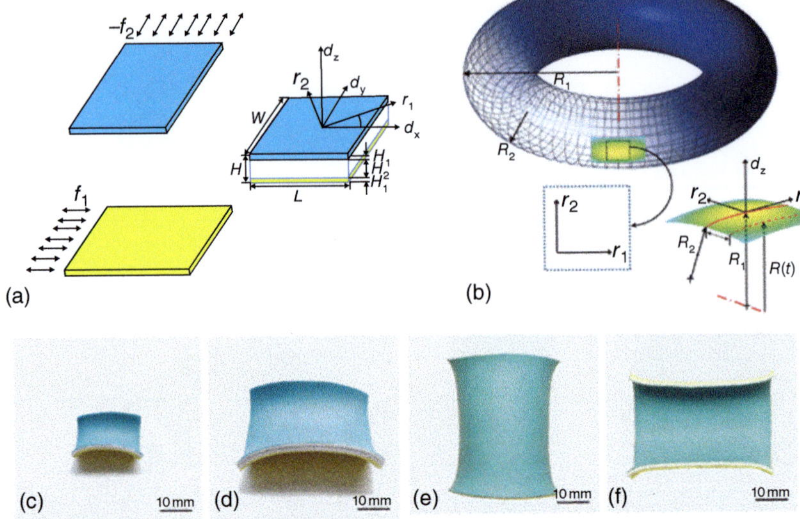

Figure 6.25 (a) An elastic strip constructed from blue and yellow latex sheets that were perpendicularly pre-stretched. The multilayer strip forms a doubly curved shape that leads a torus (b). Small (c) and large (d) thin squares form saddles, while thin strips form stable semi-cylinders (e, f). Source: Adapted from Chen et al. [24] with permission.

are illustrated in Figure 6.26a; when the parameter ξ is below a threshold value, the multilayer structure is monostable, shown in Figure 6.26b–d, while when it exceeds that critical value; the system becomes bistable (Figure 6.26e–g). The parameter ξ is roughly equivalent to the parameter $\eta \equiv W\sqrt{\kappa/H}$ [20, 24, 88], but it does not include the unknown curvature κ, which is more suitable for designing multistable structures.

Generally, neutrally stable structures are defined as zero-stiffness elastic structures [141–143]. A deformable prestressed shell structure with zero-twist rigidity is shown in Figure 6.26h. Experiments and in-depth theoretical analysis have been conducted to obtain these novel mechanical properties [141, 142]. Guest et al. [141] proposed an accessible model with assumptions of an inextensible shell and a uniform curvature throughout the shell. In such case, the midplane of the shell is simplified to lie on the surface of a cylinder and the bending of the shell is energetically equal along any axis. Seffen and Guest [142] modified the models above and obtain analytical solutions for prestressed shells having opposites- and same-sense, which present bistability and neutral stability, respectively. For isotropic materials, it is reported that when the prestress given to an elastic shell is sufficiently large the shell will enter into a neutral stability phase under same-sense prestress [144], and it has been quantitatively characterized by [141, 142]. By contrast, when the driving force is large enough applied to the shell, the opposite-sense prestress can lead to bistability of the system, similar to bifurcation occurred during the curing of unsymmetrical laminates.

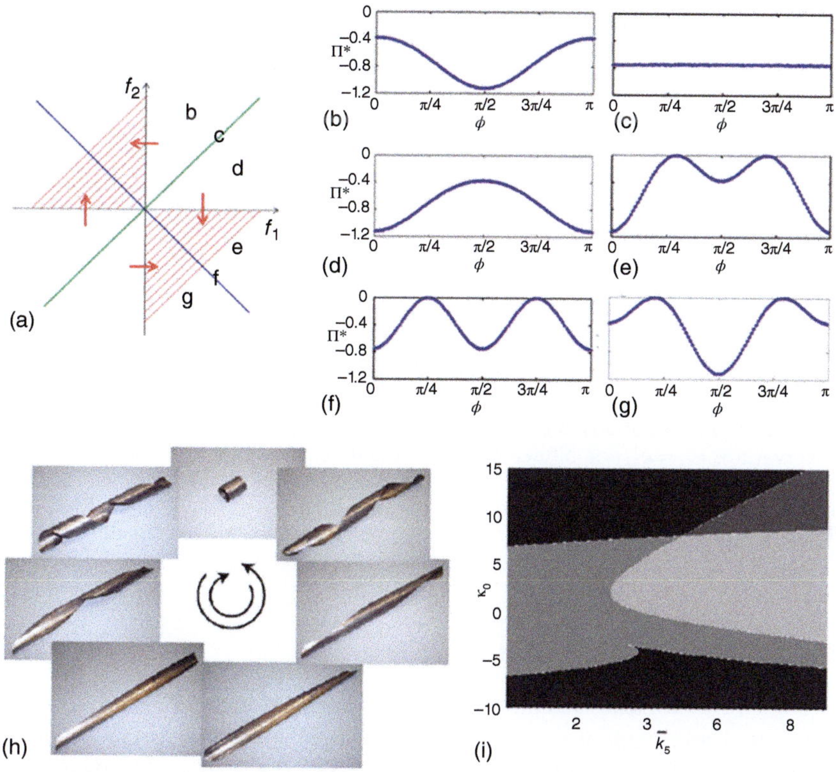

Figure 6.26 (a) Multistable shell design space. Bistability marked in red. (b–g) A plot of total strain energy versus misfit axis orientations with f_2/f_1 values of $\sqrt{3}$, 1, $1/\sqrt{3}$, $-1/\sqrt{3}$, −1, and $-\sqrt{3}$. Source: (a–g) Adapted from Chen et al. [24] with permission. (h) Different shapes of a prestressed shell structure with zero stiffness. The shell can be deformed either clockwise or anticlockwise and each shape can be kept in place simply by rubbing against the table's surface. Source: Adapted from Guest et al. [141] with permission; (i) Regions of stable equilibria for shells with combinations of same-sense prestressing and initial spherical curvature. Source: Adapted from Seffen and Guest [142] with permission.

The mechanisms of bifurcation of multistable structures have been explored by a large group of researchers [20, 142–146]. Holmes et al. [143] studied the bending of slender crosslinked polydimethylsiloxane beams where one side of the beams was immersed in one solvent and the other side was soaked in another type of solvent. They measured the resulting curvatures over time and developed analytic solutions to provide a quantitative characterization of these curvatures. It indicated that the beams bend fast to their maximum curvatures but release to their original shapes slowly. The authors further investigated axially symmetric quasi-2D discs controlled by two principal curvatures (Figure 6.27). When bifurcation occurs, the two principal curvatures show inequality in magnitude and rotation, while the material points are locked in the horizontal direction, predicted by Karman plate theory. This phenomenon was validated in experiments – during dynamic rotation, the material points only had movements in the vertical direction.

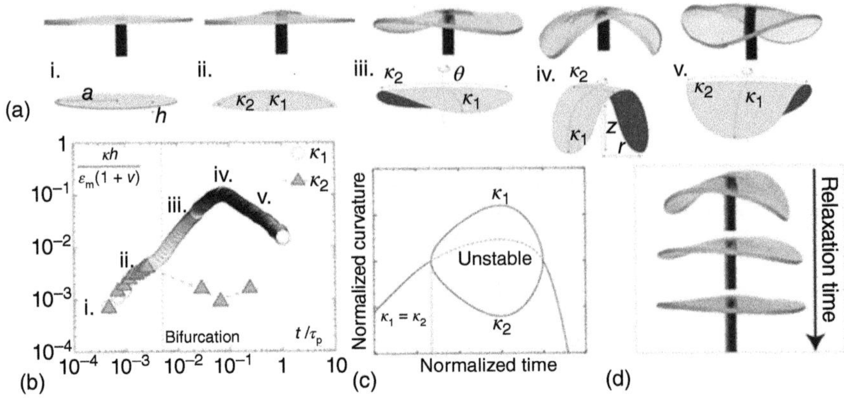

Figure 6.27 (a) Representations of circular disc buckles axisymmetrically with principal curvatures $\kappa_1 = \kappa_2$, and bifurcates with two different positive curvatures. (b) Normalized curvature as a function of nonhomogeneously swollen time. (c) A plot of normalized curvature as a function of normalized time is established when minimal total strain energy is applied and is consistent with experimental results. (d) The circular disc returns to its initial flat state. Source: Adapted from Holmes et al. [143] with permission.

6.5.3 Multistability in Bioinspired Structures

Snap-through in biological system, for example, snap-through of Venus flytrap, can be explored and explained using the mechanical instability principle [25]. Forterre et al. [25] calculated the principal curvatures of the leaves of Venus flytrap through UV fluorescent markers on their surface. The Venus flytrap was exposed to radiation via ultraviolet light and a high-speed/resolution camera was used to record the "mouth" closure process. Spatially averaged Gaussian curvature and mean curvature were used to analyze experimental data. Intrinsic principal curvatures in a closed trap of the Venus flytrap were also measured, shown in Figure 6.28a. Moreover, the elastica theory–based model was proposed to account for the coupling between bending and stretching of a leaf, mainly controlled by intrinsic principal curvatures κ_{x0} and κ_{y0}, the equilibrium shapes of leaves can be represented by K_{x0} based on the minimizer of total energy with respect to K_x and K_y. A nondimensional parameter α, related to the size and curvatures of the leaf, determines the bistability of the Venus flytrap (Figure 6.28b). Specifically, the Venus flytrap only has mono-stable state when $\alpha < \alpha_c \approx 0.8$ while when $\alpha > \alpha_c \approx 0.8$, the system shows multistability (Figure 6.28c).

Mimicking the Venus flytrap, a bionic bistable snapping surface was created by Holmes and Crosby [147] based on a mechanical buckling mechanism. Figure 6.28d shows doubly curved bistable shells where biaxial compressive stress is applied to the structure. One pre-stretched PDMS with holes is crosslinked with an uncured PDMS. This forms concave and convex shapes of bilayer PDMS are formed, which creates the essential condition for snapping through.

A similarly bioinspired snapping robot, a jumping microgel device, was developed by Lee et al. [148], shown in Figure 6.28e. Rapid actuation of this robot was realized

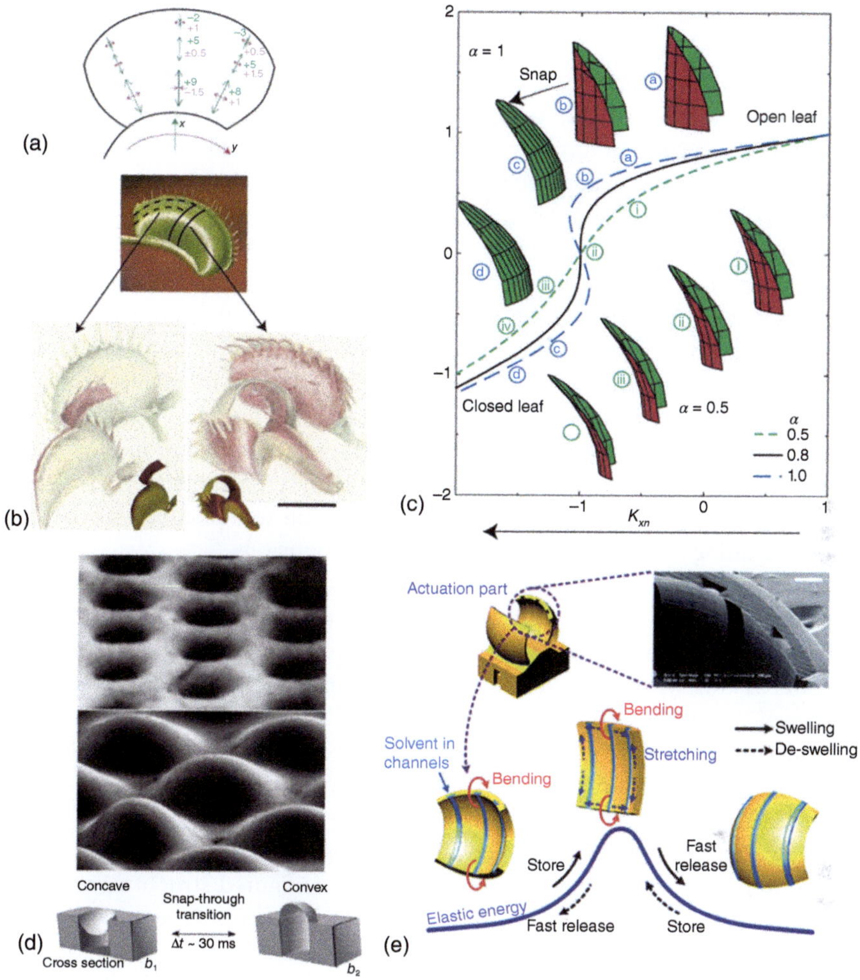

Figure 6.28 (a) Strain field measurements. (b) Cutting of the closed leaf to illustrate the natural principal curvatures. (c) Plot of mean curvature K_m versus K_{xn}. Source: (a–c) Adapted from Forterre et al. [25] with permission. (d) Snapping surface of concave microlenses simulating the Venus flytrap. Source: Adapted from Holmes and Crosby [147] with permission. (e) A swelling-induced snapping microrobot that is created by hydrogel shell having embedded microfluidic channels. Source: Adapted from Lee et al. [148] with permission.

using a dome-shaped hydrogel shell that is capable of presenting elastic instability response for external stimuli. The microgel legs of this self-jumping device can snap-buckle when swelling and achieve a rapid jump. In addition, Ravi Shankar et al. [149] devised a lightweight actuator using azobenzene-functionalized polymers to induce snap-through. The use of bistable arches stimulates a snap through, which displays the snapping speed as high as 100 mm/s. These actuators become buckled when exposed to lights, and they are suitable for remotely and precisely triggerable actuators.

6.6 Conclusion

In this chapter, the nonlinear mechanics with linearized theory and instability of ribbons, plates, and beams have been systematically studied, combining with experimental validations. A quick review of linearized Kirchhoff beam theory, classic Föppl–von Kármán plate theory, and continuum elasticity theory for helicity of ribbons was given, as well as buckling instability of shells.

Bistability and reconfiguration of nanomembrane are then studied. Bistability can dramatically improve maneuverability and expand ideas to design reconfigurable, stimuli-responsive structures. Based on the equal biaxial misfit strain between Cr and VO_2 layers and a clamped boundary condition, we successfully design a Cr/VO_2 nanomembrane with two stable shapes – one simply rolls up while the other folds itself up by generating narrow creases. More importantly, both experiments and simulations prove that bistability of the structure is greatly determined by its geometry, indicating that the configurations and functions of the self-assembled mesostructures can be precisely tuned on demand by redesigning the geometry of the 2D precursor, including the types of materials and the dimensions.

Additionally, this chapter provides a comprehensive review of state-of-the-art study for self-assembly flexible layers and multilayer structures. By introducing pre-strains into these flexible layers, they are able to exhibit predictable wrinkling, rolling, and twisting. We explain how strain-induced bending multilayers work and the process of spontaneous helix formation. We also show experimental examples of materials, such as hydrogel bi-layered structure and liquid crystal polymer. Shape transitions and multistability of helical ribbons are also included in this chapter.

Shape-morphing materials have been widely developed and used for sensors, actuators, and adaptive surfaces. These materials and reconfigured structures are preferred candidates for environment sensors, since they are passive actuators stimulated by surrounding environments, such as temperature, humidity, salinity, and pH. Compared with conventional passive sensors, structures made by flexible layers can reconfigure to any shapes thus readily adding them to substrates and devices. Moreover, their thin profile and sensitivity make them more suitable for artificial skin and could potentially replace the matrices of hard sensors currently used in the field because of their small footprint and low material cost.

Acknowledgments

This work was supported in part by the Branco Weiss–Society in Science fellowship, administered by ETH Zürich, Neukom Institute CompX Faculty award, and the Office of Naval Research (MURI N00014-20-1-2479).

References

1 Okada, K. and Shimura, Y. (1990). Reversible root tip rotation in arabidopsis seedlings induced by obstacle-touching stimulus. *Science* 250 (4980): 274–276.

2 Dawes-Hoang, R.E., Parmar, K.M., Christiansen, A.E. et al. (2005). Folded gastrulation, cell shape change and the control of myosin localization. *Development* 132 (18): 4165–4178.

3 Xu, G., Knutsen, A.K., Dikranian, K. et al. (2010). Axons pull on the brain, but tension does not drive cortical folding. *J. Biomech. Eng.* 132 (7): 071013.

4 Bunch, J.S., Van Der Zande, A.M., Verbridge, S.S. et al. (2007). Electromechanical resonators from graphene sheets. *Science* 315 (5811): 490–493.

5 Cui, Y., Wei, Q., Park, H., and Lieber, C.M. (2001). Nanowire nanosensors for highly sensitive and selective detection of biological and chemical species. *Science* 293 (5533): 1289–1292.

6 Smith, B., Zastavker, Y.V., and Benedek, G.B. (2001). Tension-induced straightening transition of self-assembled helical ribbons. *Phys. Rev. Lett.* 87 (27): 278101.

7 Zastavker, Y.V., Asherie, N., Lomakin, A. et al. (1999). Self-assembly of helical ribbons. *Proc. Natl. Acad. Sci. U.S.A.* 96 (14): 7883–7887.

8 Huang, M., Boone, C., Roberts, M. et al. (2005). Nanomechanical architecture of strained bilayer thin films: from design principles to experimental fabrication. *Adv. Mater.* 17 (23): 2860–2864.

9 Hwang, G., Dockendorf, C., Bell, D. et al. (2008). 3-D InGaAs/GaAs helical nanobelts for optoelectronic devices. *Int. J. Optomechatronics* 2 (2): 88–103.

10 Abbott, J.J., Peyer, K.E., Lagomarsino, M.C. et al. (2009). How should microrobots swim? *Int. J. Rob. Res.* 28 (11–12): 1434–1447.

11 Landau, L.D., and Lifshitz, E.M. (1965) Theory of elasticity (in Russian).

12 Atay, M.T. and Coşkun, S.B. (2009). Elastic stability of Euler columns with a continuous elastic restraint using variational iteration method. *Comput. Math. Appl.* 58 (11–12): 2528–2534.

13 Lee, S.H. and Waas, A.M. (1996). Initial post-buckling behavior of a finite beam on an elastic foundation. *Int. J. Non Linear Mech.* 31 (3): 313–328.

14 Kounadis, A.N., Mallis, J., and Sbarounis, A. (2006). Postbuckling analysis of columns resting on an elastic foundation. *Arch. Appl. Mech.* 75 (6–7): 395–404.

15 Ruocco, E. and Mallardo, V. (2016). An enhanced exponential matrix approach aimed at the stability of piecewise beams on elastic foundation. *Appl. Math. Comput.* 285: 8–25.

16 Goriely, A. and Tabor, M. (1998). Spontaneous helix hand reversal and tendril perversion in climbing plants. *Phys. Rev. Lett.* 80 (7): 1564–1567.

17 Wyczalkowski, M.A., Chen, Z., Filas, B.A. et al. (2012). Computational models for mechanics of morphogenesis. *Birth Defects Res. C Embryo Today* 96 (2): 132–152.

18 Sawa, Y., Urayama, K., Takigawa, T. et al. (2013). Shape and chirality transitions in off-axis twist nematic elastomer ribbons. *Phys. Rev. E: Stat. Nonlinear Soft Matter Phys.* 88 (2): 022502.

19 Ge, Q., Qi, H.J., and Dunn, M.L. (2013). Active materials by four-dimension printing. *Appl. Phys. Lett.* 103 (13): 131901.

20 Armon, S., Efrati, E., Kupferman, R., and Sharon, E. (2011). Geometry and mechanics in the opening of chiral seed pods. *Science* 333 (6050): 1726–1729.

21 Wang, J.S., Feng, X.Q., Wang, G.F., and Yu, S.W. (2008). Twisting of nanowires induced by anisotropic surface stresses. *Appl. Phys. Lett.* 92 (19): 191901.

22 Yu, X., Zhang, L., Hu, N. et al. (2017). Shape formation of helical ribbons induced by material anisotropy. *Appl. Phys. Lett.* 110: 091901.

23 Liu, J., Huang, J., Su, T. et al. (2014). Structural transition from helices to hemihelices. *PLoS One* 9 (4): 93183.

24 Chen, Z., Guo, Q., Majidi, C. et al. (2012). Nonlinear geometric effects in mechanical bistable morphing structures. *Phys. Rev. Lett.* 109: 114302.

25 Forterre, Y., Skotheim, J.M., Dumals, J., and Mahadevan, L. (2005). How the Venus flytrap snaps. *Nature* 433 (7024): 421–425.

26 Guo, Q., Dai, E., Han, X. et al. (2015). Fast nastic motion of plants and bioinspired structures. *J. R. Soc. Interface* 12 (110): 20150598.

27 Poppinga, S., Daber, L.E., Westermeier, A.S. et al. (2017). Biomechanical analysis of prey capture in the carnivorous Southern bladderwort (*Utricularia australis*). *Sci. Rep.* 7 (1): 1–10.

28 Kebadze, E., Guest, S.D., and Pellegrino, S. (2004). Bistable prestressed shell structures. *Int. J. Solids Struct.* 41 (11–12): 2801–2820.

29 Taffetani, M., Jiang, X., Holmes, D.P., and Vella, D. (2018). Static bistability of spherical caps. *Proc. R. Soc. A.* 474: 20170910.

30 Daynes, S., Potter, K.D., and Weaver, P.M. (2008). Bistable prestressed buckled laminates. *Compos. Sci. Technol.* 68 (15–16): 3431–3437.

31 Arrieta, A.F., Kuder, I.K., Rist, M. et al. (2014). Passive load alleviation aerofoil concept with variable stiffness multi-stable composites. *Compos. Struct.* 116 (1): 235–242.

32 Hu, N., Han, X., Huang, S. et al. (2017). Edge effect of strained bilayer nanofilms for tunable multistability and actuation. *Nanoscale* 9 (9): 2958–2962.

33 Gomez, M., Moulton, D.E., and Vella, D. (2017). Critical slowing down in purely elastic "snap-through" instabilities. *Nat. Phys.* 13 (2): 142–145.

34 Epstein, E., Yoon, J., Madhukar, A. et al. (2015). Colloidal particles that rapidly change shape via elastic instabilities. *Small* 11 (45): 6051–6057.

35 Pan, D., Ma, B., and Dai, F. (2017). Experimental investigation of broadband energy harvesting of a bi-stable composite piezoelectric plate. *Smart Mater. Struct.* 26 (3): 035045.

36 Shan, S., Kang, S.H., Raney, J.R. et al. (2015). Multistable architected materials for trapping elastic strain energy. *Adv. Mater.* 27 (29): 4296–4301.

37 Rafsanjani, A., Akbarzadeh, A., and Pasini, D. (2015). Snapping mechanical metamaterials under tension. *Adv. Mater.* 27 (39): 5931–5935.

38 Plaut, R.H. and Virgin, L.N. (2009). Vibration and snap-through of bent elastica strips subjected to end rotations. *J. Appl. Mech. Trans. ASME* 76 (4): 1–7.

39 Virgin, L.N., Guan, Y., and Plaut, R.H. (2017). On the geometric conditions for multiple stable equilibria in clamped arches. *Int. J. Non Linear Mech.* 92: 8–14.

40 Efrati, E., Sharon, E., and Kupferman, R. (2009). Elastic theory of unconstrained non-Euclidean plates. *J. Mech. Phys. Solids* 57 (4): 762–775.

41 Seffen, K.A. (2007). "Morphing" bistable orthotropic elliptical shallow shells. *Proc. R. Soc. London, Ser. A* 463 (2077): 67–83.

42 Hamouche, W., Maurini, C., Vincenti, A., and Vidoli, S. (2016). Basic criteria to design and produce multistable shells. *Meccanica* 51 (10): 2305–2320.

43 Fernandes, A., Maurini, C., and Vidoli, S. (2010). Multiparameter actuation for shape control of bistable composite plates. *Int. J. Solids Struct.* 47 (10): 1449–1458.

44 Chouaieb, N., Goriely, A., and Maddocks, J.H. (2006). Helices. *Proc. Natl. Acad. Sci. U.S.A.* 103 (25): 9398–9403.

45 Gurtin, M.E. and Ian Murdoch, A. (1975). A continuum theory of elastic material surfaces. *Arch. Ration. Mech. Anal.* 57 (4): 291–323.

46 He, J. and Lilley, C.M. (2008). Surface effect on the elastic behavior of static bending nanowires. *Nano Lett.* 8 (7): 1798–1802.

47 Li, H., Yang, Z., Zhang, Y.M., and Wen, B.C. (2010). Deflections of nanowires with consideration of surface effects. *Chin. Phys. Lett.* 27 (12): 126201.

48 Lotz, B. and Cheng, S.Z.D. (2005). A critical assessment of unbalanced surface stresses as the mechanical origin of twisting and scrolling of polymer crystals. *Polymer* 46 (3): 577–610.

49 Miller, R.E. and Shenoy, V.B. (2000). Size-dependent elastic properties of nanosized structural elements. *Nanotechnology* 11 (3): 139–147.

50 Pokroy, B., Kang, S.H., Mahadevan, L., and Aizenberg, J. (2009). Self-organization of a mesoscale bristle into ordered, hierarchical helical assemblies. *Science* 323 (5911): 237–240.

51 Zhang, L., Deckhardt, E., Weber, A. et al. (2005). Controllable fabrication of SiGe/Si and SiGe/Si/Cr helical nanobelts. *Nanotechnology* 16 (6): 655–663.

52 Chen, Z., Majidi, C., Srolovitz, D.J., and Haataja, M. (2011). Tunable helical ribbons. *Appl. Phys. Lett.* 98 (1): 011906.

53 Green, A.E. and Laws, N. (1966). A general theory of rods. *Proc. R. Soc. A: Math. Phys. Eng. Sci.* 293 (1433): 145–155.

54 O'Reilly, O.M. (1998). On constitutive relations for elastic rods. *Int. J. Solids Struct.* 35 (11): 1009–1024.

55 Stoney, G.G. (1909). The tension of metallic films deposited by electrolysis. *Proc. R. Soc. London, Ser. A* 82 (553): 172–175.

56 Tu, Z.C. and Ou-Yang, Z.C. (2008). Elastic theory of low-dimensional continua and its applications in bio- and nano-structures. *J. Comput. Theor. Nanosci.* 5 (4): 422–448.

57 Ebata, Y., Croll, A.B., and Crosby, A.J. (2012). Wrinkling and strain localizations in polymer thin films. *Soft Matter* 8 (35): 9086–9091.

58 Lee, S., Hippalgaonkar, K., Yang, F. et al. (2021). Anomalously low electronic thermal conductivity in metallic vanadium dioxide. *Science* 355 (6323): 371–374.

59 Noblin, X., Rojas, N.O., Westbrook, J. et al. (2012). The fern sporangium: a unique catapult. *Science* 335 (6074): 1322.

60 Hu, B., Zhang, Y., Chen, W. et al. (2011). Self-heating and external strain coupling induced phase transition of VO_2 nanobeam as single domain switch. *Adv. Mater.* 23 (31): 3536–3541.

61 Tian, Z., Xu, B., Hsu, B. et al. (2018). Reconfigurable vanadium dioxide nanomembranes and microtubes with controllable phase transition temperatures. *Nano Lett.* 18 (5): 3017–3023.

62 Pandey, A., Moulton, D.E., Vella, D., and Holmes, D.P. (2014). Dynamics of snapping beams and jumping poppers. *EPL* 105 (2): 24001.

63 Skotheim, J.M. and Mahadevan, L. (2005). Plant science: physical limits and design principles for plant and fungal movements. *Science* 308 (5726): 1308–1310.

64 Schmidt, O.G. and Eberl, K. (2001). Thin solid films roll up into nanotubes. *Nature* 410 (6825): 168.

65 Fedorchenko, A.I., Wang, A.B., Mashanov, V.I. et al. (2006). Strain-induced wrinkling on SiGe free standing film. *Appl. Phys. Lett.* 89 (4): 043119.

66 Brau, F., Vandeparre, H., Sabbah, A. et al. (2011). Multiple-length-scale elastic instability mimics parametric resonance of nonlinear oscillators. *Nat. Phys.* 7 (1): 56–60.

67 Mei, Y., Kiravittaya, S., Harazim, S., and Schmidt, O.G. (2010). Principles and applications of micro and nanoscale wrinkles. *Mater. Sci. Eng., R* 70 (3–6): 209–224.

68 Kim, S.P., Chew, H.B., Chason, E. et al. (2012). Nanoscale mechanisms of surface stress and morphology evolution in FCC metals under noble-gas ion bombardments. *Proc. R. Soc. London, Ser. A* 468 (2145): 2550–2573.

69 Semler, M.R., Harris, J.M., Croll, A.B., and Hobbie, E.K. (2013). Localization and length-scale doubling in disordered films on soft substrates. *Phys. Rev. E: Stat. Nonlinear Soft Matter Phys.* 88 (3): 032409.

70 Oshri, O., Brau, F., and Diamant, H. (2015). Wrinkles and folds in a fluid-supported sheet of finite size. *Phys. Rev. E: Stat. Nonlinear Soft Matter Phys.* 91 (5): 052408.

71 Zang, J., Zhao, X., Cao, Y., and Hutchinson, J.W. (2012). Localized ridge wrinkling of stiff films on compliant substrates. *J. Mech. Phys. Solids* 60 (7): 1265–1279.

72 Cendula, P., Kiravittaya, S., Mei, Y.F. et al. (2009). Bending and wrinkling as competing relaxation pathways for strained free-hanging films. *Phys. Rev. B: Condens. Matter* 79 (8): 085429.

73 Mora, T. and Boudaoud, A. (2006). Buckling of swelling gels. *Eur. Phys. J. E* 20 (2): 119–124.

74 Mei, Y., Thurmer, D.J., Cavallo, F. et al. (2007). Semiconductor sub-micro-/nanochannel networks by deterministic layer wrinkling. *Adv. Mater.* 19 (16): 2124–2128.

75 Sun, Y., Choi, W.M., Jiang, H. et al. (2006). Controlled buckling of semiconductor nanoribbons for stretchable electronics. *Nat. Nanotechnol.* 1 (3): 201–207.

76 Malachias, A., Mei, Y., Annabattula, R.K. et al. (2008). Wrinkled-up nanochannel networks: long-range ordering, scalability, and X-ray investigation. *ACS Nano* 2 (8): 1715–1721.

77 Guo, Q., Zhang, M., Xue, Z. et al. (2013). Three dimensional strain distribution of wrinkled silicon nanomembranes fabricated by rolling-transfer technique. *Appl. Phys. Lett.* 103 (26): 264102.

78 Childs, W.R., Motala, M.J., Lee, K.J., and Nuzzo, R.G. (2005). Masterless soft lithography: patterning UV/ozone-induced adhesion on poly(dimethylsiloxane) surfaces. *Langmuir* 21 (22): 10096–10105.

79 Duffy, D.C., McDonald, J.C., Schueller, O.J.A., and Whitesides, G.M. (1998). Rapid prototyping of microfluidic systems in poly(dimethylsiloxane). *Anal. Chem.* 70 (23): 4974–4984.

80 Huang, Y.Y., Zhou, W., Hsia, K.J. et al. (2005). Stamp collapse in soft lithography. *Langmuir* 21 (17): 8058–8068.

81 Choi, S., Kwon, S., Kim, H. et al. (2017). Highly flexible and efficient fabric-based organic light-emitting devices for clothing-shaped wearable displays. *Sci. Rep.* 7 (1): 1–8.

82 Cheng, H. and Yi, N. (2017). Dissolvable tattoo sensors: from science fiction to a viable technology. *Phys. Scr.* 92 (1): 013001.

83 Khang, D.-Y., Rogers, J.A., and Lee, H.H. (2009). Mechanical buckling: mechanics, metrology, and stretchable electronics. *Adv. Funct. Mater.* 19 (10): 1526–1536.

84 Genzer, J. and Groenewold, J. (2006). Soft matter with hard skin: from skin wrinkles to templating and material characterization. *Soft Matter* 2 (4): 310–323.

85 Cross, M.C. and Hohenberg, P.C. (1993). Pattern formation outside of equilibrium. *Rev. Mod. Phys.* 65 (3): 851–1112.

86 Bowden, N., Brittain, S., Evans, A.G. et al. (1998). Spontaneous formation of ordered structures in thin films of metals supported on an elastomeric polymer. *Nature* 393 (6681): 146–149.

87 Chung, S., Lee, J.H., Moon, M.-W. et al. (2008). Non-lithographic wrinkle nanochannels for protein preconcentration. *Adv. Mater.* 20 (16): 3011–3016.

88 Forterre, Y. and Dumais, J. (2011). Generating helices in nature. *Science* 333 (6050): 1715–1716.

89 Pham, J.T., Lawrence, J., Lee, D.Y. et al. (2013). Highly stretchable nanoparticle helices through geometric asymmetry and surface forces. *Adv. Mater.* 25 (46): 6703–6708.

90 Wu, Z.L., Moshe, M., Greener, J. et al. (2013). Three-dimensional shape transformations of hydrogel sheets induced by small-scale modulation of internal stresses. *Nat. Commun.* 4 (1): 1–7.

91 Bae, J., Na, J.H., Santangelo, C.D., and Hayward, R.C. (2014). Edge-defined metric buckling of temperature-responsive hydrogel ribbons and rings. *Polymer* 55 (23): 5908–5914.

92 Godinho, M.H., Canejo, J.P., Feio, G., and Terentjev, E.M. (2010). Self-winding of helices in plant tendrils and cellulose liquid crystal fibers. *Soft Matter* 6 (23): 5965–5970.

93 Canejo, J. and Godinho, M. (2013). Cellulose perversions. *Materials* 6 (4): 1377–1390.

94 Ji, X.Y., Zhao, M.Q., Wei, F., and Feng, X.Q. (2012). Spontaneous formation of double helical structure due to interfacial adhesion. *Appl. Phys. Lett.* 100 (26): 263104.

95 Matray, T.J. and Kool, E.T. (1999). A specific partner for abasic damage in DNA. *Nature* 399 (6737): 704–708.

96 Wittung, P., Nielsen, P.E., Buchardt, O. et al. (1994). DNA-like double helix formed by peptide nucleic acid. *Nature* 368 (6471): 561–563.

97 Liu, X.J., Sun, X.W., Zhang, J.J. et al. (2003). Fabrication of β-sialon powder from kaolin. *Mater. Res. Bull.* 38 (15): 1939–1948.

98 Zhang, Q., Zhao, M.-Q., Tang, D.-M. et al. (2010). Carbon-nanotube-array double helices. *Angew. Chem.* 122 (21): 3724–3727.

99 Zhao, M.Q., Huang, J.Q., Zhang, Q. et al. (2011). Stretchable single-walled carbon nanotube double helices derived from molybdenum-containing layered double hydroxides. *Carbon* 49 (6): 2148–2152.

100 Armon, S., Aharoni, H., Moshe, M., and Sharon, E. (2014). Shape selection in chiral ribbons: from seed pods to supramolecular assemblies. *Soft Matter* 10 (16): 2733–2740.

101 Sawa, Y., Ye, F., Urayama, K. et al. (2011). Shape selection of twist-nematic-elastomer ribbons. *Proc. Natl. Acad. Sci. U.S.A.* 108 (16): 6364–6368.

102 Delclos, T., Aimé, C., Pouget, E. et al. (2008). Individualized silica nanohelices and nanotubes: tuning inorganic nanostructures using lipidic self-assemblies. *Nano Lett.* 8 (7): 1929–1935.

103 Oda, R., Huc, I., Schmutz, M. et al. (1999). Tuning bilayer twist using chiral counterions. *Nature* 399 (6736): 566–569.

104 Guo, Q., Mehta, A.K., Grover, M.A. et al. (2014). Shape selection and multi-stability in helical ribbons. *Appl. Phys. Lett.* 104 (21): 211901.

105 Chen, Z. (2014). Geometric nonlinearity and mechanical anisotropy in strained helical nanoribbons. *Nanoscale* 6 (16): 9443–9447.

106 Cranford, S.W. and Buehler, M.J. (2011). Mechanical properties of graphyne. *Carbon* 49 (13): 4111–4121.

107 Lee, H.S., Cho, M., and Lee, B.J. (2011). Size dependence in self-bending mechanism of Ge/Si bilayer nanofilms. *J. Appl. Phys.* 109 (7): 074314.

108 Kit, O.O., Tallinen, T., Mahadevan, L. et al. (2012). Twisting graphene nanoribbons into carbon nanotubes. *Phys. Rev. B: Condens. Matter* 85 (8): 085428.

109 Chen, Z., Han, X., and Zheng, H. (2015). Residual stresses and Poisson's effect drive shape formation and transition of helical structures. *J. Elast.* 119: 321–333.

110 Iamsaard, S., Aßhoff, S.J., Matt, B. et al. (2014). Conversion of light into macroscopic helical motion. *Nat. Chem.* 6 (3): 229–235.

111 De Haan, L.T., Verjans, J.M.N., Broer, D.J. et al. (2014). Humidity-responsive liquid crystalline polymer actuators with an asymmetry in the molecular trigger that bend, fold, and curl. *J. Am. Chem. Soc.* 136 (30): 10585–10588.

112 Li, X. (2008). Strain induced semiconductor nanotubes: from formation process to device applications. *J. Phys. D: Appl. Phys.* 41 (19): 193001.

113 Chun, I.S. and Li, X. (2008). Controlled assembly and dispersion of strain-induced InGaAs/GaAs nanotubes. *IEEE Trans. Nanotechnol.* 7 (4): 493–495.

114 Chun, I.S., Challa, A., Derickson, B. et al. (2010). Geometry effect on the strain-induced self-rolling of semiconductor membranes. *Nano Lett.* 10 (10): 3927–3932.

115 Songmuang, R., Deneke, C., and Schmidt, O.G. (2006). Rolled-up micro- and nanotubes from single-material thin films. *Appl. Phys. Lett.* 89 (22): 223109.

116 Krause, B., Mocuta, C., Metzger, T.H. et al. (2006). Local structure of a rolled-up single crystal: an X-ray microdiffraction study of individual semiconductor nanotubes. *Phys. Rev. Lett.* 96 (16): 165502.

117 Zhen, H.L., Huang, G.S., Kiravittaya, S. et al. (2013). Light-emitting properties of a strain-tuned microtube containing coupled quantum wells. *Appl. Phys. Lett.* 102 (4): 041109.

118 Duan, P., Cao, H., Zhang, L., and Liu, M. (2014). Gelation induced supramolecular chirality: chirality transfer, amplification and application. *Soft Matter* 10 (30): 5428–5448.

119 Mateos-Timoneda, M.A., Crego-Calama, M., and Reinhoudt, D.N. (2004). Supramolecular chirality of self-assembled systems in solution. *Chem. Soc. Rev.* 33 (6): 363–372.

120 Lee, C.C., Grenier, C., Meijer, E.W., and Schenning, A.P.H.J. (2009). Preparation and characterization of helical self-assembled nanofibers. *Chem. Soc. Rev.* 38 (3): 671–683.

121 Liu, M., Zhang, L., and Wang, T. (2015). Supramolecular chirality in self-assembled systems. *Chem. Rev.* 115: 7304–7397.

122 Wang, Y., Xu, J., Wang, Y., and Chen, H. (2013). Emerging chirality in nanoscience. *Chem. Soc. Rev.* 42 (7): 2930–2962.

123 Pashuck, E.T. and Stupp, S.I. (2010). Direct observation of morphological tranformation from twisted ribbons into helical ribbons. *J. Am. Chem. Soc.* 132 (26): 8819–8821.

124 Ziserman, L., Lee, H.Y., Raghavan, S.R. et al. (2011). Unraveling the mechanism of nanotube formation by chiral self-assembly of amphiphiles. *J. Am. Chem. Soc.* 133 (8): 2511–2517.

125 Liu, C., Jin, Q., Lv, K. et al. (2014). Water tuned the helical nanostructures and supramolecular chirality in organogels. *Chem. Commun.* 50 (28): 3702–3705.

126 Ziserman, L., Mor, A., Harries, D., and Danino, D. (2011). Curvature instability in a chiral amphiphile self-assembly. *Phys. Rev. Lett.* 106 (23): 238105.

127 Selinger, R.L.B., Selinger, J.V., Malanoski, A.P., and Schnur, J.M. (2004). Shape selection in chiral self-assembly. *Phys. Rev. Lett.* 93 (15): 158103.

128 Barclay, T.G., Constantopoulos, K., Zhang, W. et al. (2012). Chiral self-assembly of designed amphiphiles: optimization for nanotube formation. *Langmuir* 28 (40): 14172–14179.

129 Gopal, A., Hifsudheen, M., Furumi, S. et al. (2012). Thermally assisted photonic inversion of supramolecular handedness. *Angew. Chem.* 124 (42): 10657–10661.

130 Yagai, S., Yamauchi, M., Kobayashi, A. et al. (2012). Control over hierarchy levels in the self-assembly of stackable nanotoroids. *J. Am. Chem. Soc.* 134 (44): 18205–18208.

131 Huang, Y., Hu, J., Kuang, W. et al. (2011). Modulating helicity through amphiphilicity – tuning supramolecular interactions for the controlled assembly of perylenes. *Chem. Commun.* 47 (19): 5554–5556.

132 Wang, X., Duan, P., and Liu, M. (2013). Organogelation-controlled topochemical [2+2] cycloaddition and morphological changes: from nanofiber to peculiar coaxial hollow toruloid-like nanostructures. *Chem. Eur. J.* 19 (47): 16072–16079.

133 Zhang, L., Wang, T., Shen, Z., and Liu, M. (2016). Chiral nanoarchitectonics: towards the design, self-assembly, and function of nanoscale chiral twists and helices. *Adv. Mater.* 28 (6): 1044–1059.

134 Wang, X., Duan, P., and Liu, M. (2014). Self-assembly of π-conjugated gelators into emissive chiral nanotubes: emission enhancement and chiral detection. *Chem. Asian J.* 9 (3): 770–778.

135 Jung, J.H., Moon, S.J., Ahn, J. et al. (2013). Controlled supramolecular assembly of helical silica nanotube-graphene hybrids for chiral transcription and separation. *ACS Nano* 7 (3): 2595–2601.

136 Shen, Z., Wang, T., Shi, L. et al. (2015). Strong circularly polarized luminescence from the supramolecular gels of an achiral gelator: tunable intensity and handedness. *Chem. Sci.* 6 (7): 4267–4272.

137 Liu, G.-F., Zhang, D., and Feng, C.-L. (2014). Control of three-dimensional cell adhesion by the chirality of nanofibers in hydrogels. *Angew. Chem. Int. Ed.* 53 (30): 7789–7793.

138 Guo, Q., Bishop, C.J., Meyer, R.A. et al. (2018). Entanglement-based thermoplastic shape memory polymeric particles with photothermal actuation for biomedical applications. *ACS Appl. Mater. Interfaces* 10 (16): 13333–13341.

139 Lu, W., Le, X., Zhang, J. et al. (2017). Supramolecular shape memory hydrogels: a new bridge between stimuli-responsive polymers and supramolecular chemistry. *Chem. Soc. Rev.* 46 (5): 1284–1294.

140 Zhang, W.M., Yan, H., Peng, Z.K., and Meng, G. (2014). Electrostatic pull-in instability in MEMS/NEMS: a review. *Sens. Actuators, A* 214: 187–218.

141 Guest, S.D., Kebadze, E., and Pellegrino, S. (2011). A zero-stiffness elastic shell structure. *J. Mech. Mater. Struct.* 6 (1–4): 203–212.

142 Seffen, K.A. and Guest, S.D. (2011). Prestressed morphing bistable and neutrally stable shells. *J. Appl. Mech. Trans. ASME* 78 (1): 0110021–0110026.

143 Holmes, D.P., Roché, M., Sinha, T., and Stone, H.A. (2011). Bending and twisting of soft materials by non-homogenous swelling. *Soft Matter* 7 (11): 5188–5193.

144 Mansfield, E.H. (1965). Bending, buckling and curling of a heated elliptical plate. *Proc. R. Soc. London, Ser. A* 288 (1414): 396–417.
145 Freund, L.B. (2000). Substrate curvature due to thin film mismatch strain in the nonlinear deformation range. *J. Mech. Phys. Solids* 48 (6): 1159–1174.
146 Guo, Q., Zheng, H., Chen, W., and Chen, Z. (2014). Modeling bistable behaviors in morphing structures through finite element simulations. *Biomed. Mater. Eng.* 24 (1): 557–562.
147 Holmes, D.P. and Crosby, A.J. (2007). Snapping surfaces. *Adv. Mater.* 19 (21): 3589–3593.
148 Lee, H., Xia, C., and Fang, N.X. (2010). First jump of microgel; actuation speed enhancement by elastic instability. *Soft Matter* 6 (18): 4342–4345.
149 Ravi Shankar, M., Smith, M.L., Tondiglia, V.P. et al. (2013). Contactless, photoinitiated snap-through in azobenzene-functionalized polymers. *Proc. Natl. Acad. Sci. U.S.A.* 110 (47): 18792–18797.

7

Nanomembranes for Cell Scaffolding and Bio-Analyses

Yue Wu, Chunyu You, Xinyi Ke, Borui Xu, and Yongfeng Mei

Fudan University, Department of Materials Science, 220 Handan Road, Shanghai 200433, P. R. China

7.1 Introduction

Membranes are an essential component for living creatures in nature. From bacteria, tardigrades to human cells, and even cancer cells, nanomembranes participate in metabolic processes and serve as the interface between cytoplasm and extracellular matrix (ECM). In native tissues, ECM supports and directs tissue organization, exerts an indispensable influence on cell behavior, including adhesion, proliferation, migration, and organization [1]. As a result, artificial membranes that combine physical geometry and chemical characters of the cell membrane and ECM had long been utilized as reasonable choices for conducting *in vitro* experiments of cells and human tissues [2, 3]. With the development of fabrication technology, these membranes had been further downsized into structures with extreme aspect ratios, which were defined as nanomembranes, and were widely employed in the field of medical and biological applications, such as tissue engineering [4], biosensing [5], biorobotics [6, 7], and regenerative medicine [8].

On the other hand, there is an increasing demand for uncovering the mysteries that exactly show how cells regulate their behavior toward stimulation from ECM, and ECM is the basis for the formation of diseases and surgery failure (such as cancer and osteoporosis). Obviously, the conventional plane substrate does not meet the need for studies of manipulating and testing cells, and efforts had been made to reform the geometry and compositions of the nanomembranes for better mimic of *in vitro* environment. Harrison et al. first observed that cells growing on a spider web follow the orientation of fibers, demonstrating that cells react to surface topography cues. From then on, several studies with various kinds of nanomembrane geometry confirmed their impact on cell behavior, such as cell adhesion, proliferation, and differentiation [9, 10]. Among all these approaches, microtubes that were fabricated by rolled-up nanotechnology offered a fine platform for encapsulating and guiding cell growth, as well as scaffolds for monitoring of single-cell behavior considering their 2D confinement of cells that resembles the environment in the native blood vessels.

Nanomembranes: Materials, Properties, and Applications, First Edition.
Edited by Yongfeng Mei, Gaoshan Huang, and Xiuling Li.
© 2022 WILEY-VCH GmbH. Published 2022 by WILEY-VCH GmbH.

In this chapter, we will discuss the diversified applications of nanomembranes in cell science and biology. As a thriving discipline, cell scaffolds enabled by nanomembranes will be introduced first. Then, utilizing the structured scaffolds, the cells can be manipulated and observed through topography and electrical regulation. And we will discuss the cell activity on nanomembranes or in nanomembrane-based structures.

7.2 Nanomembranes for Cell Scaffolding

7.2.1 Two-Dimensional Biocompatible Nanomembranes

Polymer-based substrates are traditional choices for cell scaffolding considering their flexible and soft nature that matches the elasticity of human tissues as well as the chemical composition, which resemble the components in the ECM. A main way to improve polymeric substrates for suitable cell culture platforms is tailoring their surface with different nanomembranes or nanostructures [11–14] to modify their biocompatibility, biodegradability, and biosensing functionality.

In one approach, Graeter et al. introduced a novel deblock copolymer micelle nanolithography technique to decorate polymer surfaces with metal nanostructures. Gold nanoparticles were firstly separated on an inorganic substrate and functionalized by connecting molecules. The substrate was in turn covered with different polymer solutions, which were cured by a chemical mechanism of cooling or cross-linking. By choosing linker molecules that were compatible with certain polymers, gold nanoparticles were successfully coupled to various polymer coatings, including polystyrene (PS), polydimethylsiloxane (PDMS), and polyethylene glycol diacrylate (PEGDA), forming a gold-functionalized polymer nanomembrane when separated from the substrate. Specifically, polyethylene glycol (PEG) hydrogels have previously been utilized as a biologically inert background for biosensing. After seeding fibroblasts onto the substrate, distinct attachment and spreading of the cells to the gold-decorated area of the membrane were observed as they further grew into a dense layer of fibroblasts. As a result, the biocompatibility of the sample was demonstrated as well as its potential for positioning separated, individual signaling cells [15].

Alternatively, carbon nanotubes (CNTs) were also employed to enhance cell differentiation on polyacrylic acid (PAA) films. CNT is an excellent material candidate for cell scaffolds due to its strong but flexible mechanical property and biocompatibility [16], while the surface hinders cell adhesion and proliferation [17]. By grafting CNTs with PAA via in situ polymerization with acetone, the produced sample was produced and compared to a conventional poly-*l*-ornithine (PLO) platform for neuron cell culture. The PAA-g-CNT thin film exhibited a similar cell viability rate, while both cell adhesion and neuron cell differentiation were promoted compared to PLO film [18].

Conducting polymers (CPs), regarding their semiconductor behavior, hint at potential application for biotissue sensors, neuronal stimulation, and artificial muscle actuators. However, vulnerable mechanical property and lack of biodegradability limit their biotechnological performance. To solve this problem, Armelin et al. proposed an organic–inorganic hybrid freestanding nanomembrane by blending a semiconducting poly (3-thiophene methyl acetate) (P3TMA) with poly (tetramethylene succinate) (PE44), a biodegradable polyester. The hybrid nanomembranes were degraded after being immersed in Lipase PS for one week, while the P3TMA dominant membranes showed no sign of degradation after four weeks, indicating that it successfully inherited the biodegradability of the PE44 component. The nanomembranes presented reasonable viability to culture human laryngeal epidermoid carcinoma (HEp-2) cells, and the P3TMA: PE44 50:50 blend revealed the best performance for cell adhesion and proliferation among all samples. Thus, a strong, semiconducting, biodegradable, and biocompatible polymeric platform was demonstrated [19].

In another approach, the supported lipid bilayer (SLB) was covered by polyelectrolyte multilayer (PEM) films to construct a neuron cell scaffold for adjustable cell outgrowth and differentiation. SLB is an artificial substrate with structure and fluidity that closely resembles synaptic membranes. The layer-by-layer assembled PEM nanomembranes composed of poly-L-lysine (PLL) and poly-L-glutamic acid (PLGA), which coupled through electrostatic interactions from 1 to 7.5 layers were absorbed onto the SLB surface for substrate protection and cell adhesion, as is shown in Figure 7.1. Neural stem/progenitor cells (NSPCs) outgrowth was analyzed by estimating the process length of neurons, and all samples displayed better outgrowth levels than their counterpart without SLB substrate, increasing with the number of PEM layers. Also, neurites exhibited long and straight processes on the nanomembranes with PLL as the terminal layer compared to short and distorted processes on the PLGA terminal layer. In addition, both immunostaining and western blot analysis demonstrated an encouraging differential of the NSPCs to form synapses on SLB–PEM films. The mechanism for tunable control of neuron cell growth and differential degree lies in the combined effect of SLB substrate, layer number, and surface material on mechanical property and chemical environment [20].

In addition to traditional polymeric substrates, 2D materials are also investigated for their potential for cell scaffolding. As a mostly well-known 2D material, graphene produced from graphite by either liquid-phase exfoliation (LPE) or ball milling (BM) was tested for brain cell culturing. Cell density on the graphene layer was 104 ± 11 cells/mm^2 compared to 91 ± 13 neurons/mm^2 on polyornithine-covered glass coverslips as a control group. Particularly, neither neuronal network synaptic performance nor heterogeneous spontaneous postsynaptic currents (PSCs) were disturbed by the graphene scaffold, suggesting a completely inert bio interface that possibly derives from the relatively low surface roughness (~14 or ~9 nm). Combined with its outstanding conductivity, graphene may represent an ideal scaffolding material for biosensing in medical applications [21].

Later, Rauti et al. investigated the impact of graphene oxide (GO) on cultured brain networks. Modified Hummers method produces GO flakes, which were dispersed in warm water and were separated into large GO (l-GO) and small GO (s-GO) depending on the magnitude of flake size. Neurites cultured in s-GO solution with maximum concentration up to 10 μg/ml displayed negligible difference from the control

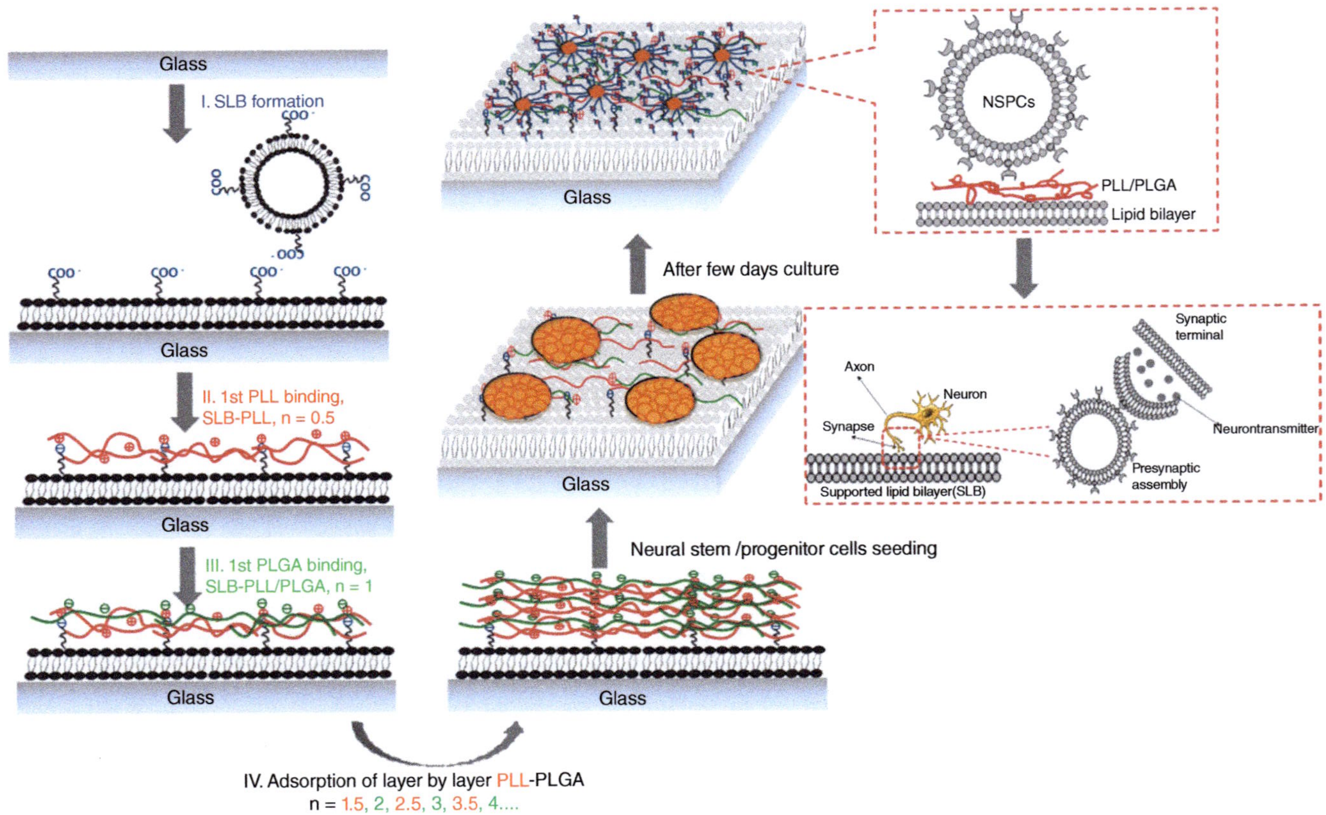

Figure 7.1 Schematic illustration (not to scale) of the layer-by-layer assembly of polypeptides adsorbed onto the SLB, the NSPC seeding, and the directed NSPC differentiation. Source: Adapted from Lee and Wu [20]. Copyright 2016, American Chemical Society.

group both in cell survival and neurite network size. By contrast, both neuron and glial cells did not survive in the l-GO medium after 6–8 days of incubation, implying cell toxicity of the sample, which suggests further examination. The regulation of s-GO on neuronal cell behavior is discussed in Section 7.3.5 [22].

More recently, $Ti_3C_2T_x$ MXene nanosheets were applied for the incubation of neural stem cells (NSCs). In comparison with graphene, MXene sheets are more hydrophobic and feasible to be surface functionalized. 2D $Ti_3C_2T_x$MXene, a biocompatible and noncytotoxic category of the MXene family, was proposed as a suitable candidate for scaffolds for neuroelectronic devices. The $Ti_3C_2T_x$ MXene was treated with laminin for better cell adhesion and subsequently seeded with NSCs. The proliferate percentage, length of the longest neurites, as well as number of branch points of cultured NSCs, all exceeded that on the tissue culture polystyrene (TCPS) control sample. For neurite differentiation, the efficiency and complexity of newborn neurons were remarkably improved [23].

7.2.2 Rolled-up Scaffolds

In the examples mentioned above, cells were cultured mainly on two-dimensional (2D) planar surfaces, while cells *in vivo* migrate and proliferate in complex three-dimensional (3D) environments of vessels and tissues. On the other hand, cell investments on 2D scaffolds mainly focus on average data of a mass of cells, which lacks meticulous study on an individual cell. To further investigate the cell behavior under different circumstances, tubular 3D microstructures were introduced as *in vitro* mimics of the biological environment in living organisms as well as scaffolds for single-cell culturing and analysis.

Rolled-up nanotechnology is a fabrication method that turns 2D nanomembranes into 3D helical structures, such as microtubes, rings, and springs [24, 25]. The principle of rolled-up nanotechnology is to deposit pre-strained nanomembrane on a sacrificial substrate. After removing the substrate layer by chemical or physical etching, membrane delaminates from the substrate and rolls up into a cylindrical structure under the initial strain gradient. This general technology enables manufacturing nanotubes from a vast variety of materials, such as biocompatible inorganic oxides, semiconductors, and biodegradable polymers. The intrinsic circular cross section resembles the geometry of *in vivo* environment, typically blood vessels. Moreover, compatibility with conventional 2D photolithography techniques gives rise to further integration with optical and electronic sensing devices for cell analysis. All the advantages mentioned above enable rolled-up microtube to be a suitable platform for cell culture scaffolds.

In 2009, Huang et al. first utilized rolled-up SiO/SiO_2 microtubes as 2D confined scaffolds to incubate individual yeast cells [26]. SiO/SiO_2 multilayer was deposited on the patterned photoresist layer by electron-beam evaporation. By removing the photoresist layer with acetone, the pre-strained multilayer was delaminated from the substrate and self-rolled up into microtubes. The tubes were subsequently coated with an atomic layer–deposited Al_2O_3 layer to improve mechanical stability and biocompatibility. The transparent nature of the microtube favored optical

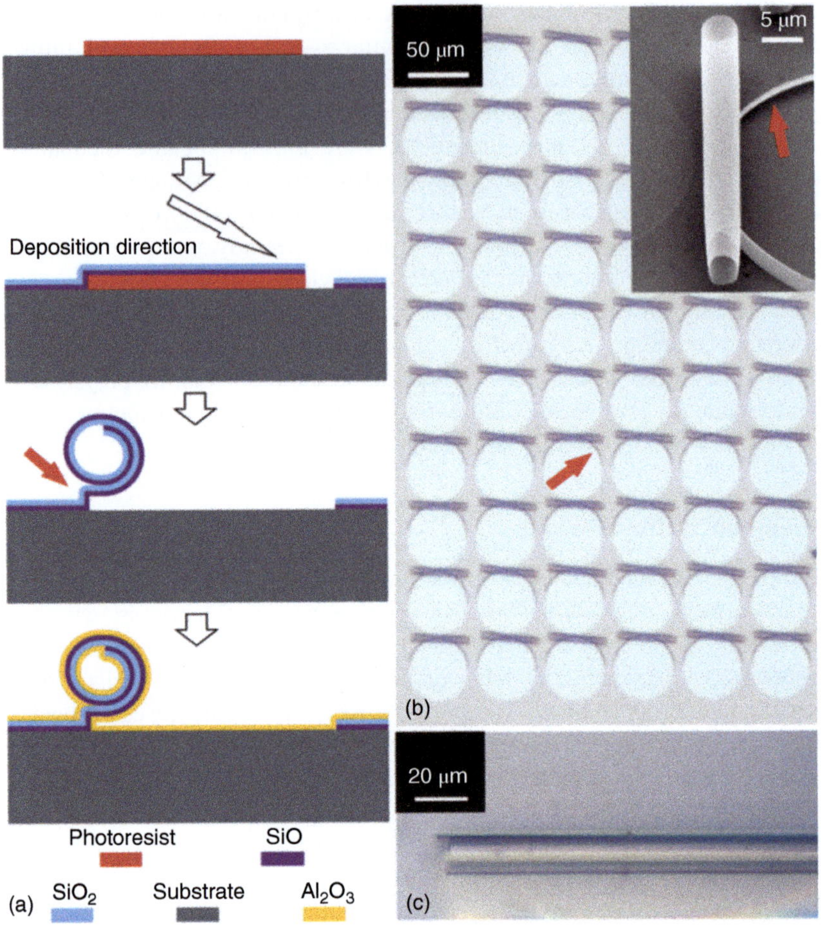

Figure 7.2 Rolled-up transparent SiO/SiO_2 microtubes. (a) Schematic illustration of the fabrication process of rolled-up SiO/SiO_2 microtubes. (b) Optical microscope image of an ordered array of rolled-up transparent microtubes with outstanding uniformity. The inset shows an SEM image of a compactly rolled microtube with vertical residual film remaining. (c) Optical microscope image of a transparent microtube after Al_2O_3 deposition. Source: (a)–(c) Huang et al. [26]. Reproduced with permission of Royal Society of Chemistry.

microscope observation of cell appearance and behavior within the scaffold, as shown in Figure 7.2. After 120 minutes of culturing, the normalized number of yeast cells inside microtubes was comparable to that on the flat surface. The confined yeast cells presented similar reproduction behavior and the growth rate showed no sign of saturation within the experiment. This indicated good biocompatibility and sufficient nutrient supply of the SiO/SiO_2 microtube scaffold.

To further improve the biocompatibility and performance of the microtube platform, both rolled-up nanomembranes and coating materials had been explored. For example, proteins such as fibronectin or poly-*d*-lysine (PDL), which promotes cell adhesion, had become a broad approach to modify tube inner surface [20, 27, 28].

In a recent publication, oxidized beta-type Ti-45Nb (TNO) membranes were bent up into microscaffolds to culture human mesenchymal stem cells (hMSCs), and their biological performance was compared to previous $SiO/SiO_2/Al_2O_3$ (SOA) microtubes [29]. Nb and its alloys are suitable for hard tissue implants due to their fine mechanical strength and low elastic modulus [30], and studies had demonstrated its ability to enhance cell adhesion and proliferation. TNO nanomembranes were deposited and rolled up into microtubes by etching of underlying Ge substrate. TNO nanomembranes possessed a lower contact angle (40°–70°) than the SOA membranes (75°–90°), hinting at a more hydrophilic property of TNO. Both reasons contributed to enhanced cell proliferation, adhesion, and migration on TNO microtubes. After 72 hours of proliferation, cell number on TNO scaffolds was 10% higher than that on SOA. Several contact points where the cell attaches to the inner surface of the tube were observed in TNO tubes, while few existed in SOA tubes. These results demonstrated the superior performance of TNO scaffolds.

Apart from inorganic oxide microtubes, semiconductor microtubes were fabricated particularly for guided neuron cell growth due to not only their physical and electrical properties that resemble natural glial cell membranes (myelin) in which axons are unsheathed but also their standard semiconductor fabrication technique that favors the integration of hybrid neural electronic/photonic system.

Arrays of semiconductor tubes were constructed by Si layer epitaxially grown on $Si_{0.8}Ge_{0.2}$ layer on top of silicon dioxide layer. Hydrofluoric (HF) etching of the sacrificial layer led to the self-assembling of the Si/SiGe bilayer under intrinsic strain induced by lattice misfit. The viability for cell culture of the scaffolds was validated by seeding cortical neurons onto the sample, as neural cells grew normally both on the planar surface and into the microtubes, even without PDL coating [31]. To accurately guide the growth of neurons into the tubular platforms, an inkjet printing technique was utilized to selectively deposit small islands of PDL adjacent to the opening of the microtubes. Seeding of a single neuron cell was obtained by scaling the diameter of the PDL pattern down to 15 μm. This technique facilitated individual axon growth into and through the microtube [31].

GaAs/InGaAs strained bilayer also served as a potential candidate for scaffolding of neurons. GaAs has several advantages that exceed Si. First, the tunable and direct bandgap of the III–V semiconductor makes it a proper material for optogenetic neural experiments. Also, higher electron velocity and mobility in GaAs indicate an improved signal-to-noise ratio when coupled with electronic devices. However, the inherent toxicity of arsenic hinders its ability for neuron cell culture. The problem was circumvented by PDMS and perylene coating of the self-rolled-up GaAs/InGaAs nanomembrane, which isolated the tube surface from neuron culture medium, as a result, preventing toxic arsenic compound from diffusing into the medium under corrosion. A live/dead staining test was conducted to analyze the viability of neuron cells grown on the treated microtubes. Survival of neurites compared to death of all cells on samples without surface treatment was observed, confirming successful suppression of the toxicity of arsenic compounds [32].

While semiconductor scaffolds strengthen electrical interaction with neuron cells, silicon nitride (SiN_x) suggested an alternative option that suits optical microscopy and stimulation of neural growth. SiN_x were deposited onto glass slides with plasma-enhanced chemical vapor deposition (PECVD) and rolled up after removing the Ge sacrificial layer with H_2O_2 solution. Primary cortical neurons were incubated on the samples to observe their guided growth and migration mechanism. For this purpose, real-time monitoring of cell morphology change was conducted, which required a fully transparent platform. SiN_x microtube scaffold on a transparent glass slides substrate served as an ideal platform for live-cell imaging experiments [33].

7.3 Cell Regulations in Nanomembranes

Cells are extremely sensitive to mechanical, chemical, topographical, and electronic cues in their extracellular environment. For a fundamental understanding of the mechanism of how cells transform the physical and chemical cues in microenvironments into a biochemical response, scaffolds, especially the ones that mimic the character of real physiological tissues, were utilized to regulate and monitor complicated cellular behaviors [34].

7.3.1 Cell Morphology and Arrangement

Apart from scaffolding materials that are mainly discussed in the previous section, the surface topography of membranes also plays an important role in cell adhesion and biocompatibility. Microstructures, such as grooves, ridges, pillars, and pores, had been adapted to tune interactions between membrane surface and cell. For instance, plasma-immersion ion implantation and deposition prepared hydrogenated amorphous carbon films were patterned by argon plasma etching with arrays of square wells of dimensions of 25 μm × 25 μm, 50 μm × 50 μm, and 100 μm × 100 μm. Contact angle tests were conducted to determine surface energy and results showed that micro-patterned samples exhibited larger contact angles, indicating a more hydrophobic surface. Consequently, human umbilical vein endothelial cells (ECs) cultured on the micro-patterned surfaces were observed to have better cell adhesion and proliferation than as-deposited samples. A tendency of the cells to adhere onto the ridges of etched wells was also observed [9]. However, these studies were still conducted on a merely 2D surface that deviated from the complex geometry environment in human tissues and organs, and the results may not be representative of common cellular behavior *in vivo* situations.

Research on cell behavior in 2D confinement was firstly conducted by incubating yeast cells in SiO/SiO_2 microtubes. In scaffolds with a diameter of ~9 μm, cells showed normal reproduction behavior of budding; however, instead of their dendritic shape on a planar surface, they grew into a "zigzag" pattern. To further investigate this phenomenon, yeast cells were seeded into rolled-up microtubes with varying diameters of 14–3.5 μm, and the results are shown in Figure 7.3a. In the largest tube of ~14 μm, cells were arranged in two rows. With the decrease of tube diameter, the arrangement changed into a "zigzag" pattern, which may

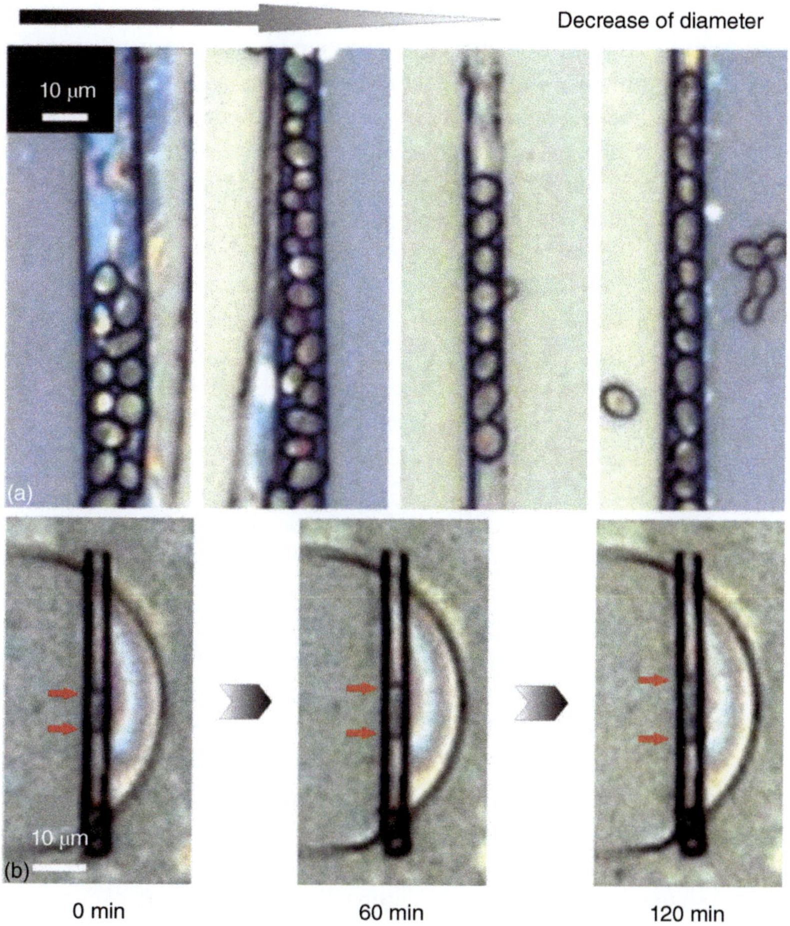

Figure 7.3 Reproduction and growth behaviors of yeast cells within SiO/SiO_2 microtubes. (a) Optical images of microtubes with different tube diameters. The yeast cells cultured in the four microtubes show different arrangements, respectively. (b) Optical image of a single yeast cell cultured in a ultra-narrow tube. The length of the yeast cell is marked by the red arrows. Source: (a) and (b) Huang et al. [26]. Reproduced with permission of Royal Society of Chemistry.

be explained by the self-rotation of the mother cell to sufficiently utilize the limited space for its budding. Further decline in tube diameter led to strict spatial confinement that limited cell rotation, resulting in the cell orientation parallel to tube axis and a straight, chain-like arrangement. As shown in Figure 7.3b, in a ~3.5 μm wide ultra-narrow microtube, cells successfully reside and grew up with a volume increase from 75 to 125 μm^3 as well as height to width aspect ratio from 2.2 to 3.7, respectively, which was smaller and more elongated than typical yeast cells, indicating tight restriction from the microtube and a suppressed bioactivity [26].

A more specific relationship between confinement and deformation of single cell and nucleus morphology was proposed by Koch et al. The width of freely growing single human osteosarcoma U2OS cells was determined to be 16 ± 2 μm; therefore, tube diameter from 25 to 4 μm was chosen for cell confinement. For tubes wider than 8 μm, only a slight impact on the aspect ratios and the nucleus width of cells was observed, and the nucleus height and length remained constant. Below 8 μm, cells exhibited a distinct increase in aspect ratio in accordance with the decrease of tube diameter as well as a substantial decrease in nucleus width and height. While the morphology change derives from the spatial confinement, the same reason may not be applied for the deformation of the nucleus as their dimensions were much smaller than microtubes, therefore indicating a direct influence from the 2D confinement of the scaffolds. Even in the narrowest microtubes below 8 μm, deoxyribonucleic acid (DNA) integrity showed negligible harm confirmed by a fluorescently labeled marker of the DNA damage response pathway in GFP-53BP1 U2OS cells. However, quantification of cell survival displayed reduced viability and the effect of mitosis was speculated [35]. A deeper look into this affliction will be discussed in Section 7.3.2.

Alternatively, a comparison of neuron cell shape under 1D and 2D confinement was made. On a planar area of the sample, prevailed spread and fibroblastic cell morphology were observed, while in the fibronectin-functionalized glass microtubes neuron cells mainly displayed a small and rounded appearance. The morphology difference was convertible as the cells migrated into and out of the channels. The distribution of actin filaments, which act as cytoskeleton for cell supporting, was investigated, and the result revealed a thickened layer of actin filament on the interface between spread cell and planar surface compared to no pronounced actin stress fibers in microtubes, which indicates tighter anchoring of the neuritis on the planar surface. In addition, trenches were fabricated with similar mechanical properties to the microtubes. Cells cultured on these trenches did not show a spherical shape, which confirmed the tube dimension, instead of curvature of the tube wall, contributed to the morphology transformation. The phenomenon that the cell volume shrank and aspect ratio increased when decreasing tube diameter mentioned before was also observed. As the diameter was reduced to about half the size of unconfined NSCs, the circularity and aspect ratio resembled that in the brain, which in turn hinted same spatial confinement *in vivo* [36].

Apart from the morphology of a single cell, feasibility to manipulate the alignment and orientation of a group of cells was also testified. Fujie et al. manufactured a scaffold for C2C12 skeletal myoblasts by patterning parallel CNT-incorporated fibronectin (Fn) ridges featuring 50 μm width and 50 μm separation as adhesive layer on ultrathin PS nanomembrane. After 24 hours of incubation, an anisotropic alignment of C2C12 myoblasts on the CNT-Fn micropatterned surfaces was observed, compared to the random orientation of the ones on non-patterned substrate. Elongation of the myoblasts was supported by a decrease in the nucleus circularity. Moreover, after eight days in differentiation medium, myotubes that were products of cell differentiation also showed a significant level of parallel alignment than the counterpart. Wrapping the membrane around a template resulted in a tubular structure with a highly organized synthetic myofiber layer in the inner surface that closely resembles the specific structure of blood vessels, hinting at its future application for biomimetic devices [7].

7.3.2 Proliferation and Mitosis

Xi et al. put emphasis on the fundamental mitosis process, which is a dictating factor in cell proliferation. Two different types of human cell lines transformed HeLa cells and RPE1 cells were chosen to be confined in rolled-up SiO/SiO_2 microtubes and studied their mitosis behavior by live-cell imaging. For HeLa cells confined in tubes with diameters of 13 μm, the chromosome arrangement after nuclear envelope breakdown (NBE) was perturbed and no congresses chromosomes near the poles were detected. Under extreme confinement of <10 μm, chromosomes failed to align into a metaphase plate throughout prometa-/metaphase. Detailed images of the process are shown in Figure 7.4. Consequently, in anaphase where the mother cell segregates their chromosomes into two daughter cells, lagging chromosomes were frequently observed and in the end formed micronucleus. A significant delay in prometa-/metaphase ranging from ~70 to ~160 minutes within microtubes ranging from 18 to 7 μm in width was measured and was deducted to be contributed to the arrangement of the distorted chromosomes, which prevented the cell from entering the spindle assembly checkpoint (SAC). Similar results were revealed for RPE1 cells, demonstrating that genomic instability in the mitosis process under severe physical restriction is a general nature for various cells [37].

A more detailed insight into mitosis of molecule level was proposed in 2016, which carefully inspected representative parameters during the separation of cells, including spindle morphology, chromosome arrangements, centrosome separation, spindle formation time, and other microstructures. In early prometaphase, normal HeLa cells possessed a typical ring-shaped chromosome formation, which is named as chromosome rosettes in the center of the cell, perpendicular to spindle axes. In contrast, cells encapsulated in microtubes of <14 μm diameters displayed a tilted plate with densely packaged chromosomes, while in the smallest tubes a cloud of chromosomes without a clear outline was observed. For spindle formation and centrosome separation, a distinct delay of separation time was testified and was explained by both the elongation of the spindle according to the elongation of the cell body and a decreased formation speed of spindle. Two microcomponents in mitosis, actin cortex and blebs, presented a remarkable enrichment especially at the cell tips, which merely appeared in free mitotic process. By adding corresponding inhibitors into the scaffolds, such structures were suppressed while the cells showed increased chromosome segregation errors (CSEs), indicating the function of actin cortex and blebbing for accurate chromosome segregation as a compensatory mechanism of the deterioration from physical confinement [38].

7.3.3 Migration and Differentiation

The phenomenon that cells seeded on rolled-up microtubes arrays voluntarily entered into the tube channels had been widely demonstrated and utilized for

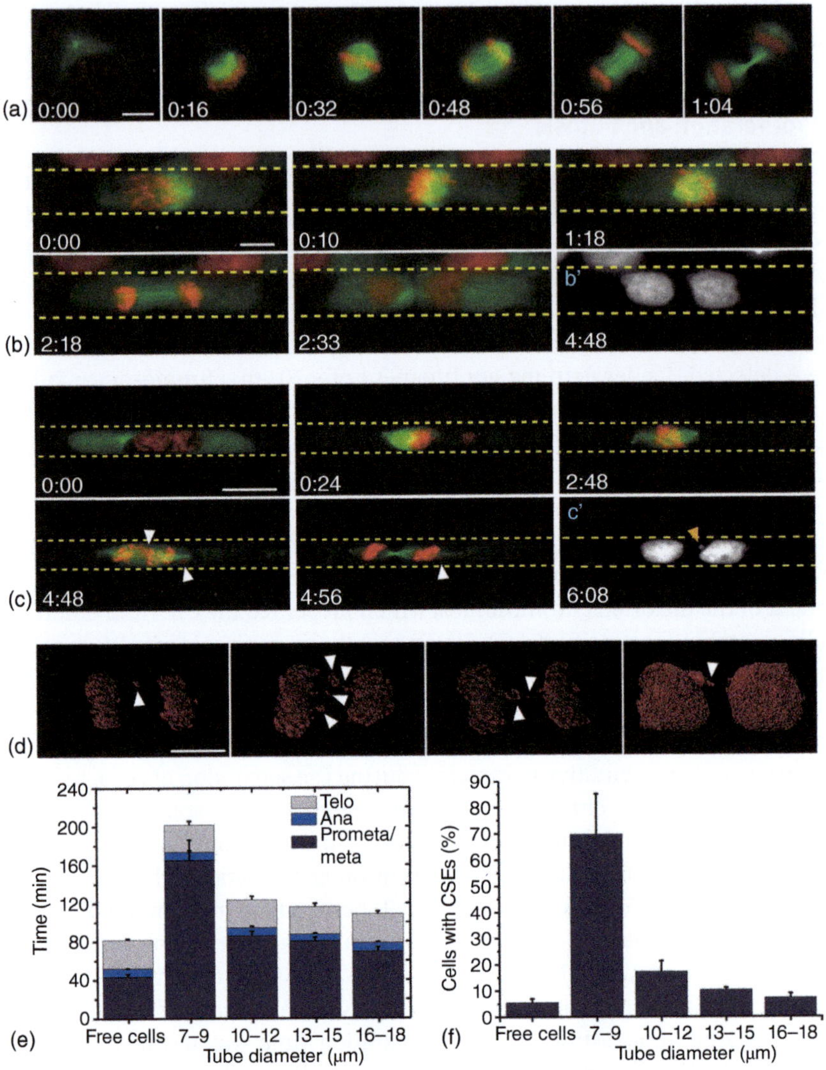

Figure 7.4 Mitotic delays during prometa-/metaphase and CSEs under spatial confinement. (a) Representative time-lapse z-stack projections of a free dividing HeLa cell coexpressing green fluorescent protein (GFP)-tubulin (green) and H2B-mCherry (red), and (b), (c) for cells dividing in microcavities with diameters of 13 μm (b) and 9 μm (c), respectively. The white arrows in (c) marked the appearance of CSEs that further developed into micronuclei marked by the orange arrow in c′. (d) The 3D reconstructions of CSEs in anaphase HeLa cells are highlighted by arrowheads. (e) The mean durations of three mitotic phases in HeLa cells. (f) Statistics of the mean percentages (±SD) of cells with CSEs during anaphase. Scale bars, 10 μm, time in hours/minutes format. Source: (a)–(f) Xi et al. [37]. Reproduced with permission of American Chemical Society.

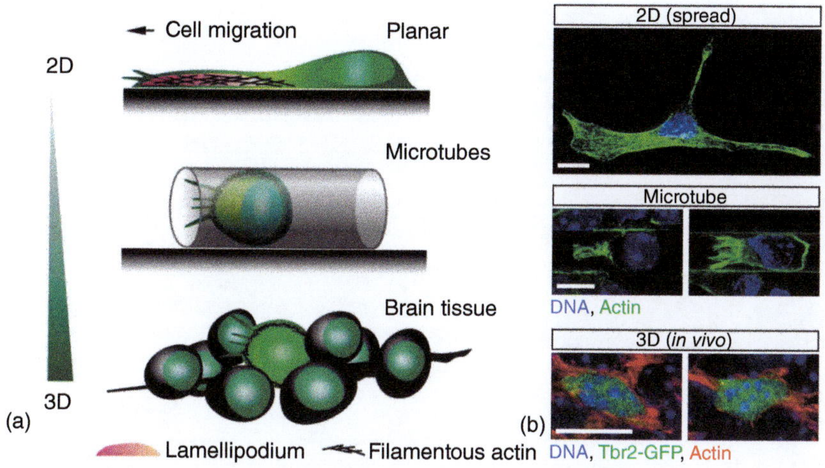

Figure 7.5 Lamellipodium and filamentous action structures of NSCs under different situations. (a) Schematic illustrations of the different appearances of NSCs under different conditions. (b) Slices of *z*-stacks for the 3D reconstruction of the immunofluorescently labeled cells under different conditions. Scale bars, 10 μm. Source: (a)–(b) Koch et al. [36]. Reproduced with permission of American Chemical Society.

cell encapsulation. In addition, how do the cells respond to cues of the external environments and adjust their migration strategy to march through complicated 3D environment was investigated in detail. There were two major components observed for cell movement, a sheet of branched actin filaments named lamellipodium and bundled actin filaments at the leading edge termed filopodia (Figure 7.5). For NSCs inside microchannels of SiO/SiO_2, finger-like filamentous actin at the cell front was quantified as the predominant structure and lamellipodium structure was scarcely detected, which contradicted the common situation of cells on the planar surface. Cell motility was quantified and the decline of migration velocity as well as a rise in arrest coefficients were detected, which had a minor correlation to tube diameter, suggesting a direct relationship with confinement dimensions. The cell structure and motility suggested a migration mode shift from mesenchymal to amoeboid migration. The critical velocity was analyzed to be 0.27 μm/min, approaching that of 0.10 ± 0.01 μm/min to 0.33 ± 0.10 μm/min of NSCs in brain tissues, confirming the microtube as a well-defined cell environment that closely mimics *in vivo* situations [36].

In the case of TNO scaffolds, hMSC) were cultured on the platform for close observation of osteogenic process. Both of the migration modes of the cell that were mentioned above were confirmed similar to previous studies. In the early stage of osteo-differentiation, the tissue-nonspecific alkaline phosphatase (TNAP) activity was measured and results showed a higher value in TNO microtubes than TNO membranes, indicating a more active differentiation potential. In addition, aligned nodule clusters were inside and along with the tubular microscaffolds (TS), which could be identified as mineralized structures. To analyze the mineralization process inside the TS, cells were tracked for five weeks with time-lapse optical microscopy.

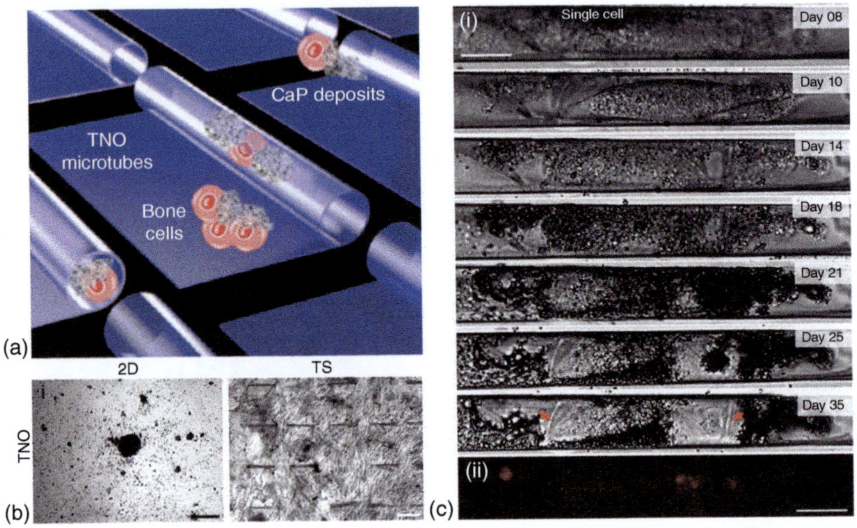

Figure 7.6 Mineralization process of bone cells onto TNO-2D and TS samples. (a) Illustration of an array of TS as scaffolds for bone cell observation. (b) Nodule generation onto TNO-2D and TS samples. (c) Observation of single cells in TNO-TS from day 8 to day 35 in DM (i). Live/dead staining of the former cell identified as dead cell (ii). Source: (a)–(c) Herzer et al. [29]. Reproduced with permission of John Wiley & Sons, Inc.

The cell body darkened until the twenty-first culture day, suggesting the death of the cells; during the following 14 days, these dead cells were substituted by a crystal-like material that was further confirmed to be a partial cell body mineralization via scanning electron microscope (SEM) and transmission electron microscope (TEM), as shown in Figure 7.6. This study overturned previous views that only big-pore diameters (larger than 50 μm) can lead to mineralized tissue, and it supports a calcification theory that dead cell bodies are able to accumulate Ca^{2+} and P_i ions and serve as nucleation sites for mineralization [29].

7.3.4 Guidance of Neuron Cell Outgrowth

Neuron cells, as a highly specialized adherent cell type, possess two types of long, rope-like unique cell protrusions, which were defined as axons and dendrites. These structures serve to establish connections to neighboring cells and to transport electrical signals. Therefore, interest had been aroused in interactions between tubular scaffolds and the neuron outgrowth.

The guiding of neurons by microtubes was first conducted within SiO/SiO_2 tubes. Motoneuron, a kind of primary cell that differentiates into long neurites after seeding, is seeded on the scaffolds, and successfully survives and extends its axons and dendrites into microtubules. The two structures were distinguished by antibody staining, and the fluorescence microscopy revealed that axons alignments were more consistent with the scaffold array topography for their longer and extremely extensive geometry. Two main guiding cues for axon outgrowth were summarized, the first was the residual of the nanomembrane on the surface, while the second was microtubes themselves. For the neurites that had entered the tube channels, they were directed by the tube walls and extend themselves to the opposite opening of the channel, and the tube array even led to the formation of complex square-shaped grid-like neurite networks, confirming the directional ability on microtubes on neuron outgrowth [39].

Later, semiconductor tubes were utilized for neurite outgrowth regulation for their potential for receiving and stimulating electronic signals with neuron cells. Yu et al. developed the SiGe platform for both neurite guidance and integration of electrical and optical functional devices. As is mentioned in the previous chapter, Si/SiGe bilayer nanomembranes were topographically patterned and released into an array of rectangular islands with buckle-delaminated 3D structures serving as interconnecting channels, as shown in Figure 7.7. The channels were fabricated toward an amplitude between 10 and 1.5 μm, which resemble the width of axons and were selectively coated with PDL for controlled cell adhesion. The channels successfully guided neurons, which interacted with each other either within or at the opening of the channels forming various neuron networks. In addition, a prominent tendency for the axons to proceed along the inner side of the channel was observed, which was contributed to the uniquely bell-shaped cross-sectional area of nanomembrane channels different from common microtubes. The electronic and optical capabilities of the membranes and channels were demonstrated, implying future applications in biosensing and optogenetics [28].

Koitmäe et al. directed a similar experiment in self-rolled-up GaAs/InGaAs microtubes. In this case, cerebellar granule neurons were specially chosen for their so-called bipolar shape with only two neurites per neuron. Coupling with a square array of microtubes and selectively coated PLL at the junctions of the channels, a clear network of guided neurons was cultured of which cell bodies

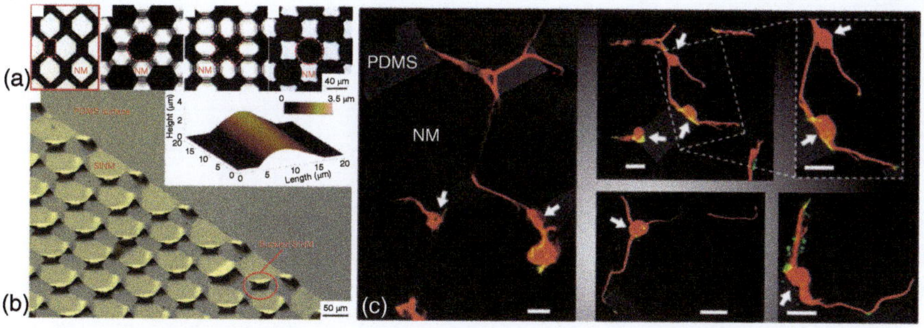

Figure 7.7 Guided neurite outgrowth within buckled channels into neuron networks. (a) Optical images of 2D arrays of various shaped islands interconnected by buckled channels with different widths from left to right, respectively. (b) Tilted SEM image showing buckled channels constructed of 72 nm Si nanomembrane, forming a near-checkerboard pattern. The inset shows an AFM image of a single SiNM buckled channel. (c) Confocal fluorescence microscope images showing buckle-delaminated channels guided neuron outgrowth, with the neuronal cell body (marked by arrows) blocked outside the channel. Scale bar in, 20 μm. Source: (a)–(c) Cavallo et al. [28]. Reproduced with permission of American Chemical Society.

adhered to the junction areas and the neurites reached out toward each other deep into the tube channels. Specifically, an SEM image of the cross section of an axon within the channel was obtained for the first time via focused ion beam, which distinctly depicted the axon adhered tightly on the side of the polymer coated on the inner surface of the microtube. By labeling the neurites with a green dye, a method of intensity line plots was introduced to distinguish an axon in the tube from a single axon and an empty tube, and the results addressed that the axons could extend for a distance of 35 μm deep into the microchannels. Finite element simulations suggested that by replacing polymer coating with biocompatible TiO_2, tube performance could be further improved for better optical probing of action potentials (APs) to understand the interactions between neurons [40].

For real-time monitoring of the neuron outgrowth process, SiN_x microtubes were utilized to incubate neuron cells. The diameter of the tubes was set to be 4.4 μm, which was slightly wider than that of a typical axon (2–3 μm). Robust guidance was confirmed as the axon bundles grew along the axial direction through the channels. Remarkably, the guidance of microtubes on axon outgrowth on the plane surface between tubes was observed for the first time, as shown in Figure 7.8. The neurites maintained the same moving orientation along the tube axis after growing out of the microtubes onto the open planar area for all the spacing from 20 to 40 μm, and the percentage of axons that proceeded along the linear direction was quantified to be around 83% for 40 μm wide tubes. A live-cell recording further confirmed the

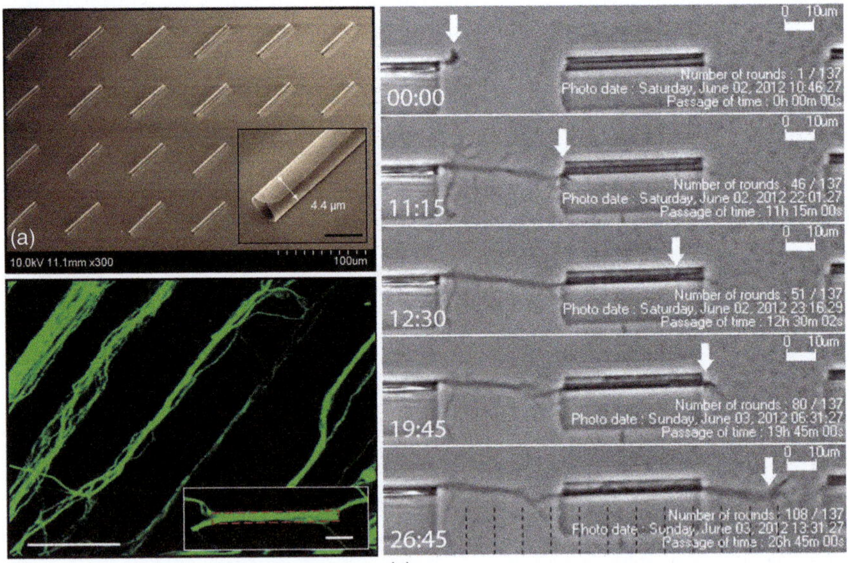

(b) (c)

Figure 7.8 Axon directional outgrowth through SiN_x channels. (a) SEM images of SiN_x microtube array. (b) Fluorescence images of corresponding immunocytochemistry of neural culture. (c) Time-lapse phase-contrast images of an axonal pathfinding process of a living cortical neuron within the tube channel. Source: (a)–(c) Froeter et al. [33]. Reproduced with permission of American Chemical Society.

outgrowth process, and interestingly, a deflection of growth path was observed as the axon turned its direction toward the opening of microtubes. This pathfinding ability was contributed to the substrate-bound nanomembranes that form a ~100 nm tall ridge, hinting at topographic cues that guided the growth cones of axons toward tubular 3D structures.

7.3.5 Mediation of Neurite Signaling

Apart from morphology regulation, nanomembranes had been explored for synaptic signaling tuning. In one instance, GO nanosheets had demonstrated their ability to downregulate neuronal signaling. Hippocampal neurons were cultured in aqueous media where s-GO were dispersed with a concentration of 10 μg/ml, and the occurrence of spontaneous PSCs was monitored to deduce synapse formation and activity after incubation. A significant decrease (0.6 ± 0.1 Hz) in PSC frequency was quantified compared to that in the control group (2.0 ± 0.1 Hz), and the probability of finding active neurons within networks was reduced while the viability, cell density, and cellular composition of survived neurons did not change, which was determined by immunofluorescence markers. All these observations indicated a direct impact of GO nanomembranes on synapses activities [22].

On the other hand, single-layer graphene (SLG) significantly improved neuron excitation levels. Hippocampal neurons were plated on SLG, multi-layer graphene (MLG), and Au-coated coverslips. PSC frequency was significantly higher for the cells on SLG platform, while those on the MLG and Au displayed no difference with the control group. To understand the mechanism of this phenomenon, a series of tests were conducted and the results excluded impacts of PSCs, morphology, and composition of neuron network and synaptic rearrangements. Instead, it was demonstrated that the restriction of mobility of K^+ ions was similar to the material surface that contributed to the mediated neuronal excitability, and the restriction was presumed to derive from the unique cation–π interactions of SLG [41].

7.4 Enhanced Therapy and Cell Analyses

7.4.1 Capturing of Cell

Most studies of cell behavior adopt a strategy that passively encapsulates cells into the scaffolds under random distribution or spontaneous migration. Nevertheless, microsystems that actively capture organisms into openings and cavities of the scaffolds are desirable for improved efficiency of cell characterization.

A microsyringe pump was introduced to "suck-up" embryonic fibroblast mouse cells suspended in the culture medium into SiO/SiO_2 microtubes. The syringe was composed of a tapered capillary with a tip diameter on the scale of tens of micrometers fixed onto an XYZ micromanipulator. The capillary was connected to a microfluidic pump with a pump rate of 0.01–0.1 μl/s. When the syringe was placed near one opening of the microtube, a cell-rich culture medium was pumped through the microtubes until a cell approached the opposite open end. The operating pump rate has subsequently risen to 0.5–1 μl/s until the oversized cell was eventually sucked up into the scaffold. The advantage of this approach is its ability to forcefully drag cells with diameters larger than the tubes (maximum $D_{cell} = 2.3\,D_{tube}$) into the scaffold [42]. However, more investigation needs to be conducted on possible damage exerted on captured cells. The rolled-up microtubes may also be coupled with a PDMS-based microfluidics platform to capture cells in a continuous flow [43].

In contrast to implanting the cells into scaffolds after they were assembled, seeding cells onto nanomembranes prior to their delamination is also a feasible method for organism encapsulation. For instance, fully biodegradable and biocompatible polysuccinimide/polycaprolactone (PSI/PCL) microtubes were produced to wrap and culture yeast cells. The polymer bilayers were spin coated on the silicon wafer substrate and were patterned by ultraviolet (UV) light illumination with a shadow mask. The bakery yeast cells were seeded onto the polymer film before the sample was incubated in PBS (0.15 M, pH = 7.4). Being immersed in the solution for nine hours, the PSI layer began to swell, triggering delamination and rolling of PSI/PCL membranes. Further incubation of the rolled-up tubes for 27 hours resulted in shrinkage of the tubes and the final tube diameter was about 80–100 μm. Number of the yeast cells wrapped in the tubes remained approximately the same during the whole PBS incubation process and increased after the scaffolds were transferred into a nutrition media, indicating successful proliferation of the yeast cells [44].

Hydrogel-based stimuli-responsive polymer microtubes were also employed to facilitate controlled capturing and releasing of neurons. A highly cross-linked hydrogel-based swelling layer and a stiff polyimide film layer were stacked on a water-soluble metal-organic sacrificial layer. When immersed into a neutral aqueous solution of sodium diethylenetriaminepentaacetic acid (DTPA), sacrificial layer was selective dissolved and the hydrogel component of remaining freestanding polymeric membrane swelled from beneath causing upward scrolling of the membrane into a Swiss roll–like structure of several windings. Tube diameter could be tuned from 50 to 500 μm by altering thickness of polyimide and hydrogel layers, which supports wrapping of nerves and nervous fasciculus of different sizes. By adding isopropanol or water into the aqueous solution, the shrinking or swelling behavior of the hydrogel layer could be manipulated and the geometry of the nanomembrane switches between planar state and Swiss roll state (Figure 7.9). Its ability to open/close under external excitation qualified intelligent cuffing and releasing of a targeted axon [45].

7.4.2 Comprehensive Cell Analysis Platform

By coupling the aforementioned nanomembranes scaffolds and sensing devices, platforms that integrated the functions of cell incubation, regulation, biosensing, and therapy enhancement had been introduced for *in vivo*, real-time monitoring for cell culturing.

In one approach, the SiO/SiO_2 microtubes were utilized as optical microcavity (μ-cavity) resonators for sensing embryonic fibroblast mouse cells by shifts in

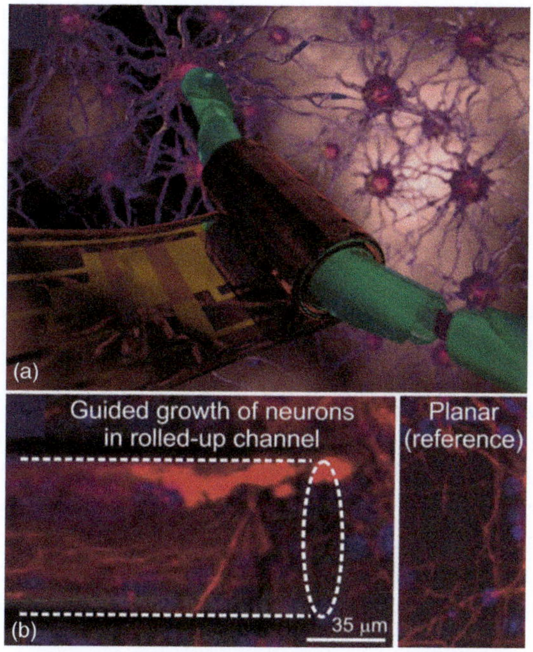

Figure 7.9 Biomimetic microelectronics for cuff implants: (a) Conceptual illustration of a rolled-up biomimetic microelectronic device for neuronal cuff-type implants. (b) A polymeric microtube guides the growing direction of differentiated neurite cells. Source: (a) and (b) Zakharchenko et al. [44]. Reproduced with permission of John Wiley & Sons, Inc.

whispering gallery modes (WGMs). The microtubes were determined to have a split-wall structure, of which a nanogap existed between different wall layers. As the cells were sucked into the tubes, the cells pressed against the inner tube wall, causing the deformation of the nanogap structure, which triggers the peak shift of the photoluminescence (PL) spectra of a μ-cavity resonator. As a result, the SiO/SiO_2 μ-cavity resonator may provide a platform for label-free *in vitro* sensing and its reproducibility was also demonstrated [46].

On the other hand, nanomembranes can be coupled with electrodes for electronic sensing of cells. For instance, Cr/Au patterned layers were deposited onto strained TiO_2 bilayer nanomembranes as electrodes for impedance analysis, and subsequently, the layers were released from the substrate and rolled up into microtubes. The fabricated sensor was placed onto a microfluidic platform and phosphate-buffered saline (PBS) containing HeLa cells flew through the cell. It was observed that a peak in tube resistance appeared when cells passed through the channel, and the peak resistance indicated a linear response with the cell concentration, with an excellent detection limit of ~5 cells/ml. In addition, a distinct resistance change of a single cell was visible in a nearly 1 million cells/ml solution [43].

Integrated with high-performance microelectronics consisting of signal amplifiers and indium gallium zinc oxide (IGZO) transistors, hydrogel-based stimuli-responsive polymer microtubes were utilized to realize signaling of ionic liquid. The flexible transistors were rolled up with the polymer microtubes and cuffed the test sample, realizing the frontend detection and amplification, which improves the sensitivity of the entire system. A glass capillary was encapsulated into the microtube and a drop of 10 pl ionic liquid with a length of 30 μm flowed through the channel, which results in the appearance of the peaks in the I–V characteristic of the transistor. The mechanism lay in the electric field in the IGZO semiconductor, which was altered by the ionic droplet. With the presence of an ionic droplet, the drain–source current raises to about 1 nA, leading to a ≈20 mV variation in gate voltage, which supported the capability of the scaffold for detecting the ionic axons captured by the scaffolds [45].

Kim et al. had successfully manufactured a multifunctional soft instrumented platform combining graphene for regulated cell culturing, Au nanofilm for electronic monitoring, and PDMS film as a flexible substrate. The modulus of PDMS was comparable to native skeletal muscle tissues, and anisotropically micropatterned graphene nanoribbons were confirmed to improve cell adhesion [47] and differentiation [48, 49], and the widths and spacing of the graphene patterns of <~5 μm were chosen for better regulation of cell alignment. C2C12 myoblasts were seeded on the scaffold for three days and exhibited a cell arrangement parallel to the aligned graphene ribbons, with the orientation ranging from −10° to 10° compared to the ribbon alignment direction, which closely resembled the morphology of the cells aligned in muscles. Measurement of impedance and temperature change of the sample during culturing process were achieved by the Au electrodes under the graphene layer, which revealed that proliferation process led to an increase in impedance, while differentiation led to a decrease of impedance, as a result, facilitating real-time monitoring of the stage of cell development. Furthermore, the device offered a platform for testing the impacts of novel nanomaterials on cells by monitoring cells viability. A summary of the integrated platform is shown in Figure 7.10 [50].

7.4.3 Nanomembranes for Tissue Mimicking and Enhanced Therapy

Apart from its biosensing ability, the nanomembrane cell-culture platform may also be employed in *in vivo* therapies. In the case of forementioned graphene scaffold, the cells sheets grown on the platform could be transfer printed onto receiving substrate (protein-coated glass) without affecting the structure and differentiation characteristics, and the transferred yields were obtained to be around 90%. The cells sheets were consequently transplanted onto scarred muscles in mice, and fluorescence microscope images of the labeled samples addressed regeneration of muscle tissues with transfer-printing treatment compared to no recovery of the control group, demonstrating the therapeutic ability of the *in vitro* cultured cell sheets on the platform [50].

Freestanding chitosan/alginate nanomembranes were also fabricated for stem cell delivery and wound healing. Multilayer nanomembranes were constructed using the layer-by-layer fabrication process, and the chitosan/alginate layers were coupled by electrostatic interactions. Results from *in vitro* tests demonstrated that the nanomembranes were biocompatible and promoted cell proliferation. To evaluate the wound-healing ability of the nanomembranes, a 5 × 5 mm full-thickness dorsal cutaneous defected wound was created on mice and treated with mesenchymal stem cell (MSC) injection and chitosan/alginate nanomembrane transplanting. Fluorescent signals indicated an accelerated migration of the cells from the fringe to the center of the created wounds under nanomembrane treatment, and the injected cell viability was also quantified to be improved on transplanted membrane surface.

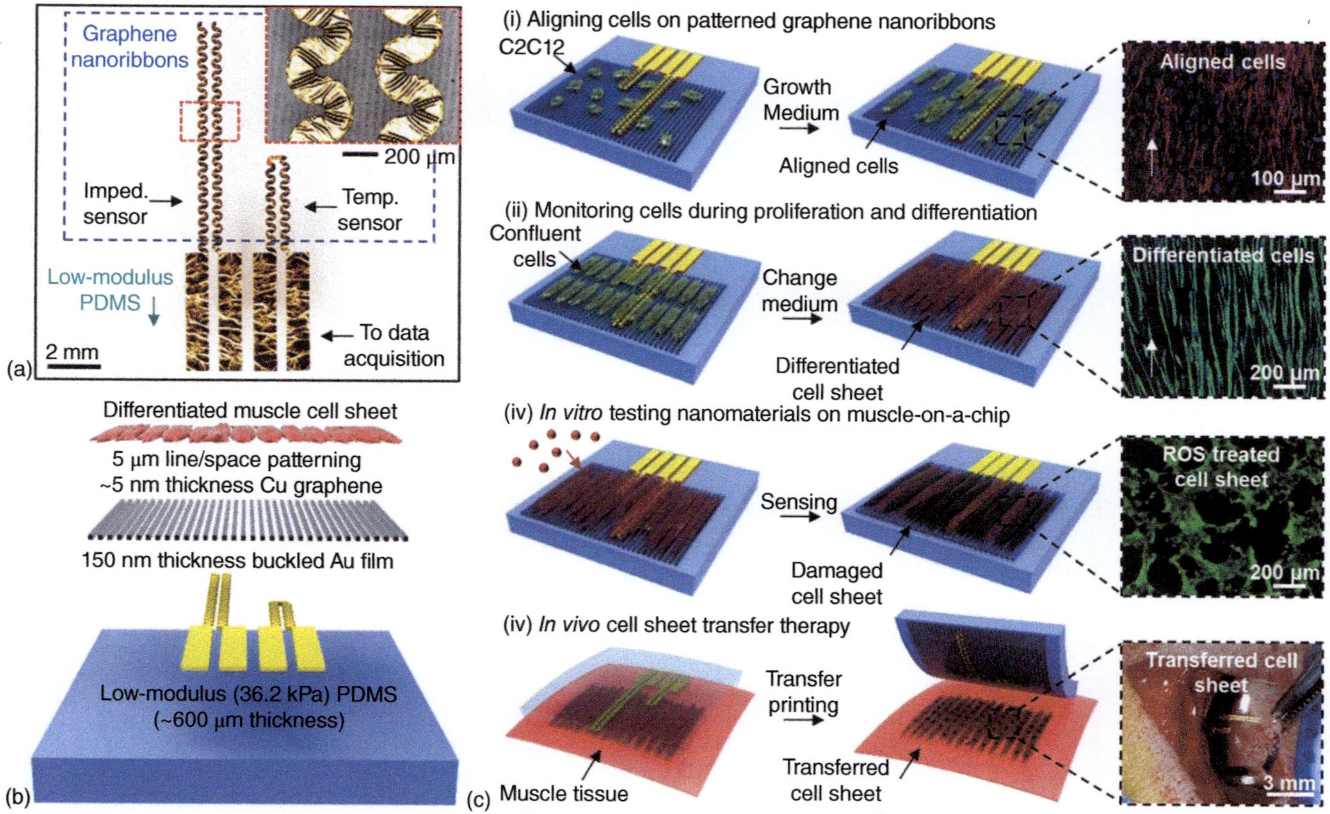

Figure 7.10 Conceptional image of the structure and key applications of the graphene-based multifunctional platform for cell culture. (a) The optical microscope image of the structure of cell-culture platform consisting of graphene nanoribbons, Au nanomembrane impedance/temperature sensors, and a low-modulus PDMS substrate. The top right inset shows a magnified optical microscope image of the buckled Au electrodes. (b) Schematic illustration of the components of the multifunctional platform. (c) Four main applications of the soft multifunctional cell-culture platform: (i) cell arrangement guiding, (ii) in situ monitoring of proliferation and differentiation, (iii) *in vitro* testing the influences of novel materials and drugs on cells, and (iv) *in vivo* cell sheet transfer printing for therapy. Source: (a)–(c) Kim et al. [50]. Reproduced with permission of American Chemical Society.

As a result, the nanomembrane was testified as an outstanding platform for cell delivery for skin tissue regeneration [51].

Regarding the intrinsic tubular nature of scaffolds fabricated by rolled-up nanotechnology, Cheng et al. demonstrated a blood vessel--mimicking microtube that was successfully transplanted into rabbits. Multilayer membranes were fabricated by combining two different kinds of polymers, poly(ε-caprolactone) (PCL) and poly(*dl*-lactide-co-glycolide) (PLGA), which mediated the mechanical performances and degradation lifecycle of the two materials. Microfluidic technique was employed to pattern three kinds of typical components in native blood vessels, ECs, smooth muscle cells (SMCs), and fibroblasts onto the membrane. By controlling the pattern width and the tube diameter, the prestressed PCL–PLGA sheet was released and rolled up into multiwall tubular structures with three layers of cell tissues between the polymer walls (Figure 7.11). To verify the self-adjust ability of

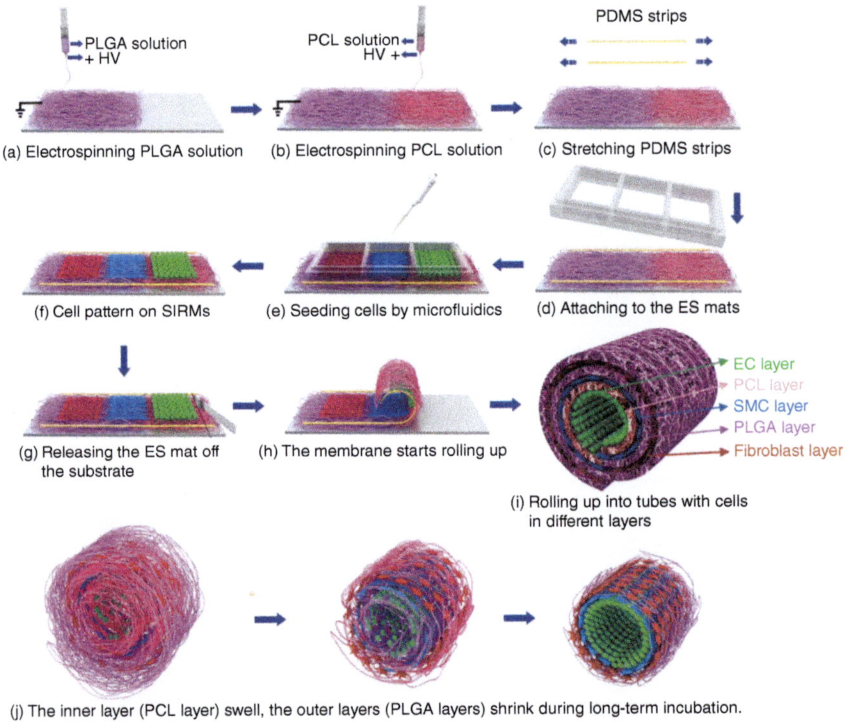

Figure 7.11 Conceptual images of the manufacturing procedure of the cell-laden multilayered PCL–PLGA tubes. (a) and (b) Electrospinning the PCL–PLGA solution onto the substrate. (c) and (d) Preparation of the PCL–PLGA SIRM. (e) and (f) Seeding three kinds of cells on planar SIRM patterned by a PDMS chip. (g) and (h) Separating the PCL–PLGA SIRM from the substrate to induce membrane rolling under intrinsic stress gradient into a tube. (i) A tube with HUVECs, SMCs, fibroblasts separated by different PCL–PLGA layers. (j) The degradation of the PCL-PLGA layers under long-term incubation while the cells merged into the polymer layers. Source: (a)–(j) Adapted from Cheng et al. [52]. Copyright 2017, John Wiley and Sons.

the scaffold, cell-laden–PCL–PLGA tubes were incubated in a culture medium for 21 days. The PCL–PLGA maintained its original shape as the polymer layer and cell layer merged into a single layer, and the tube diameter remained constant. By contrast, PCL-only tubes swelled and the PLGA tubes shrank with a sharp diameter change. During culturing period, all three bio layers presented fine viability. Thus, the fabricated sample revealed excellent degradation behavior and biocompatibility and was implanted as artificial blood vessels into rabbits for evaluation for *in vivo* performance. After three months, the PCL–PLGA microtube still allowed for proper blood flow through the channel while single component tubes collapsed in the first month. Staining images of cross sections of the microtube showed an even distribution of ECM proteins between the polymer wall layers, and immunofluorescent assays further confirm the high level as well as proper distribution of ECM, which demonstrated a transfer from artificial vessel-mimicking scaffold to fully remodeled near-native blood vessel [52].

7.5 Summary

As outlined in this chapter, wide choices of nanomembranes of polymers, inorganic nonmetallic materials, metallic oxides, semiconductors, and 2D materials had been utilized to fabricate cell scaffolding platforms. Due to the intrinsic softness of nanomembranes originating from their ultralow thickness, they are compatible with vulnerable human cells and tissues, and enormous emphasis had been put on regulating cell viability, morphology, and other behavior on those platforms by topography patterning, surface coating, or even electrical stimulation. Among different strategies, rolled-up nanotechnology was particularly investigated for its tubular nature that highly resembled the structure of native tissues, such as blood vessels; as a result, rolled-up microtubes served as an ideal platform for cell behavior regulation and observation. The adhesion, proliferation, migration, and differentiation behavior within the microtubes were all different from previous studies on the planar substrate, and clearly, those in the tubes are more creditable as they mimic the environment *in vivo*. By integrating the nanomembrane scaffolds with biosensing devices as well as microscopic imaging, scientists are able to have more fundamental learning on the molecular level details of cellular activity that happens every second in the human body. Regulation of neurite outgrowth had been demonstrated with curved scaffolds, but the control of neurite branching was not precise enough and cultured neurite network was relatively small, which needs further investigation for an artificial neurite circuit. On the other hand, artificial bionic membranes had already been explored for enhancing therapy and *in vivo* transplantation. Future lies in a fully integrated, comprehensive platform that combines cell capturing, culturing, regulation, biosensing, *in vivo* implantation, and even *in vivo* monitoring. With the continuous exploration of ultrathin materials and nanofabrication technology, nanomembranes will further expand opportunities for applications in biosensing and biomedical fields.

References

1 Sternlicht, M.D. and Werb, Z. (2001). How matrix metalloproteinases regulate cell behavior. *Annu. Rev. Cell Dev. Biol.* 17: 463–516.

2 Van Vlierberghe, S., Dubruel, P., and Schacht, E. (2011). Biopolymer-based hydrogels as Scaffolds for tissue engineering applications: a review. *Biomacromolecules* 12 (5): 1387–1408.

3 Xu, C.Y., Inai, R., Kotaki, M., and Ramakrishna, S. (2004). Aligned biodegradable nanofibrous structure: a potential scaffold for blood vessel engineering. *Biomaterials* 25 (5): 877–886.

4 Zhang, J., Chen, H., Zhao, M. et al. (2020). 2D nanomaterials for tissue engineering application. *Nano Res.* 13 (8): 2019–2034.

5 Wang, Y., Zhao, M., Ping, J. et al. (2016). Bioinspired design of ultrathin 2D bimetallic metal-organic-framework nanosheets used as biomimetic enzymes. *Adv. Mater.* 28 (21): 4149–4155.

6 Malachowski, K., Jamal, M., Jin, Q. et al. (2015). Self-folding single cell grippers. *Nano Lett.* 1: 104–110.

7 Fujie, T., Ahadian, S., Liu, H. et al. (2013). Engineered nanomembranes for directing cellular organization toward flexible biodevices. *Nano Lett.* 13 (7): 3185–3192.

8 Rethinam, S., Basaran, B., Vijayan, S. et al. (2020). Electrospun nano-bio membrane for bone tissue engineering application – a new approach. *Mater. Chem. Phys.* 249 (January): 123010.

9 Jing, F.J., Huang, N., Wang, L. et al. (2007). Behavior of human umbilical vein endothelial cells on micro-patterned amorphous hydrogenated carbon films produced by plasma immersion ion implantation & deposition and plasma etching. *Diamond Relat. Mater.* 16 (3): 550–557.

10 Flemming, R.G., Murphy, C.J., Abrams, G.A. et al. (1999). Effects of synthetic micro- and nano-structured surfaces on cell behavior. *Biomaterials* 20 (6): 573–588.

11 Maheshwari, G., Brown, G., Lauffenburger, D.A. et al. (2000). Cell adhesion and motility depend on nanoscale RGD clustering. *J. Cell Sci.* 113 (10): 1677–1686.

12 Arnold, M., Cavalcanti-Adam, E.A., Glass, R. et al. (2004). Activation of integrin function by nanopatterned adhesive interfaces. *ChemPhysChem* 5 (3): 383–388.

13 Glass, R. and M ller, M., and Spatz, J.P. (2003). Block copolymer micelle nanolithography. *Nanotechnology* 14 (10): 1153–1160.

14 Krishnamoorthy, S., Hinderling, C., and Heinzelmann, H. (2006). Nanoscale patterning with block copolymers. *Mater. Today* 9 (9): 40–47.

15 Graeter, S.V., Huang, J., Perschmann, N. et al. (2007). Mimicking cellular environments by nanostructured soft interfaces. *Nano Lett.* 7 (5): 1413–1418.

16 ZHANG, X., PRASAD, S., NIYOGI, S. et al. (2005). Guided neurite growth on patterned carbon nanotubes. *Sens. Actuators, B* 106 (2): 843–850.

17 Li, B., Ma, Y., Wang, S., and Moran, P.M. (2005). Influence of carboxyl group density on neuron cell attachment and differentiation behavior: gradient-guided neurite outgrowth. *Biomaterials* 26 (24): 4956–4963.

18 Chao, T.I., Xiang, S., Chen, C.S. et al. (2009). Carbon nanotubes promote neuron differentiation from human embryonic stem cells. *Biochem. Biophys. Res. Commun.* 384 (4): 426–430.

19 Armelin, E., Gomes, A.L., Pérez-Madrigal, M.M. et al. (2012). Biodegradable free-standing nanomembranes of conducting polymer:polyester blends as bioactive platforms for tissue engineering. *J. Mater. Chem.* 22 (2): 585–594.

20 Lee, I.C. and Wu, Y.C. (2014). Assembly of polyelectrolyte multilayer films on supported lipid bilayers to induce neural stem/progenitor cell differentiation into functional neurons. *ACS Appl. Mater. Interfaces* 6 (16): 14439–14450.

21 Fabbro, A., Scaini, D., León, V. et al. (2016). Graphene-based interfaces do not alter target nerve cells. *ACS Nano* 10 (1): 615–623.

22 Rauti, R., Lozano, N., León, V. et al. (2016). Graphene oxide nanosheets reshape synaptic function in cultured brain networks. *ACS Nano* 10 (4): 4459–4471.

23 Guo, R., Xiao, M., Zhao, W. et al. (2021). 2D $Ti_3C_2T_x$MXene couples electrical stimulation to promote proliferation and neural differentiation of neural stem cells. *Acta Biomater.* 139: 105–117, ISSN 1742-7061.

24 Xu, C., Wu, X., Huang, G., and Mei, Y. (2019). Rolled-up nanotechnology: materials issue and geometry capability. *Adv. Mater. Technol.* 4 (1): 1–26.

25 Prinz, V.Y., Seleznev, V.A., Gutakovsky, A.K. et al. (2000). Free-standing and overgrown InGaAs/GaAs nanotubes, nanohelices and their arrays. *Phys. E* 6 (1): 828–831.

26 Huang, G., Mei, Y., Thurmer, D.J. et al. (2009). Rolled-up transparent microtubes as two-dimensionally confined culture scaffolds of individual yeast cells. *Lab Chip* 9 (2): 263–268.

27 Harazim, S.M., Xi, W., Schmidt, C.K. et al. (2012). Fabrication and applications of large arrays of multifunctional rolled-up SiO/SiO_2 microtubes. *J. Mater. Chem.* 22 (7): 2878–2884.

28 Cavallo, F., Huang, Y., Dent, E.W. et al. (2014). Neurite guidance and three-dimensional confinement via compliant semiconductor scaffolds. *ACS Nano* 8 (12): 12219–12227.

29 Herzer, R., Gebert, A., Hempel, U. et al. (2021). Rolled-up metal oxide microscaffolds to study early bone formation at single cell resolution. *Small* 17: 2005527.

30 Niinomi, M., Liu, Y., Nakai, M. et al. (2016). Biomedical titanium alloys with Young's moduli close to that of cortical bone. *Regen. Biomater.* 3 (3): 173–185.

31 Yu, M., Huang, Y., Ballweg, J. et al. (2011). Semiconductor nanomembrane tubes: three-dimensional confinement for controlled neurite outgrowth. *ACS Nano* 5 (4): 2447–2457.

32 Bausch, C.S., Koitmäe, A., Stava, E. et al. (2013). Guided neuronal growth on arrays of biofunctionalized GaAs/InGaAs semiconductor microtubes. *Appl. Phys. Lett.* 103 (17): 173705.

33 Froeter, P., Huang, Y., Cangellaris, O.V. et al. (2014). Toward intelligent synthetic neural circuits: directing and accelerating neuron cell growth by self-rolled-up silicon nitride microtube array. *ACS Nano* 8 (11): 11108–11117.

34 Prinz, A.A. and Fromherz, P. (2000). Electrical synapses by guided growth of cultured neurons from the snail *Lymnaea stagnalis*. *Biol. Cybern.* 82 (4): L1–L5.

35 Koch, B., Sanchez, S., Schmidt, C.K. et al. (2014). Confinement and deformation of single cells and their nuclei inside size-adapted microtubes. *Adv. Healthc. Mater.* 3 (11): 1753–1758.

36 Koch, B., Meyer, A.K., Helbig, L. et al. (2015). Dimensionality of rolled-up nanomembranes controls neural stem cell migration mechanism. *Nano Lett.* 15 (8): 5530–5538.

37 Xi, W., Schmidt, C.K., Sanchez, S. et al. (2014). Rolled-up functionalized nanomembranes as three-dimensional cavities for single cell studies. *Nano Lett.* 14 (8): 4197–4204.

38 Xi, W., Schmidt, C.K., Sanchez, S. et al. (2016). Molecular insights into division of single human cancer cells in on-chip transparent microtubes. *ACS Nano* 10 (6): 5835–5846.

39 Schulze, B.S., Huang, G., Krause, M. et al. (2010). Morphological differentiation of neurons on microtopographic substrates fabricated by rolled-up nanotechnology. *Adv. Eng. Mater.* 12 (9): 558–564.

40 Koitmäe, A., Harberts, J., Loers, G. et al. (2016). Approaching integrated hybrid neural circuits: axon guiding on optically active semiconductor microtube arrays. *Adv. Mater. Interfaces* 3 (24): 1–6.

41 Pampaloni, N.P., Lottner, M., Giugliano, M. et al. (2018). Single-layer graphene modulates neuronal communication and augments membrane ion currents. *Nat. Nanotechnol.* 13 (8): 755–764.

42 Smith, E.J., Xi, W., Makarov, D. et al. (2012). Lab-in-a-tube: ultracompact components for on-chip capture and detection of individual micro-/nanoorganisms. *Lab Chip* 12 (11): 1917.

43 Martinez-Cisneros, C.S., Sanchez, S., Xi, W., and Schmidt, O.G. (2014). Ultra-compact three-dimensional tubular conductivity microsensors for ionic and biosensing applications. *Nano Lett.* 14 (4): 2219–2224.

44 Zakharchenko, S., Sperling, E., and Ionov, L. (2011). Fully biodegradable self-rolled polymer tubes: a candidate for tissue engineering scaffolds. *Biomacromolecules* 12 (6): 2211–2215.

45 Karnaushenko, D., Münzenrieder, N., Karnaushenko, D.D. et al. (2015). Biomimetic microelectronics for regenerative neuronal Cuff implants. *Adv. Mater.* 27 (43): 6797–6805.

46 Smith, E.J., Schulze, S., Kiravittaya, S. et al. (2011). Lab-in-a-tube: detection of individual mouse cells for analysis in flexible split-wall microtube resonator sensors. *Nano Lett.* 11 (10): 4037–4042.

47 Hong, S.W., Lee, J.H., Kang, S.H. et al. (2014). Enhanced neural cell adhesion and neurite outgrowth on graphene-based biomimetic substrates. *Biomed. Res. Int.* 2014: 1–8.

48 Park, S.Y., Park, J., Sim, S.H. et al. (2011). Enhanced differentiation of human neural stem cells into neurons on graphene. *Adv. Mater.* 23 (36): 263–267.

49 Park, J., Park, S., Ryu, S. et al. (2014). Graphene-regulated cardiomyogenic differentiation process of mesenchymal stem cells by enhancing the expression of extracellular matrix proteins and cell signaling molecules. *Adv. Healthc. Mater.* 3 (2): 176–181.

50 Kim, S.J., Cho, H.R., Cho, K.W. et al. (2015). Multifunctional cell-culture platform for aligned cell sheet monitoring, transfer printing, and therapy. *ACS Nano* 9 (3): 2677–2688.

51 Kong, Y., Xu, R., Darabi, M.A. et al. (2016). Fast and safe fabrication of a free-standing chitosan/alginate nanomembrane to promote stem cell delivery and wound healing. *Int. J. Nanomed.* 11: 2543–2555.

52 Cheng, S., Jin, Y., Wang, N. et al. (2017). Self-adjusting, polymeric multilayered roll that can keep the shapes of the blood vessel scaffolds during biodegradation. *Adv. Mater.* 29 (28): 1–8.

8

Nanomembranes for Energy Storage

Ziyu Zhang, Zhe Zhao, Ye Kong, Anni Li, Yongfeng Mei, and Gaoshan Huang

Fudan University, Department of Material Sciences, 220 Handan Road, Shanghai 200438, P. R. China

8.1 Introduction

Climate change and the growing consumption of fossil fuels are two of the most pressing issues in today's world [1–3]. In recent years, the development of low-cost, environment-friendly energy storage devices has become a pressing need to respond to economy and society's long-term development [4, 5]. At present, batteries and supercapacitors are of great importance for industry and human daily life due to their large capacitance, long cycling performance, and enhanced energy densities [6]. Thanks to the significant advances in material science and nanoscience, energy storage devices with improved performance were produced in past decades, and they have become primary power source for portable electronics in the social and industrial areas [7, 8]. Despite the achieved advances, electrochemical properties of these devices still face challenges, which should be further overcome to meet the increasing demands in more complex systems, such as high density energy storage systems for transportation systems in electric vehicles [9, 10]. Generally, electrode materials are the most important part of the whole device [10]. Several researchers have recently focused their efforts in developing and fabricating electrode materials with unique architectures [11, 12]. With the rapid development of nanotechnology, nanomaterials such as nanomembranes have been viewed as ideal energy storage materials among scientists [12, 13].

Although the concept of "nanostructure" was first raised in 1950s, due to its nanoscale characteristics and excellent controllability, nanomembrane has become one of the most promising nanostructure materials recently [14]. Nanomembrane is a new class of nanomaterials created by nanotechnologies, which are defined as quasi-two-dimensional (2D) nanostructures with a nanoscale thickness (1–100 nm) and millimeter-level or even centimeter-level lateral dimensions with complex microstructures that is freestanding or able to attach to certain substrates [15]. Besides high aspect ratios, nanomembranes also exhibit outstanding robustness, which actually may even be handled with bare hands without losing their excellent flexibility. Nanomembranes can provide abundant functions with proper designs,

Nanomembranes: Materials, Properties, and Applications, First Edition.
Edited by Yongfeng Mei, Gaoshan Huang, and Xiuling Li.
© 2022 WILEY-VCH GmbH. Published 2022 by WILEY-VCH GmbH.

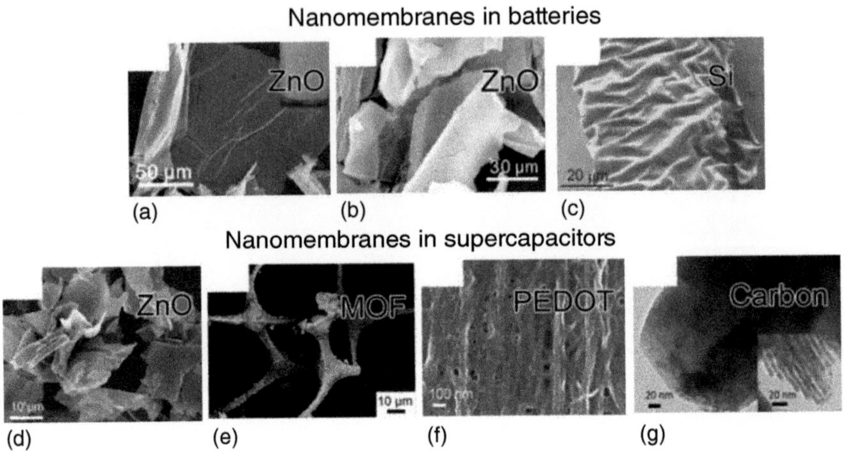

Figure 8.1 Versatile nanomembranes and their applications in (a)–(c) batteries and (d)–(g) supercapacitors. Source: (a) Zhao et al. [29]. Reproduced with permission of Royal Society of Chemistry. (b) Zhao et al. [30]. Reproduced with permission of Royal Society of Chemistry. (c) Si et al. [20]. Reproduced with permission of John Wiley & Sons, Inc. (d) Naeem et al. [13] Reproduced with permission of Elsevier. (e) Zhao et al. [32] Reproduced with permission of Elsevier. (f) Lee et al. [9] Reproduced with permission of American Chemical Society. (g) Wang et al. [25] Reproduced with permission of Royal Society of Chemistry.

and finally manifest novel and unique features [16, 17]. The control over their physical and chemical properties opens the door to a wide range of new uses, particularly in energy storage devices [18–20]. Plenty of techniques have been applied for synthesizing nanomembranes, such as atomic layer deposition (ALD) [21], adaptive ionic layer epitaxy [22], electron beam evaporation [23], layer-by-layer deposition [24], chemical vapor deposition (CVD) [25], self-assembly process [26], and so on. Various approaches endow various properties and functions of the resultant nanomembrane [27–34]. Metal oxide, composites, metal, conductive polymer (CP), inorganic materials, and metal organic frameworks (MOFs) are examples of common nanomembranes that can be used in energy storage devices, as shown in Figure 8.1. Nanomembranes in energy storage devices can be used in a variety of ways, including providing conductive support, passive layer, energy storage active materials, reducing volume expansion in the process of charging and discharging, constructing hierarchical structures and so on.

In this chapter, functions and applications of nanomembranes in energy storage devices will be introduced. The first part of the section will illustrate the applications of nanomembranes in batteries, focusing on their functions and advantages in lithium batteries, lithium–sulfur batteries and other batteries. Then the applications in supercapacitors, including double-layer capacitors and pseudocapacitors, will be demonstrated. Finally, the perspective and conclusion will discuss the application potentials and challenges of nanomembranes in energy storage devices.

8.2 Nanomembranes for Batteries

With the developing demand for energy storage, high energy density, promising safety, and long cycling life batteries have attracted researchers' interest [29, 35]. Various batteries with components of nanomembranes have so far been fabricated by utilizing the mechanical softness, better cycling stability, long-term durability, and more unique physical properties of nanomembranes [7, 36, 37]. Components such as anodes and cathodes are fabricated by nanomembranes pave the way for the fabrication of low-cost and high-capacity batteries in the future. Strategies that can precisely control the thickness of nanomembranes during fabrication processes are essential for nano-related researches, and thickness is also the key factor that affects the performance of nanomembranes [7]. With the development of thin-film fabrication processes, nanomembranes can be fabricated by more approaches [22, 33, 38–40], which paves an appealing way to fabricate ideal structures consisting of nanomembranes for batteries.

8.2.1 Nanomembranes for Li-Ion Batteries

Li-ion batteries have attracted great interest as a possible candidate for future energy storage systems due to their high energy and power densities. The high energy efficiency of Li-ion batteries allows them to be used in a variety of electric applications, ranging from portable electronics, power tools, and electric vehicles to large-scale electric grid applications, thereby improving the efficiency of energy collected from wind, solar, geo-thermal, and other green energy sources [22, 29, 35]. Nevertheless, the low theoretical capacity of graphite anode constrained the widespread use of Li-ion batteries [41–46]. Therefore, developing new electrode materials with high capacity becomes an important research direction. Non-graphic carbons such as graphene and many transition-metal oxides (ZnO [13], NiO [14], Fe_2O_3 [15], CuO [46], etc.) have been studied as ideal Li-ion battery anode materials [35].

Nanomembranes fabricated by ALD as electrodes manifest high stabilities and remarkable capacities. Zhao et al. [29] fabricated an interconnected carbon foam anchored by high mass loading of flexible ZnO nanomembranes (C/ZnO nanomembranes foam) via ALD approach [29]. The carbon foam was uniformly fixed on the ZnO nanomembranes via immersing in suspension and subsequently re-pyrolyzing. The ALD-synthesized ZnO nanomembranes are ultrathin and flexible with uniform thicknesses. Because of their enormous specific surface and contact area, two-dimensional ZnO nanomembranes are prone to adhering to the surfaces they contact. Figure 8.2a shows the synthesis procedure of C/ZnO nanomembranes foam electrode. As shown in Figure 8.2b, the C/ZnO nanomembranes foam exhibits high steady discharge capacities of 450 mAh/g at a current density of 250 mA/g. In addition, the capacities increased to 1000 mAh/g if only the mass of ZnO is included. This manifests higher capacity than the theoretical capacity of ZnO

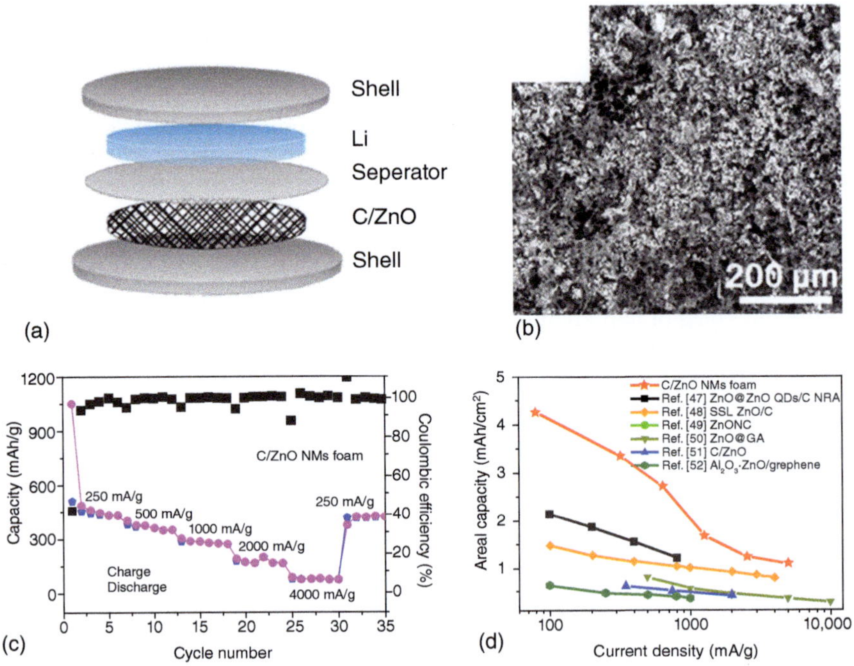

Figure 8.2 Illustration and characterization of C/ZnO nanomembranes foam. (a) Schematic of synthesis procedure of C/ZnO nanomembranes foam anode. (b) SEM image of the (b) C/ZnO nanomembranes foam composite electrode. (c) Rate performance of C/ZnO nanomembranes foam composite. (d) Areal capacity of C/ZnO nanomembranes foam composite electrode and other C/ZnO composite reported recently [51, 53]. Source: (a)–(d) Zhao et al. [29]. Reproduced with permission of Royal Society of Chemistry.

(978 mAh/g). The superior capacity can be attributed to the carbon foam, which enables fast electron transportation through the interconnective network. Besides, the porosity of carbon foam demonstrates remarkable permeability for fast Li^+ diffusion and liquid electrolyte [53]. Benefiting from the unique mechanical properties of ZnO nanomembranes, the nanostructure can release lithiation/delithiation strain through deformation, preventing performance loss caused by cracking thus extending the life of Li-ion batteries [53]. Carbon's high volumetric porosity (95%) results in a large surface area (50 m^2/g) which is suitable for significant mass loading of ZnO nanomembranes, leading to a high areal capacity of 4.3 mAh/cm^2 at 80 mA/g and meeting the requirements for practical applications.

On the other hand, Li et al. [52] deposited ZnO nanomembranes on expanded graphite (EG) by ALD as binder-free high-capacity anodes for Li-ion batteries. The preparation of ZnO/EG composite and battery assembling is shown in Figure 8.3a [52]. The charge and discharge curves of ZnO/EG with 300 cycles of ALD (ZG300) and pure ZnO at a current rate of 200 mA/g are shown in Figure 8.3b. When compared to pure ZnO anodes, which had less than 20% capacity remaining after 10 cycles, the ZG300 demonstrated extraordinary stability, with more than 80% of charge capacity remaining after 10 cycles. The superior capacity stability can be

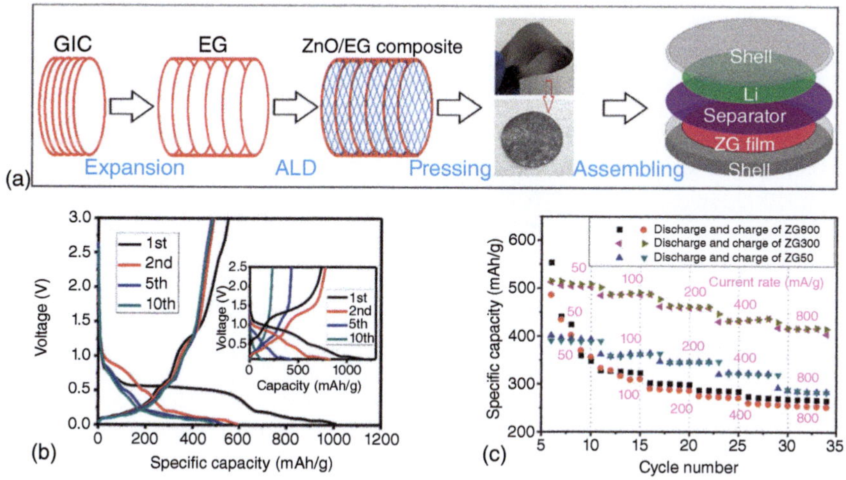

Figure 8.3 Diagram and electrical performance of ZG. (a) Diagram of ZG preparation and corresponding device fabrication. (b) Discharge/charge curves of ZG300 and pure ZnO (inset) during various cycles at 200 mA/g. (c) The rate performances of the anodes made from ZG50, ZG300, and ZG800 at various current densities. Source: (a)–(c) Adapted from Li et al. [52]. Copyright 2017, American Chemical Society.

ascribed to the structural properties of ZnO nanomembranes and EG framework. EG serves as a mechanical support for ZnO nanomembranes in the ZnO/EG composite. The ultrahigh specific surface area and numerous pores of EG provide sufficient space for ZnO nanomembranes to expand and prevent pulverization due to volume change during lithiation [48]. In Figure 8.3c, the rate performances of anodes composed of ZG50, ZG300, and ZG800 are shown, with capacity and current density calculated by using entire electrode mass instead of the mass of ZnO. As can be seen from the graph, the ZG300 composite has a substantially higher specific capacity than the ZG50 and ZG800 composites, with capacities of 510, 490, 462, 435, and 417 mAh/g at current rates of 50, 100, 200, 500, and 800 mA/g, respectively. This is due to the better coverage of ZnO nanomembranes layer and full utilization of the exposed surface area in ZG300. In addition, during charge and discharge process at a constant current rate of 200 mAh/g, the ZG300 has the greatest capacity of 436.4 mAh/g and 98% Coulombic efficiency.

Nanomembranes with multi-layered structure have also become a promising design for Li-ion applications and therefore have drawn considerable attention. Zhao et al. [30] used template-assisted ALD method with pyrolysis process to assemble sandwiched porous carbon/ZnO nanomembrane/porous carbon (CZC) trilayer nanosheets as anodes for Li-ion batteries [54]. Sandwiched structure of CZC nanosheets can be observed by transmission electron microscope (TEM) images presented in Figure 8.4a,b. The porous carbon on both sides of ZnO nanomembranes was deposited by secondary pyrolysis of hydrocarbons under high temperature, which formed an electric contact network around ZnO nanomembranes and enabled fast Li^+ movement [55]. Meanwhile, the increased surface area in the composition of porous carbon layers and nanomembrane results in a shorter

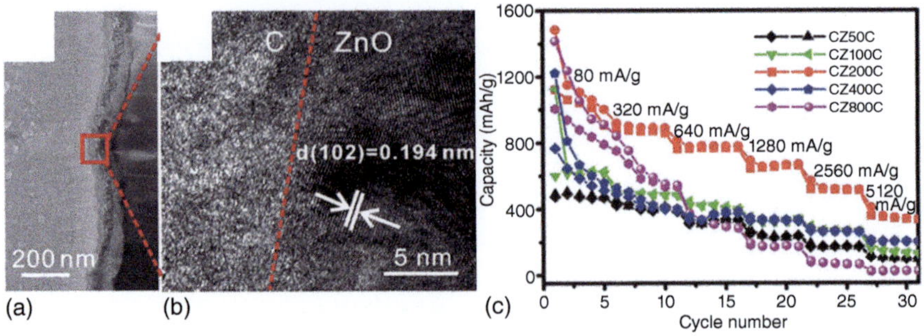

Figure 8.4 Morphological and electrochemical properties of CZC nanosheet. (a) Cross-sectional TEM image of CZC nanosheet. (b) High-resolution TEM image of the C/ZnO interface. (c) Galvanostatic discharge/charge capacities of CZ50C, CZ100C, CZ200C, CZ400C, CZ800C at different current densities. Source: (a)–(c) Zhao et al. [30]. Reproduced with permission of Royal Society of Chemistry.

ion-diffusion length in the inner ZnO nanomembrane. At current densities of 80, 320, 640, 1280, and 2560 mA/g, CZC nanosheets with 200 cycles of ALD deposition process (CZ200C) had the maximum reversible capacity of 1081, 889, 780, 668, and 520 mAh/g. respectively. Furthermore, CZ200C is still able to exhibit outstanding cycling performance when cycled at 5000 mA/g for more than 1450 cycles. In Figure 8.4c, there was no obvious capacity decline in CZ200C after 200 cycles at 4500 mA/g.

Moreover, rolled-up nanotechnology can combine with the deposition technique to achieve enhanced performance of Li-ion batteries. This is an advanced method for rearranging the geometry of nanomembranes using strain release [55–58]. Deng et al. [55] used rolled-up nanotechnology to form C/Si/C trilayer nanomembranes as anodes for Li-ion batteries. The illustration of the rolled-up C/Si/C microtubes preparation process is shown in Figure 8.5a. Scanning electron microscope (SEM) images of nanomembranes are illustrated in Figure 8.5b. After 82 cycles of charge and discharge, the C/Si/C nanomembranes still retain their rolled-up structures, indicating that it is an ideal anode material for Li-ion batteries. Furthermore, when the current rate is reduced to 500 mA/g, the Si nanomembranes covered with double carbon layers show a reverse capacity of roughly 1000 mAh/g. It aids in the improvement of conductivity and mechanical stability (Figure 8.5c).

On the other hand, Huang et al. [56] fabricated TiO_x/Si/TiO_x trilayer nanomembranes microtubes by rolled-up technique. Here, TiO_x and Si nanomembranes were deposited on polymethyl methacrylate (PMMA) sacrificial layer in order by electron beam evaporation. The TiO_x layers with Ti^{3+} and oxygen vacancies are deposited by precisely controlled deposition, which are conducive to the transportation of electrons and Li^+ ions through the whole electrode [36]. The morphology characterizations of TiO_x/Si/TiO_x nanomembranes observed by SEM are illustrated in Figure 8.6a–c. As shown in Figure 8.6c, the TiO_x/Si/TiO_x nanomembranes show remarkable stability: their integrity is maintained after 50 cycles of Li^+ insertion/extraction. Furthermore, the Si nanomembranes with TiO_x coating lower

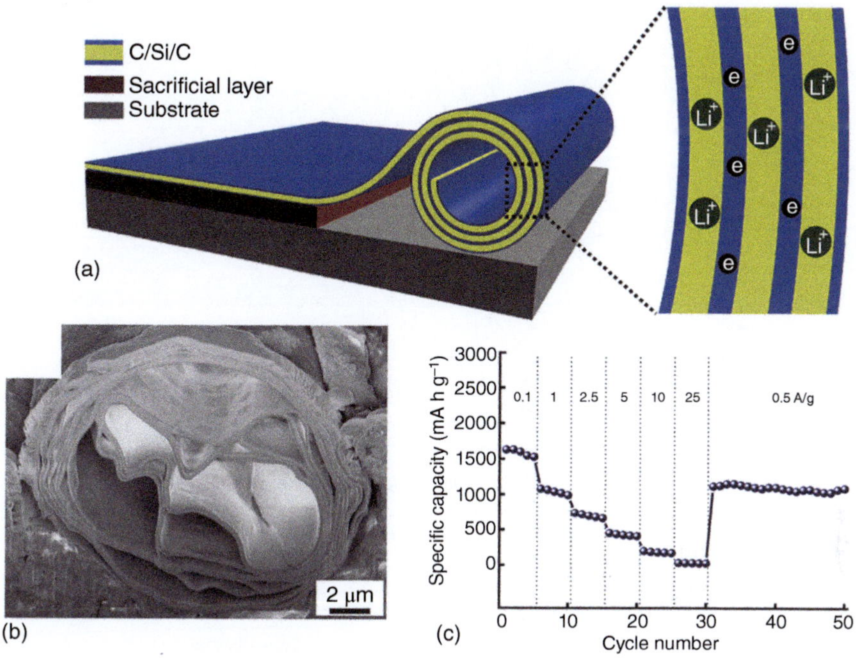

Figure 8.5 Schematic and characterization of rolled-up C/Si/C multilayer microtubes. (a) Fabrication of naturally rolled-up C/Si/C multilayer microtubes. (b) High-magnification SEM image of cross-section of rolled-up tube after the 82nd charge. (c) Specific capacity at various current densities. Source: (a)–(c) Deng et al. [55]. Reproduced with permission of John Wiley & Sons, Inc.

the battery's internal resistance and exhibit great stability: retaining 94.5% of highest capacity after 500 cycles (Figure 8.6d).

8.2.2 Nanomembranes for Na-Ion Batteries and Zn-Ion Batteries

Na-ion batteries possess relatively high energy density and feature low environmental pollution. The advantages give them the potential to become promising energy storage devices. Furthermore, the advantages of Na-ion batteries also lie in their lower cost, higher abundance of sodium than lithium but comparable energy density to Li-ion batteries [59, 60]. Thanks to the outstanding flexibility, ultrathin thickness and large surface area of nanomembranes, electrodes consisting of nanomembranes can be fabricated to enhance the performance of Na-ion batteries. By growing SnS_2 nanomembrane arrays on carbon cloth (CC) and covering the SnS_2 nanomembranes with graphene nanomembranes, Xu et al. [61] constructed SnS_2@graphene nanomembrane arrays on carbon cloth as anodes. As illustrated in Figure 8.7a, the SnS_2 nanomembrane arrays homogeneously grow on the carbon, and the gaps of SnS_2 nanomembrane arrays are wrapped with graphene [61–63]. In the novel structure, both SnS_2 nanomembrane arrays and graphene work as

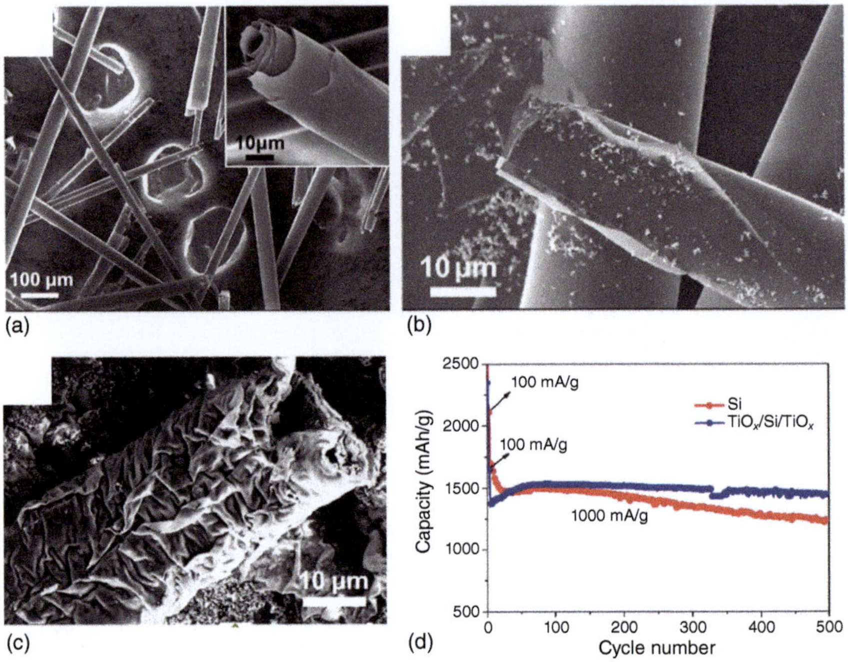

Figure 8.6 SEM appearances and cycling performance of $TiO_x/Si/TiO_x$ nanomembranes. (a) Low and high magnification (inset) SEM images of $TiO_x/Si/TiO_x$ nanomembranes. SEM images of $TiO_x/Si/TiO_x$ nanomembranes (b) before Li^+ insertion and (c) after 50 cycles of Li^+ insertion/extraction at 100 mA/g. (d) Cycling performance of $TiO_x/Si/TiO_x$ nanomembranes at 1000 mA/g. Source: (a)–(d) Huang et al. [56]. Reproduced with permission of Elsevier.

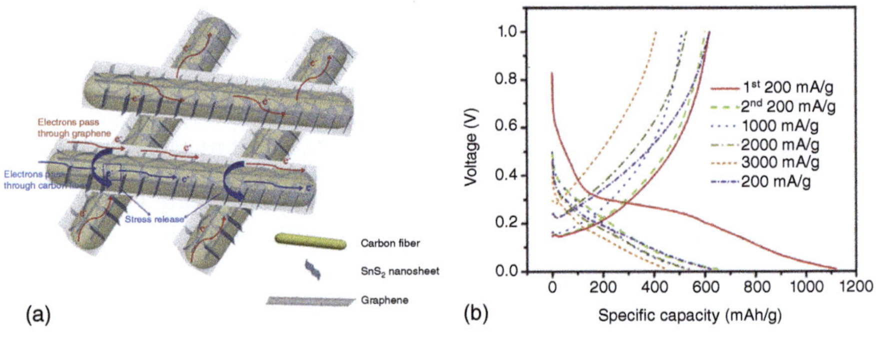

Figure 8.7 Diagram and electrical properties of SnS_2@graphene nanomembrane arrays. (a) Schematic illustration of the SnS_2@graphene nanomembrane arrays. (b) The charge/discharge curves at various current rates. Source: (a) and (b) Adapted from Xu et al. [61]. Copyright 2016, Elsevier BV.

conductivity materials to enhance the conductivity of Na-ion batteries, allowing faster electron transportation in the electrodes. Besides, the structural stability and integrity are improved since the stress caused by volume changes during cycling can be effectively relieved by graphene and SnS_2 nanomembranes [64, 65]. Figure 8.7b shows charge/discharge curves of SnS_2@graphene nanomembrane arrays. When

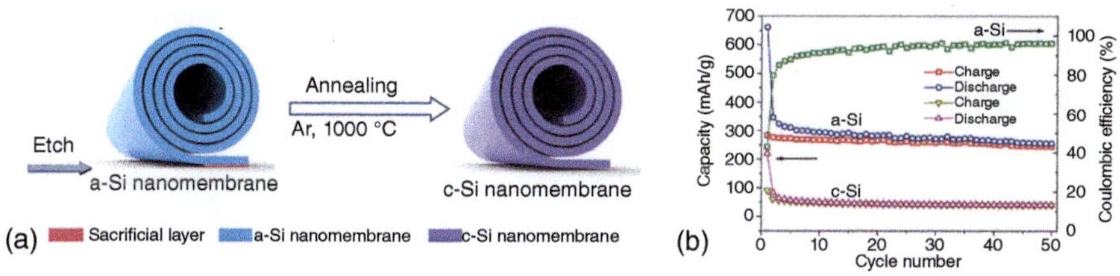

Figure 8.8 Diagram and cycling performance of a-Si and c-Si nanomembranes. (a) Schematic diagram of fabricating rolled-up a-Si and c-Si nanomembranes. (b) Comparison of cycling performance between a-Si and c-Si nanomembranes at a current density of 100 mA/g. Source: (a) and (b) Adapted from Huang et al. [66]. Copyright 2018, Wiley-VCH.

cycled at 200 mA/g, SnS_2@graphene nanomembrane arrays show a discharge and charge capacity of 1121 mAh/g and 643 mAh/g in the first cycle, respectively, and a discharge capacity of 652 mAh/g in the second cycle.

It is also worth noting that some nanomembranes contribute to a higher capacity of Na-ion batteries [57, 58]. Huang et al. [66] fabricated monolayer amorphous Si (a-Si) nanomembranes and monolayer crystalline Si (c-Si) nanomembranes by using rolled-up nanotechnology and physical vapor deposition (Figure 8.8a). The rate capability of a-Si nanomembranes and c-Si nanomembranes at current density of 100 mAh/g is compared in Figure 8.8b, with a-Si nanomembranes showing a substantially greater rate capability of (255 mAh/g) than c-Si nanomembranes (41 mAh/g) after 50 cycles. The a-Si nanomembranes contain more dangling bonds than the c-Si nanomembranes due to the disordered structures in a-Si nanomembranes. Therefore, the Na^+ ions are prone to diffuse into the inner part of the a-Si nanomembranes resulting in high sodium storage capacity.

Zn-ion batteries are attracting more and more attention because of their inexpensive cost and the abundance of zinc [67–69]. The unique properties of nanomembranes make them an ideal solution to overcome the shortcomings of poor cycle stabilities and limited energy densities. He et al. [68] synthesized layered VS_2 nanomembranes via hydrothermal reaction and used them as cathode material for Zn-ion batteries (Figure 8.9a). During the intercalation/deintercalation process of Zn^{2+}, there is an expansion of 1.73% along the c-axis and a slight shrink of a- and b-axes of VS_2 nanomembranes. In long-term cycle tests, the interlayer spacing of VS_2 nanomembranes provides sufficient space and a large surface area for reversible zinc ion insertion/extraction (0.74 Å), and produces a significant capacity of 190.3 mAh/g at a current density of 500 mA/g (Figure 8.9b).

Due to the high surface area to volume ratio, nanomembranes are also suitable materials for electrodes with atomic vacancies [7]. Zeng et al. [70] designed nickel cobaltite ($NiCo_2O_4$) nanomembranes with oxygen-vacancy and phosphate-ion modulation as ideal materials for Zn-ion batteries that synthesized by solvothermal method. The SEM images of P-$NiCo_2O_{4-x}$ nanomembranes are shown in Figure 8.10a, which resulted from the insertion of phosphate ions and oxygen vacancies. Here, the nanomembranes enable more oxygen vacancies and phosphate ions than bulk materials due to nanomembranes' higher specific surface area.

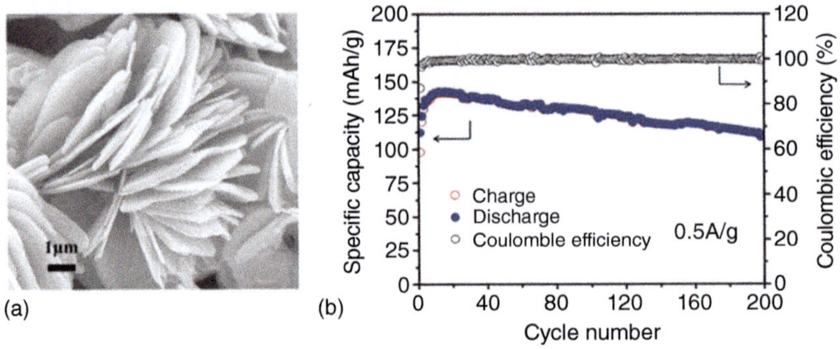

Figure 8.9 Morphologic characterization and cycling performance in layered VS_2 nanomembranes. (a) SEM image of layered VS_2 nanomembranes. (b) Long-term cyclic properties at a current density of 500 mA/g. Source: (a) and (b) He et al. [68]. Reproduced with permission of John Wiley & Sons, Inc.

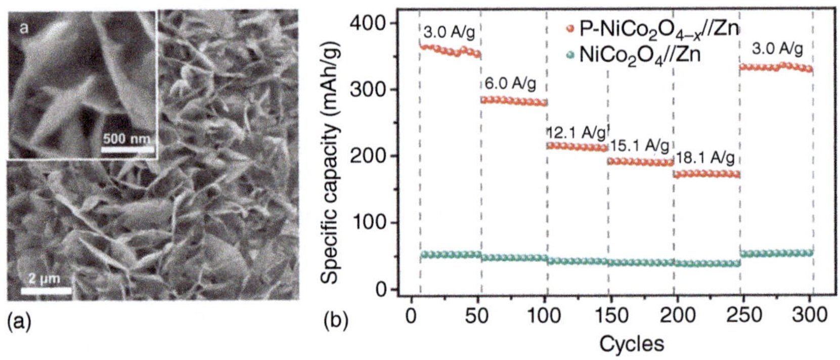

Figure 8.10 Morphologic characteristics and electrical performance of P-$NiCo_2O_{4-x}$ nanomembranes. (a) SEM images of the P-$NiCo_2O_{4-x}$ nanomembranes. (b) Rate performances of the $NiCo_2O_4$//Zn and P-$NiCo_2O_{4-x}$//Zn batteries. Source: (a) and (b) Zeng et al. [70]. Reproduced with permission of John Wiley & Sons, Inc.

They display improved conductivity, excellent rate performance, and extraordinary capacity. After 250 cycles, the as-fabricated P-$NiCo_2O_{4-x}$//Zn battery has a superior specific capacity of 332.4 mAh/g, which is approximately ten times larger than the pristine $NiCo_2O_4$//Zn battery [71] (Figure 8.10b).

8.2.3 Nanomembranes for Li-S Batteries

With the development of industry, the requirements for energy storage are getting higher and higher. As high-energy lithium batteries, Li-S batteries have drawn attention due to their high energy density and high cost effectiveness [72–75].

A Li-S battery usually consists of a lithium anode, a sulfur cathode, and an electrolyte [72]. The cathode material is generally composed of sulfur and other highly conductive materials. This is mainly because sulfur itself does not conduct electricity. When discharging, lithium in the negative electrode loses electrons and becomes lithium ions, while in the positive electrode, sulfur reacts with lithium ions and electrons to generate sulfide. Under the applied voltage, the positive and negative electrodes of the lithium–sulfur battery have opposite reactions, which is the charging process.

Lithium–sulfur batteries have the advantages such as light weight [76], good conductivity [77], low cost, abundance of sulfur [78], long endurance, and high stability [79]. Although lithium–sulfur batteries have many of the above-mentioned advantages, there are still many fatal problems that need to be solved. According to the above-mentioned reaction mechanism of Li-S batteries and many research results, it is shown that lithium–sulfur batteries have following main problems during charge/discharge process. The first problem is the formation, dissolution, and migration of lithium polysulfide in electrolyte [80]. The second problem is the changes in volume and morphology of sulfur cathodes during cycling [81]. Moreover, the shuttles of polysulfide between the electrodes during the cycle is also a problem [82]. The above-mentioned problems interact and cause capacity loss of the lithium–sulfur battery [83]. Even though lithium–sulfur batteries have attractive prospects, there is still a long way before wide range of practical applications. And the improvement of lithium–sulfur battery performance from nanomembrane modification is a noteworthy direction.

Generally, the improvement of lithium–sulfur batteries by nanomembrane can be divided into two ways, namely the improvement of Li metal anode and S cathode.

Graphene-based material is popular in the modification of S cathodes to achieve high performance. For example, reduced graphene oxide and sulfur (RGO@S) is a classic structure applied in the modification of S cathodes. Liu et al. [84] prepared flexible large-area porous RGO@S nanomembrane by subsequent reduction shown in Figure 8.11a. The discharge capacity of RGO@S composite structure reached 1381 mAh/g at 0.1C. The structure also showed an excellent cyclic stability (Figure 8.11b).

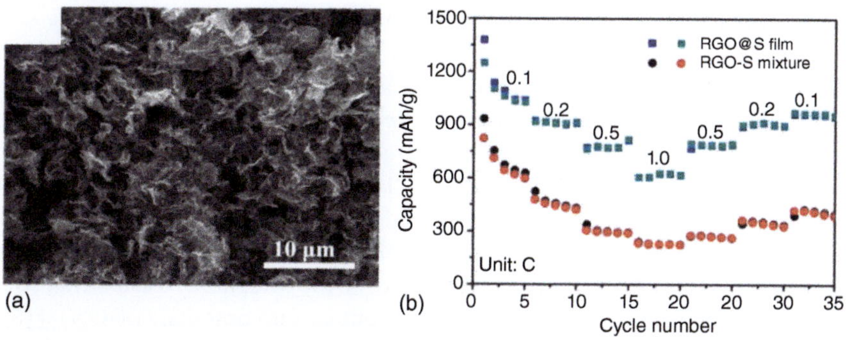

Figure 8.11 Schematic and performance of RGO@S composite nanomembrane. (a) The cross-sectional SEM images of RGO@S composite nanomembrane. (b) Rate performance of RGO@S composite nanomembrane and RGO@S mixture. Source: (a) and (b) Liu et al. [84]. Reproduced with permission of Elsevier.

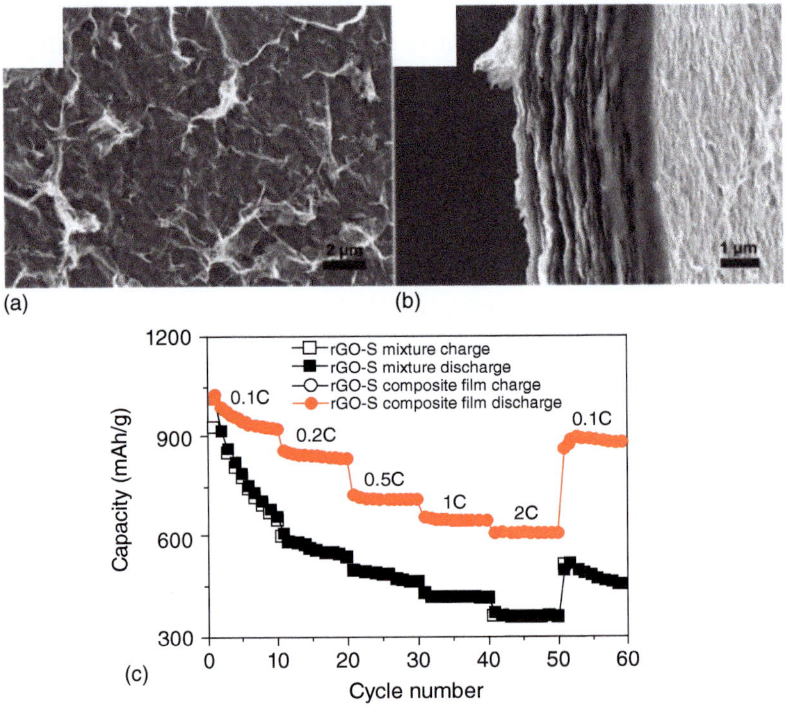

Figure 8.12 Morphology and performance of rGO@S composite nanomembrane. (a) Surface and (b) cross-sectional SEM images of the rGO@S composite nanomembrane. (c) Rate capability performances of the rGO@S composite nanomembrane and rGO–S mixture cathodes. Source: (a)–(c) Luo et al. [85]. Reproduced with permission of Royal Society of Chemistry.

In addition, Luo et al. [85] fabricated a freestanding RGO@S nanomembrane. The composite was achieved by the combination of a sulfur solution with a solvated rGO membrane through the freeze-drying technique. Figure 8.12a,b show the surface and the cross-sectional morphology of the RGO@S composite nanomembrane, respectively. This paper-like structure was cut into slices with suitable size and served as cathodes of Li-S batteries, which have shown relatively stable performance. The discharge capacity after 100, 200, and 300 cycles are 807, 721, and 668 mAh/g, respectively. As shown in Figure 8.12(c), the rGO@S composite nanomembrane cathode showed a high charge/discharge capability, which was on average 953 mA h g^{-1} at 0.1C, 845 mA h g^{-1} at 0.2C, 714 mA h g^{-1} at 0.5C, 650 mA h g^{-1} at 1C, and 612 mA h g^{-1} at 2C. It can be observed that the average discharge capacity of the rGO@S composite nanomembrane is higher than that of the rGO@S mixture between 0.1C and 2C. Furthermore, Chen et al. [86] prepared freestanding rGO nanotubes-wrapped sulfur nanoparticles (different sulfur contents) (RGONTs@S), and the nanomembrane composites exhibited an excellent cyclic stability and rate capability in comparison with RGO@S mixture electrodes.

Modification on Li anodes is another important way for the improvement of Li-S batteries. Zhang [87] investigated the effect of $LiNO_3$ on Li anode and sulfur cathode, and this work revealed that the $LiNO_3$ played opposite role when applied to Li anode and sulfur cathode. When the $LiNO_3$ was applied to Li anode, it helped to form a

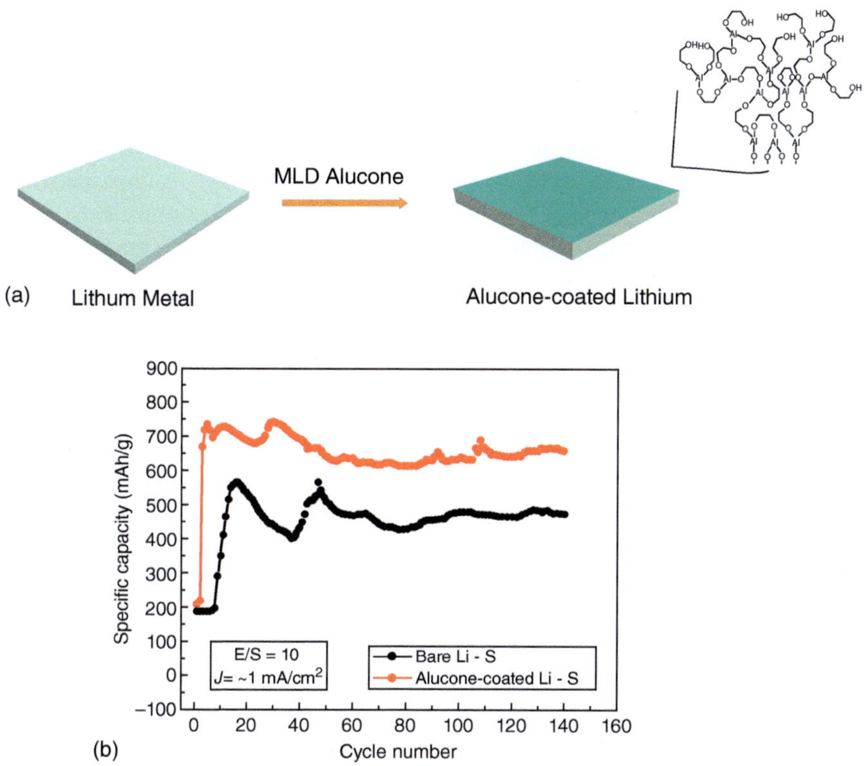

Figure 8.13 Schematic and performance of the alucone-coated Li-S battery. (a) Schematic of the MLD coating modification. (b) Cyclic performance of the bare Li and alucone-coated Li with sulfur content of ~5 mg/cm^2 and E/S = 10 at 0.1 °C. Source: (a) and (b) Adapted from Chen et al. [88]. Copyright 2018, American Chemical Society.

stable passivation layer, and with the consumption of $LiNO_3$ on the Li anode, the formed passivation layer was stable and grew infinitely with the time, leading to the promotion of the Li-S battery performance.

Molecular layer deposition (MLD) is an efficient method to deposit nanomembrane on all kinds of substrates. Chen et al. [88] used MLD to form alucone coating on Li as Figure 8.13a shows. The organic/inorganic hybrid nanomembrane with a highly cross-linked structure helps stabilize the lithium, prevent dendritic growth, and reduce side reactions as far as possible. At a high current rate of ~1 mA/cm, the capacity of the flavonoid-coated Li-S battery was 657.7 mAh/g after 140 cycles, which is 39.5% higher than that of the bare Li-S battery (Figure 8.13b).

8.3 Nanomembranes for Supercapacitors

8.3.1 Nanomembranes for Pseudocapacitors

Supercapacitor is a new type of energy storage device [89]. Its energy density is much higher than traditional capacitors, and its power density is much higher than that of batteries. [90]. There are two main types of supercapacitors according to their difference in energy storage mechanisms: electric double-layer capacitor (EDLC) and pseudocapacitor [91, 92]. The pseudocapacitors could provide more outstanding energy density values than EDLCs [93–96].

Conway et al. [88, 97] proposed the concept of "pseudocapacitance" and believed that the rapid electrochemical reaction on the electrode surface can contribute to energy storage. Pseudocapacitors are not like traditional capacitors (based on charge storage), but more like batteries (mainly based on redox reactions). Compared with traditional capacitors, supercapacitors have higher power density, longer cycle life, wider operating temperature limits, and are maintenance-free and environment friendly. Currently, the most studied pseudocapacitors are mainly based on the redox reaction of transition-metal compounds [98–100], and the doping and dedoping process of conductive polymers [101–103].

Although the introduction of pseudocapacitors increases the storage density of charges, it also brings a series of problems, including a reduced performance at high charge/discharge speed and a reduced cycle life. Compared with the electrostatic effect of EDLCs, the rate of redox process is much slower [104]. Therefore, in the case of rapid charging/discharging and high current charging, charges and ions diffusion in the pseudocapacitor system is more difficult, which then affects the performance of the pseudocapacitor [105]. The stability of the electrode materials is currently the primary limiting factor for most of the redox capacitors [106]. For transition-metal compounds, the dissolution of metal ions is the most important factor leading to electrode degradation. For conductive polymers, the charge/discharge process is accompanied with the migration of ions, and this causes repeated expansion/contraction of molecular chains. Such repeated process leads to permanent changes in the structure. Since the internal resistance of the pseudocapacitor material is relatively large, it affects the overall progress of the electrochemical reaction. The heat generated during the charge/discharge process also affects the stability and safety of the devices.

Nanomembrane is an ideal modification for the pseudocapacitors. Improvements of the performance of pseudocapacitors are benefited from nanomembrane's mature preparation process, broad material source, and excellent conformality [36, 107, 108].

Yan et al. [109] fabricated the flower-like $FeCo_2S_4$/rGO nanomembrane multilayer structure as the active material (Figure 8.14a,b). From the cross-sectional image, an $FeCo_2S_4$/rGO nanomembrane multilayer structure can be observed. Due to the synergy between conductive reduced graphene oxide film (RGOF) and active $FeCO_2S_4$, the structure was applied as an improved positive electrode with rich redox reactions in this high-performance supercapacitor. The rGO nanomembrane was prepared by a cost-effective and mass-produced self-assembly method. The obtained specific capacitance of the $FeCo_2S_4$/RGOF reached 2487.97 F/g at a current density of 5 mA/cm^2. The capacitance retention rate after 5000 cycles was 95.9%, as shown in Figure 8.14c.

Liu et al. [110] developed an efficient way to fabricate amorphous FeOOH quantum dots (QDs) on graphene nanosheets to assemble mesoporous nanomembrane (Figure 8.15a), which provided a large surface area and a high electron mobility. Due to the heterostructure of the hybrid nanosheets, the FeOOH QDs

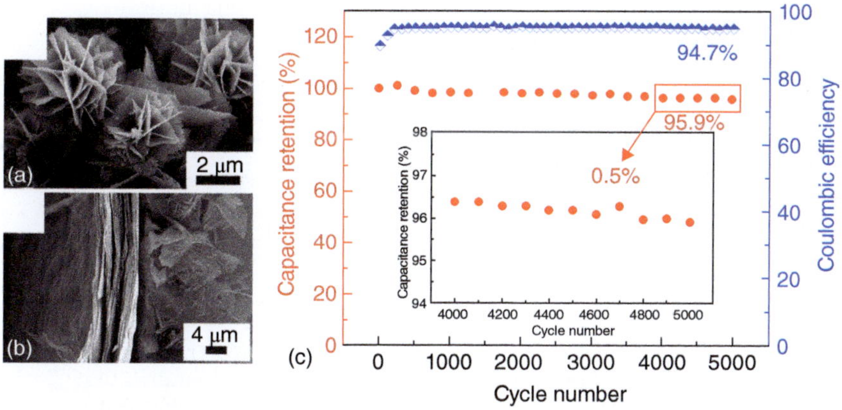

Figure 8.14 Morphology from different angles and performance of $FeCo_2S_4$/RGOF. SEM images of the $FeCo_2S_4$/RGOF. (a) Surface view and (b) cross-section view. (c) Coulombic efficiency and cyclic stability of $FeCo_2S_4$/RGOF at 20 mA/cm^2 after over 5000 cycles. Source: (a)–(c) Yan et al. [109]. Reproduced with permission of Elsevier.

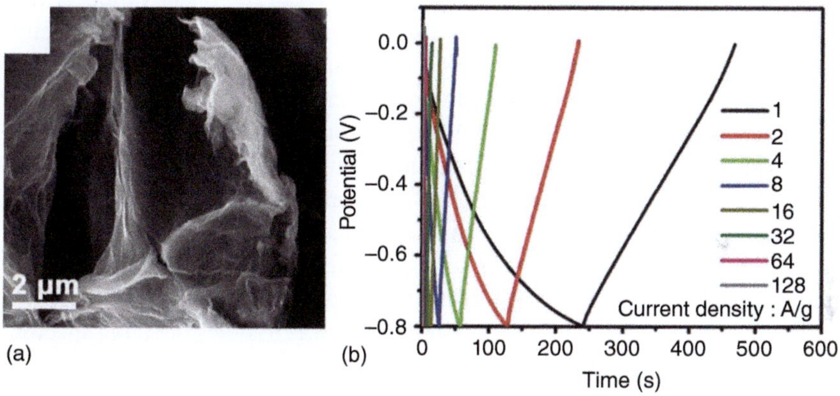

Figure 8.15 Morphology and performance of FeOOH QDs nanomembrane/graphene hybrid nanosheets. (a) SEM image of the FeOOH QDs nanomembrane/graphene hybrid nanosheets. (b) The charge/discharge curves of the FeOOH/graphene hybrid nanosheets electrode at different current densities. Source: (a) and (b) Liu et al. [110]. Reproduced with permission of John Wiley & Sons, Inc.

nanomembrane/graphene hybrid nanosheets exhibited excellent pseudocapacitance performance. It exhibited a large capacitance of about 365 F/g from −0.8 to 0 V relative to Ag/AgCl electrode. Its capacitance retention rate after 20,000 cycles was 89.7% and the rate capability was 189 F/g when the current density was 128 A/g, as shown in Figure 8.15b.

Guan et al. [111] deposited Co_3O_4 nanomembranes on flexible carbon nanotube/carbon cloth (CC) substrates by ALD as illustrated in Figure 8.16a. The ALD coating of Co_3O_4 nanomembranes on the CNT surface enabled a better electrical contact and increased the overall capacitance due to the Faradic reaction.

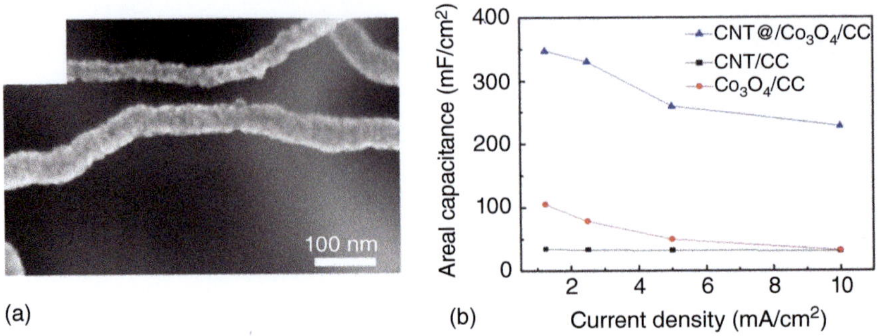

Figure 8.16 Morphology and performance of the Co_3O_4 coated CC. (a) The morphology of Co_3O_4 nanomembrane on a flexible CC. (b) Capacitance at different current densities. Source: (a) and (b) Guan et al. [111]. Reproduced with permission of IOP Publishing.

As Figure 8.16b shows, for CNT@2400-Co_3O_4/CC, the highest capacitance obtained was 416.7 mF/cm^2, and this improvement mainly originated from the flexibility of CC/CNTs substrate, uniformly coated Co_3O_4 nanoparticles, and unique hierarchical structure. The conformality of ALD coating contributed in the preservation of the hierarchical structure, and it showed an ultrastable cyclic ability, and no capacity decay was observed even after 50 000 cycles at high current.

Vanadium oxide is a viable way for adding pseudocapacitive layers to porous carbon electrodes. Parsons et al. [112] grew V_2O_5 film on activated carbon powder by ALD process to form meso- and microporous carbon electrodes, and achieved a pseudocapacitance of 540 F/g. Figure 8.17a is a schematic of the ALD modification process. By applying V_2O_5 nanomembrane to the carbon powder, V_2O_5 increased the pseudocapacitance under low ALD cycles. A thicker V_2O_5 layer increased the total charge capacity, while a thinner V_2O_5 layer required a shorter charge diffusion time, which stored Faraday charges faster. Figure 8.17b shows the large specific capacitance and good rate performance.

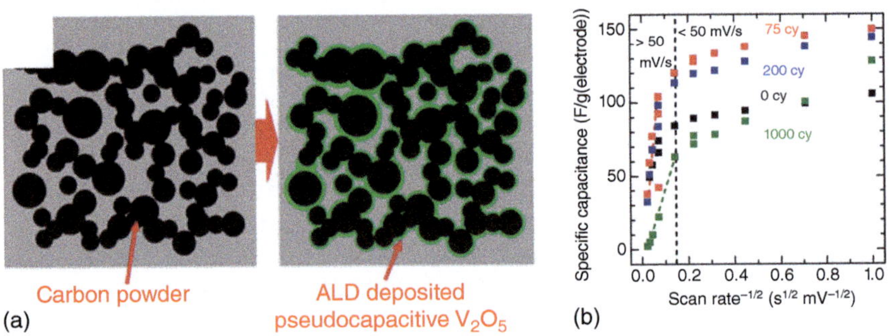

Figure 8.17 Schematic and performance of the V_2O_5-coated carbon powders. (a) Schematic image of V_2O_5 nanomembrane deposited by ALD technique on active carbon powders. (b) Specific capacitance at different scan rates. Source: (a) and (b) Adapted from Daubert et al. [112]. Copyright 2015, American Chemical Society.

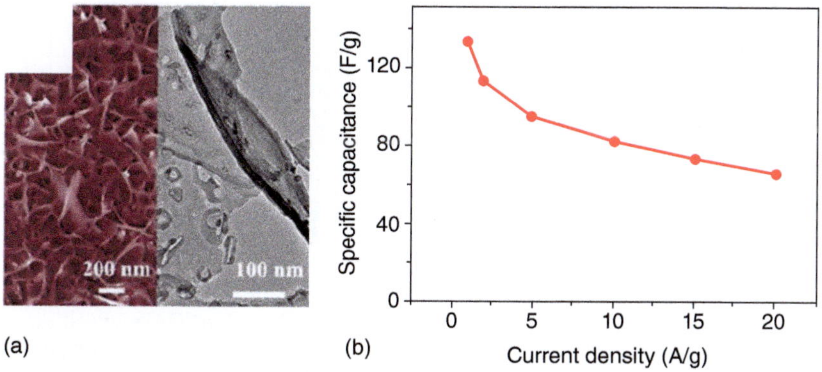

Figure 8.18 Morphology and performance of NiCoP@NF. (a) SEM and TEM images of NiCoP@NF. (b) Specific capacitance versus discharge current density. Source: (a) and (b) Lan et al. [113]. Reproduced with permission of Royal Society of Chemistry.

Lan et al. [113] prepared a nickel–cobalt phosphide nanosheet array anchored on nickel foam (NiCoP @ NF) and used it as a self-supporting electrodes (Figure 8.18a). Combining the composition characteristics and the advantages of the array architecture, the NiCoP @ NF electrode supported by foamed nickel exhibited excellent electrochemical performance compared with Ni-Co (layered double hydroxides) LDH@NF and $NiCoO_2$@NF electrodes. In addition, a binder-free supercapacitor device was designed on this basis. Its energy density reached 27 Wh/kg at a power density of 647 W/kg as shown in Figure 8.18b.

Apart from transition-metal oxide films, other kinds of nanomembrane also have great influences on improvement of pseudocapacitors. Conductive polymers (CPs) are most likely to become the basis of flexible supercapacitors as the pseudocapacitor materials. Due to their natural flexibility suitable for flexible supercapacitor applications, they may promote the development of the next-generation flexible devices [114].

Li et al. [115] used the wave-structured $Ti_3C_2T_x$ MXene film as the negative electrode and the graphene/carbon nanotube/polyaniline (rGO/CNT/PANI) nanomembrane, which is shown in Figure 8.19a. The wavy MXene negative electrode in this work was prepared by mechanically compressing 3D templated porous $Ti_3C_2T_x$ film, and has metallic conductivity and redox-active surface. The wavy $Ti_3C_2T_x$ nanomembrane provided better ion transport compared with the other one prepared by vacuum filtration. As shown in Figure 8.19b, the device outputs both high volume energy density of 70 W/h and high volumetric power density of 111 kW/L, which exceeds most of the symmetric devices. The pseudocapacitance characteristics of the device were verified by the galvanostatic charge-discharge (GCD) curves at different current densities in Figure 8.19c.

Zhou et al. [116] fabricated an asymmetric supercapacitor by using an array of activated nanoporous graphene sheets as the positive electrode, and introduced a conductive poly-(ethylenedioxythiophene) nanomembrane-coated CNT as the negative electrode. The conductive polymer nanomembrane improved the charge

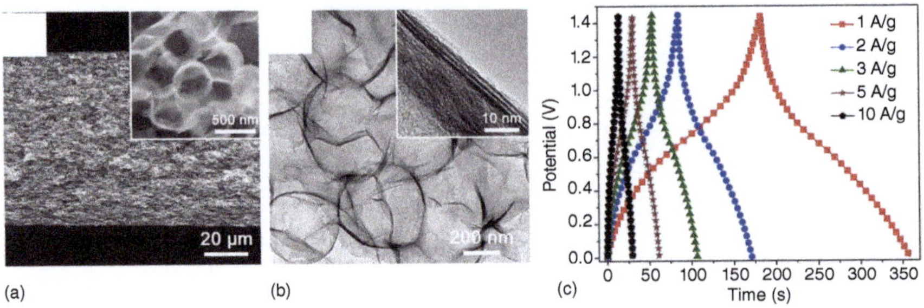

Figure 8.19 Morphology of $Ti_3C_2T_x$ film and performance of $Ti_3C_2T_x$/rGO/CNT/PANI asymmetric device. (a) SEM images of the electrodes. (b) TEM micrographs of the negative electrode. (c) Nonlinear GCD curves of the device in 3 M H_2SO_4 electrolyte. Source: (a)–(c) Li et al. [115]. Reproduced with permission of Elsevier.

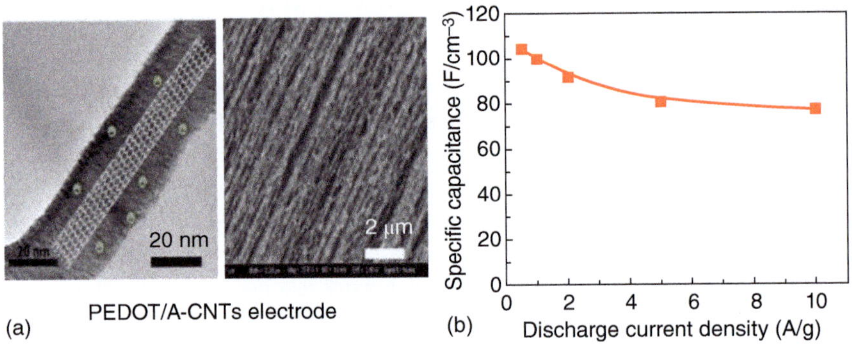

Figure 8.20 Morphology and performance of the nanostructured PEDOT/A-CNT electrode. (a) TEM images with different magnifications of the polymer nanomembrane-coated negative electrode. (b) Capacity of the polymer-modified battery at different discharge current densities. Source: (a) and (b) Zhou et al. [116]. Reproduced with permission of Royal Society of Chemistry.

storage ability of the negative electrode. The whole structure combined the ion transport pathways and enhanced charge storage capability, resulting in a promoted performance. The morphology of the structure is shown in Figure 8.20a,b. The asymmetric configuration of the supercapacitor contributed to the ability of the electrodes to be tailorable. As shown in Figure 8.20b, the volume capacitance of the battery under different discharge currents decreases slightly as the current increases and then tends to be stable.

Nejati et al. [117] used a one-step polymerization and coating process to synthesize unsubstituted polythiophene nanomembrane using oxidative chemical vapor deposition (oCVD). The coating was conformally coated in a porous nanostructure as shown in Figure 8.21a. The ultra-thin conformal polythiophene coating deposited by oCVD technique has good shape preservation ability, and its capacitance was 1.6 times larger than that of the plane membranes, which may be attributed to the absence of confinement in nanotubes. Experiments found that the ultra-thin polythiophene coating can maintain the morphology of the nanostructure, and the

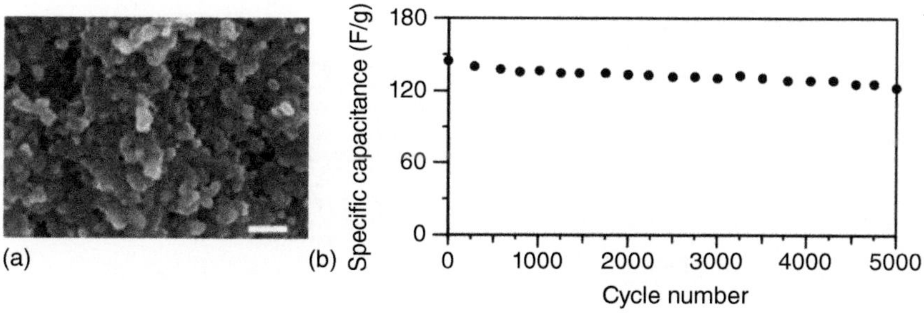

Figure 8.21 Morphology and performance of the polythiophene-coated porous structure. (a) SEM images of the polythiophene nanomembrane-coated porous structure (scale bar is 200 nm). (b) Capacitance of the sample during up to 5000 cycles (at 100 mV/s). Source: (a) and (b) Nejati et al. [117]. Reproduced with permission of American Chemical Society.

charge amount was significantly increased [117]. Compared with bare activated carbon, the specific capacitance and volume capacitance of the pseudocapacitor composed of ultra-thin polythiophene coated with activated carbon increased by 50% and 250%, respectively. The device showed excellent cyclic performance during 5000 test cycles, while the capacitance was only reduced by 10% as shown in Figure 8.21b.

8.3.2 Nanomembranes for Electric Double-Layer Capacitors

EDLCs is a family of energy storage devices with a number of advantages [118]. Electrostatic charge dispersion and accumulation at the interface of electrode and electrolyte are considered to explain the mechanism of EDLCs [118, 119]. Although pseudocapacitors could deliver appealing values of energy density, various tough issues with pseudocapacitive materials, such as excessive cost, unsatisfactory cyclic behaviors, and poor rate performances, are urgent to be solved before putting EDLCs into practical applications [13].

Despite the fact that EDLCs have a lower energy density than pseudocapacitors, their exceptional cyclic stability and superior rate performance indicate the potential of application in high-power energy storage systems [120, 121]. MOF materials have been widely studied since their discovery because of the features, for example, large specific surface area, highly controllable metal ion doping, and stable organic source [25, 119]. With the technology of ALD induction, some researchers have recently designed many promising approaches for constructing stable and uniform MOF nanomembranes [31, 32, 118]. By carbonizing the ALD induced zeolitic imidazolate framework (ZIF)-8-film-coated carbon foam, Zhao et al. [32] developed an effective approach for growing uniform flexible N-doped carbon films on carbon foam. The whole process can be summarized as follows: ZnO nanomembrane was deposited on the surface of carbon foam skeleton by ALD technique first; then the ZnO nanomembrane was converted to ZIF-8 layer and assembled on the skeleton with firm attaching. With the subsequent carbonization process, ZIF-8 transformed into N-doped carbon films. Meanwhile, the flexibility of carbon foam still

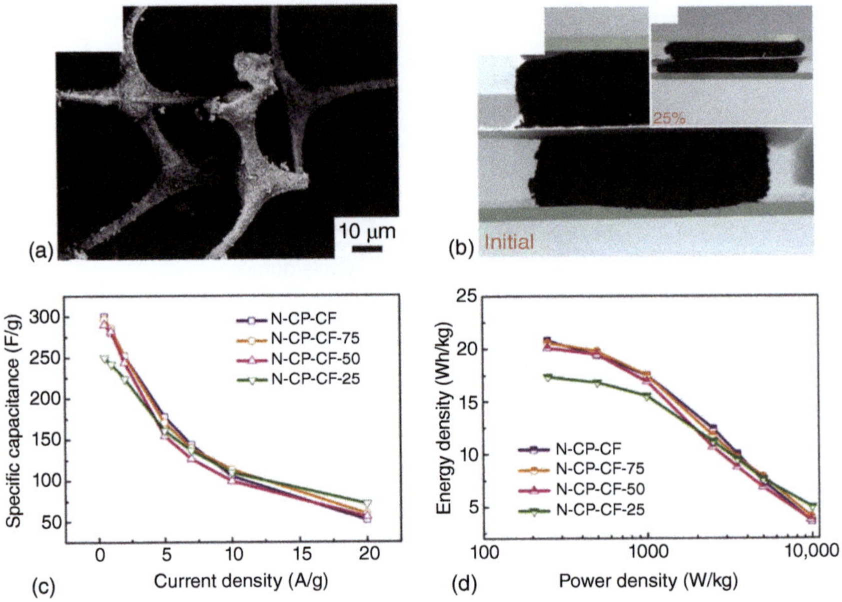

Figure 8.22 Morphology and performance of ZIF-8 nanomembrane on carbon foam. (a) The SEM image of ZIF-8 nanomembrane on carbon foam. (b) Photographs of the pyrolysis ZIF-8 nanomembrane on carbon foam (N-CP-CF) with various compression rates. (c) The specific capacitances of electrode under different compression rates as function of current density. (d) Energy densities of N-CP-CF under different power densities. Source: (a)–(d) Zhao et al. [32]. Reproduced with permission of Elsevier.

remained. The uniform growth of ZIF-8 film can be seen clearly from the SEM image in Figure 8.22, and the image displays the composite in its initial state, which exhibits excellent flexibility that can rebound after applying pressure. The capacitor assembled from N-CP-CF demonstrates a large capacitance of 300 F/g at initial state as well as a stable remaining capacitance of 250 F/g at a large compress strain of 25%. At the initial condition, a high energy density of 20.8 Wh/kg and a substantial energy density of 17.36 Wh/kg are achieved at quarter of original thickness under a great deformation.

Wang et al. [25] used Friedel–Crafts cross-linking of phenyl self-assembled monolayers (SAMs) on silica nanospheres (SNSs) to develop the three-dimensional (3D) carbon nanomembranes (CNMs). The TEM images of foam-like structures, 3D CNMs, with phenylfunctional degrees between 10% and 20% are presented in Figure 8.23a. A structurally ill-defined problem arises when this number reaches 20%. It's possible that the problem comes from the agglomerate of phenyltriethoxysilane (PTES) self-condensate. By modifying the sizes of the SNSs templates, the CNMs' pore sizes can be easily tuned. As demonstrated in Figure 8.23b, CNMs exhibit a large surface area (>1500 m^2/g) and excellent conductivity. The specific capacitance as a function of CNM current density is shown in Figure 8.23c. Specific capacitance of CNMs decreased from 202 to 132 F/g with a capacitance retention of 65.3% as current density increased from 0.5 to 20 A/g. Long-term cycle stability

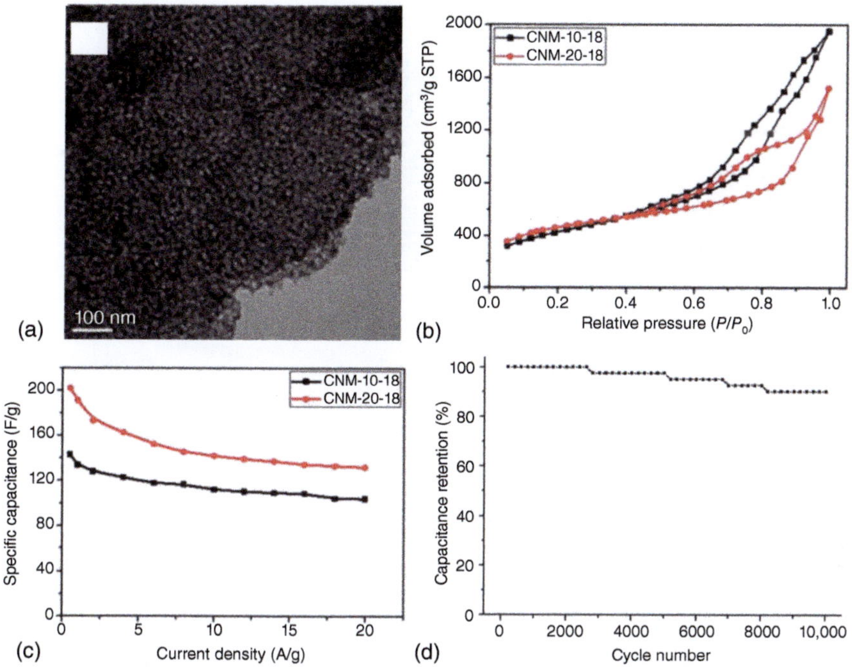

Figure 8.23 Morphology and performance of carbon nanomembrane. (a) TEM image of CNM. (b) Nitrogen adsorption and desorption isotherms of CNM. (c) Specific capacitance as a function of current density ranging from 0.5 to 20 A/g. (d) Capacitance retention of carbon nanomembrane measured in a two-electrode system with a 6 M KOH aqueous solution as the electrolyte at a current density of 4 A/g. Source: (a)–(d) Wang et al. [25]. Reproduced with permission of Royal Society of Chemistry.

as a result of CNM at a current density of 4 A/g in a 6 M KOH solution is shown in Figure 8.23d. After 10 000 cycles, CNM can still retain up to 90% of its initial capacitance. The ability of supercapacitors based on 3D porous CNMs to retain most of capacitance at high current density indicates that they are applicable to energy storage equipment that for long-term use.

Nanomembranes also can be applied in flexible all-in-one supercapacitor. For example, Karbhal et al. [118] developed an all-solid-state planar micro-supercapacitor (MSC) integrated with boron carbon nitride (BCN) electrodes with excellent performance. The electrodes were fabricated by a scalable and fast laser patterning technique. Heteroatom doping is a promising method for enhancing carbon electrode electrochemical activity. Figures 8.24a,b shows the morphology characteristics of a device built on an indium tin oxide (ITO)/polyethylene glycol terephthalate (PET) platform by using BCN as an interdigital electrode. As illustrated in Figure 8.24c, the BCN exhibits a very high power density of 487 mW/cm^2 at 1 mA/cm^2, as well as a high energy density of 10 mWh/cm^2 at 0.15 mA/cm^2, both of which are superior to other MSC devices compared in the plot. BCN-based MSC devices have significantly higher energy and power densities than other devices, indicating that BCN is a promising electrode material for MSCs. Figure 8.24d demonstrates the device's cycling stability at a current density of 1 mA/cm^2, with

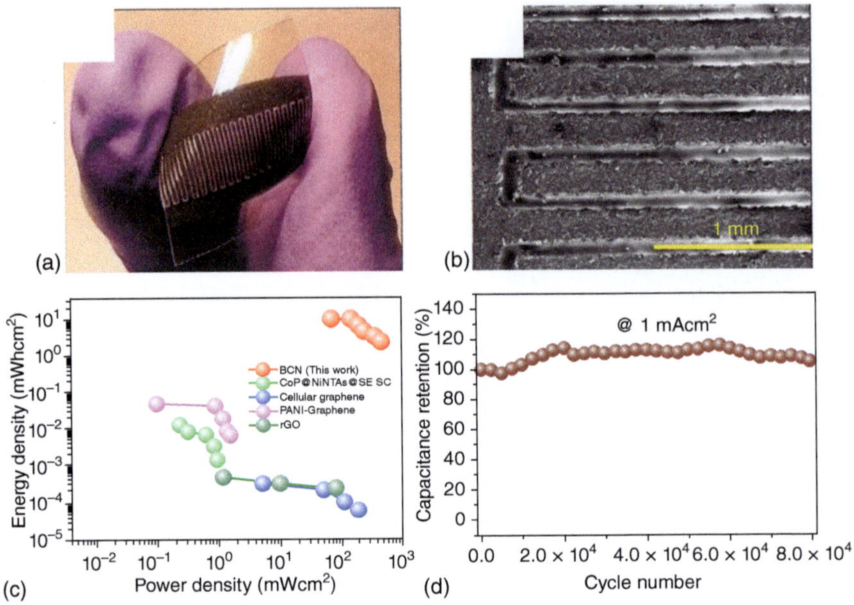

Figure 8.24 Morphology and performance of the BCN device. (a) Optical image of the device. (b) Top-view SEM images of the device. (c) Ragone plot of BCN (1 : 1 : 1)-based MSC compared with the other MSCs. (d) Cycle stability plot at 1 mA/cm^2. Source: (a)–(d) Karbhal et al. [118]. Reproduced with permission of Elsevier.

nearly 100% capacity retentions up to 80 000 cycles. The specific capacity of BCN nanosheets increased slightly and then stabilized, which was attributed to enhanced ion transportation in BCN nanosheets during the rapid process of charging and discharging. BCN is able to gradually increase the diffusion and interaction between electrolyte ions and electrode material during cycling since it is a type of 2D layered materials. At the electrode interface, the improved wettability benefits from the boron and nitrogen, resulting in increased capacitance during cyclic process. The specific capacity becomes saturated after a few cycles and then stabilizes.

8.4 Summary

In summary, we have summarized the recent progresses in application of outstanding performance energy storage devices based on nanomembranes, including Li-ion batteries, Na-ion batteries, Zn-ion batteries, Li-S batteries, and supercapacitors (EDLCs and pseudocapacitors). In particular, rolled-up tubular micro/nanoarchitectures and 2D nanosheets fabricated using the nanomembrane technique have been used in the design of battery anodes, leading to significant improvements in specific capacity and cyclability. The advanced fabrication strategies and unique structural properties of nanomembranes are mainly responsible for the enhanced battery performance. Composites based on ALD-pretreated nanomembranes have also been applied in anodes, and their conformal and uniform deposition of active layer results in exceptionally improved properties. Furthermore, a novel strategy to prepare MOF membrane via ALD induction technique has also been proposed and applied in high-performance supercapacitors. Due to its high mass loading of MOF membrane and high specific surface area with hierarchical porous structure, enhanced specific capacitance is achieved. Plenty of materials, including metal, metal oxide, organic composites, and inorganic materials, are prepared into nanomembranes by versatile techniques, and the nanomembranes have been applied in energy storage devices successfully. Diversity of material selections in nanomembranes technology has been proved to be applicable to the manufacture of ultra-compact capacitors with higher performance. This also increases the possibility of on-chip integration of microsized energy storage systems and microelectronic devices. Despite the success of utilizing nanomembranes in energy storage devices, the research on on-chip integrated and scalable energy storage systems is still in the initial stage. Replacing liquid electrolytes with solid or gel polymer electrolytes is one of the most difficult challenges in the fabrication process. There is still considerable work to be done in terms of developing better solid electrolytes and encapsulating liquid electrolytes in nanomembrane-based energy storage systems. Flexible power devices, on the other hand, are essential for flexible and wearable electronic applications. We stress the potential of utilizing nanomembrane for flexible energy storage devices due to the remarkable robustness and extraordinary flexibility of them. All in all, the nanomembranes pave the way for a new generation of energy storage applications.

References

1 Armand, M., Endres, F., MacFarlane, D.R. et al. (2009). Ionic-liquid materials for the electrochemical challenges of the future. *Nat. Mater.* 8 (8): 621–629. https://doi.org/10.1038/nmat2448.

2 Armand, M. and Tarascon, J.M. (2008). Building better batteries. *Nature* 451 (7179): 652–657. https://doi.org/10.1038/451652a.

3 Bufon, C.C.B., Gonzalez, J.D.C., Thurmer, D.J. et al. (2010). Self-assembled ultra-compact energy storage elements based on hybrid nanomembranes. *Nano Lett.* 10 (7): 2506–2510. https://doi.org/10.1021/nl1010367.

4 Etacheri, V., Marom, R., Elazari, R. et al. (2011). Challenges in the development of advanced Li-ion batteries: a review. *Energy Environ. Sci.* 4 (9): 3243–3262. https://doi.org/10.1039/c1ee01598b.

5 Goodenough, J.B. and Kim, Y. (2010). Challenges for rechargeable Li batteries. *Chem. Mater.* 22 (3): 587–603. https://doi.org/10.1021/cm901452z.

6 Goodenough, J.B. and Park, K.-S. (2013). The Li-ion rechargeable battery: a perspective. *J. Am. Chem. Soc.* 135 (4): 1167–1176. https://doi.org/10.1021/ja3091438.

7 Huang, G. and Mei, Y. (2018). Assembly and self-assembly of nanomembrane materials-from 2D to 3D. *Small* 14 (14): 1703665. https://doi.org/10.1002/smll.201703665.

8 Ji, X., Lee, K.T., and Nazar, L.F. (2009). A highly ordered nanostructured carbon-sulphur cathode for lithium-sulphur batteries. *Nat. Mater.* 8 (6): 500–506. https://doi.org/10.1038/nmat2460.

9 Lee, J.A., Shin, M.K., Kim, S.H. et al. (2012). Hybrid nanomembranes for high power and high energy density supercapacitors and their yarn application. *Acs Nano* 6 (1): 327–334. https://doi.org/10.1021/nn203640a.

10 Li, D., Mueller, M.B., Gilje, S. et al. (2008). Processable aqueous dispersions of graphene nanosheets. *Nat. Nanotechnol.* 3 (2): 101–105. https://doi.org/10.1038/nnano.2007.451.

11 Liu, C., Li, F., Ma, L.-P. et al. (2010). Advanced materials for energy storage. *Adv. Mater.* 22 (8): E28. https://doi.org/10.1002/adma.200903328.

12 Lu, X., Deng, J., Si, W. et al. (2015). High-performance Li-O_2 batteries with tri-layered Pd/MnO_x/Pd nanomembranes. *Adv. Sci.* 2 (9): 1500113. https://doi.org/10.1002/advs.201500113.

13 Naeem, F., Naeem, S., Zhao, Z. et al. (2020). Atomic layer deposition synthesized ZnO nanomembranes: a facile route towards stable supercapacitor electrode for high capacitance. *J. Power Sources* 451: 227740. https://doi.org/10.1016/j.jpowsour.2020.227740.

14 Ozgur, U., Alivov, Y.I., Liu, C. et al. (2005). A comprehensive review of ZnO materials and devices. *J. Appl. Phys.* 98 (4): 041301. https://doi.org/10.1063/1.1992666.

15 Poizot, P., Laruelle, S., Grugeon, S. et al. (2000). Nano-sized transition-metaloxides as negative-electrode materials for lithium-ion batteries. *Nature* 407 (6803): 496–499. https://doi.org/10.1038/35035045.

16 Reina, A., Jia, X., Ho, J. et al. (2009). Large area, few-layer graphene films on arbitrary substrates by chemical vapor deposition. *Nano Lett.* 9 (1): 30–35. https://doi.org/10.1021/nl801827v.

17 Rinzler, A.G., Hafner, J.H., Nikolaev, P. et al. (1995). Unraveling nanotubes – field-emission from an atomic wire. *Science* 269 (5230): 1550–1553. https://doi.org/10.1126/science.269.5230.1550.

18 Scrosati, B. and Garche, J. (2010). Lithium batteries: status, prospects and future. *J. Power Sources* 195 (9): 2419–2430. https://doi.org/10.1016/j.jpowsour.2009.11.048.

19 Shanmugam, M., Durcan, C.A., and Yu, B. (2012). Layered semiconductor molybdenum disulfide nanomembrane based Schottky-barrier solar cells. *Nanoscale* 4 (23): 7399–7405. https://doi.org/10.1039/c2nr32394j.

20 Si, W., Moench, I., Yan, C. et al. (2014). A single rolled-up Si tube battery for the study of electrochemical kinetics, electrical conductivity, and structural integrity. *Adv. Mater.* 26 (47): 7973. https://doi.org/10.1002/adma.201402484.

21 Simon, P. and Gogotsi, Y. (2008). Materials for electrochemical capacitors. *Nat. Mater.* 7 (11): 845–854. https://doi.org/10.1038/nmat2297.

22 Tarascon, J.M. and Armand, M. (2001). Issues and challenges facing rechargeable lithium batteries. *Nature* 414 (6861): 359–367. https://doi.org/10.1038/35104644.

23 Trotta, R., Atkinson, P., Plumhof, J.D. et al. (2012). Nanomembrane quantum-light-emitting diodes integrated onto piezoelectric actuators. *Adv. Mater.* 24 (20): 2668–2672. https://doi.org/10.1002/adma.201200537.

24 Vendamme, R., Onoue, S.-Y., Nakao, A. et al. (2006). Robust free-standing nanomembranes of organic/inorganic interpenetrating networks. *Nat. Mater.* 5 (6): 494–501. https://doi.org/10.1038/nmat1655.

25 Wang, Y., Xiong, R., Dong, L. et al. (2014). Synthesis of carbon nanomembranes through cross-linking of phenyl self-assembled monolayers for electrode materials in supercapacitors. *J. Mater. Chem. A* 2 (15): 5212–5217. https://doi.org/10.1039/c3ta15311h.

26 Winter, M. and Brodd, R.J. (2004). What are batteries, fuel cells, and supercapacitors? *Chem. Rev.* 104 (10): 4245–4269. https://doi.org/10.1021/cr020730k.

27 Xu, K. (2004). Nonaqueous liquid electrolytes for lithium-based rechargeable batteries. *Chem. Rev.* 104 (10): 4303–4417. doi: 10.1021/cr030203g.

28 Zhang, Y., Yan, Z., Nan, K. et al. (2015). A mechanically driven form of Kirigami as a route to 3D mesostructures in micro/nanomembranes. *Proc. Natl. Acad. Sci. U.S.A.* 112 (38): 11757–11764. https://doi.org/10.1073/pnas.1515602112.

29 Zhao, Y., Huang, G., Li, Y. et al. (2018). Three-dimensional carbon/ZnO nanomembrane foam as an anode for lithium-ion battery with long-life and high areal capacity. *J. Mater. Chem. A* 6 (16): 7227–7235. https://doi.org/10.1039/c8ta00940f.

30 Zhao, Y., Huang, G., Wang, D. et al. (2018). Sandwiched porous C/ZnO/porous C nanosheet battery anodes with a stable solid- electrolyte interphase for fast and long cycling. *J. Mater. Chem. A* 6 (45): 22870–22878. https://doi.org/10.1039/c8ta07848c.

31 Zhao, Z., Kong, Y., Lin, X. et al. (2020). Oxide nanomembrane induced assembly of a functional smart fiber composite with nanoporosity for an ultra-sensitive flexible glucose sensor. *J. Mater. Chem. A* 8 (48): 26119–26129. https://doi.org/10.1039/d0ta09211h.

32 Zhao, Z., Kong, Y., Liu, C. et al. (2020). Atomic layer deposition-induced integration of N-doped carbon particles on carbon foam for flexible supercapacitor. *J. Materiomics* 6 (1): 209–215. https://doi.org/10.1016/j.jmat.2020.01.011.

33 Zhao, Z., Kong, Y., Zhang, Z. et al. (2020). Atomic layer-deposited nanostructures and their applications in energy storage and sensing. *J. Mater. Res.* 35 (7): 701–719. https://doi.org/10.1557/jmr.2019.329.

34 Zhao, Z., Zhang, Z., Zhao, Y. et al. (2019). Atomic layer deposition inducing integration of Co, N codoped carbon sphere on 3D foam with hierarchically porous structures for flexible hydrogen producing device. *Adv. Funct. Mater.* 29 (48): 1906365. https://doi.org/10.1002/adfm.201906365.

35 Nitta, N., Wu, F., Lee, J.T. et al. (2015). Li-ion battery materials: present and future. *Mater. Today* 18 (5): 252–264. https://doi.org/10.1016/j.mattod.2014.10.040.

36 Huang, S., Zhang, L., Lu, X. et al. (2017). Tunable pseudocapacitance in 3D TiO_2-delta nanomembranes enabling superior lithium storage performance. *ACS Nano* 11 (1): 821–830. https://doi.org/10.1021/acsnano.6b07274.

37 Lu, X., Yin, Y., Zhang, L. et al. (2019). 3D Ag/NiO-Fe_2O_3/Ag nanomembranes as carbon-free cathode materials for Li-O_2 batteries. *Energy Storage Mater.* 16: 155–162. https://doi.org/10.1016/j.ensm.2018.05.002.

38 Yang, R., Gong, Z., Zhang, X. et al. (2019). Single-walled carbon nanotubes (SWCNTs) and Poly(3,4-ethylenedioxythiophene) nanocomposite microwire-based electronic biosensor fabricated by microlithography and layer-by-layer nanoassembly. *J. Nanosci. Nanotechnol.* 19 (12): 7591–7595. https://doi.org/10.1166/jnn.2019.16752.

39 Nazari, M., Kashanian, S., and Mohammadi, R. (2019). Electrodeposition of anionic, cationic and nonionic surfactants and gold nanoparticles onto glassy carbon electrode for catechol detection. *J. Nanoanal.* 6 (1): 48–56. https://doi.org/10.22034/jna.2019.664401.

40 Oviroh, P.O., Akbarzadeh, R., Pan, D. et al. (2019). New development of atomic layer deposition: processes, methods and applications. *Sci. Technol. Adv. Mater.* 20 (1): 465–496. https://doi.org/10.1080/14686996.2019.1599694.

41 Wang, D., Li, Y., Zhao, Y. et al. (2019). Cycling-induced capacity increase of graphene aerogel/ZnO nanomembrane composite anode fabricated by atomic layer deposition. *Nanoscale Res. Lett.* 14 (1): 69. https://doi.org/10.1186/s11671-019-2900-7.

42 Thackeray, M.M., Wolverton, C., and Isaacs, E.D. (2012). Electrical energy storage for transportation—approaching the limits of, and going beyond, lithium-ion batteries. *Energy Environ. Sci.* 5 (7): 7854–7863. https://doi.org/10.1039/c2ee21892e.

43 Huang, X.H., Xia, X.H., Yuan, Y.F. et al. (2011). Porous ZnO nanosheets grown on copper substrates as anodes for lithium ion batteries. *Electrochim. Acta* 56 (14): 4960–4965. https://doi.org/10.1016/j.electacta.2011.03.129.

44 Hong, Y., Yang, J., Xu, J. et al. (2019). Template-free synthesis of hierarchical NiO microtubes as high performance anode materials for Li-ion batteries. *Curr. Appl Phys.* 19 (6): 715–720. https://doi.org/10.1016/j.cap.2019.03.019.

45 Li, X., Qiao, L., Li, D. et al. (2013). Three-dimensional network structured α-Fe_2O_3 made from a stainless steel plate as a high-performance electrode for lithium ion batteries. *J. Mater. Chem. A* 1 (21): 6400–6406. https://doi.org/10.1039/c3ta10821j.

46 Hu, Z. and Liu, H. (2015). Three-dimensional CuO microflowers as anode materials for Li-ion batteries. *Ceram. Int.* 41 (6): 8257–8260. https://doi.org/10.1016/j.ceramint.2015.03.010.

47 Zhang, G., Zhang, H., Zhang, X. et al. (2015). Solid-solution-like ZnO/C composites as excellent anode materials for lithium ion batteries. *Electrochim. Acta* 186: 165–173. https://doi.org/10.1016/j.electacta.2015.10.133.

48 Zhou, Z., Zhang, K., Liu, J. et al. (2015). Comparison study of electrochemical properties of porous zinc oxide/N-doped carbon and pristine zinc oxide polyhedrons. *J. Power Sources* 285: 406–412. https://doi.org/10.1016/j.jpowsour.2015.03.124.

49 Sun, H., Mei, L., Liang, J. et al. (2017). Three-dimensional holey-graphene/niobia composite architectures for ultrahigh-rate energy storage. *Science* 356 (6338): 599–604. https://doi.org/10.1126/science.aam5852.

50 Liu, J., Li, Y., Ding, R. et al. (2009). Carbon/ZnO nanorod array electrode with significantly improved lithium storage capability. *J. Phys. Chem. C* 113 (13): 5336–5339. https://doi.org/10.1021/jp900427c.

51 Yu, M., Wang, A., Wang, Y. et al. (2014). An alumina stabilized ZnO-graphene anode for lithium ion batteries via atomic layer deposition. *Nanoscale* 6 (19): 11419–11424. https://doi.org/10.1039/c4nr02576h.

52 Li, Y., Zhao, Y., Huang, G. et al. (2017). ZnO nanomembrane/expanded graphite composite synthesized by atomic layer deposition as binder-free anode for lithium ion batteries. *ACS Appl. Mater. Interfaces* 9 (44): 38522–38529. https://doi.org/10.1021/acsami.7b11735.

53 Zhang, G., Hou, S., Zhang, H. et al. (2015). High-performance and ultra-stable lithium-ion batteries based on MOF-derived ZnO@ZnO quantum dots/C core-shell nanorod arrays on a carbon cloth anode. *Adv. Mater.* 27 (14): 2400–2405. https://doi.org/10.1002/adma.201405222.

54 Pan, S., Zhao, Y., Huang, G. et al. (2015). Highly photocatalytic TiO_2 interconnected porous powder fabricated by sponge-templated atomic layer deposition. *Nanotechnology* 26 (36): 364001. https://doi.org/10.1088/0957-4484/26/36/364001.

55 Deng, J., Ji, H., Yan, C. et al. (2013). Naturally rolled-up C/Si/C trilayer nanomembranes as stable anodes for lithium-ion batteries with remarkable cycling performance. *Angew. Chem. Int. Ed.* 52 (8): 2326–2330. https://doi.org/10.1002/anie.201208357.

56 Huang, S., Zhang, L., Liu, L. et al. (2018). Rationally engineered amorphous TiO_x/Si/TiO_x nanomembrane as an anode material for high energy lithium ion battery. *Energy Storage Mater.* 12: 23–29. https://doi.org/10.1016/j.ensm.2017.11.010.

57 Huang, G. and Mei, Y. (2012). Thinning and shaping solid films into functional and integrative nanomembranes. *Adv. Mater.* 24 (19): 2517–2546. https://doi.org/10.1002/adma.201200574.

58 Mei, Y., Huang, G., Solovev, A.A. et al. (2008). Versatile approach for integrative and functionalized tubes by strain engineering of nanomembranes on polymers. *Adv. Mater.* 20 (21): 4085–4090. https://doi.org/10.1002/adma.200801589.

59 Kundu, D., Talaie, E., Duffort, V. et al. (2015). The emerging chemistry of sodium ion batteries for electrochemical energy storage. *Angew. Chem. Int. Ed.* 54 (11): 3431–3448. https://doi.org/10.1002/anie.201410376.

60 Yabuuchi, N., Hara, R., Kajiyama, M. et al. (2014). New O_2/P_2-type Li-excess layered manganese oxides as promising multi-functional electrode materials for rechargeable Li/Na batteries. *Adv. Energy Mater.* 4 (13): https://doi.org/10.1002/aenm.201301453.

61 Xu, W., Zhang, K., Zhang, L. et al. (2016). SnS_2@graphene nanosheet arrays grown on carbon cloth as freestanding binder-free flexible anodes for advanced sodium batteries. *J. Alloys Compd.* 654 (5): 357–362. https://doi.org/10.1016/j.jallcom.2015.09.050.

62 Liu, B., Zhang, J., Wang, X. et al. (2012). Hierarchical three-dimensional $ZnCo_{(2)}O_{(4)}$ nanowire arrays/carbon cloth anodes for a novel class of high-performance flexible lithium-ion batteries. *Nano Lett.* 12 (6): 3005–3011. https://doi.org/10.1021/nl300794f.

63 Zhang, Y., Xie, J., Zhang, S. et al. (2015). Ultrafine tin oxide on reduced graphene oxide as high-performance anode for sodium-ion batteries. *Electrochim. Acta* 151: 8–15. https://doi.org/10.1016/j.electacta.2014.11.009.

64 Zhao, Y., Feng, J., Liu, X. et al. (2014). Self-adaptive strain-relaxation optimization for high-energy lithium storage material through crumpling of graphene. *Nat. Commun.* 5: 4565. https://doi.org/10.1038/ncomms5565.

65 Qu, B., Ma, C., Ji, G. et al. (2014). Layered SnS_2-reduced graphene oxide composite--a high-capacity, high-rate, and long-cycle life sodium-ion battery anode material. *Adv. Mater.* 26 (23): 3854–3859. https://doi.org/10.1002/adma.201306314.

66 Huang, S., Liu, L., Zheng, Y. et al. (2018). Efficient sodium storage in rolled-up amorphous Si nanomembranes. *Adv. Mater.* 30 (20): e1706637. https://doi.org/10.1002/adma.201706637.

67 Song, M., Tan, H., Chao, D. et al. (2018). Recent advances in Zn-Ion batteries. *Adv. Funct. Mater.* 28 (41): 1802564. https://doi.org/10.1002/adfm.201802564.

68 He, P., Yan, M., Zhang, G. et al. (2017). Layered VS_2 nanosheet-based aqueous Zn ion battery cathode. *Adv. Energy Mater.* 7 (11): 1601920. https://doi.org/10.1002/aenm.201601920.

69 Xu, C., Li, B., Du, H. et al. (2012). Energetic zinc ion chemistry: the rechargeable zinc ion battery. *Angew. Chem. Int. Ed.* 51 (4): 933–935. https://doi.org/10.1002/anie.201106307.

70 Zeng, Y.X., Lai, Z.Z., Han, Y. et al. (2018). Oxygen-vacancy and surface modulation of ultrathin nickel cobaltite nanosheets as a high-energy cathode for advanced Zn-ion batteries. *Adv. Mater.* 30 (33): 1802396. https://doi.org/10.1002/adma.201802396.

71 Zhang, H., Zhang, X., Li, H. et al. (2018). Flexible rechargeable Ni//Zn battery based on self-supported $NiCo_2O_4$ nanosheets with high power density and good cycling stability. *Green Energy Environ.* 3 (1): 56–62. https://doi.org/10.1016/j.gee.2017.09.003.

72 Bruce, P.G., Freunberger, S.A., Hardwick, L.J. et al. (2012). LiO_2 and Li-S batteries with high energy storage. *Nat. Mater.* 11 (1): 19–29. https://doi.org/10.1038/nmat3191.

73 Manthiram, A., Fu, Y.Z., Chung, S.H. et al. (2014). Rechargeable lithium-sulfur batteries. *Chem. Rev.* 114 (23): 11751–11787. https://doi.org/10.1021/cr500062v.

74 Manthiram, A., Chung, S.H., and Zu, C.X. (2015). Lithium-sulfur batteries: progress and prospects. *Adv. Mater.* 27 (12): 1980–2006. https://doi.org/10.1002/adma.201405115.

75 Fang, R.P., Zhao, S.Y., Sun, Z.H. et al. (2017). More reliable lithium-sulfur batteries: status, solutions and prospects. *Adv. Mater.* 29 (48): 1606823. https://doi.org/10.1002/adma.201606823.

76 Zang, Y., Pei, F., Huang, J.H. et al. (2018). Large-area preparation of crack-free crystalline microporous conductive membrane to upgrade high energy lithium-sulfur batteries. *Adv. Energy Mater.* 8 (31): 1802052. https://doi.org/10.1002/aenm.201802052.

77 Wei, B.B., Shang, C.Q., Wang, X. et al. (2020). Highly conductive VC embedded in carbon matrix as effective trapper and catalyst for Li-S batteries. *Chem. Commun.* 56 (91): 14295–14298. https://doi.org/10.1039/d0cc04741d.

78 Yang, C., Li, P., Yu, J. et al. (2020). Approaching energy-dense and cost-effective lithium-sulfur batteries: from materials chemistry and price considerations. *Energy* 201: 117718. https://doi.org/10.1016/j.energy.2020.117718.

79 Peng, H.J., Huang, J.Q., Cheng, X.B. et al. (2017). Review on high-loading and high-energy lithium-sulfur batteries. *Adv. Energy Mater.* 7 (24): 1700260. https://doi.org/10.1002/aenm.201700260.

80 Wang, L.L., Ye, Y.S., Chen, N. et al. (2018). Development and challenges of functional electrolytes for high-performance lithium-sulfur batteries. *Adv. Funct. Mater.* 28 (38): 1800919. https://doi.org/10.1002/adfm.201800919.

81 Canas, N.A., Wolf, S., Wagner, N. et al. (2013). In-situ X-ray diffraction studies of lithium-sulfur batteries. *J. Power Sources* 226: 313–319. https://doi.org/10.1016/j.jpowsour.2012.10.092.

82 Zheng, D., Wang, G.W., Liu, D. et al. (2018). Systematic and rapid screening for the redox shuttle inhibitors in lithium-sulfur batteries. *Electrochim. Acta* 282: 687–693. https://doi.org/10.1016/j.electacta.2018.06.117.

83 Hofmann, A.F., Fronczek, D.N., and Bessler, W.G. (2014). Mechanistic modeling of polysulfide shuttle and capacity loss in lithium-sulfur batteries. *J. Power Sources* 259: 300–310. https://doi.org/10.1016/j.jpowsour.2014.02.082.

84 Liu, Y., Yao, M.J., Zhang, L.L. et al. (2019). Large-scale fabrication of reduced graphene oxide-sulfur composite films for flexible lithium-sulfur batteries. *J. Energy Chem.* 38: 199–206. https://doi.org/10.1016/j.jechem.2019.03.034.

85 Luo, S.W., Yao, M.J., Lei, S. et al. (2017). Freestanding reduced graphene oxide-sulfur composite films for highly stable lithium-sulfur batteries. *Nanoscale* 9 (14): 4646–4651. https://doi.org/10.1039/c7nr00999b.

86 Chen, K., Cao, J., Lu, Q.Q. et al. (2018). Sulfur nanoparticles encapsulated in reduced graphene oxide nanotubes for flexible lithium-sulfur batteries. *Nano Res.* 11 (3): 1345–1357. https://doi.org/10.1007/s12274-017-1749-2.

87 Zhang, S.S. (2012). Role of $LiNO_3$ in rechargeable lithium/sulfur battery. *Electrochim. Acta* 70: 344–348. https://doi.org/10.1016/j.electacta.2012.03.081.

88 Chen, L., Huang, Z.N., Shahbazian-Yassar, R. et al. (2018). Directly formed alucone on lithium metal for high-performance Li batteries and Li-S batteries with high sulfur mass loading. *ACS Appl. Mater. Interfaces* 10 (8): 7043–7051. https://doi.org/10.1021/acsami.7b15879.

89 Poonam, S.K., Arora, A. et al. (2019). Review of supercapacitors: materials and devices. *J. Energy Storage* 21: 801–825. https://doi.org/10.1016/j.est.2019.01.010.

90 Gonzalez, A., Goikolea, E., Barrena, J.A. et al. (2016). Review on supercapacitors: technologies and materials. *Renewable Sustainable Energy Rev.* 58: 1189–1206. https://doi.org/10.1016/j.rser.2015.12.249.

91 Wei, L., Sevilla, M., Fuertes, A.B. et al. (2011). Hydrothermal carbonization of abundant renewable natural organic chemicals for high-performance supercapacitor electrodes. *Adv. Energy Mater.* 1 (3): 356–361. https://doi.org/10.1002/aenm.201100019.

92 Chen, X.L., Qiu, L.B., Ren, J. et al. (2013). Novel electric double-layer capacitor with a coaxial fiber structure. *Adv. Mater.* 25 (44): 6436–6441. https://doi.org/10.1002/adma.201301519.

93 Sruthi, T. and Kartick, T. (2019). Route to achieving enhanced quantum capacitance in functionalized graphene based supercapacitor electrodes. *J. Phys. Condens. Matter* 31 (47): 475502. https://doi.org/ARTN 475502.

94 Liu, S.P. and Pan, X.X. (2020). An aqueous hybrid supercapacitor employing honeycomb-shaped MoS2 nanosheets as anode. *J. Nanosci. Nanotechnol.* 20 (2): 1020–1027. https://doi.org/10.1166/jnn.2020.16907.

95 Feng, Y.Y., Li, Y.J., Yang, W. et al. (2020). Facile synthesis of nickel cobalt layered double hydroxide nanosheets intercalated with sulfate anion for high-performance supercapacitor. *J. Nanosci. Nanotechnol.* 20 (2): 1260–1268. https://doi.org/10.1166/jnn.2020.16983.

96 Fan, M.Q., Ren, B., Yang, X.D. et al. (2019). NiO@NiO and NiO@Co_3O_4 hollow core/shell composites for high-performance supercapacitor electrodes. *J. Nanosci. Nanotechnol.* 19 (12): 7785–7789. https://doi.org/10.1166/jnn.2019.16857.

97 Conway, B.E., Birss, V., and Wojtowicz, J. (1997). The role and utilization of pseudocapacitance for energy storage by supercapacitors. *J. Power Sources* 66 (1-2): 1–14. https://doi.org/10.1016/s0378-7753(96)02474-3.

98 Augustyn, V., Simon, P., and Dunn, B. (2014). Pseudocapacitive oxide materials for high-rate electrochemical energy storage. *Energy Environ. Sci.* 7 (5): 1597–1614. https://doi.org/10.1039/c3ee44164d.

99 Low, W.H., Khiew, P.S., Lim, S.S. et al. (2019). Recent development of mixed transition metal oxide and graphene/mixed transition metal oxide based hybrid nanostructures for advanced supercapacitors. *J. Alloys Compd.* 775: 1324–1356. https://doi.org/10.1016/j.jallcom.2018.10.102.

100 Wang, T., Chen, H.C., Yu, F. et al. (2019). Boosting the cycling stability of transition metal compounds-based supercapacitors. *Energy Storage Mater.* 16: 545–573. https://doi.org/10.1016/j.ensm.2018.09.007.

101 Snook, G.A., Kao, P., and Best, A.S. (2011). Conducting-polymer-based supercapacitor devices and electrodes. *J. Power Sources* 196 (1): 1–12. https://doi.org/10.1016/j.jpowsour.2010.06.084.

102 Wang, Y.K., Chen, F., Liu, Z.X. et al. (2019). A highly elastic and reversibly stretchable all-polymer supercapacitor. *Angew. Chem. Int. Ed.* 58 (44): 15707–15711. https://doi.org/10.1002/anie.201908985.

103 Abdah, M.A.A.M., Azman, N.H.N., Kulandaivalu, S. et al. (2020). Review of the use of transition-metal-oxide and conducting polymer-based fibres for high-performance supercapacitors. *Mater. Design* 186: 108199. https://doi.org/ARTN 108199.

104 Conway, B.E. and Pell, W.G. (2003). Double-layer and pseudocapacitance types of electrochemical capacitors and their applications to the development of hybrid devices. *J. Solid State Electrochem.* 7 (9): 637–644. https://doi.org/10.1007/s10008-003-0395-7.

105 Lee, H., Kumbhar, V.S., Lee, J. et al. (2020). Highly reversible crystal transformation of anodized porous V_2O_5 nanostructures for wide potential window high-performance supercapacitors. *Electrochim. Acta* 334: 135618. https://doi.org/ARTN 135618.

106 Liu, C.Y., Chen, Y.F., Dong, Z.Y. et al. (2019). Tuning Mn^{2+} additive in the aqueous electrolyte for enhanced cycling stability of birnessite electrodes. *Electrochim. Acta* 298: 678–684. https://doi.org/10.1016/j.electacta.2018.12.123.

107 Gui, Q.Y., Wu, L.X., Li, Y.Y. et al. (2019). Scalable wire-type asymmetric pseudocapacitor achieving high volumetric energy/power densities and ultralong cycling stability of 100 000 times. *Adv. Sci.* 6 (10): 1802067. 10.1002/advs.201802067.

108 Zhi, J., Deng, S., Wang, Y.F. et al. (2015). Highly ordered metal oxide nanorods inside mesoporous silica supported carbon nanomembranes: high performance electrode materials for symmetrical supercapacitor devices. *J. Phys. Chem. C* 119 (16): 8530–8536. 10.1021/acs.jpcc.5b01230.

109 Yan, S.X., Luo, S.H., Feng, J. et al. (2020). Rational design of flower-like $FeCo_2S_4$/reduced graphene oxide films: Novel binder-free electrodes with ultra-high conductivity flexible substrate for high-performance all-solid-state pseudocapacitor. *Chem. Eng. J.* 381: 122695. https://doi.org/ARTN 122695.

110 Liu, J.Q., Zheng, M.B., Shi, X.Q. et al. (2016). Amorphous FeOOH quantum dots assembled mesoporous film anchored on graphene nanosheets with superior electrochemical performance for supercapacitors. *Adv. Funct. Mater.* 26 (6): 919–930. 10.1002/adfm.201504019.

111 Guan, C., Qian, X., Wang, X.H. et al. (2015). Atomic layer deposition of Co_3O_4 on carbon nanotubes/carbon cloth for high-capacitance and ultrastable supercapacitor electrode. *Nanotechnology* 26 (9): 094001. https://doi.org/Artn 094001.

112 Daubert, J.S., Lewis, N.P., Gotsch, H.N. et al. (2015). Effect of meso- and micro-porosity in carbon electrodes on atomic layer deposition of pseudocapacitive V_2O_5 for high performance supercapacitors. *Chem. Mater.* 27 (19): 6524–6534. https://doi.org/10.1021/acs.chemmater.5b01602.

113 Lan, Y.Y., Zhao, H.Y., Zong, Y. et al. (2018). Phosphorization boosts the capacitance of mixed metal nanosheet arrays for high performance supercapacitor electrodes. *Nanoscale* 10 (25): 11775–11781. https://doi.org/10.1039/c8nr01229f.

114 Yang, S.S., Sun, L.J., An, X.H. et al. (2020). Construction of flexible electrodes based on ternary polypyrrole@cobalt oxyhydroxide/cellulose fiber composite for supercapacitor. *Carbohydr. Polym.* 229: 115455. https://doi.org/ARTN 115455.

115 Li, K., Wang, X.H., Wang, X.F. et al. (2020). All-pseudocapacitive asymmetric MXene-carbon-conducting polymer supercapacitors. *Nano Energy* 75: 104971. https://doi.org/ARTN 104971.

116 Zhou, Y., Lachman, N., Ghaffari, M. et al. (2014). A high performance hybrid asymmetric supercapacitor via nano-scale morphology control of graphene, conducting polymer, and carbon nanotube electrodes. *J. Mater. Chem. A* 2 (26): 9964–9969. https://doi.org/10.1039/c4ta01785d.

117 Nejati, S., Minford, T.E., Smolin, Y.Y. et al. (2014). Enhanced charge storage of ultrathin polythiophene films within porous nanostructures. *ACS Nano* 8 (6): 5413–5422. https://doi.org/10.1021/nn500007c.

118 Karbhal, I., Basu, A., Patrike, A. et al. (2021). Laser patterning of boron carbon nitride electrodes for flexible micro-supercapacitor with remarkable electrochemical stability/capacity. *Carbon* 171: 750–757. https://doi.org/10.1016/j.carbon.2020.09.053.

119 Ji, Z., Ma, D., Dai, W. et al. (2021). Anchoring nitrogen-doped carbon quantum dots on nickel carbonate hydroxide nanosheets for hybrid supercapacitor applications. *J. Colloid Interface Sci.* 590: 614–621. https://doi.org/10.1016/j.jcis.2021.01.102.

120 Sun, Y., Zhang, J., Liu, S. et al. (2021). An enhancement on supercapacitor properties of porous CoO nanowire arrays by microwave-assisted regulation of the precursor. *Nanotechnology* 32 (19): 195707. https://doi.org/10.1088/1361-6528/abe264.

121 Sun, Z., Fang, L., and Cai, J. (2021). A hierarchical composites constructed by ZIF-8 derived carbon core and Mn_3O_4/N-doped carbon shell as efficient polysulfide entrapment host for Li-S batteries. *J. Alloys Compd.* 860: 158461. https://doi.org/10.1016/j.jallcom.2020.158461.

9

Nanomembrane Robotics

Alexander A. Solovev

Fudan University, Department of Materials Science, 220 Handan Road, Shanghai 200433, People's Republic of China

9.1 Introduction

The term "(nano)-membrane" depends on the field where it is used. Modern nano/microtechnology is a diverse field, and its subdisciplines often operate with different terminologies, which can be often a discrepancy. For instance, a nanomembrane can be considered a porous two-dimensional (2D) millimeter-thick structure that contains nanometer-sized pores. In fact, nanomembranes, by definition, do not exclude the existence of nanopores. However, unless specified, our nanomembranes or nanomembots, discussed here, do not contain nanopores. Unlike a conventional definition of a membrane, i.e. a selective layer that allows certain substances/molecules to pass through but stops others, the definition of a nanomembrane in this book is different. Here, the term "nanomembrane" is used to describe a free-standing thin film with a thickness below 100 nm and with at least a few orders of magnitude surface-to-volume aspect ratio. Nanomembranes with quasi-2D layer/wall thicknesses below 10 nm can integrate unprecedented quantum, electronic, physical, mechanical, and chemical nanomembrane properties [1]. For example, optical resonators, radial magnetization, cell-containing containers, and stimuli-responsive structures. When the wall/layer of nanomembrane is comparable to the electron's wavelength, quantum confinement effects arise with modifications of electron energy gaps, energy levels, conduction bands, valence bands, and potential wells [2]. Rolled-up tubes made of inorganic nanomembranes technology enable integrating various materials' properties, effects, and functions. In nanotechnology approach to science, one can take a designer point of view by tailoring the materials' combinations, the nanoscale thickness of the layers, multiple rotations, an aspect ratio of tubes (length, diameter); we are pulling together elements from different disciplines, materials science, physics, electronics, chemistry, and sometimes biology. As a result, scientists and engineers create novel combinations of devices (with properties that often do not exist in nature) [1, 2]. A newly discovered method of "rolled-up nanotechnology" on polymers enabled the fabrication of highly

Nanomembranes: Materials, Properties, and Applications, First Edition.
Edited by Yongfeng Mei, Gaoshan Huang, and Xiuling Li.
© 2022 WILEY-VCH GmbH. Published 2022 by WILEY-VCH GmbH.

parallel arrays of strain-engineered tubular nanomembots with tunable layers' thicknesses, diameters, and aspect ratios [3, 4]. If catalysts are used, previously demonstrated for nanorods [5, 6], nano/microtubes, and other particles, catalytic layers/segments decompose chemical fuel and microstructures self-propel in their fuel solutions. In this book, the term "nanomembot" denotes exclusively a mobile nano/microstructure (that is robust enough to ensure self-propulsion) with a large aspect ratio [7]. Nanomembranes can roll up into tubes, be twisted into helical structures, be deposited on microparticles, or self-assemble in biomimetic capsules.

Engineers and scientists discovered how to construct nanoscale machines and robots, which can help explore the nano/microworld [8]. Initially, the nano/micromachinery research field had stalled due to inherent difficulty to answer the following questions: From where does the power come to tetherless/wireless nanobots? How can nanobots overcome Brownian diffusion and highly viscous environments at a very low Reynolds number? How to control the nanobots' motion? Which tasks can nanobots perform? It was recognized that environments, forces, and interactions in the nano/microworld can be quite different from the world where humans developed intuition. For example, if we take a reductionist approach to fabricate the nanobot – a construction that mimics macroscale robots is typically impractical. The nano/micromachinery must be redesigned [9]. Due to a high surface-to-volume ratio and negligible weight, nanostructures are critically affected by interfacial, electrostatic, Van der Waals, and viscous forces. As a result, nanomembots can "stick" to the surface if located in the air. However, if nano/–structures are placed in an aqueous solution, the electrochemical double layer forms at the solid–liquid interface. It helps to keep a nanoscale gap between the nanobot and the solid surface. Simultaneously, due to reduced weight, nanobots can move in fluids if they generate sufficient propelling forces. However, a robot with dimensions below 1 μm is highly influenced by the Brownian diffusion. The mean-squared displacement of Brownian particles changes linearly with time $<x^2> = 2Dt$, where D is the diffusion coefficient and t is time [10]. Biological cells steer their internal components by Brownian motion rather than fighting against it (with few examples of biological motor proteins, which overcome diffusion at the solid–liquid interfaces). Biological components diffuse, meet/collide, and react with each other more often than expected. For instance, in a cell with micron dimensions, two proteins can meet approximately twice a second. Likewise, diffusion of oxygen molecule across 10-μm distance takes less than a second, while across 1 cm distance it takes 6.6 hours. Nano/microbots with sizes $10–10^5$ times larger than oxygen molecules move even slower, and it would take a long time for them to travel over longer distances. Subsequently, it would be more practical if nanobots integrate a motor or engine. One prominent example of a biological micro-object that over-swims diffusion is a bacterium [11]. A bacterium swims, tastes its environment, and navigates in chemotactic gradients. Subsequently, the dynamic behavior of bacteria can be recorded and studied, such as displacement, speed, and trajectories of motion. For example, a 1-m-long neuron cell in a brain would take approximately 1000 years to deliver specific proteins by diffusive transport. However, motor proteins can deliver cargo in about one week [12]. Initially, it was complicated to integrate a nanoscale motor or engine on the board of a nano/microscale machine, because classical motors require a "tank" with the fuel. Nevertheless, at the submicron scale, carrying fuels in tanks would not last long due to a rapid diffusion or "leakage" of fuel molecules out of the tank. For this reason, biological nanomachines discovered strategies to convert local chemical energy available in their local environments into autonomous movements.

Scaling laws are applied to understand the origin of the differences between the macro- and the micro-objects and their properties. For example, scaling laws illustrate that shrinking a body size leads to a size reduction and different modifications of physical, mechanical, fluidic, electromagnetic, thermodynamic, optical, and biological effects [13]. Quantities of physical interest have dimensions, and the scaling laws predict a relationship between them. Examples discussed by Rogers, Adams, and Pennathur illustrate how small-scale robots can be significantly stronger (relatively) than their macroscopic counterparts [12]. For example, the effect of scaling on the strength-to-weight ratio is apparent in biological creatures. If we consider men's gold medal-winning lifts from the Olympic Games as a percentage of the competitors' weights, the smaller the competitor, the greater the percentage of his/her bodyweight he/she can lift. Furthermore, that can be explained by scaling laws. The strength of humans' muscles depends on the cross section (2D) area, and weight depends on the muscles' volume (3D). Based on the strength-to-weight ratio determined above, it can be explained why it can support several thousands of its body weight, but an elephant can only support around 1/10 its body weight. A nanobot is a million times smaller than ants. Subsequently, a large variety of phenomena that change with robot size can be calculated.

Figure 9.1 shows examples of various nanomembots with planar, helical, tubular, gripper-like, and spherical bodies' architectures. Figure 9.1a illustrates a thin-film robot self-assembled by the origami method [14]. Robot joints consist of plastic that

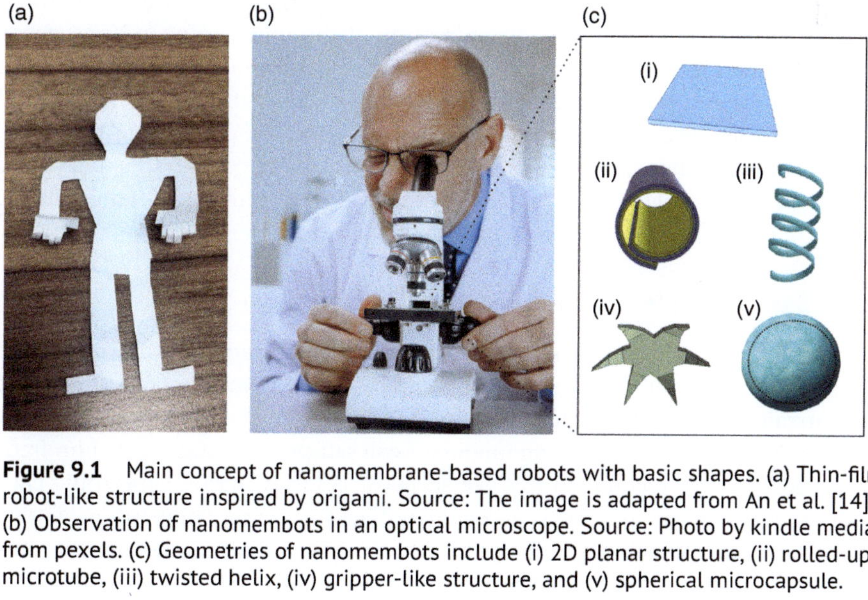

Figure 9.1 Main concept of nanomembrane-based robots with basic shapes. (a) Thin-film robot-like structure inspired by origami. Source: The image is adapted from An et al. [14]. (b) Observation of nanomembots in an optical microscope. Source: Photo by kindle media from pexels. (c) Geometries of nanomembots include (i) 2D planar structure, (ii) rolled-up microtube, (iii) twisted helix, (iv) gripper-like structure, and (v) spherical microcapsule.

heats up from 55 to 120 °C, leading to contracting and folding in minutes. Typically, an optical microscope is used to observe nanomembots, as shown in Figure 9.1b. Different nanomembranes have been fabricated to demonstrate multifunctional nanomembots with various body shapes: (i) planar, (ii) tubular, (iii) helical, (iv) gripper-like, and (v) spherical structures, for example (Figure 9.1c) [15], [16]. Recent achievements in the field of man-made nano/micromachines [17] led to demonstrations of autonomously and remote-controlled robots at the nano/microscale capable of performing specific tasks, such as "on-the-fly" delivery and assembly of cargo [18], detection of DNA hybridization [19], capture and removal/drilling of cancer cells [20], and the isolation of pathogens in complex media. Helical nanomembots mimic the motion of biological flagellum and bacteria with corkscrew-like shapes [21]. Emergent collective behaviors have also been reported for these synthetic catalytic nanomotors, in some cases mimicking the chemotaxis and phototaxis phenomena occurring in living systems [22]. A rich variety of nanomembots can be fabricated using 2D and 3D origami [23, 24], kirigami/origami [25], and 4D printing [26] methods. Besides standard microfabrication methods involving atomic layer, electron beam, and thermal depositions, nanomembots can be prepared using microfluidics. Recently, high progress has been achieved to explore spherical nanomembots enabled by nanoparticle-shelled microbubbles [27] and polymer/hydrogel [28, 29] microcapsules fabricated using microfluidic techniques. In fact, in biology, nanomembranes are essential building blocks of cells, encapsulating reactions' networks, organelles, genetic and hereditary information.

9.2 Motive Mechanisms

When nanomembots are released from the substrate, they require energy and integrated motors to propel and perform tasks. In most cases, macroscale swimming strategies cannot be applied to achieve motion at the microscale, and one needs to redesign the machinery. Biological microorganisms invented efficient working mechanisms of motion. And human nanorobots can adapt these mechanisms and at the same time discover novel nonbiological principles of propulsion. Here, two strategies are considered: (i) chemical-to-mechanical energy conversion and (ii) an application of external fields (magnetic field, electric field, ultrasound, temperature, and a light source) to power/control nanobots.

9.2.1 Basic Principles of Motion at the Microscale

Different forces and interactions with an environment dominate the motion of robots at different scales. If nanomembots can beat the kT limit imposed by nature and generate forces on the order of pico-to-micro-Newtown, they can successfully overcome Brownian diffusion. Nanomembots can self-propel ultrafast, i.e. hundreds of body lengths per second (in relative speed, it is faster than a supersonic car). Moreover, comparison of speeds research initiated a racing competition among various nano/micromachines with optimal catalysts types, size, design, aspect ratio, reaction–diffusion processes, and fuels' compositions [30]. It is a well-known fact that large rockets gain momentum due to expelled mass ($p = mv$, you can throw away more heavy mass or expel it at a faster speed to accelerate based on inertia, $F = ma$). Since inertia is negligible for the nanomembot, the main issue is a low Reynolds number, i.e. extremely high fluid viscosity. Various methods have been utilized to quantify the nanobots' efficiency and predict optimal parameters to achieve ultrafast motion at very low Reynolds numbers. Let us consider a swimmer, shown in Figure 9.2a. The conservation of momentum tells us that the momentum swimmer gives to a body going forward is the same as the swimmer pulls the water backward. If we consider a scallop (Figure 9.2b), it swims by the same principle using conservation of momentum. Inertia depends on the object's mass or volume (in three dimensions). However, the viscous force depends on surface area (in two dimensions). If we reduce the size of the scallop by 10 000 times, it turns out the microscallop cannot swim by the same mechanism. Moreover, that is a very

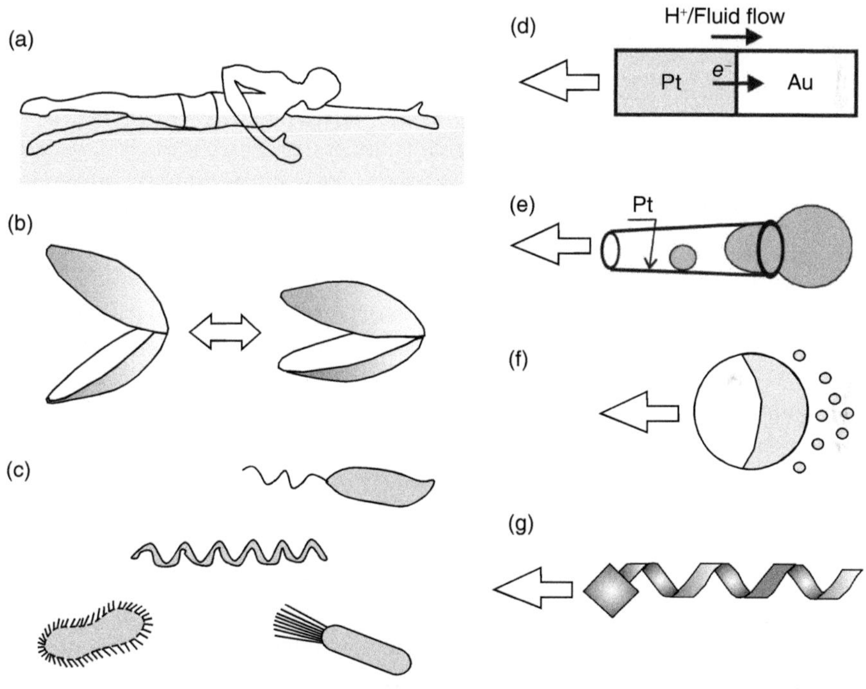

Figure 9.2 Illustrations of biological and synthetic swimmers at different scales. (a) A schematic image of a swimmer and forces acting on the body: drag, thrust, weight, buoyancy. (b) A schematic image of the scallop's motion by clapping its shells together and pumping a jet of water past the shell hinges. (c) Examples of bacteria with various motility strategies: nonreciprocal cilia beating, helical body motion, flagellum rotation, and comet tails, i.e. polymerization reactions. (d) A schematic of bimetallic Pt–Au catalytic nanorod-based nanomotor driven by the self-electrophoresis, i.e. by oxidation–reduction reaction in hydrogen peroxide solutions. (e) A schematic image of a catalytic tubular micromotor self-propelled by oxygen microbubble recoil and fluid/fuel self-pumping during decomposition of hydrogen peroxide. (f) Illustration of a Janus micromotor driven by the self-diffusiophoresis. (g) Biomimetic magnetic nanomembot that resembles motion of bacteria with helical shapes.

low Reynolds number that measures a ratio between inertial and viscous forces. Reynolds number becomes very small, and subsequently, a viscous force dominates. The Reynolds number is determined using the following equation:

$$Re = \frac{\rho \vartheta L}{\mu} \tag{9.1}$$

where ρ is the density of the fluid, ϑ is the object's velocity, L is the characteristic dimension of an object, and μ is the fluid's dynamic viscosity. Typically, laminar flow corresponds to the Reynolds number below 2000, and turbulent flow appears if Reynolds numbers are greater than 3500. For a nanomembot with a size of bacteria (Figure 9.2c), the Reynolds number is on order of 10^{-6}, and the fluid is highly viscous. Using Eq. (9.1), one can also estimate the viscosity that humans would feel if bodies are reduced to the size of the bacterium at a low Reynolds number. Humans at the scale of bacteria would feel the viscosity of water is similar to the viscosity of honey.

Nature discovered how energy-rich molecules (e.g. ATP) and facile fuels, including gradients of protons and salts, can power bio-nanomotors. In his seminal paper, Purcell proposed a "scallop theorem" stating that a reciprocal motion (a deformation with time-reversal symmetry) cannot lead to net propulsion at low Reynolds numbers [31]. At the microscale, time does not matter and the motion pattern is the same, whether an object moves slow or fast, forward or backward in time. For example, microorganisms achieve their propulsion by propagating traveling waves along their flagella (or rotating them) to break the time-reversibility requirement. Such microorganisms escape the scallop theorem's constraints. Purcell also estimated the time required for transport by stirring and time for transport by diffusion. It was demonstrated that to out-swim the diffusion, an object should swim a minimum distance determined as a constant diffusion ratio and the object's velocity. If a nanobot does not swim that far it has not gone anywhere. Subsequently, Brownian diffusion randomizes the motion of nanobots, particularly with a body size below 1 μm. The Peclet number (Pe) is a dimensionless number that measures the ratio of the convective over the diffusion transport rates

$$Pe = \frac{\vartheta L}{D} \tag{9.2}$$

where ϑ is the object velocity, L is the distance of swimming, and D is the diffusion coefficient. Subsequently, sub-microscale nanomembots must generate a high enough propelling force to overcome Brownian motion. The Peclet number measures the effectiveness of motion by stirring, i.e. a ratio between the advective and diffusive transport rates. Thus, to outswim the diffusion, the nanomembot must swim a minimum distance determined by the ratio $L \geq D/\vartheta$. For example, accepting the diffusion coefficient of hydrogen peroxide 10^{-5} cm^2/s, the nanomembot moving at speed 30 μm/s must swim at least 30 μm distance. According to Purcell, if the microswimmer does not go that far, it has not gone anywhere.

Figure 9.3a shows an object trying to move in a cycle (the solid line shows object shape; when it changes shape, the dotted contour moves). During conventional swimming, cyclic deformations and reciprocal motion are applied. Strikingly, swimming principles by reciprocal motion do not work at the microscale because inertial

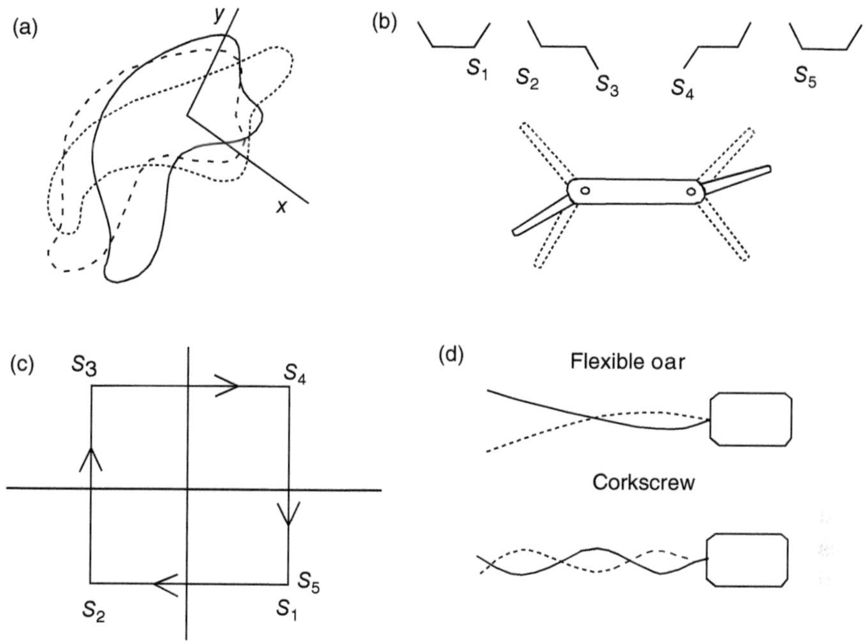

Figure 9.3 Examples of swimming strategies at a low Reynolds number considered by Purcell. (a) A hypothetical swimmer with four hinges. (b) An imaginary swimmer with two hinges. (c) A hypothetical swimmer going through sequences of configurations $S_1 \rightarrow S_5$. (d) Application of flexible oar and corkscrew to achieve motive force. Source: Reproduced from Purcell [31], with the permission of the American Association of Physics Teachers.

terms in the Navier–Stokes equation are neglected. Figure 9.3b demonstrates an imaginary animal with two hinges that can go through the sequences of configurations $S_1 \rightarrow S_5$ around the loop and it can swim (Figure 9.3c) [31]. Figure 9.3d is an example of micromotors consisting of a flexible oar and a corkscrew. These types of motions are not reciprocal in configuration space. Helical structures can rotate and translate microorganisms. Biological cells and bacteria propel themselves using nonreciprocal mechanisms of motion at low *Re*. However, Fischer noted that most biomedical fluids are non-Newtonian, where Purcell's scallop theorem is not valid. Fischer's lab demonstrated magnetically actuated nanobots that mimic a scallop reciprocal periodic motion. The nanobot can swim in shear-thickening and shear-thinning (non-Newtonian) fluids, i.e. by modulating fluid viscosity upon changing the shear rate [32].

Next, the dynamic parameters and forces of nanomembots are discussed. For a self-propelling nanomembot with a spherical shape, the drag force can be calculated using the Stokes equation, $F_d = 6\pi\mu r\vartheta$, where r is the sphere radius, μ is the fluid viscosity, and ϑ is the sphere velocity. Accepting the viscosity of water as 10^{-3} Ns/m^2, the nanomembot radius 5×10^{-7} m and velocity 3×10^{-5} m/s the drag force equals 2.8×10^{-13} N. Since the motive force equals mass m times acceleration a, one can estimate the rate at which the robot (weighing approximately 1 pg) would stop. According to the equation, $a = F_d/m$, one can calculate the rate at which the robot stops, where m is the mass of the nanomembot. Accepting the nanomembot mass 1×10^{-15} kg, the stopping rate is 280 m/s^2. Subsequently, the time the nanomembot would stop equals $t = \Delta\vartheta/a$ or 1×10^{-7} s. Thus, the distance the nanomembot travels while coasting to a halt is 2×10^{-12} m [12]. Theoretically, it is smaller than the diameter of a hydrogen atom (1.06×10^{-10} m). Ultimately, nanobots could potentially move with atomic precision if there would be no constant collision with surrounding water molecules, i.e. Brownian diffusion. Subsequently, the movements of nanobots are very different from the motion of macroscale machines. Suppose a flying airplane switches off its jet engines, then it continuously flies by inertia. However, a nanoscale machine immediately stops.

9.2.2 Chemically Powered NanomemBots

To self-propel, nanomembots require the continuous generation of a motive force due to a highly viscous environment. In chemistry, a catalyst is a substance that lowers the energy barrier and increases the rate of chemical reactions. Biological enzymes utilize catalytic reactions without themselves being altered or consumed during the reaction. Catalytically powered micromachines convert local chemical energy into motive forces using catalysts [33]. For instance, platinum metal is a perfect catalyst for hydrogen peroxide decomposition into water and oxygen, according to the reaction $2\ H_2O_2 \rightarrow 2H_2O + O_2$. Moreover, it was found that catalytic microparticles immersed in the hydrogen peroxide fuel self-propel by different mechanisms: self-electrophoresis, self-diffusiophoresis, dynamic surface tension, pumping of fluids, and a gaseous bubble recoil [34, 35]. Today, smart materials contribute significantly to the progress of nanomembots' development [36]. Various methods of fabrication of nanomembots have been shown, including electron-beam deposition [37], microfluidics [27], and electrochemistry [38].

Figure 9.4a shows bimetallic catalytic Pt/Au nanorods in hydrogen peroxide solution. These nanobots, independently discovered by Paxton et al. [5] and Ozin et al. [6], consist of 2 μm-long gold–platinum rods propelling themselves by self-electrophoresis. In this example, nanorods utilize simultaneous oxidation–reduction reactions. The gold surface catalyzes a reaction that takes protons, electrons, and H_2O_2 to produce H_2O. The reaction calls for protons and electrons to move in from the platinum side. Protons drag water and electrons flow inside the rod, and equal a number of protons, which move along the rod's surface. The protons' positive charges attract water molecules, which get dragged along the surface of the nanorod. Protons are produced and the platinum surface continuously generates the needed protons by catalyzing the breakdown of H_2O_2 into electrons, protons, and O_2. The nanorod operates like a motor (driven by electron/proton transfer), while the nanotubular structure is related to an engine (accumulates and expels gas). Another example, shown in Figure 9.4b, a concentration gradient of molecules around a Janus colloidal motor can cause the nanomembot spontaneous motion. The motive mechanism is known as self-diffusiophoresis – the concentration gradient of a molecular solute that acts on the surface of the micromachine [42].

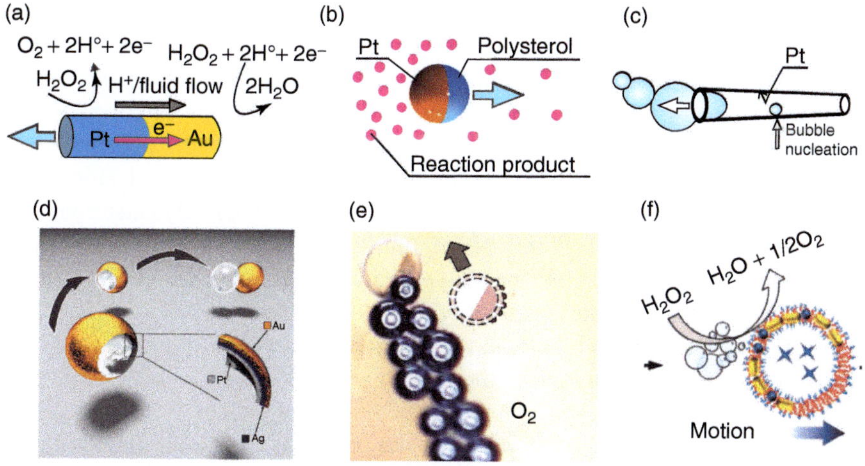

Figure 9.4 (a) Self-propelled catalytic nanomotor by self-electrophoresis due to chemical decomposition of hydrogen peroxide. (b) A schematic image of Janus nanomembot driven by self-diffusiophoresis. (c) A schematic image of catalytic tubular micromotor powered by oxygen bubbles recoil/pumping. (d) A schematic image of micromotor with hollow catalytic shells. Source: Reprinted with permission from Huang et al. [39]. Copyright 2013 American Chemical Society. (e) Optical micrograph of soft hydrogel micromotor with coated catalytic hemisphere. IOP Publishing. Source: Reproduced with permission from Zhu et al. [40]. All rights reserved. (f) A synthetic vesicle-like structure with enclosed catalytic nanoparticles. Source: Reproduced with permission from Wang et al. [41]. Copyright 2015, Wiley, VCH.

One of the most efficient motion mechanisms is based on the mass exchange, e.g. driven by gas evolution, interfacial surface tension, or microbubble recoils. Nanomembots are fabricated using rolled-up nanomembranes (Figure 9.4c) [1, 2], spherical nanoshell (Figure 9.4d) [39], hydrogel microcapsule with an externally coated catalyst (Figure 9.4e) [40], and vesicle-like nanomembot containing nanoparticles enclosed in the lipid membrane (Figure 9.4f) [41]. Catalysts' type, size, curvature, materials, and fuels' composition have crucial effects on gas bubbles formation in heterogeneous processes on catalytic surfaces. A successful generation of gas bubbles from bubble nuclei requires a gas supersaturation in/above the catalytic defect/cavity. For example, convex Janus microparticles generate bubbles only if the particle size is larger than 10 μm. In contrast, concave nanoshell microstructures with sizes below 2 μm generate and eject bubbles at high rates. An efficient bubble-propelled motion leads to ultrahigh velocities. Tubular polyaniline/Pt nanomembots with 2 μm diameter self-propel at 350 body lengths/s in 0.2% H_2O_2 solution [43]. Increasing the aspect ratio of tubes (length/diameter), molecular diffusion is confined, creating favorable conditions for bubble nucleation. For instance, by increasing the aspect ratio of tubes from 2 to 100 a substantial increase of the O_2 bubbles evolution is observed [44, 45]. The diffusion equation is used to estimate the time when molecules reach the opposite wall of the tube $t_{mix} = D^2/d$, where D is the diameter of the tube and d is the diffusion coefficient (if $d = 2.1 \times 10^{-9}$ m^2/s for O_2 in water and $D = 10$ μm, $t = 50$ ms). Nanotubes with sizes 20 nm do not generate bubbles; such nanotubes self-propel by self-diffusiophoresis [46]. Theoretical models predict the propulsion mechanisms [30, 47–49]. Importantly, during the operation of bubble-driven nanomembots, surfactants can be added into the fuel solution to decrease surface tension and stabilize microbubbles. The bubble recoil induces the pumping of chemical fuel/fluid into the rolled-up microtube. Despite a low chemical-to-mechanical energy conversion efficiency, on the order of 10^{-10}, nanomembots achieve high speeds at a very low Reynolds number [50]. However, a further increase in the concentration of chemical fuel does not lead to faster motion. Catalysts turnover number sets the maximum number of chemical conversions of substrate molecules per second that a single catalytic site can execute. In the future, optimization of nanomembots' geometry, reaction–diffusion processes, fuels can lead to higher efficiencies that will enable novel applications.

Figure 9.5a shows an array of fabricated tubular nanomembots using the rolled-up nanomembrane method. Figure 9.5b indicates the mechanism of oxygen microbubbles nucleation in the tubular microcavity: the bubble is absent (i), nucleates (ii), migrates (iii), and recoils (iv) from the microtube. Figure 9.5c shows a schematic image of bubbles nucleation and fluid pumping in a rolled-up microtube during decomposition of hydrogen peroxide. Figure 9.5d shows optical micrographs of self-propelled microtubes in hydrogen peroxide solution containing surfactants. Tubular nanomembots self-propel in straight, circular, and three-dimensional helical trajectories.

One of the main difficulties of tubular nanomembots is that a laminar fluid pumping through the microtube costs a lot of energy. Resistance to a laminar flow in the microtube can be calculated using the Hagen–Poiseuille flow derived from the Navier–Stokes equation $R = 8\mu L/\pi r^4$ [52], where μ is a fluid viscosity, L is the channel length, and r is the channel radius. Subsequently, the flow resistance depends on the fourth fold of diameter. And even a small variation in the diameter of the tubular nanomembot leads to a dramatic change in the fluid flow resistance. The pressure gradient ΔP to pump fluid along the tube is $\Delta P = 12v\mu/d^2$, where v is the average velocity of the fluid, μ is the viscosity of the fluid, and d is the diameter of the tube. Assuming the tube is 10 μm in diameter 2×10^8 Pa/m [12], at the microscale, reducing the tube radius by a factor of 10 leads to a 1000-fold increase in the pressure drop per unit length. To achieve pumping through microtubes, electrokinetic phenomena are typically applied, i.e. electroosmosis and electrophoresis. Assuming that the channel height is 100 nm and the fluid flow is 0.5 mm/s, (accepting the electrochemical double layer 3 nm, zeta potential −100 mV, the relative permittivity of water 78.3) it can be determined that the needed current is on the order of *fA*. Due to the significance of interfacial forces at the microscale, "pressurized" growing bubbles in microchannels cause an imbalance of interfacial forces. While migrating/recoiling bubbles in channels lead to efficient pumping mechanisms, such nanomembots self-propel by generating gas bubbles that eject and pump (hydrogen peroxide) fuel solutions.

One of the fundamental questions concerns the chemical-to-mechanical energy conversion efficiencies of man-made nanomembots (e.g. self-electrophoretic, bubble-propelled, ultrasonic, magnetic, and thermophoretic) [53, 54]. Inspired by the high energy conversion efficiency of biological nanomotors, Mallouk's lab

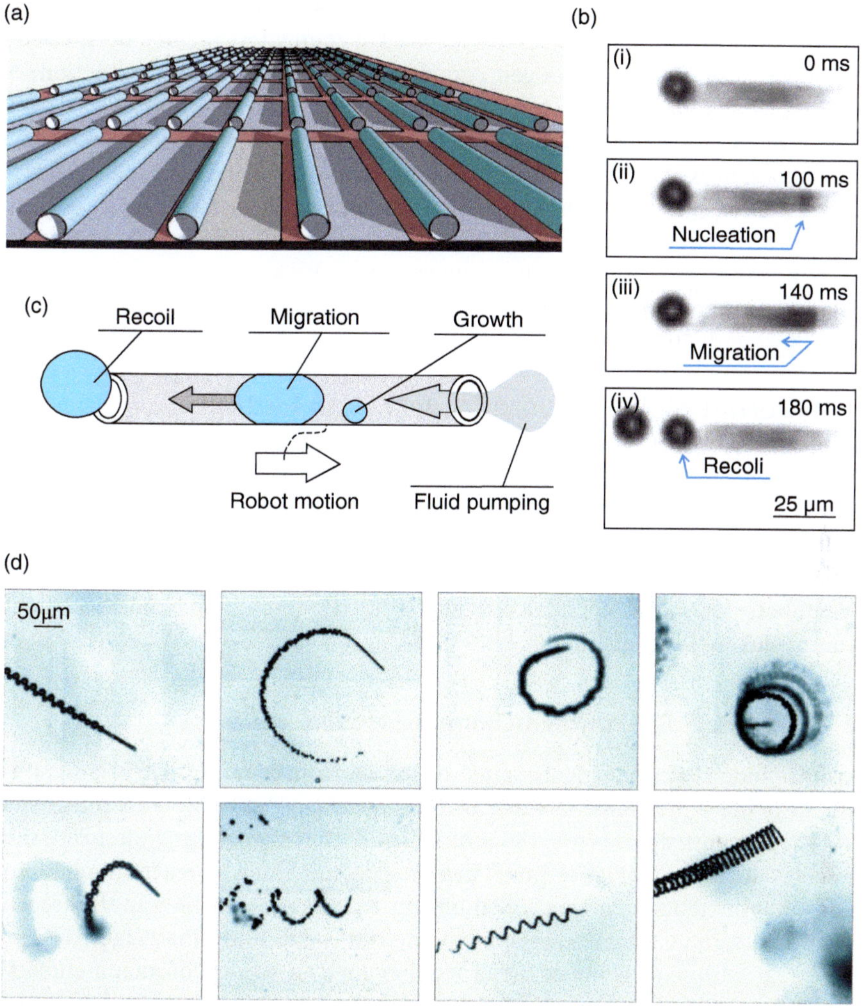

Figure 9.5 Nanomembot based on rolled-up catalytic nanomembranes. (a) An array of fabricated catalytic microtubes using the rolled-up angular deposition method. Source: Reproduced with permission from Solovev et al. [18] Copyright 2010, Wiley, VCH. (b) Bubble recoil mechanism of individual catalytic microtube by the recoil of oxygen bubbles (i–iv) during decomposition of hydrogen peroxide. (c) A schematic image of fuel/fluid self-pumping into the microtube due to nucleation, growth, and ejection of oxygen microbubble. (d) Self-propelled motion of tubular nanomembots in straight, circular, and three-dimensional helical trajectories. Source: Images (b)–(d) are adopted from the dissertation of Solovev, "Chemnitz University of Technology" [51]. Reproduced under the terms of the CC-BY Creative Commons Attribution 4.0 International License (https://creativecommons.org licenses/by/4.0/).

estimated efficiencies of self-electrophoretic and bubble propelled nanomembots to be on order of 10^{-9}–10^{-10} [50]. Potential ways of energy loss include nonelectrochemical decomposition of hydrogen peroxide, use of a very exoergic reaction to achieve a little potential drop along the nanobot surface, the fraction of electrical energy converted into mechanical work, and the electroosmotic flow that is opposite to the direction of electrophoretic propulsion. Another intriguing question concerns the discussion of how small the nanobot can be and if the dynamics of biological molecules can be better understood using perspectives of nano/micromachines. Recently, a controversy of diffusion enhancement by biological enzymes has been the subject of intense debates [55].

9.3 External Field–Powered/Controlled Motion

Nanomembots containing additional segments/layers can be powered using external sources and fields. Alternatively, chemically propelled nanomembots can be controlled or modulated using magnetic, electric, acoustic fields, electrochemical potential, light, and heat sources. Versatile control enables applications of nanomembots for a variety of tasks, including lab-on-a-chip, biomedical, and environmental applications.

9.3.1 Magnetic Field–Powered/Controlled Motion

A ferromagnetic (Fe) layer incorporated in tubular rolled-up geometry is demonstrated to achieve magnetic control of nanomembots [18, 56]. First magnetic control experiments started from the positioning of the magnet bar underneath the microscope, as shown in Figure 9.6a. Then, magnetized catalytic reaction–powered tubular nanomembots were controlled during their motion to pick up, transport, deliver, and assemble micro-objects, including beads and thin films. Later, a more sophisticated setup consisting of the closed-loop precise control of nanomembots' motion in three dimensions has been developed (Figure 9.6b). Nanomembots move straight using catalytic decomposition of hydrogen peroxide, while the orientation of nanomembots is controlled using magnetic torque. The closed-loop system is designed based on the current-magnetic force map. The U-turn technique is applied to characterize the magnetic dipole moment of the nanomembots (estimated to be 1.3×10^{-10} A/m^2 at speed of 100 μm/s and magnetic field of 2 mT) [57]. Figure 9.6c shows a typical helical structure nanomembot that is often referred to as "artificial flagellum" moving in the corkscrew type of motion using an external magnetic field. For example, using rapid polymerization, flow lithography combined with microfluidic spinning was demonstrated to fabricate helical nanomembots [58]. Subsequently, metal-coated polymer magnetic flagellum-like spermbots are demonstrated to carry (motion deficient) cells for fertilization-assisted operations [59].

Unlike conventional solid-state microbots, which cannot change their shape and motility type after fabrication, recently, reprogrammable shape-morphing

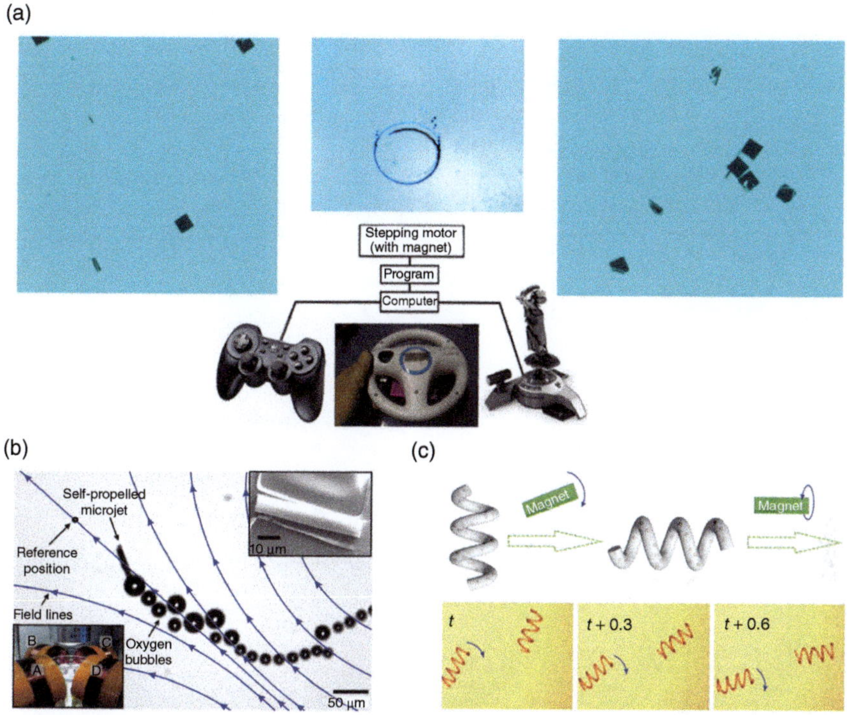

Figure 9.6 External control of nanomembots using a magnetic field. (a) Control of catalytic tubular nanomembot using an external magnetic field to transport and assemble micro-objects at the air-fluid interface. (b) A more sophisticated setup is shown using 3D coils for precise positioning, directionality, and speed control of nanomembots [57]. Source: Khalil et al. [57]PLOS ONE/CC BY 4.0. (c) Helical magnetic nanomembot actuated using an external magnetic field. Source: Reproduced with permission from Yu et al. [58]. Copyright 2015, Wiley, VCH.

nanomembots with embedded magnetic particles has attracted great attention. Encoding multi-shape-morphing nanomembots is particularly attractive for minimally invasive medicine, smart structures, and metamaterials. For instance, the Curie temperature effect can be used to heat and subsequently reorient magnetic domains by keeping the orientation of the magnetic field during cooling. Surface walking nanomembot and griper have been demonstrated [60]. Figure 9.7a shows a shape morphing: reorientation of ferromagnetic CrO_2 domains of PDMS elastomer using heating above the Curie temperature point (118 °C) and by applying an external magnetic field during cooling. Figure 9.7b shows the shortest heating–cooling cycle in a 1.3 mm diameter structure, demonstrated in 5.7 seconds. Figure 9.7c illustrates the demagnetization of the same material by heating again above the Curie point (w/o magnetic field). A laser (808 nm) is used for local heating of elastomer, followed by demagnetization of heated spots using a 15 mT external magnetic field (90% magnetization efficiency). Six-leg body robot is shown with fabricated legs, wings, and tail (Figure 9.7d–h). Parts of the robot can be demagnetized in different directions (Figure 9.7d,e). Measured magnetic flux density, i.e. magnetization

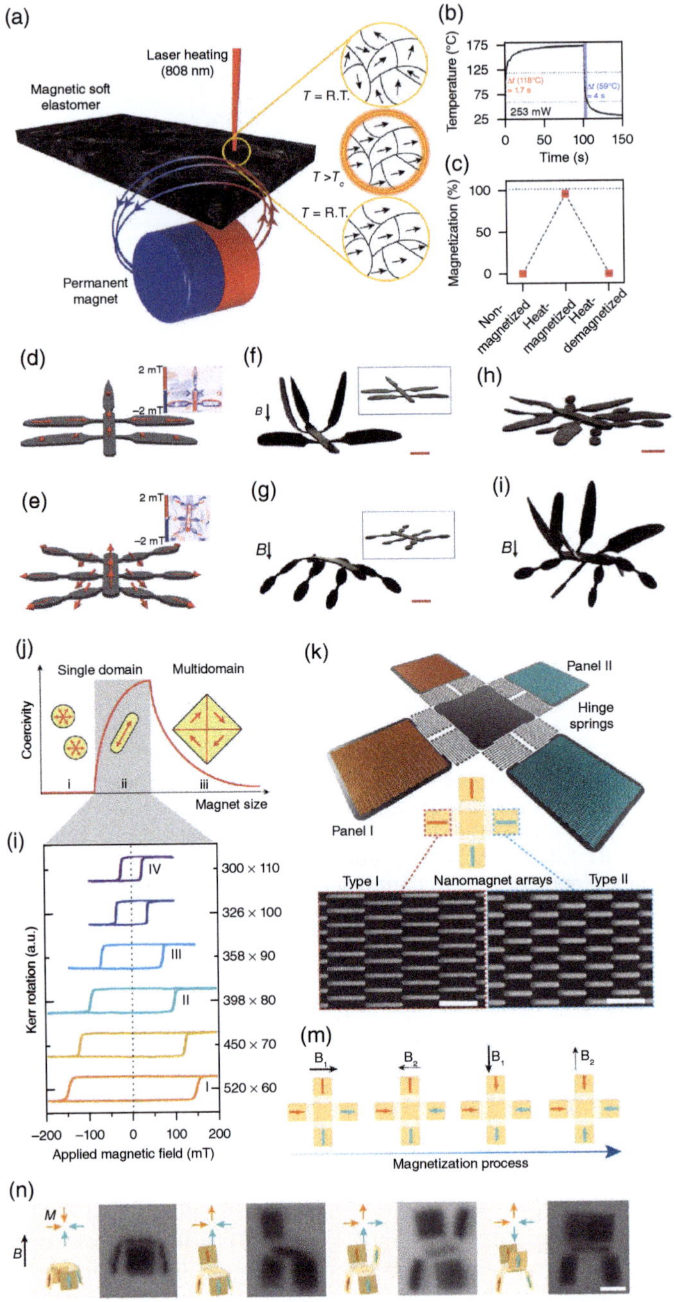
(a)
Laser heating
(808 nm)
Magnetic soft
elastomer
T = R.T.
T > Tc
T = R.T.
Permanent
magnet
(b)
Temperature (°C)
175
125
75
25
253 mW
0
50
100
150
Time (s)
(c)
Magnetization (%)
100
50
0
Non-magnetized
Heat-magnetized
Heat-demagnetized
(d)
2 mT
−2 mT
(e)
2 mT
−2 mT
(f)
B
(g)
B
(h)
(i)
B
(j)
Single domain
Multidomain
Coercivity
i
ii
iii
Magnet size
(i)
IV
III
II
I
300 × 110
326 × 100
358 × 90
398 × 80
450 × 70
520 × 60
Kerr rotation (a.u.)
−200
−100
0
100
200
Applied magnetic field (mT)
(k)
Panel II
Hinge
springs
Panel I
Type I
Nanomagnet arrays
Type II
(m)
B1
B2
B1
B2
Magnetization process
(n)
M
B

directions, is indicated in insets of Figure 9.7d,e. A magnetic field (60 mT) is applied perpendicular to the plane, creating a magnetic torque on elastomer structures, which have different directions of magnetizations, shown in Figure 9.7f,g. It is noteworthy to mention that various components can be stacked together to assemble the complex configurations nanomembot (Figure 9.7h–l).

In another example, Cui et al. demonstrated more complex programmable modular microbots consisting of arrays of nanomagnets (Figure 9.7j, region ii) [61]. Small magnets (between 100 and 500 nm) lead to the appearance of a single domain with a stable remanent magnetization and tunable magnetic anisotropy. Figure 9.7k shows a nanomembot design consisting of four panels with 60 nm thick decodable magnetic panels and a central nonmagnetic structure. The bottom image of Figure 9.7k illustrates the SEM image. For instance, magnets with different magnetic switching fields can be integrated on opposite panels, e.g. panels I and II (keeping the same volume, but varying the aspect ratios of the magnetic material). Figure 9.7l shows the magnetic hysteresis loops, where the coercive fields B_c (for switching the magnets) start from 30 mT for 300–110 nm magnets and 140 mT for 520–60 nm ratio magnets. Subsequently, it is feasible to encode the nanomembot by magnetic fields, represented in Figure 9.7m. When a magnetic field ($B_1 > B_c$(I)) is applied along the x-direction, it magnetizes both I and II magnets. A lower field (B_c(I) > B_2 > B_c(II)) is applied to magnetize the opposite direction panels (orange and turquoise horizontal arrows). As a result, if fields B_1 and B_2 are applied in sequence along the y-direction, it leads to the head-to-head magnetization of the other two panels. Various sequences of B_1 and B_2 fields can be used to achieve different configurations of the nanomembot. Finally, the nanomembot is released from

Figure 9.7 Reprogrammable shape-morphing structures effected by heat and magnetism. (a) Laser heating above Curie temperature of an elastomer containing CrO_2 particles in PDMS material. (b) Elastomer heating above the Curie temperature (118 °C) for 1.7 seconds and cooling to achieve half of the initial temperature. (c) Elastomer magnetization with 90% efficiency by magnetization and demagnetization using Curie temperature effect (w/o external magnetic field). (d)–(g) Specifically designed shape of elastomer nanomembot. Red arrows indicate magnetization directions and color bars show magnetic flux density strength. (f) and (g) Actuation of parts using a magnetic field. The scale bar is 2 mm. (h) and (i) Wings and legs parts are stacked to produce a three-dimensional hierarchical structure using a magnetic field (60 mT) applied in the directions pointed out by black arrows. The scale bar is 2 mm. (j) Engineering of four-panel shape-morphing nanomembot: magnetic states dependence on size, (i) superparamagnetic, (ii) stable single domain, and (iii) multidomain state. Red arrows show directions of magnetization. The red line represents the size relation of the coercivity of the magnets. Shown nanomembots employ region *ii* nanomagnets (a gray area). (k) Design of individual nanomembot with dimensions 520–60 nm (type I) magnets on panel I and 398–80 nm (type II) magnets on panel II. (l) Magneto-optical Kerr hysteresis loops measurement of a single domain magnets (the same volume, but with six different aspect ratios; sizes are shown in the right corner). (m) Configurations of the encoding of the nanomembot using two magnetic fields. (n) Magnetic positioning using type I and type II magnets, leading to four different shapes of the nanomembot under applied magnetic field $B = 15$ mT. Scale bars are 500 nm (b) and 10 μm (images in [e]). Source: (a)–(i) Adapted from Alapan et al. [60]. Images (j)–(n) are reproduced with permission from Cui et al. [61] Copyright 2019, Springer Nature.

the surface and powered by magnetic torque, $\tau = m \times B$, where B is the magnitude of an applied magnetic field and m is the total magnetic moment of the magnetic array on a panel. Shape-morphing nanomembot is demonstrated in Figure 9.7n movable by controlling actuation fields (<15 mT) and switching (30–140 mT) fields for I–IV magnets. Interestingly, Cui et al. showed how a nanomembot with a bird-like configuration could exhibit more complex movements, such as hovering, side-slipping, flapping, and turning (not shown here) [61]. The discussed techniques will undoubtedly help develop more sophisticated morphable 2D and 3D nanomembots for a wide variety of applications.

9.3.2 Light-Powered Motion

A large growing body of literature has focused on the light-powered nanomembots, which are of special interest to explore novel motion mechanisms and applications. Essential questions have arisen, including how to drive the mechanism of light actuation, choose and optimize the elemental composition of nanomaterials, maximize light-harvesting properties, utilize oxidation–reduction potential, thermal and plasmonic effects. Light absorption must be maximized while providing sufficient energy to facilitate molecular reduction and oxidation. For example, a light-powered motion of TiO_2 [62, 63], ZnO [64], and quantum-dot-sensitized motors [65] has been demonstrated. The catalytic tubular nanomembot power was switched "on" and "off" on-demand in hydrogen peroxide solutions using a white light source, specific laser wavelength, and intensity [66]. In another example, superfast near-infrared light-powered polymer microrockets showed specific promise [67]. Light-powered motion can lead to efficient ways for precise motion control of nanomembots [68].

Figure 9.8a–d shows main mechanisms, which can be applied to light-driven nanomembots (i.e. gas evolution, photothermal, osmotic, and electrophoretic) using different light sources (NIR, Vis, and UV) [69]. Figure 9.8a shows a Janus nanomembot driven by the photothermophoresis powered by NIR irradiation. Light energy can be absorbed and dissipated as heat, forming a temperature gradient around the robot surface. Subsequently, the light intensity can be used to control the velocity of the nanomembot. Figure 9.8 illustrates individual nanomembot driven by self-electrophoresis due to photocatalytic distribution of generated electrons, protons, and ions. A combination of TiO_2 and SiO_2 can be used, whereupon UV irradiation electrons are transferred from the valence band to the conduction band of the nanomembot aiding in photocatalytic decomposition of hydrogen peroxide, diffusion of protons, and propulsion toward the TiO_2 end of the nanomembot. However, it is mentioned that light-driven self-electrophoretic robots cannot reach high speeds. Using another example, the schematic image of Figure 9.8c demonstrates a tubular bubble-driven nanomembot. The bubble-propelled mechanism is well known for previously discussed chemical fuel (hydrogen peroxide)–driven tubular micromotors. However, light-powered nanomembots, driven by hydrogen peroxide, can be implemented with power conversion efficiency approximately 10 times higher than convention bubble-propelled hydrogen peroxide–powered nanomembots. Since amorphous TiO_2 generates peroxide complexes under light

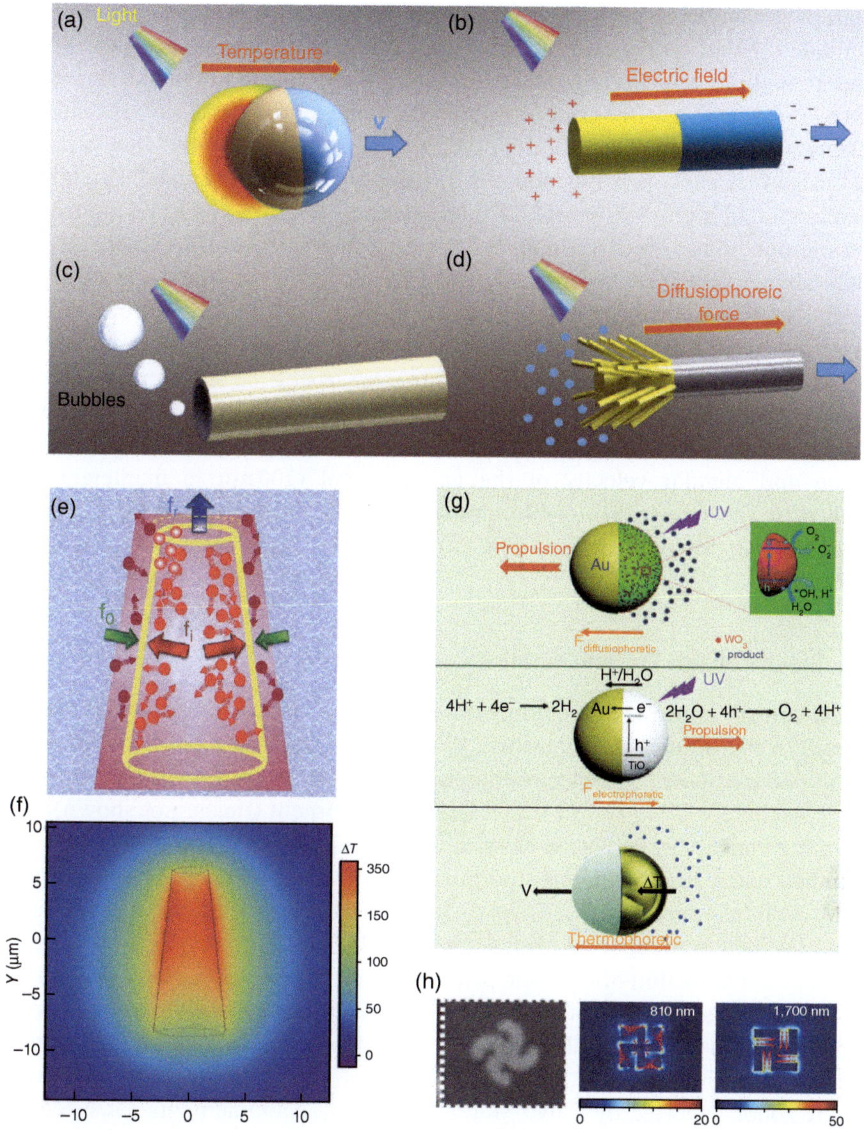

Figure 9.8 Light-powered motion of nanomembots using catalytic, thermal, plasmonic, diffusion, and oxidation–reduction properties of materials. (a) A photothermally driven nanomembot moving by variation of the temperature gradient. (b) A self-electrophoresis-powered nanomembot, driven by the self-induced electric field and fuel pumping. (c) A tubular nanomembot powered by bubbles. (d) The nanomembot based on asymmetric brunches driven by diffusiophoresis. (e) and (f) A polymer tubular microbot driven by light-induced thermophoresis. (g) A Janus nanomembot powered by photocatalytic reactions. (h) A nanomembot structure with rotations induced by light. Source: Images (a)–(d) Zhan et al. [69] are reproduced under the terms of the non-commercial-NoDerivs 3.0 unported (CC BY-NC-ND 3.0) attribution. Images (e) and (f) are reproduced with permission from Wu et al. [67]. Copyright 2015, Wiley, VCH. Image (g) is reprinted with permission from Zhang et al. [70]. Copyright (2015) American Chemical Society. Image (h) is reproduced with permission from Liu et al. [71] Copyright 2010, Springer Nature.

irradiation, generated electrons are transferred toward peroxide complexes, leaving accumulated holes, enhancing the decomposition of hydrogen peroxide (that produces O_2 bubbles). Figure 9.8d shows a self-diffusiophoretic nanomembot driven by the diffusive solute concentration difference, such as ·OH, O^{2-} radicals. If O_2 is generated during the decomposition of hydrogen peroxide on the TiO_2 surface, the diffusive flux of O_2 is responsible for the nanomembot movement. The lab of He demonstrated a polymeric tubular nanomembot with decorated Au nanoshell (AuNSs) propelling at speeds as high as 160 μm/s, shown in Figure 9.8e,f [67]. The nanomembot is powered by plasma resonance absorption of metallic structure using NIR light, leading to the appearance of a heat gradient and thermophoretic forces. Ren's group fabricated light-powered 1-μm diameter Au–WO_3@C nanomembot propelling by diffusiophoresis as fast as 16 μm/s using 40 mW/cm^2 UV light (in water), illustrated in Figure 9.8g [70]. Zhang and coworkers realized nanoscale plasmonic robot powered by polarized light, shown in Figure 9.8h [71]. The direction and angular velocity of the nanomembot (100 nm in diameter) are controlled by the wavelength of illuminated light via different plasmonic modes (photons linear, angular momentum influence on adsorbing, and scattering processes).

9.3.3 Electrical and Electrochemical Control

Electrochemical control of nanomembots is demonstrated using electrodes immersed in the solution. For instance, Wang's lab showed motion modulation of catalytic nanorod-based self-electrophoretic nanomembots powered by hydrogen peroxide decomposition [72]. In the described experiment (image not shown), an Au working electrode was placed close to nanorods. The motion of nanomembots was studied using 10-second potential pulses of +1.0 and −0.4 V (versus Ag/AgC, 100 mM KCl reference electrode) in a 5 wt% H_2O_2 fuel solution. Accelerating nanorods' velocities is reported from 4 μm/s (−0.4 V) to 22 μm/s (+1.0 V). Authors proved that the reduction of O_2 is the primary factor responsible for acceleration at negative potentials, while, in other experiments, electrical control of nanomembots is implemented using external electrodes, typically located outside the solution, where nanomembots operate. Figure 9.9a shows the electrical control of the nanowire robot, reported by Calvo-Marzal et al. [73]. Nanomembots consisted of polypyrrole–cadmium (PPy–Cd) nanowire hybrids fabricated using electropolymerization of PPy and electrodeposition of Cd. A spatially uniform AC square-wave electric field is applied to propel the nanowire at 17.2 μm/s by electroosmotic flow. Velev's group developed theory and experiment to power semiconductor particles using external fields considering motors' size, shape, electric polarizability, coatings, and electrokinetics: induced charge electrophoresis, dielectrophoresis, related effects. Figure 9.9b shows one example of the electrical circuit diagram of a p–n junction diode in water [74]. When AC electric field is applied, the diode is propelled (large red and gray spheres are positive and negative charges moving

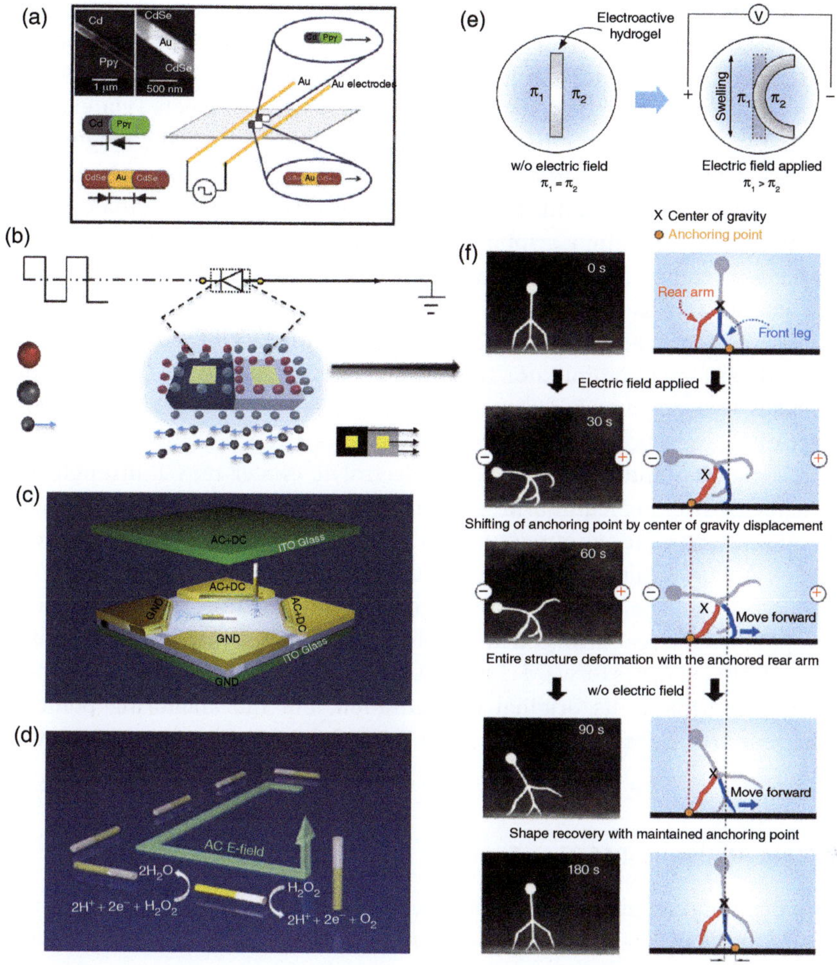

Figure 9.9 Electrical/electrochemical control of nanomembots. (a) Propulsion of nanowire diode-based nanomembot powered by an external electrical field. Source: Copyright 2010, The Royal Society of Chemistry. Calvo-Marzal et al. [73] (b) Electrical circuit diagram of a p–n junction diode located in water. Source: Reproduced under the terms of the CC-BY Creative Commons Attribution 4.0 International License (https://creativecommons.org licenses/by/4.0/). Ohiri et al. [74] (c) Setup to control catalytically driven nanomembot in an electric field. Source: Image is reproduced with permission from Guo et al. [75] Copyright 2018, American Chemical Society. (d) Schematic of trajectory change of catalytically powered nanomembot-based robot. Source: Image is reproduced with permission from Guo et al. [75] Copyright 2018, American Chemical Society. (e) Schematic image of polymer actuation using an electric field. Source: Image is reproduced with permission from Han et al. [76] Copyright 2018, American Chemical Society. (f) Walking of alien-like structure using an electric field. Source: Han et al. [76]. Reproduced with permission of American Chemical Society.

across the diode, small gray spheres with blue arrows are negative counter ions in the fluid, correspondingly). Fan's group demonstrated catalytic nanorods' AC and DC electric field guidance using a 3D orthogonal microelectrode setup (Figure 9.9c,d) [75]. An AC field is used to align the nanomembots (via electric torque on the induced dipoles). In contrast, the DC field is used to change their speed (via electrophoretic, electroosmotic effects). Recently, electroactive hydrogels (EAHs) showed great promise in designing electric field–powered actuators, for example, for directional motion, gripping, and delivery of objects (Figure 9.9e) [76]. In the discussed example, the photocurable precursor solution for EAH contains 13.3 M of acrylic acid as a monomer, 145 mM poly (ethylene glycol) diacrylate (PEGDA 700) as a cross-linker, and 43 mM of phenylbis(2,4,6-trimethylbenzoyl) phosphine oxide as the photoinitiator. Subsequent placement of EAH in electrolyte solutions leads to a polymer bending under electric field due to EAH ionization in the electrolyte, osmotic pressure is created by cations concentrations difference between the material and the surrounding electrolyte (water moves through the EAH to achieve swelling equilibrium). For instance, EAH film bends toward the cathode and the bending curvature can be controlled using an electric field (Figure 9.9e). A mobile alien-like EAH structure was recorded using manipulations performed by the electric field. The structure moves toward the rear, shifting the anchoring point to the rear arm (red arm) under the electric field. As the structure deforms, the front leg slides forward (Figure 9.9f). When an electric field is switched off, the robot returns to its original configuration, but the anchoring point is still on the rear arm. Subsequently, the front leg slides forward, moving the entire structure.

9.3.4 Other Methods of Motion Control

The motion of nanomembots can be programmed using designed shapes, which influence motive and drag forces balance. For instance, using rolled-up nanomembranes' straight, circular, and three-dimensional helical autonomous movements can be achieved. Mei's group designed grating structured walls of tubular nanomembranes as guided empennages [77]. Heat pulses have been applied to achieve thermal modulation of catalytic nanomembot' speeds [78]. Another essential method to power/control motion of nanomembots is ultrasound-modulated propulsion [79–82]. For instance, an air bubble has been included in a soft swimmer. When ultrasound is applied, the bubble oscillates and propels the nanomembot [83]. Cylindrical channels that trap bubbles with a diameter around 60 μm are demonstrated to alternate flows of intake and discharge through the opening, propelling micromachines. By increasing the Reynolds number, the difference between the intake and discharge streaming flows achieves higher values, creating a motive force [84]. To maximize the oscillating amplitude of the bubbles, a resonant frequency is used. In another example, acoustically powered nanoshells are driven by the acoustic streaming stress over the asymmetric surface [85]. Ultrasound-powered nanomembots do not require chemical fuels, and simultaneously, they can be combined with state-of-the-art biomedical imaging methods (see Section 9.4).

9.4 Potential Applications

The first generation of nanomembots is used to perform simplified tasks related to their property to overcome diffusion, generate motive forces, and transport nano/micro-objects. Since the discovery of nano/microbots, engineers and scientists envisioned their multiple potential applications, such as environmental remediation, on-the-fly transport and assembly of micro-objects, minimally invasive surgery, biomedical cargo (drugs, cells) delivery, artificial fertilization, roving biosensors, isolation of pathogens, cancer cells, ultrasound-driven imaging, and theranostics. Below our discussion is focused on biomedical and environmental nanomembots.

9.4.1 Biomedical Applications

A large and growing body of literature has investigated biomedical applications of nanomembots [86]. Bio-applications of nanomembots include isolation of bacteria [87], microsurgical operations [88], cells transportation [89], delivery and release of drugs [90, 91]. Another essential direction is utilization of nanomembots in biosensing [92]. For instance, in immunoassays, antibody-functionalized robots can be used to detect the cortisol target analyte [93]. Nucleic acids are extracted from solutions, such as employing cationic polyethyleneimine (PEI)-functionalized robots using carbodiimide chemistry with inner Ni/Pt layers and an outer cationic PEI layer [94]. Ultrasound-driven gold nanowires functionalized with graphene oxide and dye-labeled single-stranded DNA are used to detect human papillomavirus (HPV16 E6 mRNA transcripts) [95].

Figure 9.10a shows externally controlled multimodal magnetoelastic nanomembot with several degrees of freedom [96]. The nanomembot moves in a stomach phantom and can switch between different modes, i.e. swim, climb, roll, jump, crawl, and walk. Figure 9.10b shows a rolled-up nanomembrane microbot with a sharp end that drills into cancer cell material [97]. Microbots with sharp parts have significantly smaller contact areas. These microbots can exert a large contact pressure to penetrate soft tissue. However, for motion in/through tissue and depending on bioapplication: (i) the mass of the nanomembot must be high enough to increase the penetration force and/or (ii) the size of the nanomembot must be small enough. Figure 9.10c shows an example of helical nanomembots with a size below 100 nm, which can penetrate the dense body of the eye, propelling toward the retina [98]. A coating with specific molecules is used to decrease the friction of nanomembots with the dense tissue of the eye. Clinical optical coherence tomography was used to monitor the movement of helical nanobots. In another promising approach, an actuation of spermbots was demonstrated using an external magnetic field. Spermbots are capable of sperm cells delivery for potential *in vivo* reproduction, shown in Figure 9.10d [59]. A versatile magnetic control is used to manipulate stimuli-responsive magnetic microgrippers through channels, shown in Figure 9.10e. Such thermobiochemically driven microgrippers have specific promise toward biopsy and related biomedical procedures [99]. Figure 9.10f,g demonstrates

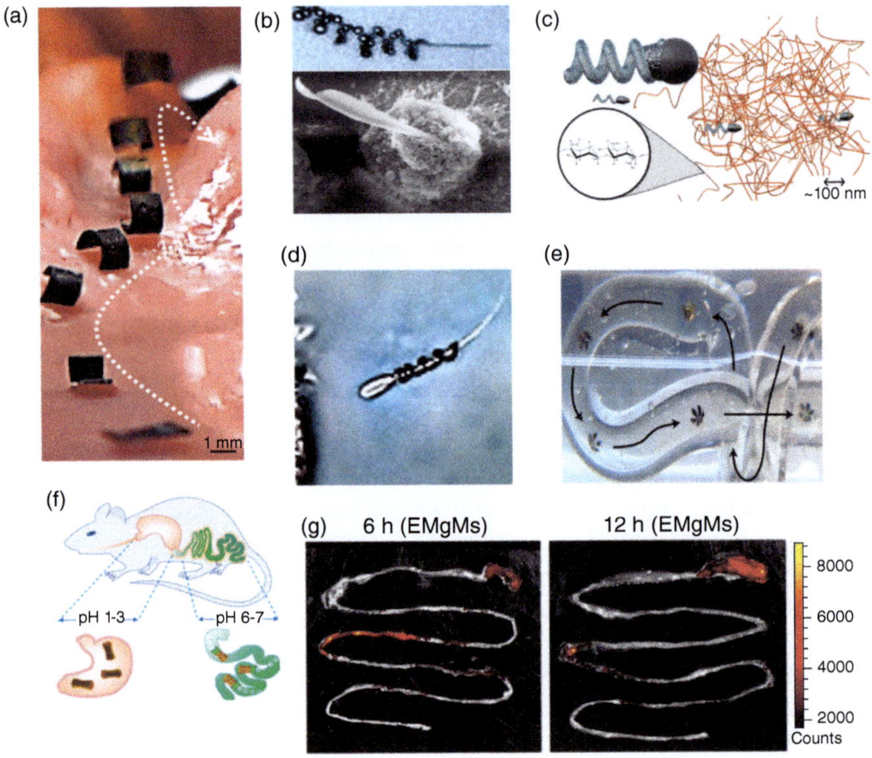

Figure 9.10 Examples of biomedical applications of microbots. (a) Movement of magnetic elastic multimodal robot in stomach phantom. Source: Hu et al. [96]. Reproduced with permission of Springer Nature. (b) Self-propelled nanotubes with sharp end drilling cancer cell material. Source: Solovev et al. [97]. Reproduced with permission of American Chemical Society. (c) Helical nanobot operating that can propel through the dense tissue of the biological eye. Source: Modified from Wu et al. [98]. (d) Magnetic spermbot delivering sperm cell during fertilization. Source: Medina-Sánchez et al. [59]. Reproduced with permission of American Chemical Society. (e) Magnetic microgrippers operating in a complex geometry of three-dimensional channels. Source: Leong et al. [99]. Reproduced with permission of PNAS. (f) *In vivo* micromotor-enhanced drug delivery in the GI tract of mice. Source: Image is reproduced with permission from Li et al. [100] Copyright 2016, American Chemical Society. (g) Superimposed fluorescent images of mouse GI tracts after 6 and 12 hours post-administration of magnesium microbots loaded with the dye rhodamine 6G. Source: Li et al. [100]. Reproduced with permission of American Chemical Society.

magnesium-based microbots' precise positioning and controllable retention in the specific locations of the mouse gastrointestinal tract [100]. Microbots deliver cargo at the target site due to the dissolution of pH-sensitive PEDOT coating. Figure 9.10g indicates superimposed fluorescence images of *in vivo* distribution of nanomembots in the tract of mice after 6 and 12 hours post-administration (red color shows a fluorescent signal).

Research on the subject has been mostly restricted to solving challenges of biocompatible fuels. Today, by taking interdisciplinary approach a fully biocompatible materials can be applied to prepare bio-nanomembots in combination with

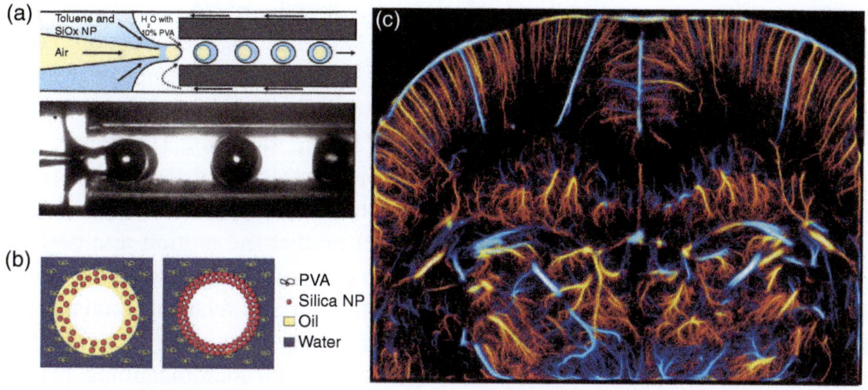

Figure 9.11 Applications of bubble-based nanomembots for ultrasound imaging. (a) The capillary-based microfluidic device generating nanoparticle-shelled microbubbles. Source: Adams et al. [27]. Reproduced with permission of John Wiley & Sons, Inc. (b) A schematic image showing the bubbles' shell stabilization process by nanoparticles when oil evaporates. Source: Reproduced with permission from Adams et al. [27]. Copyright 2020, Wiley, VCH. (c) Superresolution ultrasound imaging of the (animal) brain using microbubble contrast agents. Source: Hingot et al. [104]. Reproduced with permission of Springer Nature.

fluorescent tomography, optical coherence tomography, surface-enhanced Raman scattering, magnetic resonance, photo-acoustics, and ultrasound imaging [101]. Today, ultrasound at hospitals can be used to observe organs and large arteries. However, suppose a patient requires detailed research of the cardiovascular system or microcapillaries. In that case, a conventional ultrasound technique shows little contrast due to very similar mechanical properties of tissue and blood. Microbubble-based nanomembots (with sizes below the smallest capillaries, i.e. sub-10 μm) with desired gaseous core and mechanically elastic shells can oscillate, scatter ultrasound, and drastically enhance *in vivo* imaging contrast of the smallest capillaries [102, 103]. Figure 9.11a shows microfluidics-based fabrication of silica microbubbles using air-in-oil-in-water (A/O/W) emulsions [27]. Figure 9.11b shows a schematic image of microbubble shell stabilization after drying of oil. Due to the hydrophobic property of silica nanoparticles, they form rigid shells. Figure 9.11c illustrates a super-resolution (colorful) image of blood flow in rat brain using ultrasound-driven noninvasive imaging modality [104]. Subsequently, principles of nanomembots control of precise motion, speed, and position can be applied to enhance biomedical imaging and theranostics.

9.4.2 Water Cleaning and Environmental Remediation

Wastewater can contain various contaminants, including microbes, viruses, nanoparticles, organic molecules, heavy metal ions, and other trace elements. Different mechanisms must be used to adsorb, separate, decompose, and decontaminate various pollutants. Environmental remediation and cleaning of polluted water using mobile nanomembot, including high surface area materials and photocatalysts, is a new approach that started only a decade ago [105–108]. Nanomembots increase mobility and reduce time of cleaning operations, in comparison to static materials that rely only on diffusion [109–111]. For instance, tubular TiO_2 nanomembots with inner and outer surface Pt coatings were demonstrated to achieve light-induced decomposition of rhodamine B [112]. Several strategies are implemented using light-assisted water cleaning, including (i) chemically driven selective degradation and adsorption, (ii) chemically driven with photoinduced remediation, and (iii) light-driven photoinduced remediation [113]. A machine-learning algorithm was used to model/predict the motion and performance of water cleaning, leading to a prediction of efficient catalytic nanomembots [114]. Well-established methods to remove pollutants using such mechanisms as molecular absorption and oxidation are implemented, combining technologies of catalytic nanomembranes [115] and dynamic nano/micromachines [112]. Figure 9.12a shows catalytic hydrogel $KMnO_4$ nanomembot operated in hydrogen peroxide solution by oxygen bubble recoil [116]. Besides, $KMnO_4$ catalyzes hydrogen peroxide, producing MnO_2 and hydroxyl radicals, which react to organic pollutants. Eskandarloo et al. discussed several mechanisms of the pollutants decontamination, oil droplets, heavy metals, and organic compounds removal using NMs-enabled

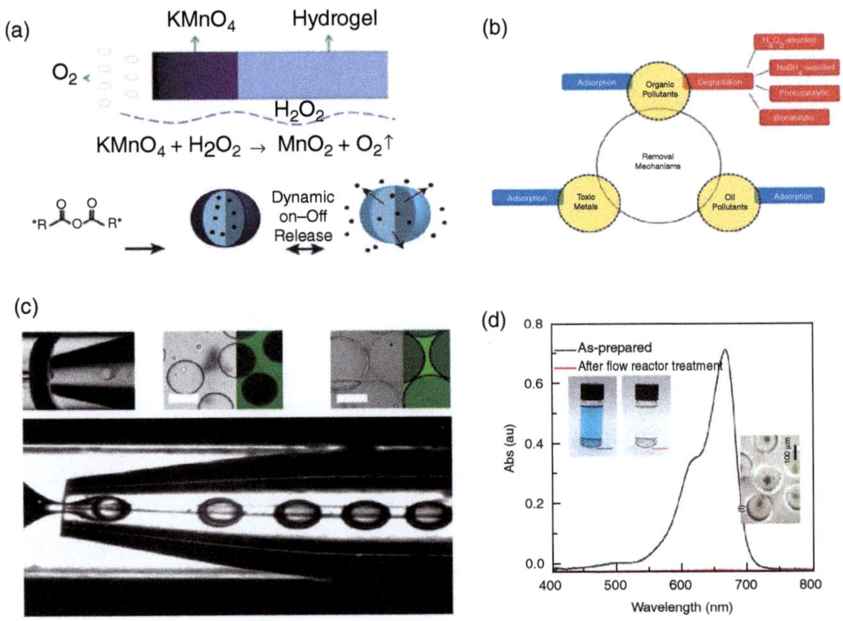

Figure 9.12 Water cleaning using nanomembots. (a) $KMnO_4$/hydrogel nanomembot self-propelled in a hydrogen peroxide solution. Source: Image is reproduced with permission from Wang et al. [116] Copyright 2016, American Chemical Society. (b) Pollutants removal using different mechanisms. Source: Eskandarloo et al. [117]. Copyright 2017, The Royal Society of Chemistry. (c) Fabrication of hydrogel microcapsules with encapsulated photocatalysts. Images are reproduced with permission. Source: Werner et al. [118]. Reproduced with permission of American Chemical Society. (d) Water cleaning using hydrogel microcapsule-based nanomembot removing methylene blue pollutant. Source: Liu et al. [119]. Reproduced with permission of Royal Society of Chemistry.

adsorption and degradation, shown in Figure 9.12b [117]. Degradation mechanism can be used to decontaminate stable organic pollutants from water, for example, by H_2O_2, $NaBH_4$, chemo-/photo-/bio-/catalytic degradation. To detoxify metals, such as Pb, As, Cd, Tl, Cr, and Hg, graphene can be used as excellent absorbers of heavy metal ions. Other approaches utilize ligand-modified nanomembots, such as functionalized Mg/Au with SAM of meso-2,3-dimercaptosuccinic acid, to remove Pb, Cd, and Zn ions. Activated carbon can be used to remove such organic compounds as organophosphorus nerve agents and nitroaromatic explosives. Charged molecules can be absorbed, transported, and separated by polymer motors [120, 121]. Fenton reaction was used due to generated ·OH hydroxyl radicals, where H_2O_2 is used both as a fuel for NMs and as a reagent for the Fenton oxidation. In another approach, Prussian blue (PB)-sodium dodecyl sulfate (SDS)-reduced GO hydrogel, polyacrylamide (PAM) hydrogel, and emulsion hydrogel soft motors can be used to remove methylene blue as a model organic pollutant. Light absorption with the energy of photons higher than the bandgap of semiconductor photocatalysts (e.g. TiO_2) results in electron–hole pairs, which perform chemistry on surfaces due to reactive ·OH radicals, electrons, and holes. If combined with additional plasmonic heating, e.g. Au–Pt/TiO_2, the structure can remove large mixtures of organic pollutants (e.g. methyl orange, rhodamine B, and methylene blue). Other materials for pollutants removal by *adsorption* include (oil drops) Au/Ni/PEDOT/Pt, $MnFe_2O_4$/OA, (heavy metals) Pt/Ni/GO_x, DMSA/Mg/Ti/Au, (organic compounds) rGO/Fe_2O_3/SiO_2–Pt. Methods of pollutants removal by *degradation* are (H_2O_2 assisted): (heavy metals/organic compounds) activated carbon/Pt, (methyl paraoxon – ethyl paraoxon – bis(4-nitriphenyl) phosphate) polymer/Pt, (rhodamine 6G) Fe/Pt, (methylene blue) PB/SDS/rGO/hydrogel; $KMnO_4$/PAM hydrogel. Methods to remove pollutants by *degradation* consist of ($NaBH_4$-assisted photocatalytic): (methylene blue – rhodamine 6G) MnO_2, (methylene blue – rhodamine B – methyl orange) Au/Pt/TiO_2, (Bis(4-nitrophenyl) phosphate-methylparaoxon-*Bacillus globigii* spore) TiO_2/Au/Mg, (rhodamine B) TiO_2/Pt; biocatalytic: (2-amino-4-chlorophenol-eriochorome black-T) laccase/SDS, (guaicaol–catechol–2-amino-4-chlorophenol) enzyme-rich tissue.

However, suspended nanoparticles, i.e. nanomembots, in rivers and lakes are complicated to separate/filtrate after the water-cleaning operation is accomplished. Subsequently, one strategy consists of embedding nanoparticles into water-permeable supportive materials, such as hydrogels – materials with high water-swelling capacity and unique properties including controllable reaction, diffusion, separation, purification, and delivery of molecular species. Hydrogels are highly absorbent polymeric materials that can absorb a large amount of water and swell with broad applications, including forward osmosis, desalination, wastewater cleanup, removal of pollutants, microbial, chemical, and biohazardous threats. Polymer nanocomposites are shown to have a high potential to avoid the agglomeration of high surface area nanoparticles for water-cleaning tasks [122]. Permeability of such hydrogel films can be easily controlled, for example, by optimization of the composition ratio of PEGDA and RNH3Cl monomers (e.g. ~200 l/mm/h/atm or above for 80% w/w water in the pre-polymerized PEGDA and RNH_3Cl) [123].

Graphene-based hydrogels with agarose (AG) as a stabilizer and reducing agent and the polydopamine-functionalized graphene hydrogel (PDA-GH) are shown to enhance degradation and removal of toxic dye pollutants, organic solvents, heavy metal ions (Cd_2^+, Pb_2^+, and Cu_2^+), and oils [124]. Ultrahigh surface area nanomaterials, i.e. MOF in the hydrogel was demonstrated for water cleaning. A high surface area absorbers/catalysts, such as MnO_2, Ag, Pt, Carbon, TiO_2 ZrO_2–GO/Pt, FePt, metal–organic framework (i.e. MOF with catalytic Fe), can be encapsulated in hydrogel particles using the microfluidic device at kHz rates. Thus, multiple filtering layers consisting of different nanoparticle-loaded composite hydrogels can be constructed for smart photodegradation, separation, adsorption, sensing, and water quality monitoring [130]. Functional groups of hydrogels (e.g. –OH, –COOH, and $-NH_2$) are highly effective in binding and removing selective ions.

Figure 9.12c,d shows novel hydrogel microcapsules synthesized using methacrylic anhydride [118]. Microcapsules exhibit tailored functionality for size-selective uptake and release of molecular species and for on-demand encapsulation of aqueous photocatalysts for water-cleaning applications. Permeability of capsule's shell shut off upon deswelling in acidic conditions and swollen in neutral and alkaline solution. In our example, microcapsules with dispersed photocatalytic (TiO_2, ZnO) nanoparticles in an aqueous core are used for the removal of methylene blue pollutants from an aqueous solution using an adsorption–oxidation mechanism [119]. A prototype flow microreactor was assembled to demonstrate a controllable water purification approach in a short time using photocatalytic hydrogel capsules. Compared to catalytic nanoparticles that are fixed statically in the bulk hydrogel, diffusion in the aqueous core of microcapsules is approximately two times faster. These parameters can be used to optimize water-cleaning procedures in the future.

9.5 Conclusion and Future Prospects

Biological molecular nanomachines widely use the chemical energy conversion approach to move and perform complex operations. Man-made nanomembots represent a paradigm shift from static to dynamic nanotechnology. Autonomous nanomembots are released from the substrate and they can swim, sense, interact, assemble, and perform tasks. Today, nanomembots can operate with great precision, high strength, and ultrafast speeds (relative). Subsequent essential research has started optimizing velocities, materials, efficiencies, biocompatibility, fuels, fabrication, and motion control methods toward potential applications. Nano- and micro-mechanical devices utilize advanced dynamic properties of materials, which can be further explored in the future: (i) ultrafast relative motion, (ii) ultrahigh-strength to weight ratio, (iii) ultra-precise atomistic movements (due to high viscous drag force and negligible inertia), (iv) increased performance due to ultra-low mass (e.g. enzymes convert a million of molecules per second), (v) compactness, i.e. positioning of a large number of nanomembots in a small volume space (e.g. direction, location, and steps), (vi) ability to move through narrow channels (e.g. for biomedical cargo delivery or minimally invasive surgery), (vii) explore novel biomimetic ways of clean energy transductions, and (viii) swarming behaviors of entities, which aggregate together or act as a single ensemble.

References

1 Gulina, L.B., Tolstoy, V.P., Solovev, A.A. et al. (2020). Gas-solution interface technique as a simple method to produce inorganic microtubes with scroll morphology. *Progr. Nat. Scie.: Mater. Int.* 30 (3): 279–288.

2 Huang, G.S. and Mei, Y.F. (2018). Assembly and self-assembly of nanomembrane materials-from 2D to 3D. *Small* 14 (14): e1703665.

3 Solovev, A.A., Mei, Y.F., Ureña, E.B. et al. (2009). Catalytic microtubular jet engines self-propelled by accumulated gas bubbles. *Small* 5: 1688–1692.

4 Mei, Y.F., Solovev, A.A., Sanchez, S., and Schmidt, O.G. (2011). Rolled-up nanotech on polymers: from basic perception to self-propelled catalytic micro-engines. *Chem. Soc. Rev.* 40: 2109–2119.

5 Paxton, W.F., Kistler, K.C., Olmeda, C.C. et al. (2004). Catalytic nanomotors: autonomous movement of striped nanorods. *J. Am. Chem. Soc.* 126 (41): 13424–13431.

6 Ozin, G.A., Manners, I., Fournier-Bidoz, S., and Arsenault, A. (2005). Dream nanomachines. *Adv. Mater.* 17 (24): 3011–3018.

7 Mei, Y.F., Huang, G.S., Solovev, A.A. et al. (2008). Versatile approach for integrative and functionalized tubes by strain engineering of nanomembranes on polymers. *Adv. Mater.* 20 (21): 4085–4090.

8 Mujtaba, J., Liu, J.R., Dey, K.K. et al. (2021). Micro-bio-chemo-mechanical-systems: micromotors, microfluidics, and nanozymes for biomedical applications. *Adv. Mater.* 33 (22): 2007465.

9 Whitesides, G.M. (2001). The once and future nanomachine. *Scientific American* 285: 70–75.

10 Einstein, A. (1905). Über die von der molekularkinetischen Theorie der Wärme geforderte Bewegung von in ruhenden Flüssigkeiten suspendierten Teilchen. *Annalen der Physik* 17: 549–560.

11 Zhang, L., Peyer, K. E., Petit, T., et al. (2010) Motion control of artificial bacterial flagella. 10th IEEE Conference on Nanotechnology (IEEE-NANO), 11772690, 893–896. Ilsan, Korea (South): Institute for Electrical and Electronics Engineers (IEEE).

12 Rogers, B., Adams, J., and Pennathur, S. (2015). *Nanotechnology Understanding Small Systems*, 3e. CRC Press ISBN-13: 978-1482211726; ISBN-10: 9781482211726.

13 Wautelet, M. (2001). Scaling laws in the macro-, micro- and nanoworlds. *Eur. J. Phys.* 22: 601–611.

14 An, B., Miyashita, S., Ong, A. et al. (2018). An end-to-end approach to self-folding origami structures. *IEEE Transac. Robot.* 34 (6): 1409–1424.

15 Ning, H., Zhang, Y., Zhu, H. et al. (2018). Geometry design, principles and assembly of micromotors. *Micromachines* 9 (2): 75.

16 Xi, W., Kong, F., Yeo, J.C. et al. (2017). Soft tubular microfluidics for 2D and 3D applications. *PNAS* 114 (40): 10590–10595.

17 Solovev, A.A., Sanchez, S., and Schmidt, O.G. (2013). Collective behaviour of self-propelled catalytic micromotors. *Nanoscale* 5: 1284–1293.

18 Solovev, A.A., Sanchez, S., Pumera, M. et al. (2010). Magnetic control of tubular catalytic microbots for the transport, assembly, and delivery of micro-objects. *Adv. Funct. Mater.* 20: 2430–2435.

19 Kagan, D., Campuzano, S., Balasubramanian, S. et al. (2011). Functionalized micromachines for selective and rapid isolation of nucleic acid targets from complex samples. *Nano Lett.* 11: 2083.

20 Balasubramanian, S., Kagan, D., Hu, C.-M.J. et al. (2011). Micromachine-enabled capture and isolation of cancer cells in complex media. *Angew. Chem. Int. Ed.* 50: 4161.

21 Shi, X., Liu, J.R., Kong, Y. et al. (2021). A strain-engineered helical structure as a self-adaptive magnetic microswimmer. *ChemNanoMat.* 7 (6): 607–612.

22 Sen, A., Ibele, M., Hong, Y., and Velegol, D. (2009). Chemo and phototactic nano/microbots. *Faraday Discuss.* 143: 15–27.

23 Rogers, J., Huang, Y., Schmidt, O.G., and Gracias, D.H. (2016). Patterning via self-organization and self-folding. *MRS Bulletin* 41 (2): 123–129.

24 Liu, Y., Shaw, B., Dickey, M.D., and Genzer, J. (2017). Sequential self-folding of polymer sheets. *Sci. Adv.* 3 (3): e1602417.

25 Chen, S., Chen, J., Zhang, X. et al. (2020). Kirigami/origami: unfolding the new regime of advanced 3D microfabrication/nanofabrication with "folding". *Light: Sci. & Appl.* 9: 75.

26 Lui, Y.S., Sow, W.T., Tan, L.P. et al. (2019). 4D printing and stimuli-responsive materials in biomedical aspects. *Acta Biomater.* 1 (92): 19–36.

27 Adams, L.L.A., Lee, D., Mei, Y.F. et al. (2020). Nanoparticle-shelled catalytic bubble micromotor. *Adv. Mater. Interfaces* 7 (4): 1901583.

28 Lin, X., Xu, B., Zhu, H. et al. (2020). Requirement and development of hydrogel micromotors towards biomedical applications. *Research* 2020: 1–15, 7659749. https://doi.org/10.34133/2020/7659749.

29 Ortiz-Rivera, I., Mathesh, M., and Wilson, D.A. (2018). A supramolecular approach to nanoscale motion: polymersome-based self-propelled nanomotors. *Acc. Chem. Res.* 51 (9): 1891–1900.

30 Li, L., Wang, J., Li, T. et al. (2014). Hydrodynamics and propulsion mechanism of self-propelled catalytic micromotors: model and experiment. *Soft Matter* 10: 7511–7518.

31 Purcell, E.M. (1977). Life at low Reynolds number. *Am. J. Phys.* 45: 3–11.

32 Qiu, T., Lee, T.C., Mark, A.G. et al. (2014). Swimming by reciprocal motion at low Reynolds number. *Nat. Commun.* 5: 5119.

33 Ismagilov, R.F., Schwartz, A., Bowden, N., and Whitesides, G.M. (2002). Autonomous movement and self-assembly. *Angew. Chem. Int. Ed.* 41: 652–654.

34 Gibbs, J. and Zhao, Y. (2011). Catalytic nanomotors: fabrication, mechanism, and applications. *Front. Mater. Sci.* 5: 25–39.

35 Liu, M. and Zhao, K. (2021). Engineering active micro and nanomotors. *Micromachines* 12 (6): 687.

36 Soto, F., Karshalev, E., Zhang, F. et al. (2021). Smart materials for microrobots. *Chem. Rev.* 122 (5): 5365–5403. https://doi.org/10.1021/acs.chemrev.0c00999.

37 Sun, J., Tan, H., Lan, S. et al. (2021). Progress on the fabrication strategies of self-propelled micro/nanomotors. *J. Colloid Interface Sci.* 2: 100011.

38 Serrà, A. and García-Torres, J. (2021). Electrochemistry: a basic and powerful tool for micro- and nanomotor fabrication and characterization. *Appl. Mater. Today* 22: 100939.

39 Huang, W., Manjare, M., and Zhao, Y. (2013). Catalytic nanoshell micromotors. *J. Phys. Chem. C* 117: 21590–21596.

40 Zhu, H., Nawar, S., Werner, J.G. et al. (2019). Hydrogel micromotors with catalyst-containing liquid core and shell. *Journal of Physics: Condens. Matter Phys.* 31 (21): 214004.

41 Wang, L., Liu, Y., He, J. et al. (2015). Continuous microfluidic self-assembly of hybrid janus-like vesicular motors: autonomous propulsion and controlled release. *Small* 11 (31): 3762–3767.

42 Liu, W., He, R., Zhu, H. et al. (2010). Ultrafast nanotube based diffusiophoresis nanomotors. *Appl. Phys. Lett.* 96: 053114.

43 Gao, W., Sattayasamitsathit, S., Orozco, J., and Wang, J. (2011). Highly efficient catalytic microengines: template electrosynthesis of polyaniline/platinum microtubes. *J. Am. Chem. Soc.* 133: 11862–11864.

44 Solovev, A.A., Sanchez, S., Mei, Y.F., and Schmidt, O.G. (2011). Tunable catalytic tubular micropumps operating at low concentration of hydrogen peroxide. *Phys. Chem. Chem. Phys.* 13: 10131.

45 Sanchez, S., Solovev, A.A., Harazim, S. et al. (2011). The smallest man-made jet engine. *Chem. Records* 11 (6): 367.

46 Li, J., Liu, W., Wang, J. et al. (2017). Nanoconfined atomic layer deposition of TiO_2/Pt nanotubes: toward ultrasmall highly efficient catalytic nanorockets. *Adv. Funct. Mater.* 27: 1700598.

47 Li, L., Wang, J., Li, T. et al. (2015). A unified model of drag force for bubble-propelled catalytic micro/nano-motors with different geometries in low Reynolds number flows. *J. Appl. Phys.* 117: 104308; *RSC Adv.*, **6**, 65624.

48 Man, V.H., Li, M.S., Wang, J. et al. (2019). Nonequilibrium atomistic molecular dynamics simulation of tubular nanomotor propelled by bubble propulsion. *J. Chem. Phys.* 151: 024103.

49 Liu, L., Bai, T., Chi, Q. et al. (2017). How to make a fast, efficient bubble-driven micromotor: A mechanical view. *Micromachines* 8: 267.

50 Wang, W., Chiang, T.Y., Velegol, D., and Mallouk, T.E. (2013). Understanding the efficiency of autonomous nano- and microscale motors. *J. Am. Chem. Soc.* 135: 10557–10565.

51 Solovev, A. A. (2012) Catalytic tubular microjet engines. Ph.D. Thesis, TU Chemnitz, IFW Dresden, Germany.

52 Hudson, S.D. (2010). Poiseuille flow and drop circulation in microchannels. *Rheol Acta* 49: 237–243.

53 Lin, Y., Geng, X., Chi, Q. et al. (2019). Driving forces of the bubble-driven tubular micromotor based on the full life-cycle of the bubble. *Micromachines* 10: 415.

54 Wang, Y., Tu, Y., and Peng, F. (2021). The energy conversion behind micro-and nanomotors. *Micromachines* 12 (2): 222.

55 Zhang, Y. and Hess, H. (2021). Chemically-powered swimming and diffusion in the microscopic world. *Nat. Rev. Chem.* 5: 500–510.

56 Zhao, G. and Pumera, M. (2013). Magnetotactic artificial self-propelled nanojets. *Langmuir* 29: 7411–7415.

57 Khalil, S.M., Magdanz, V., Sanchez, S. et al. (2014). Wireless magnetic-based closed-loop control of self-propelled microjets. *PLoS One* 9 (2): e83053.

58 Yu, Y., Shang, L., Gao, W. et al. (2017). Microfluidic lithography of bioinspired helical micromotors. *Angew. Chem. Int. Ed.* 56: 12127.

59 Medina-Sánchez, M., Schwarz, L., Meyer, A.K. et al. (2016). Cellular cargo delivery: toward assisted fertilization by sperm-carrying micromotors. *Nano Lett.* 16 (1): 555–561.

60 Alapan, Y., Karacakol, A.C., Guzelhan, S.N. et al. (2020). Reprogrammable shape morphing of magnetic soft machines. *Sci. Adv.* 6 (38): eabc6414.

61 Cui, J., Huang, T.-Y., Luo, Z. et al. (2019). Nanomagnetic encoding of shape-morphing micromachines. *Nature* 575: 164–168.

62 Enachi, M., Guix, M., Postolache, V. et al. (2016). Light-induced motion of microengines based on microarrays of TiO_2 nanotubes. *Small* 12: 5497–5505.

63 Mou, F., Li, Y., Chen, C. et al. (2015). Single-component TiO_2 tubular microengines with motion controlled by light-induced bubbles. *Small* 11: 2564–2570.

64 Dong, R., Wang, C., Wang, Q. et al. (2017). ZnO-based microrockets with light-enhanced propulsion. *Nanoscale* 9: 15027–15032.

65 María Hormigos, R., Jurado Sánchez, B., and Escarpa, A. (2019). Multi-light-responsive quantum dot sensitized hybrid micromotors with dual-mode propulsion. *Angew. Chem. Int. Ed.* 58: 3128.

66 Solovev, A.A., Smith, E., BofBufon, C.C. et al. (2011). Light-controlled propulsion of catalytic microengines. *Angew. Chem. Int. Ed.* 50 (46): 10875–10878.

67 Wu, Z., Si, T., Gao, W. et al. (2016). Superfast near-infrared light-driven polymer multilayer rockets. *Small* 12: 577–582.

68 Zhan, Z., Wei, F., Zheng, J. et al. (2018). Recent advances of light-driven micro/nanomotors: toward powerful thrust and precise control. *Nanotechnol. Rev.* 7 (6): 555–581.

69 Zhan, Z., Wei, F., Zheng, J. et al. (2018). Recent advances of light driven micro/nanomotors: towards powerful thrust and precise control. *Nanotech. Rev.* 7 (6): 555–581.

70 Zhang, Q., Dong, R., Wu, Y. et al. (2017). Light-driven Au-WO_3@C janus micromotors for rapid photodegradation of dye pollutants. *Appl. Mater. Interfaces* 9 (5): 4674–4683.

71 Liu, M., Zentgraf, T., Liu, Y. et al. (2010). Light-driven nanoscale plasmonic motors. *Nature Nanotech.* 5: 570–573.

72 Calvo-Marzal, P., Manesh, K.M., Kagan, D. et al. (2009). Electrochemically-triggered motion of catalytic nanomotors. *Chem. Commun.* 4509–4511.

73 Calvo-Marzal, P., Sattayasamitsathit, S., Balasubramanian, S. et al. (2010). Propulsion of nanowire diodes. *Chem. Commun.* 2010 (46): 1623–1624.

74 Ohiri, U., Shields, C.W., Han, K. et al. (2018). Reconfigurable engineered motile semiconductor microparticles. *Nat. Commun.* 9: 1791.

75 Guo, J., Gallegos, J.J., Tom, A.R., and Fan, D. (2018). Electric-field-guided precision manipulation of catalytic nanomotors for cargo delivery and powering nanoelectromechanical devices. *ACS Nano* 12 (2): 1179–1187.

76 Han, D., Farino, C., Yang, C. et al. (2018). Soft robotic manipulation and locomotion with a 3D printed electroactive hydrogel. *ACS Appl. Mater. Interfaces* 10 (21): 17512–17518.

77 Huang, G.S., Wang, J., Liu, Z. et al. (2017). Rocket-inspired tubular catalytic microjets with grating-structured walls as guiding empennages. *Nanoscale* 9: 18590–18596.

78 Balasubramanian, S., Kagan, D., Manesh, K.M. et al. (2009). Thermal modulation of nanomotor movement. *Small* 5 (13): 1569–1574.

79 Xu, T., Soto, F., Gao, W. et al. (2014). Ultrasound-modulated bubble propulsion of chemically powered microengines. *J. Am. Chem. Soc.* 136: 8552–8555.

80 Ren, L., Wang, W., and Mallouk, T.E. (2018). Two forces are better than one: combining chemical and acoustic propulsion for enhanced micromotor functionality. *Acc. Chem. Res.* 51: 1948–1956.

81 Wang, W., Duan, W., Zhang, Z. et al. (2015). A tale of two forces: simultaneous chemical and acoustic propulsion of bimetallic micromotors. *Chem. Commun.* 51: 1020–1023.

82 Rao, K.J., Li, F., Meng, L. et al. (2015). A force to be reckoned with: a review of synthetic microswimmers powered by ultrasound. *Small* 11 (24): 2836–2846.

83 Ahmed, D., Dillinger, C., Hong, A., and Nelson, B.J. (2017). Artificial acousto-magnetic soft microswimmers. *Adv. Mater. Technol.* 2: 1700050.

84 Feng, J., Yuan, J., and Cho, S.K. (2015). Micropropulsion by an acoustic bubble for navigating microfluidic spaces. *Lab Chip* 15: 1554–1562.

85 Soto, F., Wagner, G.L., Garcia-Gradilla, V. et al. (2016). Acoustically propelled nanoshells. *Nanoscale* 8: 17788–17793.

86 Jiang, W., Ma, L., and Xu, X. (2018). Recent progress on the design and fabrication of micromotors and their biomedical applications. *Bio-Des. Manuf.* 1: 225–236.

87 Zhang, B., Huang, G.S., Wang, L. et al. (2019). Rolled-up monolayer graphene tubular micromotors: enhancedperformance and antibacterial property. *Chem. Asian J.* 14: 2479–2484.

88 Xi, W., Solovev, A.A., Ananth, A.N. et al. (2013). Magnetic micro-drillers: towards minimally invasive surgery. *Nanoscale* 5: 1294–1297.

89 Sanchez, S., Solovev, A.A., Schulze, S., and Schmidt, O.G. (2011). Controlled manipulation of multiple cells using catalytic microbots. *Chem. Commun.* 47: 698–700.

90 Wu, Z., Lin, X., Zou, X. et al. (2015). Biodegradable protein-based rockets for drug transportation and light-triggered release. *ACS Appl. Mater. Interfaces* 7: 250–255.

91 Wang, L., Chen, J., Feng, X. et al. (2016). Self-propelled manganese oxide-based catalytic micromotors for drug delivery. *RSC Adv.* 6: 65624–65630.

92 Jurado-Sánchez, B. (2018). Nanoscale biosensors based on self-propelled objects. *Biosens.* 8: 59.

93 Esteban-Fernández de Ávilaa, B., Zhao, M., Campuzano, S. et al. (2017). Rapid micromotor-based naked-eye immunoassay. *Talanta* 167: 651–657.

94 Luo, M., Jiang, Y., Su, J. et al. (2019). Surface charge-reversible tubular micromotors for extraction of nucleic acids in microsystems. *Chem. Asian J.* 14: 2503–2511.

95 Qualliotine, J.R., Bolat, G., Beltrán-Gastélum, M. et al. (2019). Acoustic nanomotors for detection of human papillomavirus–associated head and neck cancer. *Otolaryngology–Head and Neck Surgery* 161 (5): 814–822.

96 Hu, W., Lum, G., Mastrangeli, M., and Sitti, M. (2018). Small-scale soft-bodied robot with multimodal locomotion. *Nature* 554: 81–85.

97 Solovev, A.A., Xi, W., Gracias, D.H. et al. (2012). Self-propelled nanotools. *ACS Nano* 6 (2): 1751–1756.

98 Wu, Z., Troll, J., Jeong, H.-H. et al. (2018). A swarm of slippery micropropellers penetrates thevitreous body of the eye. *Sci. Adv.* 4: eaat4388.

99 Leong, T.G., Randall, C.L., Benson, B.R. et al. (2009). Tetherless thermobiochemically actuated microgrippers. *Proc. Natl. Acad. Sci.* 106 (3): 703–708.

100 Li, J., Thamphiwatana, S., Liu, W. et al. (2016). Enteric micromotor can selectively position and spontaneously propel in the gastrointestinal tract. *ACS Nano* 10 (10): 9536–9542.

101 Erol, O., Pantula, A., Liu, W., and Gracias, D.H. (2019). Transformer hydrogels: a review. *Adv. Mater. Technol.* 4: 1900043.

102 Angilè, F.E., Vargo, K.B., Sehgal, C.M. et al. (2014). Recombinant protein-stabilized monodisperse microbubbles with tunable size using a valve-based microfluidic device. *Langmuir* 30 (42): 12610–12618.

103 Pulsipher, K.W., Hammer, D.A., Lee, D., and Sehgal, C.M. (2018). Engineering theranostic microbubbles using microfluidics for ultrasound imaging and therapy: a review. *Ultrasound Med. Biol.* 44 (12): 2441–2460.

104 Hingot, V., Errico, C., Heiles, B. et al. (2019). Microvascular flow dictates the compromise between spatial resolution and acquisition time in ultrasound localization microscopy. *Sci. Rep.* 9: 2456.

105 Liu, W., Ge, H., Chen, X. et al. (2019). Fish-scale-like intercalated metal oxide-based micromotors as efficient water remediation agents. *ACS Appl. Mater. Interfaces* 11 (17): 16164–16173.

106 Srivastava, S.K., Guix, M., and Schmidt, O.G. (2016). Wastewater mediated activation of micromotors for efficient water cleaning. *Nano Lett.* 16 (1): 817–821.

107 Ye, H., Wang, Y., Xu, D. et al. (2021). Design and fabrication of micro/nano-motors for environmental and sensing applications. *Appl. Mater. Today* 23: 101007.

108 Zarei, M. and Zarei, M. (2018). Self-propelled micro/nanomotors for sensing and environmental remediation. *Small* 14 (30): e1800912.

109 Gao, W. and Wang, J. (2014). The environmental impact of micro/nanomachines: a review. *ACS Nano* 8 (4): 3170–3180.

110 Guo, J., Moo, S., and Pumera, M. (2015). Chemical energy powered nano/micro/macromotors and the environment. *Chem. Eur. J.* 21: 58–72.

111 Minh, T.D., Ncibi, M.C., Srivastava, V. et al. (2021). Micro/nano-machines for spilled-oil cleanup and recovery: a review. *Chemosphere* 271: 129516.

112 Wang, S., Jiang, Z., Ouyang, S. et al. (2017). Internally/externally bubble-propelled photocatalytic tubular nanomotors for efficient water cleaning. *ACS Appl. Mater. Interfaces* 9 (28): 23974–23982.

113 Safdar, M., Simmchen, J., and Jänis, J. (2017). Light-driven micro- and nanomotors for environmental remediation. *Environ. Sci. Nano* 4: 1602–1616.

114 Zeng, M., Yuan, S., Huang, D., and Cheng, Z. (2019). Accelerated design of catalytic water-cleaning nanomotors via machine learning. *ACS Appl. Mater. Interfaces* 11 (43): 40099–40106.

115 Liang, C., Zhan, C., Zeng, F. et al. (2018). Bilayer tubular micromotors for simultaneous environmental monitoring and remediation. *ACS Appl. Mater. Interfaces* 10 (41): 35099–35107.

116 Wang, H., Gu, X., and Wang, C. (2016). Self-propelling hydrogel/emulsion-hydrogel soft motors for water purification. *ACS Appl. Mater. Interfaces* 8 (14): 9413–9422.

117 Eskandarloo, H., Kierulfa, A., and Abbaspourrad, A. (2017). Nano- and micromotors for cleaning polluted waters: focused review on pollutant removal mechanisms. *Nanoscale* 9: 13850–13863.

118 Werner, J.G., Nawar, S., Solovev, A.A., and Weitz, D.A. (2018). Hydrogel microcapsules with dynamic ph-responsive properties from methacrylic anhydride. *Macromolecules* 51 (15): 5798–5805.

119 Liu, J.R., Chen, H., Shi, X. et al. (2020). Hydrogel microcapsules with photocatalytic nanoparticles for removal of organic pollutants. *Environ. Sci.: Nano* 7: 656–664.

120 Xuan, M., Lin, X., Shao, J. et al. (2015). Motion-based, high-yielding, and fast separation of different charged organics in water. *Chem. Phys. Chem.* 16: 147–151.

121 Wani, O.M., Safdar, M., Kinnunen, N., and Janis, J. (2016). Dual effect of manganese oxide micromotors: catalytic degradation and adsorptive bubble separation of organic pollutants. *Chemistry* 22: 1244–1247.

122 Pandey, N., Shukla, S.K., and Singh, N.B. (2017). Water purification by polymer nanocomposites: an overview. *Nanocomposites* 3: 47–66.

123 La, Y.-H., McCloskey, B.D., Sooriyakumaran, R. et al. (2011). Bifunctional hydrogel coatings for water purification membranes: improved fouling resistance and antimicrobial activity. *J. Membr. Sci.* 372: 285–291.

124 Lu, H., Zhang, S., Guo, L., and Lic, W. (2017). Applications of graphene-based composite hydrogels: a review. *RSC Adv.* 7: 51008–51020.

10

Nanomembranes Technology for Microrobots: from Origami to 4D Construction

Xinyi Lin, Chunyan Qu, Jizhai Cui, and Yongfeng Mei

Fudan University, Department of Materials Science, 220 Handan Road, Shanghai 200433, P. R. China

10.1 Introduction

Origami, as the art of paper folding, dates back at least 400 years, with tens of thousands of folding patterns documented [1–3]. Created mainly for aesthetics at the beginning, origami now has been applied in mechanical engineering [4, 5] and space applications [6], allowing engineers and scientists to take advantage of the bright ideas people have had over the centuries on how to effectively fold a paper [7–11]. There are several advantages of origami. First, it transforms a flat sheet into 3D intricate shapes with relatively little processing [12]. The process is also often straightforward without sacrificing structural integrity. Second, the principles are scalable, enabling the miniaturization of devices for functioning in the small-scale environment.

The static structure of origami could be further equipped with deformation capability to satisfy the needs of special applications, where the motion or transformation of the device is also important. Recently, origami has been used to design deployable structures that can fold and unfold reversibly under control, which we call 4D origami in this book [13–16]. The additional dimension is defined as time due to a transformation in shape over time under specific external stimuli [16]. The concept of 4D origami is applied to overcome many challenges in engineering, such as packing large solar panels to reduce their physical size. For medical applications, small-scale 4D origami forceps can be used as a medical instrument for minimally invasive surgery inside the human body, which can morph and manipulate biological tissues. The transformation of 4D origami ensures a small incision, which offers significant advantages [17].

In the future, noninvasive surgeries are believed to be the first choice of patients [18]. How could we apply 4D origami devices to achieve noninvasive surgeries? One of the essential steps is to make it smaller. When we want to build milli-size or micro-size devices with 4D origami, stimuli-responsive free-standing thin films, i.e. nanomembranes, are often one of the best choices. Nanomembrane in the micro-world is like thin paper in the macro-world, which can be folded into more compact and smaller forms.

Nanomembranes: Materials, Properties, and Applications, First Edition.
Edited by Yongfeng Mei, Gaoshan Huang, and Xiuling Li.
© 2022 WILEY-VCH GmbH. Published 2022 by WILEY-VCH GmbH.

In this chapter, we introduce the 4D origami based on nanomembrane engineering. We discuss the basic fabrication method of 4D origami and various responsive mechanisms of nanomembrane materials for 4D origami according to different external stimuli. In the end, a brief discussion is presented for better combining nanomembrane materials into 4D origami constructions for the next generation of microrobots.

10.2 Fabrication of Smart Nanomembrane Origami Devices: From 2D to 4D

In recent years, with the demand for smaller, cheaper, faster electronic devices, 3D devices have been widely studied, and significant progress has been reported during the past decades [19]. According to Moore's law, the performance enhancement of electronic devices depends on the reduction in size. Over 50 years since the law was proposed in 1965, traditional 2D electronic devices are proving to approach the limit in size. Compared with 2D devices, 3D devices usually have higher integration densities within the 3D space, which means a smaller size, so that they are more suitable for many applications, such as implantable devices. 3D micro- and nanostructures and their devices have been used widely in fields of biomedical devices [20–24], microelectromechanical systems (MEMS) [25–27], energy storage platforms [28–33], optoelectronic components [34, 35], and electronics [15, 36–42]. The traditional fabrication methods of on-chip 3D micro- and nanostructures rely mainly on top-down (subtractive manufacturing) and bottom-up (additive manufacturing) strategies, including layer-by-layer lithography/stacking [43], 3D translational writing [8, 44], 3D printing [45, 46], two- and multiphoton lithography [47–51], large-area projection micro-stereolithography (LAPμSL) [52, 53], and template-assisted deposition [54–56]. Although being very mature, highly precise, and widely compatible, these techniques are now approaching the bottleneck of fundamental law limits [8]. Also, most of the methods mentioned above follow a "linear" rule, i.e. the fabrication volume grows linearly with the fabrication time. If micro- and nanostructure arrays fabrication or mass production is required, the fabrication time becomes extremely long. This limits the fabrication efficiency of the desired 3D micro- and nanostructures. Moreover, for devices such as capacitors, inductors, RF antennas, and magnetic devices, the electrical performance is directly related to the specific 3D geometry, which in turn determines the distribution of the respective physical fields. With the development of 3D micro- and nanodevices, many nonplanar shapes, such as tubes, helices, and other complex geometries, have shown that they are essential for these devices. As methods like 3D printing have challenges in preparing these structures agilely and conveniently, fabrication methods that build 3D structures according to their 2D precursors show their advantages, for example, origami based on nanomembranes. There are at least three advantages: (i) nanomembranes or sheets can be mass-produced inexpensively; (ii) nanomembranes or sheets are compatible with most planar processes and a few patterning techniques; (iii) planar structures can be stacked efficiently for

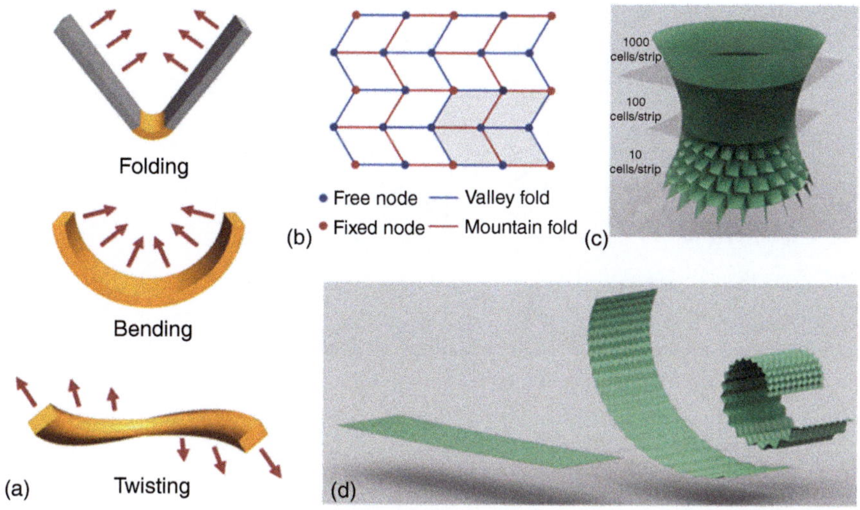

Figure 10.1 Three forms of deformation from 2D to 3D. (a) Schematic illustrations of folding, bending, and twisting. (b) Mountain/valley fold orientations and patterns of fixed/free nodes in a Miura-ori pattern. (c) Hyperboloid constructed by employing different densities of facets. (d) Generalized cylindrical Miura-ori patterns. Source: Adapted from Chen et al. [8]. Copyright 2020, Springer Nature.

storage, transport, or remote deployment [57]. As a result, origami and related techniques have shown emerging applications in devices in the space industry [6], microelectromechanical and nanoelectromechanical systems [58], energy storage systems [30], biomedical devices [59], and mechanical and photonic materials [4, 8, 35, 60]. 3D origami micro- and nanostructures have a promising future in fields of mechanical, electronic, magnetic, and optical applications [15] (Figure 10.1).

The fabrication technologies of 3D origami structures based on planar structures mainly include three forms of deformation. The first is rigid folding of subunits along a flexible hinge, and the second scheme is gradual bending [61], in which the whole subunit is deformed [8]. The third type is multidirectional twisting [62]. This kind of deformation involves folding or bending actions in opposite directions, which does not apply to traditional 3D fabrication [8]. Rigid folding and gradual bending are the main forms of deformations used in origami. Rigid folding is usually used in structures with hinges like boxes or microgrippers, while gradual bending can be used to build tubes and helices. Pattern designing on planar structure is a fundamental process in the fabrication of desired 3D structures through a rigid folding, and a few concepts should be defined and understood before that. In origami, the "mountain fold" refers to the crease protruding from the paper and can be represented with a solid line. On the contrary, "valley fold" means the crease is folded inwardly and represented with a dashed line. There is also a concept of "kirigami," which requires an extra step of cutting compared with origami. There are also steps of folding in most kirigami structures, similar to origami. Thus, in this chapter, no matter whether the film structures went through the cutting step or not, as long as it

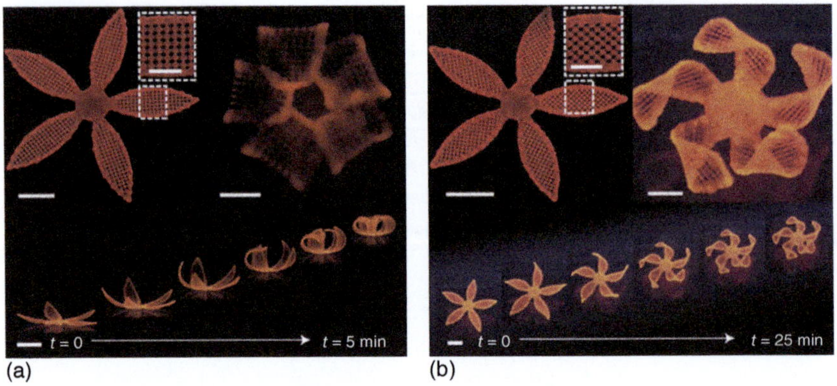

Figure 10.2 4D origami flowers blooming with time. (a) Simple flowers fabricated by 4D printing, which is composed of 90°/0° and (b) −45°/45° bilayers that are oriented with respect to the long axis during the swelling process. Source: Gladman et al. [63]. Adapted with permission of Springer Nature.

conformed to the method of preparing 2D structure first and forming 3D structure based on it by folding/bending, it can be collectively called "origami" (Figure 10.2).

Based on 3D origami, researchers have developed 4D origami to obtain more complex and changeable structures. The morphology of 3D structures evolved according to the change in environment or external stimulation in the dimension of "time" [56] so that the fabrication method was named as "4D" origami. 4D structures are more suitable for most in-site, real-time, and controllable applications, for example, the drug release in the body. Commonly, the realization of 4D origami relies mainly on the response to stimuli, such as thermal/temperature, light, electric, magnetic, chemical, and ultrasonic.

As a result, to fabricate the 4D origami micro- and nanostructures, three main processes are necessary. First, the fabrication and patterning of nanomembranes or sheets. Next, the folding/bending of the nanomembranes. Finally, the morphology changes vary over time in response to specific stimuli. The next sections mainly summarize the fabrication method of origami devices and the way they transform to 4D structures.

10.2.1 3D Fabrication and Design

The fabrication, design, and patterning of 2D structures are based on the traditional nanomembrane fabrication planar processes. According to the ways to form the structures on a 2D substrate, most of the conventional fabrication methods can be classified into additive or subtractive processes. The two classes of processes are the most common ways to fabricate independent or stacked nanomembranes and structures, and in most cases, the complete set fabrication process is finished with their combination.

Additive processes are a series of "bottom-up" methods, where nanomembranes are deposited or coated on a substrate with or without a mask. In this way, patterns form directly on the nanomembranes, before the formation of films. The typical and common additive processes include deposition, spin coating, and epitaxy growth.

Subtractive processes are "top-down" methods, and all of them rely on the pre-prepared nanomembranes or structures, that is, patterns form after film preparation. The technologies of lithography, dry and wet etching, cutting, etc. belong to the subtractive processes.

The transformation from planar designs to 3D structures is the essential stage for origami. Usually, this stage can be finished according to the following four mechanisms: (i) apply external forces to the nanomembranes, the folding or bending is driven by mechanical force, for example, the capillary force and surface tension caused by liquid drops; (ii) driven by the internal competition in the nanomembranes materials, for example, the residual stress that makes nanomembrane rolling or folding after etching or the process of the focused ion beam (FIB), or competition caused by the swelling properties difference between bilayer materials; (iii) the response of materials to specific stimuli, for example, shape-memory alloy (SMA) and shape-memory polymers (SMPs). Mechanisms 1 and 2 rely on the force applied, while mechanism 3 relies on the materials used (Figure 10.3).

Origami driven by external and internal forces is usually observed in all kinds of structures whether with rigid folding or gradual bending. They are the most used methods to fabricate origami structures.

The capillary force is derived from the surface tension of a liquid when the liquid molecules have greater attraction to each other than to the molecules of the air [61, 64, 65]. Due to favorable downscaling with length, at small scales, capillary forces become extremely large relative to forces that scale with volumes, such as gravity or inertia, and to forces that scale with the area, such as elasticity [64]. As a result, origami structures can be actuated by the surface tension or capillary forces on nanomembranes at small size scales, and the method is suitable for different kinds of materials. As mentioned above, there are two main types of deformations in origami,

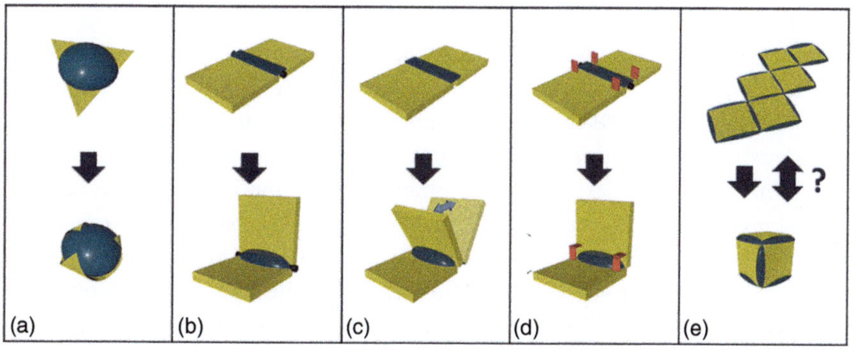

Figure 10.3 Classification of capillary self-folding mechanisms. (a) Continuous folding. (b) Self-folding with mechanical hinges. (c) Hingeless self-folding with multiple energetically equivalent final states. (d) Hingeless self-folding with mechanical locking. (e) Hingeless self-folding with capillary locking, capillary self-alignment, and potential dynamic reversibility. Source: Adapted from Kwok et al. [64]. Copyright 2020, Wiley-VCH.

Figure 10.4 Ion-beam actuation of a Ti/Al/Cr multilayer cubic assembled origami. Source: Chalapat et al. [71]. Reprinted with permission of John Wiley & Sons, Inc.

i.e. gradual bending and rigid bending, which can be achieved with surface tension or capillary forces. For gradual bending, flat nanomembranes with different shapes of square, triangular, and circular have been bent into square, tetrahedron, and "empanada" packets, respectively [66]. For rigid folding, structures were fabricated based on hinges, for example, shaped nets were folded into flowers, cubes, tetrahedrons, etc. [67]. The surface tension– or capillary forces–actuated origami method works with different materials. As reported in previous work, rigid inorganic membranes [65, 68, 69], soft polymer membranes [70], and materials membranes [66] are suitable for this mechanism (Figure 10.4).

Among all the actuation mechanisms mentioned above, origami driven by FIB process is a new real-time and on-site way of deformation. Origami structures fabricated by FIB usually have nanometer accuracy, and therefore FIB technology is suitable for the fine processes of nano-origami. Moreover, different from conventional schemes, the FIB-triggered origami provides a simple one-step method for nanostructures fabrication, as the predesign of multiple materials or multistep processes that are necessary for most other origami methods can be omitted. Several achievements have been made on FIB process-actuated micro-origami in recent years [62, 71–82]. For example, Liu et al. [76] introduced a FIB-actuated one-step and on-site nano-origami of gold film. With the irradiation of ion beam, versatile buckling, rotation, and twisting of nanostructures were simultaneously or selectively achieved. It was because of the combination of the tensile stress generated at the vacancies when gold atoms sputtered away and the compressive stress induced by the implanting of gallium ions into the film. The versatile method was proposed to be used in the fabrications of functional structures in areas of plasmonics, nanophotonics, optomechanics, MEMS/NEMS, etc.

10.2.2 4D construction

As mentioned above, 4D origami is based on 3D origami structures, and the dimension of time is introduced, as the morphology of 3D structures changes with time. The change relies mainly on the morphologies and structures change of responsive materials in response to stimuli. The deformations are usually reversible in response to the specific stimulus, and the folding angles can often be tuned [63, 83]. Thus, there is no clear boundary between 3D and 4D structures in practical examples published, and the 4D microstructures can be observed during the folding and unfolding transition processes from 2D to 3D. As a result, the changing processes from 3D to 4D are not underlined in subsequent sections, and the actuation mechanisms are classified according to the stimuli types.

10.3 4D Origami Actuated by Different Stimuli

The nanomembranes that respond to small changes in the environment can be preprogrammed to deform precisely in response to the surrounding specific stimuli. Generally speaking, the deformation is between two stable shapes and multistable shapes. The external stimuli can be classified into two main types: physical stimuli and chemical stimuli. In this section, we discuss how this nanomembrane-based origami responds to external stimuli. In Sections 10.3.1–10.3.5, we mainly discuss physical stimuli such as light, ultrasound, temperature, electric field, and magnetic field. In Section 10.3.6, we mainly introduce chemical stimuli, including pH, moisture, and gas.

10.3.1 Thermal/Temperature-Responsive Origami

The most common actuation mechanism of origami is the structures' responses to thermal stimuli [84]. According to the mechanism, the thermal/temperature-responsive origami can be divided into two types. The first type is the folding that relies on the differential strains, like the difference in swelling properties between multilayer films or the change of volume and properties under different temperatures, which applies force to the films and therefore cause the folding or bending. The second type is the folding that relies on temperature-responsive materials, such as SMA and SMP (Figure 10.5).

10.3.1.1 Swelling- or Shrinking-Actuated Passive Origami

The different strain between neighboring objects in response to thermal/temperature caused by the difference in swelling properties between different materials is a conventional actuation mode [83, 85–87]. A contraction/expansion force is generated between the bilayer or multilayer composite materials to perform the desired fold. Essentially, the mechanism of folding is derived from the deformation caused by stress.

Swelling property relying on actuation mechanism can be observed in bilayer or multilayer films with different materials. The thermo-responsive hydrogel poly

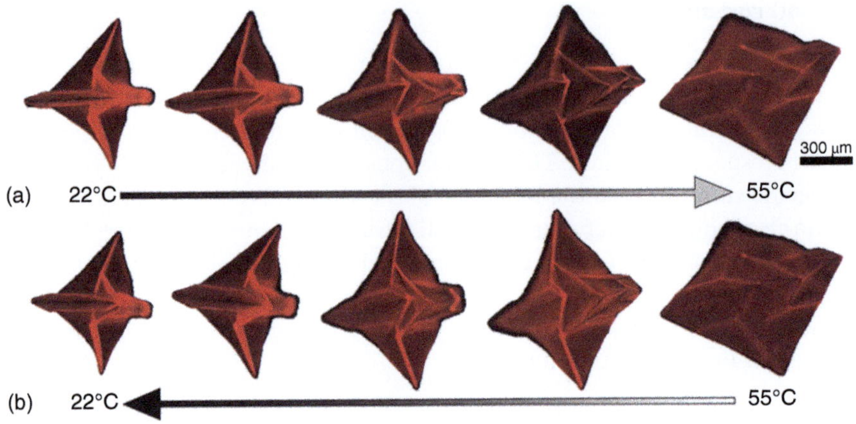

Figure 10.5 Thermal-actuated origami. (a) The unfolding of structure (deswell of polymer layer) increased with temperature increase from 22 to 55 °C. (b) The fold of structure (reswell of polymer layer) decreased with temperature decrease from 55 to 22 °C. Source: Adapted from Na et al. [83]. Copyright 2015, Wiley-VCH.

(*N*-isopropyl acrylamide) (PNIPAM) and its copolymers are commonly used materials because of their swelling and shrinking at reduced and elevated temperatures, respectively [88]. The fold of structures along with the increasing temperature is consistent with the reversible switching behavior of PNIPAM from collapsed to swollen state at a certain transition temperature. The so-called low critical solution temperature (LCST) of PNIPAM can be tuned by controlling the constituent monomer units that are used to copolymerize. Na et al. [83] published a reversible origami structure with a rigid copolymer of poly(*p*-methylstyrene) (PpMS) and a thermo-responsive hydrogel PNIPAM trilayer. As the PNIPAM layer shrank when heated, the PpMS/PNIPAM/PpMS trilayer folded when heated from 22 to 55 °C in aqueous media, and unfolded in the reverse process. Open stripes of defined width were placed in one of the PpMS layers, so the mountain and valley folds were defined, and the bending angles were controlled. With this simple approach, reproducible and reversible complex microscale origami patterns can be fabricated. The platform was expected to be used in the fields of microrobotics, biomedical devices, and mechanical metamaterials. Xu et al. [87] fabricated an ultrathin thermo-responsive self-folding origami with a polydopamine (PD)-functionalized graphene layer and a PNIPAM active layer. The patterned graphene flowers, dumbbells, and boxes folded when heated from 35 to 45 °C, and the extent of folding can be tuned by temperature. Also, as reported in the article, only those functionalized regions folded upon heating, while the 3D microstructures can be controlled by selective functionalization. Monolayer nanomembranes can also be actuated by swelling differences. This structure was used to build nonlinear resistors and creased transistors, and can be used to encapsulate and deliver cells or other biologics. Yoon et al. [89] demonstrated a PNIPAM self-folding system with flexible hinges and rigid panels. With selective UV exposure, the thicker panels were fully cross-linked by high-energy UV, while the thinner flexible hinge has a cross-link gradient induced by low-energy UV, which leads to a swelling gradient at the same temperature. Due to the swelling gradient, the hinges bend while being heated. Applications for this structure are predicted in optics, electronics, and robotics.

Not only heating, but folding of structures during cooling has also been proposed. In fact, the folding or unfolding processes of structures in response to temperature are dependent on the patterns and structures designed. Zakharchenko et al. [85] and Stoychev et al. [86] researched on the reversible edge rolling of hydrophobic polycaprolactone (PCL) and PNIPAM bilayers theoretically and experimentally. Different from the structures mentioned above, the PNIPAM layer was the under layer, while the comparatively rigid PCL was used to fabricate the upper layer. With the swelling of the PNIPAM layer at reduced temperature, the bilayer nanomembrane folded due to the uneven deformation, as the PCL layer restricted the swelling of the active PNIPAM layer. In this example, the encapsulation, delivery, and release of cells and drugs are performed.

Breger et al. [90] developed a bi-directional folding microgripper, which can change from open to close either when heated or cooled. The polypropylene fumarate (PPF) solution was coated on the PNIPAM film to fabricate the patterned bilayer structure. The original bilayer was flat, or the gripper was open. When the temperature was beyond the LCST (36 °C), the PNIPAM layer shrank, which caused the closing of the microgripper with the PNIPAM layer facing inward. When the temperature was below the LCST, however, the PNIPAM turned to swell, causing the microgripper to open and then close in the opposite direction, i.e. the PNIPAM layer faced outward.

10.3.1.2 Temperature-Responsive Materials-Actuated Active Origami

In this method, the natural properties of temperature-responsive materials change with temperature; thus, the folding takes place. Through the pre-programming of planar structures, the changes in the properties of materials at different temperatures can be observed, and macroscopic folding can be achieved. SMAs and SMPs are common materials used in this actuation mode; however, most of them are of a millimeter or centimeter scale [91, 92]. Kuribayashi et al. [59] developed an origami stent graft for biomedical application with laminates SMAs (Ti and Ni) membranes. The stent graft was devised, by cooling through liquid nitrogen, to be folded in a fully martensitic state and unfolded at near the human body temperature, as the martensitic transformation temperature of the TiNi film was adjustable to the required temperature. The as-prepared stent graft was believed to be useful for minimum invasive surgery, such as vascular surgery, using an endoscope.

The change of phase and mechanical properties of materials in response to temperature can also cause the folding of 2D structures in some cases. For example, the decrease of modulus with the increasing temperature or the melt of materials can also be an actuation approach of origami. Leong et al. [20] developed an actuation mechanism that used trilayer joints composed of a polymer and a stressed bimetallic thin film patterned between rigid phalanges. The prepared microgripper without polymer supports was closed spontaneously due to internal stress. After the rigid polymer layer was added, the trilayer joint formed and the microgripper was forced to open. However, once the polymer was softened, the microgripper tended to get closed again. It was thus possible to control the closing of the microgripper as the elastic modulus of polymer decreased when heated, and there was a negative correlation between the elastic modulus and the predicted joint angle of the microgripper. When heated at >40 °C, the microgripper closed completely. The capillary force caused by the melt and reflow of polymers and metals can also actuate the origami processes [69, 70]. For example, Joung et al. [70] reported self-assembled cubic graphene and graphene oxide boxes with polymer SU-8 frames and polymer (SPR 220) hinges. When heated to about 100 °C in water, the polymer melted and reflowed, which induced the transformation from 2D nets into 3D structures. The 3D open and closed boxes generate highly confined electric fields due to plasmon–plasmon coupling at each of the faces and can be used as a sensor with high sensitivity to detect targeted substances while maintaining their integrity due to the impermeability of the graphene membranes [61].

10.3.2 Light-Responsive Origami

Folding in response to light is another common triggering method that has been used in origami and can be roughly divided into two categories according to the different driving mechanisms. One is to use the light–heat conversion effect. First, convert the incident light into heat, then the folding occurs due to the thermal-driven mechanism. The other is to use photo-responsive materials, such as photo-deformable polymers or composite materials, to produce deformations.

10.3.2.1 Light–Thermal Conversion-Actuated Origami

The optothermal actuation scheme was one of the earliest optical drive methods and was the only optical actuation mechanism found in the literature in the 1990s, before all the researches of other mechanisms [93]. A millimeter- or centimeter-scale example has been published since then [94–98]. Nanometer-scale examples have also been researched, as nanomembranes have been introduced into the fabrication of light–thermal conversion-actuated origami structures. Carbon materials such as graphene [99] and carbon nanotubes (CNTs) [100, 101] have been used widely, considering their ability to absorb light (near-infrared ray, NIR) and convert it into heat. Fusco et al. [99] reported the fabrication of an untethered, self-folding, soft microrobotic platform with a bilayer system. The GO-mixed PNIPAM layer has a transition temperature of about 40 °C, which ensured a slightly higher temperature than a human body's. As a result, the structure did not respond to physiological body temperature conditions and can allow external stimuli to control the actuation of the microrobot. In this process, GO can help absorb NIR and turn it into heat quickly. The fast switch of the microrobot from closed to open ensures that it can be used in the load, delivery, and release of drugs.

More materials have been used in light–thermal conversion-actuated origami. For example, dopamine and its derivatives have also been applied to actuate origami structures with the light–thermal mechanism, as dopamine–melanin colloidal nanospheres can convert light into heat with an efficiency 100 times larger than

(a) (b)

Figure 10.6 (a) SEM pictures of a ring of micro-mirrors and (b) close-up of a hinge. Source: Ocampo et al. [104]. Adapted with permission of Elsevier.

that of CNTs [102]. Inspired by this outstanding light–thermal conversion property of these nanospheres, Li et al. [103] used a polydopamine (PDA) coating to replace the carbon materials as a photothermal layer in origami structures. The ultrathin active layer (normally less than 50 nm) was able to achieve enough heat to trigger a much thicker film (Figure 10.6).

10.3.2.2 Photostriction-Actuated Origami

The photostriction actuated deformation relies mainly on the uneven thermal expansion of the different component layers during laser heating, and the accumulation of photo-generated carriers which cause a photo-induced stress. Ocampo et al. [104, 105] reported on the optical actuation of GaAs micro-origami mirrors. When illuminated with a high-power Ar laser, the mirror angle, with respect to the substrate, was deflected proportionally to the beam intensity. The inclination angle increased with the increasing laser power, that is, there was a positive relationship between the angle and the power density. The process was reversible, as the mirror returned to its original angle once the illumination stopped [104] (Figure 10.7).

10.3.3 Ultrasonic-Responsive Origami

10.3.3.1 Ultrasonic-Assisted Origami

Ultrasonic can be used to assist thermal actuated origami, as its vibration energy can reduce static friction, which makes the thermokinetic force great enough to support the folding process [107]. The use of ultrasonic to reduce static friction has been reported early. At first, ultrasonic was used to reduce the friction only, which made the folding of films easier. Later, people found that ultrasonic can also provide stresses that can help the origami-folding processes.

Kaajakari and Lai [108] reported the first-ever use of ultrasonic pulses for probeless release and mass actuation of micromachined devices in 1999. They successfully actuated several micromachines in the desired direction by applying an ultrasonic pulse. They also proposed a surface-micromachined hinged structure using thermokinetic forces with ultrasonic vibration energy used to

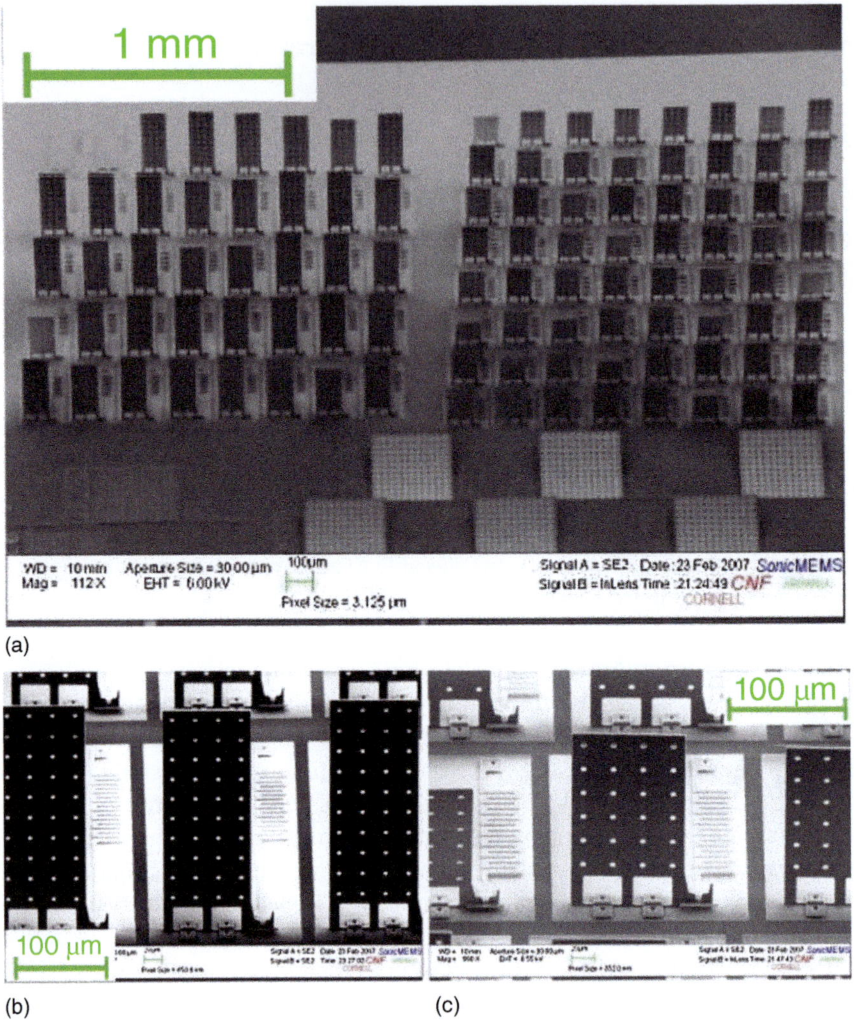

Figure 10.7 (a) SEM pictures of two designs (left and right) of 8 × 8 ultrasound enhanced electrostatic batch assembly microhinges arrays. (b) Close-up view of the left array. (c) Close-up view of the right array. Source: Ardanuc and Lal [106]. Reprinted with permission of Elsevier.

reduce the static friction [107]. Later, Ardanuc and Lai [106] fabricated 8 × 8 arrays of microhinges to prove the electrostatic batch assembly for MEMS enhanced by ultrasonic. An off-chip, bulk-piezoelectric ceramic was used to actuate the system, and the ultrasonic actuation was used as the motion-starting or lubricating agent of the folding. In the mentioned arrays, the yield of assembly can be up to 100%. The ultrasonic-assisted approach was pointed out to be useful for the one-time assembly of large arrays of simple structures, such as microneedles or electron-emitting tips.

10.3.3.2 Ultrasonic-Actuated Rolling

As reported in previous work, ultrasonic can drive nanomembranes to roll. However, this method focuses mainly on 2D materials, such as graphene [109, 110] and graphene oxide [111, 112]. After being treated with a short duration and high-energy ultrasonic, the 2D material nanomembranes could fold into carbon nanoscrolls [56]. Viculis et al. [110] studied the mechanism in detail. The donor-type graphite intercalation compounds KC8 were prepared and reacted with ethanol. The highly exothermic reaction causes exfoliation. Upon sonication, such graphene sheets would roll up into carbon nanoscrolls with a converting rate over 80%, while a much smaller degree of scrolling was observed (<10%) for those without sonication.

10.3.3.3 Ultrasonic-Heat Conversion–Actuated Origami

Ultrasonic-induced folding caused by sound–heat conversion has been published [113, 114]. High-intensity focused ultrasound (HIFU) was used to heat the SMPs spatially localized, which led to the folding. However, it is worth noting that to our knowledge, till now, there are still only studies on the macroscale films and structures that fold according to this mechanism. Micro-origami based on this method is still to be studied and developed.

10.3.4 Electric-Responsive Origami

Electrical stimuli–responsive 4D origami is fabricated by using conductive polymer nanomembranes, such as polyaniline, polyacetylene, and their derivatives [115]. As shown in Figure 10.8a, humidity-sensitive polypyrrole nanomembrane exhibits a significant expansion in the air under an electrical field. This deformation is associated with the desorption of water vapor due to Joule heating by electricity. Furthermore, as presented in Figure 10.8b–d, a biomorphic origami robot is made, moving with a caterpillar-like motion by repeated expansion and contraction. The deformation is due to changes in the elastic modulus of the conductive nanomembrane through reversible sorption and desorption of water vapor [116]. Another type of electrical stimuli-responsive origami exploits electrostatic attraction, insulating fluids, and dielectrophoresis to deliver robotic devices. Taghavi et al. introduced an electrostatic active origami made from any combination of conducting and insulating materials. The simplest type of electro-origami is an electro-ribbon actuator that can lift 1000 times its weight. More complex electro-origami devices, including robotic cilia, locomoting robots, and origami artificial muscles, also have been investigated [117].

10.3.5 Magnetic-Responsive Origami

Magnetic stimulus–responsive 4D origami is often composed of magnetic nanomembranes [118]. In most cases, these magnetic nanomembranes are either metal nanomembranes composed of ferromagnetic elements such as Co [119] or polymer film with magnetic particles inside [120, 121]. The latter kind of composite nanomembranes are programmable and flexible. After being fabricated

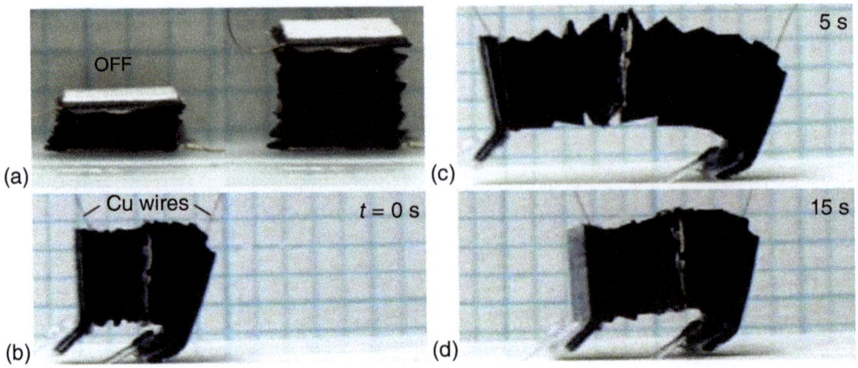

Figure 10.8 Electro-active origami microrobot. (a) Origami microrobot under 0 V (left) and 2 V(right). (b)–(d) Time profile of biomorphic origami robot in action. Source: Palleau et al. [116]. Adapted with permission of John Wiley & Sons, Inc.

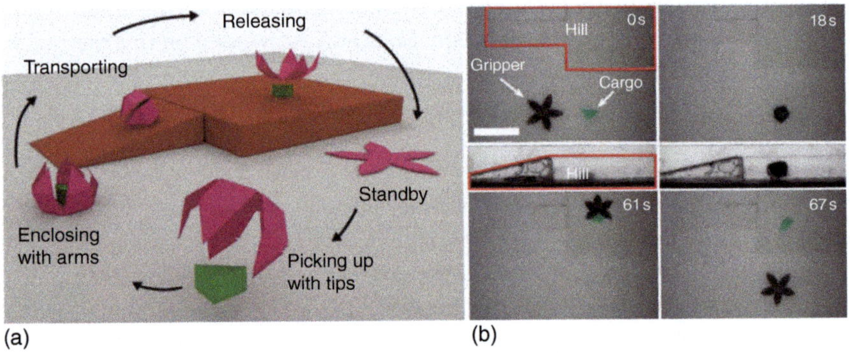

Figure 10.9 Magnetic-responsive origami microgripper. (a) Working scheme of the magnetic microgripper. (b) Time profile of the cargo transportation task in silicone oil. Source: Xu et al. [120]. Adapted with permission from AAAS.

into the desired 2D pattern and applied to specific magnetization, the origami often has different magnetization directions at different parts of the origami with hinges or creases as the boundaries. We give a few examples of magnetic 4D origami. As seen in Figure 10.9a, the flexible magnetic millimeter-scale robot could transform reversibly into complex shapes under an applied magnetic field, achieving object manipulation. The NdFeB permanent magnetic particles were encased by cured polymer nanomembrane. The microgripper made by the magnetic nanomembrane consists of a base and several arms to catch the cargo. As illustrated in Figure 10.9b, after the microrobot encapsulates the cargo, it starts to roll to the targeted places and releases the cargo. In another example, a 4 mm long thin magnetic origami can manage the different modes of locomotion. It can walk, roll, jump, swing, and move from tiny spaces under magnetic control. Magnetic particles are embedded in a thin layer of elastomer. The flexibility of this magnetic origami makes it possible to accomplish complex tasks such as targeted cargo delivery [122]. Yoonho Kim and his coworkers printed ferromagnetic domains in soft materials, introducing domain patterns and magnetization strength into the fabrication of 4D origami. Under applied magnetic fields, magnetic origami shows fast shape-shifting characteristics, suggesting its new possibilities in flexible electronics and origami robotics [123]. By combining magnetic particles into polymer nanomembranes, the size of magnetic origami fabricated by the nanomembranes came down to around 100 μm. To fabricate smaller magnetic origami and make the origami programable, Jizhai Cui et al. developed a strategy to encode shape-morphing instructions into origami microrobots. Two types of single domain nanomagnetic arrays have different magnetic switching fields due to the unit bars having various length–width ratios. The two kinds of arrays are set on the connected panels and are programed by applying a specific sequence of magnetic fields with tailored switching fields, resulting in specific shape transformations. An origami bird has been engineered to mimic the different flying modes of a real bird, which is achieved with predesigned arrangements of the nanomagnet arrays of different body parts, such as head, neck, body, tail, and wings. By encoding the nanomagnetic arrays with different configurations, four types of transformation, such as flapping, turning, hovering, and side-slipping, have been achieved. As presented in Figure 10.10, the origami bird flaps its wings with different frequencies under an applied magnetic field. Magnetic particles can also be embedded in inorganic nanomembrane materials. Park et al. put Fe_3O_4 magnetic nanoparticles inside 2D graphene. The morphology of 2D graphene can be tuned in the presence of an external magnetic field [125].

Besides solid magnetic materials, liquid magnetic materials are also used to fabricate 4D origami, which is called capillary origami. Capillary origami is the wrapping of a fluid drop by a planar elastic nanomembrane due to the interplay between elastic forces and capillary. When the fluid drop is a drop of magnetic fluid, the shape of the origami is strongly dependent on the applied magnetic field [126]. In other words, by precisely manipulating the external magnetic field, this special kind of origami could change its shape into the desired one.

Except for being used for magnetic actuation, magnetic materials could act as a material to generate heat. Tang et al. proposed a strategy to precisely control the shape of magnetothermal-sensitive hydrogels. They used an alternating magnetic field to heat PNIPAM with the help of magnetic materials, and then PNIPAM could deform with the temperature change [127].

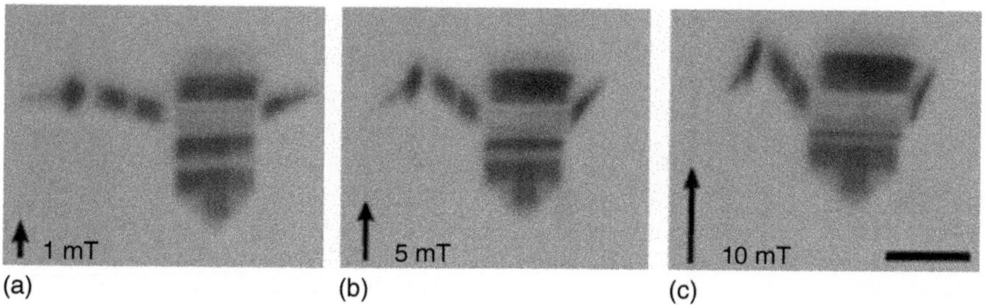

Figure 10.10 Nanomagnetic-encoded origami-flapping bird. The origami-flapping bird is under the control of a magnetic field. Scale bar, 30 μm. Source: Reprinted from Cui et al. [124]. Copyright 2019, Springer Nature.

10.3.6 Chemical-Responsive Origami

10.3.6.1 Solution-Responsive Origami

pH-responsive nanomembranes are often polymers that can respond to pH variations. This kind of polymer has ionic pendant groups to accept and/or donate protons in response to an environmental pH alteration [128]. As the pH in the solution changes, the rapid change of ionized pendant groups causes electrostatic repulsive forces between groups and thus the volume transition, creating an osmotic swelling force. Two types of pH-responsive materials are found: anionic and cationic polymer. The anionic polymer has pendant groups like sulfonic acid. Deprotonation occurs and the pendant groups are ionized when the surrounding pH is higher than a specific pH, i.e. pKa. The cationic polymer has pendant groups, including amine groups that would ionize when the pH is under another specific pH, i.e. pKb. As the pH changes, pH-responsive nanomembrane exhibits different swelling behavior, which is applied in the transformation of 4D origami [129, 130]. Karnaushenko et al. incorporated pH-sensitive hydrogel nanomembrane into thin-film electronics [131]. The swelling of the hydrogel, which can be manipulated by the environmental pH, can trigger the rolling of the thin-film device and achieve mechanically adaptive microchannels of a cuff-type implant. The origami device can roll and unroll repeatedly (Figure 10.11). For instance, an alginate hydrogel monolayer designed like starfish can shrink or stretch its body under stimulation by changes in the pH. The monolayer is an inhomogeneous hydrogel network with different swelling behaviors that can cause heterogeneous deformation. When absorbed into solution, heterogeneous deformation of hydrogel network can lead to environmental stimulus drug-releasing or folding motion [132].

Various polymers show various swelling capabilities. Two layers of polymers with different cross-linking degrees are often adopted as a chemical actuation strategy, i.e. a two-layer deformation strategy [133–135]. When put into a solvent, these two layers of polymers swell differently, causing stress difference and thus the 3D transformation. After removing the solvent, these two layers could be back to the 2D shape. For example, one layer of cross-linked polymer poly(dimethylglyoxime-urethane) (PDOU) and one layer of uncross-linked PDOU can form a simple bilayer origami actuator [136]. The adsorption and desorption of solution can repeatedly trigger the programmed deformation. The general strategy of two layers of origami can also

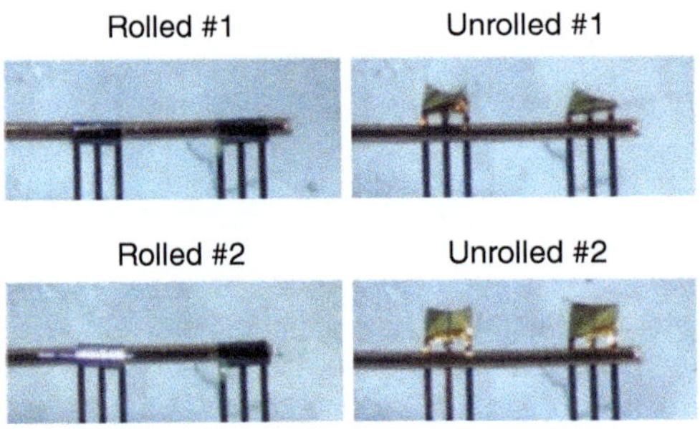

Figure 10.11 Hydrogel microtube with reconfigurable properties caused by swelling effect. The rolled/unrolled state is controlled by the content of isopropanol in a water solution. Source: Karnaushenko et al. [131]. Reprinted with permission of John Wiley & Sons, Inc.

consist of one polymer layer and one non-polymer layer. Weiming Li et al. fabricated metal/polyvinyl alcohol (PVA)–poly (acrylic acid) (PAA) microtube origami that can roll/unroll under the control of environmental fluids. In aqueous media, the PVA–PAA polymer layer has a stronger water absorption compared to the metal nanomembrane, which causes the volume change of the hydrogel and the increasing stress differences between the metal layer and hydrogel layer. With the stress differences increasing, rolling behavior occurs [134].

10.3.6.2 Gas-Responsive Origami

The interaction between nanomembranes and reactive gases is also a way to construct 4D origami. As shown in Figure 10.12a, the surface of Cu within Cr/Cu bilayer can be modified on exposure to oxidative or reductive environments [137]. The oxidation of Cu is achieved by heating the device in an oxidation environment, and the reduction of Cu is implemented in a hydrogen environment upon heating. It is proved that the oxidation of Cu could cause volume change of the film and increases tensile stresses. The stress direction would reverse when the Cu layer transformed from oxidation state to reduction state, and *vice versa*. The Cr/Cu bilayer microgripper opens up in one direction on heating in air, and opens up and then closes in the opposite direction on heating in hydrogen (Figure 10.12b).

Xu et al. applied hydrogen to stimuli nanomembranes structure as 4D origami (Figure 10.13a) [138]. During hydrogen adsorption, Pd in Pd/Cr/Ti nanomembranes exhibits volume expansion and strain change, which produces an obvious rolling of the nanomembrane. As illustrated in Figure 10.13b, before and after hydrogen adsorption, the nanomembrane origami shows a distinct morphology. After hydrogen cut off, Pd/Cr/Ti nanomembranes transform from rolling state to 2D compact nanomembrane. When put in a hydrogen environment, the deformation of Pd/Cr/Ti nanomembrane arrays could have more than 50% transmittance change, which can be observed with naked eyes. From another perspective, this 4D origami also made

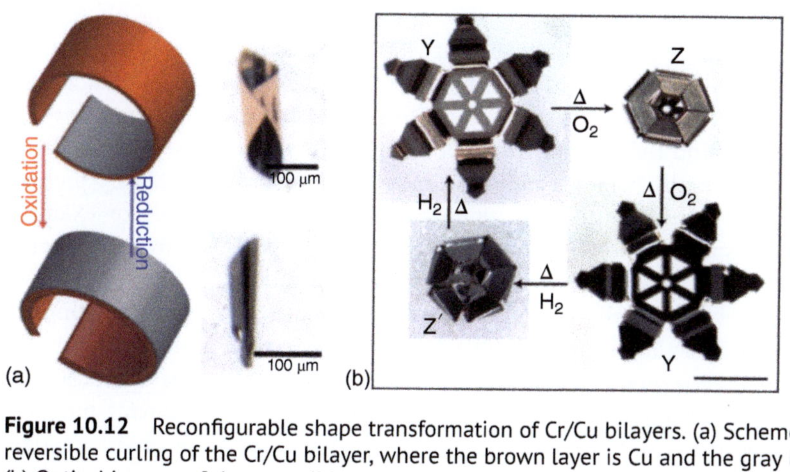

Figure 10.12 Reconfigurable shape transformation of Cr/Cu bilayers. (a) Scheme of the reversible curling of the Cr/Cu bilayer, where the brown layer is Cu and the gray layer is Cr. (b) Optical images of the reversible transition of the Cr/Cu bilayer microgripper. Scale bar, 500 μm. Source: Randhawa et al. [137]. Adapted with permission of John Wiley & Sons, Inc.

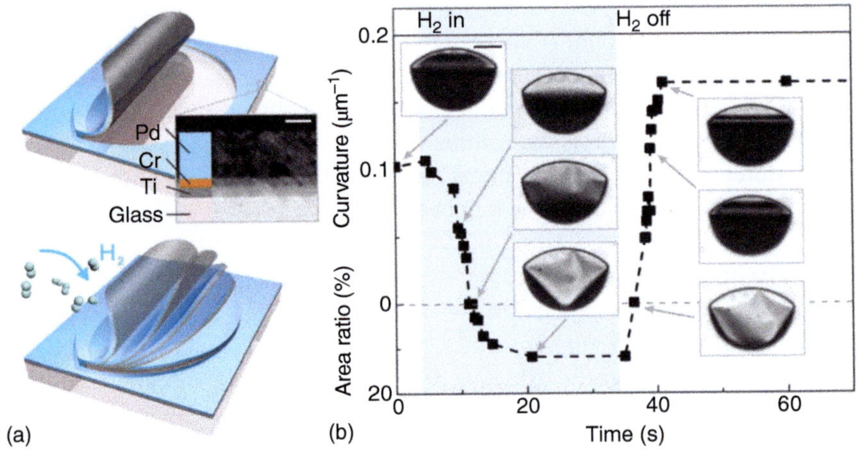

Figure 10.13 4D origami response to hydrogen gas environment. (a) Scheme illustrating spontaneous shape changing in hydrogen due to volume expansion of Pd. (b) Diagram depicting the origami change with and without 2% hydrogen stimuli. Optical images of the corresponding origami are shown in the inset. Scale bar, 25 μm. Source: Xu et al. [138].

good use of the strategy of two-layer deformation. Pd layer is the stress regulation layer that could regulate, and Cr/Ti layer acts as an inert layer. The external stimuli change mainly the volume of the Pd layer and thus increase the stress between Pd and Cr/Ti layers.

10.3.7 Discussion

For chemical-responsive 4D origami, we can observe that the double-layer deformation strategy is the basic and one of the most adopted methods. According to this strategy, one origami device mainly consists of two kinds of layers. One is the active layer, while the other kind is the inertial layer. When exposed under some specific environmental stimuli, the active layer induces volume change due to the abrupt change of swelling capability or vacancies in the lattice and so on. The different volume changes in the active layer and the inertial layer would cause stress differences and thus construct the 4D origami structure. If we go one step further, we may find that except magnetic stimuli origami, double-layer deformation strategies are also widely applied in other 4D origami actuated by physical fields (light field, ultrasound field, temperature field, and so on). For example, temperature-responsive poly(*N*-isopropylacryl-amide) (PNIPAM) layer is attached to platinum (Pt)/PCL layers. When the temperature is above 32 °C, the active-layer PNIPAM changes its swelling ability and absorbs more solution, increasing the stress between layers and causing Pt/PCL/PNIPAM nanomembranes to roll [139].

For magnetic-responsive 4D origami, there are mainly three strategies for fabricating the shape-reconfigurable devices. One is using permanent magnetic particles, as we discussed in Section 10.3.5. The magnetic particles are magnetized as the polymer cross-linked to create ferromagnetic domains. Various 2D planar structures with programmed ferromagnetic domains together could create a blaze of shape-changing origami [140, 141]. The permanent magnetic particle has a size limit of around 5 μm, which limits the size of 4D origami using this strategy. Besides using permanent magnetic particles and designing multidomains to program the origami, using soft magnetic materials combined with shape anisotropy is another strategy. Single-domain soft magnetic bar arrays (Figure 10.10) are a typical example. Each bar unit has an easy axis and a hard axis for magnetization according to the shape. It is the shape effect of bar arrays that determines the encoding of the origami's motion. One of the most distinct differences between these two strategies is that the magnetization direction could be erased and re-encoded using the latter method. In most circumstances, if a 4D origami has differently oriented magnetic moments, they tend to attract each other without external magnetic stimulation, which means that magnetic origami fabricated using the two strategies could hardly be down to micro-size with no aggregation issue. The third strategy is to use superparamagnetic particles inside polymers [141]. By fabricating magnetically anisotropic micro-origami with superparamagnetic materials, the micro-origami can align its magnetic moments with the applied magnetic field direction. This strategy needs no spontaneous magnetization. Each part of the micro-origami is designed to have a different magnetic anisotropy so that the origami could move in multidirections. There is one common fabrication method in the strategies using magnetic particles – adding ferromagnetic/superparamagnetic particles into a fast-curing polymer, i.e. digital light processing. In this way, the magnetization of ferromagnetic particles or the desired position of superparamagnetic particles can be fixed when the polymer is cross-linked.

10.4 Future 4D Origami Microrobots Fabricated from Nanomembrane Platforms

The concept of origami robotics has so far been successfully applied in the design, fabrication, and control of prototypes with the scale from millimeter to centimeter and even meter. Origami robots have already reached the market [142]. However, micro-size origami is still being investigated. Here we want to discuss the potential challenges for microrobots made from 4D origami based on a nanomembrane system.

The first challenge is to make the 4D origami more intellectual at the microscale. For 4D origami in the macro-world, integration of sensing, control, actuation, and power system could be easily achieved by adding sensing components, computational components, and actuation systems into the main origami robotic body. On the other hand, in the microworld there are few micro-origami with an integrated system from sensing to processing and that react precisely toward the surroundings due to these components' large size. In most cases, micro-origami is designed using only one or two kinds of smart nanomembranes. To draw an analogy, 4D micro-origami now lacks the "brain" to make the best decisions and reactions when facing environmental changes. In 2020, Oliver G. Schmidt's team boldly attempted by incorporating microelectronic circuits, engines, actuators, sensors, power supplies, and controllers on one platform. This micro-origami is driven by wireless power transfer, and its direction and local shape could be altered by heating temperature-sensitive PNIPAM nanomembranes [143]. With the development of ultrathin and flexible electronics, it is promising to combine microelectronic systems into micro-origami for better-accomplishing tasks. Secondly, the external stimuli are often limited to one or two types due to only one or two kinds of environmental stimuli materials being used. To tackle *in vivo/ex vivo* complex conditions, multistimuli-responsive origami is required.

The advantage of 4D micro-origami compared to other microrobots is that it may switch from a compact soft nanomembrane structure into semi-rigid or rigid 3D structures. With the flexibility in rigidity and several transformed stable structures, 4D origami microrobots are more likely to achieve multitasks that need different rigidity.

Last but not least, we could notice that for all the 4D origami except magnetic-responsive origami, the deformation is rough and often is simply rolling/unrolling or stretching/closing with the changes of external stimuli. 4D origami microrobot with rough deformation has relatively limited motion behaviors. Accurate and diverse localized mechanical deformation responding to light, temperature, ultrasound, or chemicals is important when facing complex working environment. One possible method is to use discontinuous nanomembranes instead of one single sheet of environmental stimulus material. This discontinuous nanomembrane is composed of multiple material segments. In each segment, the material-responsive capability may have slight differences. For example, in each segment of a kind of polymer, the cross-linking states are different, which would cause different segments to swell differently when the external conditions changed. This may cause various deformation modes and motion modes. From the perspective of control precision and deformation diversification, magnetically responsive origami is superior compared with other mechanisms for the application of multidomain patterns, as discussed in Section 10.3.5.

In conclusion, we discussed 4D origami fabricated by a top-down approach using environmental stimuli nanomembranes. By tailoring the cross-linking state of polymer nanomembranes, functional additives in nanomembranes, and nanomembrane combinations as well as 2D patterns, origami can achieve reversible deformation and various motion behaviors under specific environmental stimuli. To better fabricate and apply 4D origami microrobots based on nanomembrane systems in the future, the integration of an intelligent system, the diversification of responding mechanisms, and the minimization of the control unit are the main challenges. We believed that with the smarter nanomembrane platform, 4D origami microrobot could show their capabilities from a wide range of applications from medical treatment to space exploration.

References

1 Ning, X., Wang, X.J., Zhang, Y. et al. (2018). Assembly of advanced materials into 3D functional structures by methods inspired by origami and kirigami: a

review. *Adv. Mater. Interfaces* 5 (13): 1800284. https://doi.org/180028410.1002/admi.201800284.

2 Felton, S., Tolley, M., Demaine, E. et al. (2014). A method for building self-folding machines. *Science* 345 (6197): 644–646. https://doi.org/10.1126/science.1252610.

3 Howlett, R. (1998). Ecology – spring origami. *Nature* 391 (6668): 641–641. https://doi.org/10.1038/35512.

4 Silverberg, J.L., Evans, A.A., McLeod, L. et al. (2014). Using origami design principles to fold reprogrammable mechanical metamaterials. *Science* 345 (6197): 647–650. doi: 10.1126/science.1252876.

5 Li, S.G., Vogt, D.M., Rus, D. et al. (2017). Fluid-driven origami-inspired artificial muscles. *Proc. Natl. Acad. Sci. U.S.A.* 114 (50): 13132–13137. doi: 10.1073/pnas.1713450114.

6 Zirbel, S.A., Lang, R.J., Thomson, M.W. et al. (2013). Accommodating thickness in origami-based deployable arrays. *J. Mech. Des.* 135 (11): 111005. https://doi.org/10.1115/1.4025372.

7 Lazarus, N., Smith, G.L., and Dickey, M.D. (2019). Self-folding metal origami. *Adv. Intell. Syst.* 1 (7): 1900059.

8 Chen, S., Chen, J., Zhang, X. et al. (2020). Kirigami/origami: unfolding the new regime of advanced 3D microfabrication/nanofabrication with "folding". *Light Sci. Appl.* 9 (1): 75. https://doi.org/10.1038/s41377-020-0309-9.

9 Onal, C.D., Wood, R.J., and Rus, D. (2013). An origami-inspired approach to worm robots. *IEEE/ASME Trans. Mechatron.* 18 (2): 430–438. https://doi.org/10.1109/Tmech.2012.2210239.

10 Kim, S.J., Lee, D.Y., Jung, G.P. et al. (2018). An origami-inspired, self-locking robotic arm that can be folded flat. *Sci. Robot.* 3 (16): eaar2915. https://doi.org/eaar291510.1126/scirobotics.aar2915.

11 Boyvat, M., Koh, J.S., and Wood, R.J. (2017). Addressable wireless actuation for multijoint folding robots and devices. *Sci. Robot.* 2 (8): https://doi.org/eaan154410.1126/scirobotics.aan1544.

12 Xu, W.N. and Gracias, D.H. (2019). Soft three-dimensional robots with hard two-dimensional materials. *ACS Nano* 13 (5): 4883–4892. https://doi.org/10.1021/acsnano.9b03051.

13 Zhou, W.X., Qiao, Z.G., Zare, E.N. et al. (2020). 4D-printed dynamic materials in biomedical applications: chemistry, challenges, and their future perspectives in the clinical sector. *J. Med. Chem.* 63 (15): 8003–8024. https://doi.org/10.1021/acs.jmedchem.9b02115.

14 Peraza-Hernandez, E.A., Hartl, D.J., Malak, R.J. Jr., et al. (2014). Origami-inspired active structures: a synthesis and review. *Smart Mater. Struct.* 23 (9): 094001.

15 Zhang, Y., Zhang, F., Yan, Z. et al. (2017). Printing, folding and assembly methods for forming 3D mesostructures in advanced materials. *Nat. Rev. Mater.* 2 (4): 17019. https://doi.org/10.1038/natrevmats.2017.

16 Huang, G.S. and Mei, Y.F. (2018). Assembly and self-assembly of nanomembrane materials-from 2D to 3D. *Small* 14 (14): https://doi.org/170366510.1002/smll.201703665.

17 Ng, C. S. X., Tan, M. W. M., Xu, C. Y. et al. (2020). Locomotion of miniature soft robots. *Adv. Mater.* 33, 2003558 https://doi.org/200355810.1002/adma.202003558

18 Miyashita, S., Guitron, S., Yoshida, K. et al. (2016). Ingestible, controllable, and degradable origami robot for patching stomach wounds. In: *2016 IEEE International Conference on Robotics and Automation (ICRA)*, 909–916. IEEE.

19 Karnaushenko, D., Kang, T., and Schmidt, O.G. (2019). Shapeable material technologies for 3D self-assembly of mesoscale electronics. *Adv. Mater. Technol.* 4 (4): https://doi.org/180069210.1002/admt.201800692.

20 Leong, T.G., Randall, C.L., Benson, B.R. et al. (2009, Jan 20). Tetherless thermobiochemically actuated microgrippers. *Proc. Natl. Acad. Sci. U.S.A.* 106 (3): 703–708. https://doi.org/10.1073/pnas.0807698106.

21 Feiner, R., Engel, L., Fleischer, S. et al. (2016, Jun). Engineered hybrid cardiac patches with multifunctional electronics for online monitoring and regulation of tissue function. *Nat. Mater.* 15(6): 679https://doi.org/10.1038/nmat4590

22 Froeter, P., Huang, Y., Cangellaris, O.V. et al. (2014). Toward intelligent synthetic neural circuits: directing and accelerating neuron cell growth by self-rolled-up silicon nitride microtube array. *ACS Nano* 8 (11): 11108–11117. https://doi.org/10.1021/nn504876y.

23 Tian, B., Liu, J., Dvir, T. et al. (2012). Macroporous nanowire nanoelectronic scaffolds for synthetic tissues. *Nat. Mater.* 11 (11): 986–994. https://doi.org/10.1038/nmat3404.

24 Yu, M., Huang, Y., Ballweg, J. et al. (2011). Semiconductor nanomembrane tubes: three-dimensional confinement for controlled neurite outgrowth. *ACS Nano* 5 (4): 2447–2457. https://doi.org/10.1021/nn103618d.

25 Bishop, D., Pardo, F., Bolle, C. et al. (2012). Silicon micro-machines for fun and profit. *J. Low Temp. Phys.* 169 (5–6): 386–399. https://doi.org/10.1007/s10909-012-0654-z.

26 Wood, R.J. (2014). The challenge of manufacturing between macro and micro. *Am. Sci.* 102 (2): 124–131. https://doi.org/10.1511/2014.107.124.

27 Piyawattanametha, W., Patterson, P.R., Hah, D. et al. (2005). Surface- and bulk-micromachined two-dimensional scanner driven by angular vertical comb actuators. *J. Microelectromech. Syst.* 14 (6): 1329–1338. https://doi.org/10.1109/jmems.2005.859073.

28 Zhang, H., Yu, X., and Braun, P.V. (2011). Three-dimensional bicontinuous ultrafast-charge and -discharge bulk battery electrodes. *Nat. Nanotechnol.* 6 (5): 277–281. https://doi.org/10.1038/nnano.2011.38.

29 Pan, L., Yu, G., Zhai, D. et al. (2012). Hierarchical nanostructured conducting polymer hydrogel with high electrochemical activity. *Proc. Natl. Acad. Sci. U.S.A.* 109 (24): 9287–9292. https://doi.org/10.1073/pnas.1202636109.

30 Deng, J., Ji, H., Yan, C. et al. (2013). Naturally rolled-up C/Si/C trilayer nanomembranes as stable anodes for lithium-ion batteries with remarkable cycling performance. *Angew. Chem. Int. Ed.* 52 (8): 2326–2330. https://doi.org/10.1002/anie.201208357.

31 Wu, H., Yu, G., Pan, L. et al. (2013). Stable Li-ion battery anodes by in-situ polymerization of conducting hydrogel to conformally coat silicon nanoparticles. *Nat. Commun.* 4: 1943. https://doi.org/10.1038/ncomms2941.

32 Pikul, J.H., Zhang, H.G., Cho, J. et al. (2013). High-power lithium ion microbatteries from interdigitated three-dimensional bicontinuous nanoporous electrodes. *Nat. Commun.* 4: 1732. https://doi.org/10.1038/ncomms2747.

33 Sun, K., Wei, T.-S., Ahn, B.Y. et al. (2013). 3D printing of interdigitated Li-ion microbattery architectures. *Adv. Mater.* 25 (33): 4539–4543. https://doi.org/10.1002/adma.201301036.

34 Fan, Z., Razavi, H., Do, J.-W. et al. (2009). Three-dimensional nanopillar-array photovoltaics on low-cost and flexible substrates. *Nat. Mater.* 8 (8): 648–653. https://doi.org/10.1038/nmat2493.

35 Lamoureux, A., Lee, K., Shlian, M. et al. (2015). Dynamic kirigami structures for integrated solar tracking. *Nat. Commun.* 6: 8092. https://doi.org/10.1038/ncomms9092.

36 Ahn, B.Y., Duoss, E.B., Motala, M.J. et al. (2009). Omnidirectional printing of flexible, stretchable, and spanning silver microelectrodes. *Science* 323 (5921): 1590–1593. https://doi.org/10.1126/science.1168375.

37 Huang, W., Yu, X., Froeter, P. et al. (2012). On-chip inductors with self-rolled-up SiN_x nanomembrane tubes: a novel design platform for extreme miniaturization. *Nano Lett.* 12 (12): 6283–6288. https://doi.org/10.1021/nl303395d.

38 Huang, W., Koric, S., Yu, X. et al. (2014). Precision structural engineering of self-rolled-up 3D nanomembranes guided by transient quasi-static FEM modeling. *Nano Lett.* 14 (11): 6293–6297. https://doi.org/10.1021/nl5026369.

39 Grimm, D., Bufon, C.C.B., Deneke, C. et al. (2013). Rolled-up nanomembranes as compact 3D architectures for field effect transistors and fluidic sensing applications. *Nano Lett.* 13 (1): 213–218. https://doi.org/10.1021/nl303887b.

40 Xu, S., Yan, Z., Jang, K.-I. et al. (2015). Assembly of micro/nanomaterials into complex, three-dimensional architectures by compressive buckling. *Science* 347 (6218): 154–159. https://doi.org/10.1126/science.1260960.

41 Yan, Z., Zhang, F., Liu, F. et al. (2016). Mechanical assembly of complex, 3D mesostructures from releasable multilayers of advanced materials. *Sci. Adv.* 2 (9): e1601014. https://doi.org/10.1126/sciadv.1601014.

42 Yu, X., Huang, W., Li, M. et al. (2015). Ultra-small, high-frequency, and substrate-immune microtube inductors transformed from 2D to 3D. *Sci. Rep.* 5: 9661. https://doi.org/10.1038/srep09661.

43 Liu, N., Guo, H., Fu, L. et al. (2008, Jan). Three-dimensional photonic metamaterials at optical frequencies. *Nat. Mater.* 7 (1): 31–37. doi: 10.1038/nmat2072.

44 Gansel, J.K., Thiel, M., Rill, M.S. et al. (2009). Gold helix photonic metamaterial as broadband circular polarizer. *Science* 325 (5947): 1513–1515. https://doi.org/10.1126/science.1177031.

45 Skylar-Scott, M.A., Gunasekaran, S., and Lewis, J.A. (2016). Laser-assisted direct ink writing of planar and 3D metal architectures. *Proc. Natl. Acad. Sci. U.S.A.* 113 (22): 6137–6142. https://doi.org/10.1073/pnas.1525131113.

46 Destino, J.F., Dudukovic, N.A., Johnson, M.A. et al. (2018). 3D printed optical quality silica and silica-titania glasses from sol-gel feedstocks. *Adv. Mater. Technol.* 3 (6): 1700323. https://doi.org/10.1002/admt.201700323.

47 Vyatskikh, A., Delalande, S., Kudo, A. et al. (2018). Additive manufacturing of 3D nano-architected metals. *Nat. Commun.* 9: 593. https://doi.org/10.1038/s41467-018-03071-9.

48 Cumpston, B.H., Ananthavel, S.P., Barlow, S. et al. (1999). Two-photon polymerization initiators for three-dimensional optical data storage and microfabrication. *Nature* 398 (6722): 51–54.

49 Kawata, S., Sun, H.B., Tanaka, T. et al. (2001). Finer features for functional microdevices - Micromachines can be created with higher resolution using two-photon absorption. *Nature* 412 (6848): 697–698. https://doi.org/10.1038/35089130.

50 LaFratta, C.N., Fourkas, J.T., Baldacchini, T. et al. (2007). Multiphoton fabrication. *Angew. Chem. Int. Ed.* 46 (33): 6238–6258. https://doi.org/10.1002/anie.200603995.

51 Chu, W., Tan, Y., Wang, P. et al. (2018). Centimeter-height 3D printing with femtosecond laser two-photon polymerization. *Adv. Mater. Technol.* 3 (5): 1700396. https://doi.org/10.1002/admt.201700396.

52 Zheng, X., Smith, W., Jackson, J. et al. (2016). Multiscale metallic metamaterials. *Nat. Mater.* 15(10): 1100-+. https://doi.org/10.1038/nmat4694

53 Zheng, X., Lee, H., Weisgraber, T.H. et al. (2014). Ultralight, ultrastiff mechanical metamaterials. *Science* 344 (6190): 1373–1377. https://doi.org/10.1126/science.1252291.

54 Gao, W., Sattayasamitsathit, S., Orozco, J. et al. (2011). Highly efficient catalytic microengines: template electrosynthesis of polyaniline/platinum microtubes. *J. Am. Chem. Soc.* 133 (31): 11862–11864. https://doi.org/10.1021/ja203773g.

55 Li, J., Rozen, I., and Wang, J. (2016). Rocket science at the nanoscale. *ACS Nano* 10 (6): 5619–5634. https://doi.org/10.1021/acsnano.6b02518.

56 Xu, C., Wu, X., Huang, G. et al. (2019). Rolled-up nanotechnology: materials issue and geometry capability. *Adv. Mater. Technol.* 4 (1): 1800486. https://doi.org/10.1002/admt.201800486.

57 Liu, Y., Genzer, J., and Dickey, M.D. (2016, Jan). "2D or not 2D": Shape-programming polymer sheets. *Prog. Polym. Sci.* 52: 79–106. https://doi.org/10.1016/j.progpolymsci.2015.09.001.

58 Rogers, J., Huang, Y., Schmidt, O.G. et al. (2016). Origami MEMS and NEMS. *MRS Bull.* 41 (2): 123–129. https://doi.org/10.1557/mrs.2016.2.

59 Kuribayashi, K., Tsuchiya, K., You, Z. et al. (2006). Self-deployable origami stent grafts as a biomedical application of Ni-rich TiNi shape memory alloy foil. *Mater. Sci. Eng., A* 419 (1-2): 131–137. https://doi.org/10.1016/j.msea.2005.12.016.

60 Wang, Z., Jing, L., Yao, K. et al. (2017). Origami-based reconfigurable metamaterials for tunable chirality. *Adv. Mater.* 29 (27): 1700412. https://doi.org/10.1002/adma.201700412.

61 Liu, Z., Cui, A., Li, J. et al. (2019). Folding 2D structures into 3D configurations at the micro/nanoscale: principles, techniques, and applications. *Adv. Mater.* 31 (4): 1802211. https://doi.org/10.1002/adma.201802211.

62 Li, J. and Liu, Z. (2018). Focused-ion-beam-based nano-kirigami: from art to photonics. *Nanophotonics* 7 (10): 1637–1650. https://doi.org/10.1515/nanoph-2018-0117.

63 Gladman, A. S., Matsumoto, E. A., Nuzzo, R. G. et al. (2016). Biomimetic 4D printing. *Nat. Mater.* 15(4): 413-+. https://doi.org/10.1038/nmat4544

64 Kwok, K.S., Huang, Q., Mastrangeli, M. et al. (2020). Self-folding using capillary forces. *Adv. Mater. Interfaces* 7 (5): 1901677. https://doi.org/10.1002/admi.201901677.

65 Cho, J.-H., Azam, A., and Gracias, D.H. (2010). Three dimensional nanofabrication using surface forces. *Langmuir* 26 (21): 16534–16539. https://doi.org/10.1021/la1013889.

66 Reynolds, M.F., McGill, K.L., Wang, M.A. et al. (2019). Capillary origami with atomically thin membranes. *Nano Lett.* 19 (9): 6221–6226. https://doi.org/10.1021/acs.nanolett.9b02281.

67 van Honschoten, J.W., Berenschot, J.W., Ondarcuhu, T. et al. (2010). Elastocapillary fabrication of three-dimensional microstructures. *Appl. Phys. Lett.* 97 (1): 014103. https://doi.org/10.1063/1.3462302.

68 Guo, X., Li, H., Ahn, B.Y. et al. (2009). Two- and three-dimensional folding of thin film single-crystalline silicon for photovoltaic power applications. *Proc. Natl. Acad. Sci. U.S.A.* 106 (48): 20149–20154. https://doi.org/10.1073/pnas.0907390106.

69 Cho, J.-H., Keung, M.D., Verellen, N. et al. (2011). Nanoscale origami for 3D optics. *Small* 7 (14): 1943–1948. https://doi.org/10.1002/smll.201100568.

70 Joung, D., Nemilentsau, A., Agarwal, K. et al. (2017). Self-assembled three-dimensional graphene-based polyhedrons inducing volumetric light confinement. *Nano Lett.* 17 (3): 1987–1994. https://doi.org/10.1021/acs.nanolett.6b05412.

71 Chalapat, K., Chekurov, N., Jiang, H. et al. (2013). Self-organized origami structures via ion-induced plastic strain. *Adv. Mater.* 25 (1): 91–95. https://doi.org/10.1002/adma.201202549.

72 Arora, W.J., Sijbrandij, S., Stern, L. et al. (2007). Membrane folding by helium ion implantation for three-dimensional device fabrication. *J. Vac. Sci. Technol., B* 25 (6): 2184–2187. https://doi.org/10.1116/1.2779049.

73 Arora, W.J., Smith, H.I., and Barbastathis, G. (2007). Membrane folding by ion implantation induced stress to fabricate three-dimensional nanostructures. *Microelectron. Eng.* 84 (5-8): 1454–1458. https://doi.org/10.1016/j.mee.2007.01.182.

74 Cui, A., Liu, Z., Li, J. et al. (2015). Directly patterned substrate-free plasmonic "nanograter" structures with unusual Fano resonances. *Light Sci. Appl.* 4: e308. https://doi.org/10.1038/lsa.2015.81.

75 Jia, P., Zuber, K., Guo, Q. et al. (2019). Large-area freestanding gold nanomembranes with nanoholes. *Mater. Horiz.* 6 (5): 1005–1012. https://doi.org/10.1039/c8mh01302k.

76 Liu, Z., Du, H., Li, J. et al. (2018). Nano-kirigami with giant optical chirality. *Sci. Adv.* 4 (7): eaat4436. https://doi.org/10.1126/sciadv.aat4436.

77 Liu, Z., Du, H., Li, Z.-Y. et al. (2018). Invited article: nano-kirigami metasurfaces by focused-ion-beam induced close-loop transformation. *APL Photonics* 3 (10): 100803. https://doi.org/10.1063/1.5043065.

78 Mao, Y., Zheng, Y., Li, C. et al. (2017). Programmable bidirectional folding of metallic thin films for 3D chiral optical antennas. *Adv. Mater.* 29 (19): 1606482. https://doi.org/10.1002/adma.201606482.

79 Park, B.C., Jung, K.Y., Song, W.Y. et al. (2006). Bending of a carbon nanotube in vacuum using a focused ion beam. *Adv. Mater.* 18 (1): 95–98. https://doi.org/10.1002/adma.200501223.

80 Rajput, N.S., Banerjee, A., and Verma, H.C. (2011). Electron- and ion-beam-induced maneuvering of nanostructures: phenomenon and applications. *Nanotechnology* 22 (48): 485302. https://doi.org/10.1088/0957-4484/22/48/485302.

81 Xia, L., Wu, W.G., Xu, J. et al. (2006). 3D nanohelix fabrication and 3D nanometer assembly by focused ion beam stress-introducing technique. In: *Mems 2006: 19th IEEE International Conference on Micro Electro Mechanical Systems, Technical Digest*, 118–121.

82 Yoshida, T., Baba, A., and Asano, T. (2005). Fabrication of micro field emitter tip using ion-beam irradiation-induced self-standing of thin films. *Jpn. J. Appl. Phys., Part 1* 44 (7B): 5744–5748. https://doi.org/10.1143/jjap.44.5744.

83 Na, J.-H., Evans, A.A., Bae, J. et al. (2015). Programming reversibly self-folding origami with micropatterned photo-crosslinkable polymer trilayers. *Adv. Mater.* 27 (1): 79–85. https://doi.org/10.1002/adma.201403510.

84 Jin, B.J., Song, H.J., Jiang, R.Q. et al. (2018). Programming a crystalline shape memory polymer network with thermo- and photo-reversible bonds toward a single-component soft robot. *Sci. Adv.* 4 (1): https://doi.org/ARTN eaao386510.1126/sciadv.aao3865.

85 Zakharchenko, S., Puretskiy, N., Stoychev, G. et al. (2010). Temperature controlled encapsulation and release using partially biodegradable thermo-magneto-sensitive self-rolling tubes. *Soft Matter* 6 (12): 2633–2636. https://doi.org/10.1039/c0sm00088d.

86 Stoychev, G., Zakharchenko, S., Turcaud, S. et al. (2012). Shape-programmed folding of stimuli-responsive polymer bilayers. *ACS Nano* 6 (5): 3925–3934. https://doi.org/10.1021/nn300079f.

87 Xu, W., Qin, Z., Chen, C.-T. et al. (2017). Ultrathin thermoresponsive self-folding 3D graphene. *Sci. Adv.* 3 (10): e1701084. https://doi.org/10.1126/sciadv.1701084.

88 Zhao, Q., Liang, Y.H., Ren, L. et al. (2018). Bionic intelligent hydrogel actuators with multimodal deformation and locomotion. *Nano Energy* 51: 621–631. https://doi.org/10.1016/j.nanoen.2018.07.025.

89 Yoon, C., Xiao, R., Park, J. et al. (2014). Functional stimuli responsive hydrogel devices by self-folding. *Smart Mater. Struct.* 23 (9): 094008. https://doi.org/10.1088/0964-1726/23/9/094008.

90 Breger, J.C., Yoon, C., Xiao, R. et al. (2015). Self-folding thermo-magnetically responsive soft microgrippers. *ACS Appl. Mater. Interfaces* 7 (5): 3398–3405. https://doi.org/10.1021/am508621s.

91 Felton, S.M., Becker, K.P., Aukes, D.M. et al. (2015). Self-folding with shape memory composites at the millimeter scale. *J. Micromech. Microeng.* 25 (8): 085004. https://doi.org/10.1088/0960-1317/25/8/085004.

92 Ge, F. and Zhao, Y. (2017). A new function for thermal phase transition-based polymer actuators: autonomous motion on a surface of constant temperature. *Chem. Sci.* 8 (9): 6307–6312. https://doi.org/10.1039/c7sc01792h.

93 Sulfridge, M., Saif, T., Miller, N. et al. (2002). Optical actuation of a bistable MEMS. *J. Microelectromech. Syst.* 11 (5): 574–583. https://doi.org/10.1109/jmems.2002.803417.

94 Liu, Y., Boyles, J.K., Genzer, J. et al. (2012). Self-folding of polymer sheets using local light absorption. *Soft Matter* 8 (6): 1764–1769. https://doi.org/10.1039/c1sm06564e.

95 Liu, Y., Mailen, R., Zhu, Y. et al. (2014). Simple geometric model to describe self-folding of polymer sheets. *Phys. Rev. E* 89 (4): 042601. https://doi.org/10.1103/PhysRevE.89.042601.

96 Liu, Y., Miskiewicz, M., Escuti, M.J. et al. (2014). Three-dimensional folding of pre-strained polymer sheets via absorption of laser light. *J. Appl. Phys.* 115 (20): 204911. https://doi.org/10.1063/1.4880160.

97 Liu, Y., Shaw, B., Dickey, M.D. et al. (2017). Sequential self-folding of polymer sheets. *Sci. Adv.* 3 (3): e1602417. https://doi.org/10.1126/sciadv.1602417.

98 Tang, Z., Gao, Z., Jia, S. et al. (2017). Graphene-based polymer bilayers with superior light-driven properties for remote construction of 3D structures. *Adv. Sci.* 4 (5): 1600437. https://doi.org/10.1002/advs.201600437.

99 Fusco, S., Sakar, M.S., Kennedy, S. et al. (2014). An integrated microrobotic platform for on-demand, targeted therapeutic interventions. *Adv. Mater.* 26 (6): 952–957. https://doi.org/10.1002/adma.201304098.

100 Deng, J., Li, J., Chen, P. et al. (2016). Tunable photothermal actuators based on a pre-programmed aligned nanostructure. *J. Am. Chem. Soc.* 138 (1): 225–230. https://doi.org/10.1021/jacs.5b10131.

101 Zhang, X., Yu, Z., Wang, C. et al. (2014). Photoactuators and motors based on carbon nanotubes with selective chirality distributions. *Nat. Commun.* 5: 2983. https://doi.org/10.1038/ncomms3983.

102 Liu, Y.L., Ai, K.L., Liu, J.H. et al. (2013). Dopamine-melanin colloidal nanospheres: an efficient near-infrared photothermal therapeutic agent for in vivo cancer therapy. *Adv. Mater.* 25 (9): 1353–1359. https://doi.org/10.1002/adma.201204683.

103 Li, Z., Zhang, X., Wang, S. et al. (2016). Polydopamine coated shape memory polymer: enabling light triggered shape recovery, light controlled shape reprogramming and surface functionalization. *Chem. Sci.* 7 (7): 4741–4747. https://doi.org/10.1039/c6sc00584e.

104 Ocampo, J.M.Z., Vaccaro, P.O., Kubota, K. et al. (2004). Characterization of GaAs-based micro-origami mirrors by optical actuation. *Microelectron. Eng.* 73-4: 429–434. https://doi.org/10.1016/j.mee.2004.03.012.

105 Ocampo, J.M.Z., Vaccaro, P.O., Fleischmann, T. et al. (2003). Optical actuation of micromirrors fabricated by the micro-origami technique. *Appl. Phys. Lett.* 83 (18): 3647–3649. https://doi.org/10.1063/1.1622800.

106 Ardanuc, S.M. and Lal, A. (2013). Ultrasound enhanced electrostatic batch assembly for MEMS. *Sens. Actuators, A* 197: 136–149. https://doi.org/10.1016/j.sna.2013.03.017.

107 Kaajakari, V. and Lal, A. (2003). Thermokinetic actuation for batch assembly of microscale hinged structures. *J. Microelectromech. Syst.* 12 (4): 425–432. https://doi.org/10.1109/jmems.2003.811747.

108 Kaajakari, V. and Lal, A. (1999). Pulsed ultrasonic actuation of polysilicon surface micromachines. In: *1999 IEEE Ultrasonics Symposium Proceedings*, vol. 1 and 2 (ed. S.C. Schneider, M. Levy and B.R. McAvoy), 643–646. IEEE.

109 Calvaresi, M., Quintana, M., Rudolf, P. et al. (2013). Rolling up a graphene sheet. *ChemPhysChem* 14 (15): 3447–3453. https://doi.org/10.1002/cphc.201300337.

110 Viculis, L.M., Mack, J.J., and Kaner, R.B. (2003). A chemical route to carbon nanoscrolls. *Science* 299 (5611): 1361–1361. https://doi.org/10.1126/science.1078842.

111 Wang, S., Tang, L.A.L., Bao, Q. et al. (2009). Room-temperature synthesis of soluble carbon nanotubes by the sonication of graphene oxide nanosheets. *J. Am. Chem. Soc.* 131 (46): 16832–16837. https://doi.org/10.1021/ja905968v.

112 Wang, X., Yang, D.-P., Huang, G. et al. (2012). Rolling up graphene oxide sheets into micro/nanoscrolls by nanoparticle aggregation. *J. Mater. Chem.* 22 (34): 17441–17444. https://doi.org/10.1039/c2jm32810k.

113 Li, G., Fei, G., Liu, B. et al. (2014). Shape recovery characteristics for shape memory polymers subjected to high intensity focused ultrasound. *RSC Adv.* 4 (62): 32701–32709. https://doi.org/10.1039/c4ra04586f.

114 Li, G., Yan, Q., Xia, H. et al. (2015). Therapeutic-ultrasound-triggered shape memory of a melamine-enhanced poly(vinyl alcohol) physical hydrogel. *ACS Appl. Mater. Interfaces* 7 (22): 12067–12073. https://doi.org/10.1021/acsami.5b02234.

115 Palleau, E., Morales, D., Dickey, M.D. et al. (2013). Reversible patterning and actuation of hydrogels by electrically assisted ionoprinting. *Nat. Commun.* 4: https://doi.org/ARTN 225710.1038/ncomms3257.

116 Okuzaki, H., Kuwabara, T., Funasaka, K. et al. (2013). Humidity-sensitive polypyrrole films for electro-active polymer actuators. *Adv. Funct. Mater.* 23 (36): 4400–4407. https://doi.org/10.1002/adfm.201203883.

117 Taghavi, M., Helps, T., and Rossiter, J. (2018). Electro-ribbon actuators and electro-origami robots. *Sci. Robot.* 3 (25): eaau9795. https://doi.org/ARTN eaau979510.1126/scirobotics.aau9795.

118 Miyashita, S., Guitron, S., Li, S.G. et al. (2017). Robotic metamorphosis by origami exoskeletons. *Sci. Robot.* 2 (10): eaao4369. https://doi.org/ARTN eaao436910.1126/scirobotics.aao4369.

119 Han, K., Shields, C.W., Diwakar, N.M. et al. (2017). Sequence-encoded colloidal origami and microbot assemblies from patchy magnetic cubes. *Sci. Adv.* 3 (8): e170110. https://doi.org/ARTN e170110810.1126/sciadv.1701108.

120 Xu, T.Q., Zhang, J.C., Salehizadeh, M. et al. (2019). Millimeter-scale flexible robots with programmable three-dimensional magnetization and motions. *Sci. Robot.* 4 (29): eaav4494. https://doi.org/ARTN eaav449410.1126/scirobotics.aav4494.

121 Lum, G.Z., Ye, Z., Dong, X.G. et al. (2016). Shape-programmable magnetic soft matter. *Proc. Natl. Acad. Sci. U.S.A.* 113 (41): E6007–E6015. https://doi.org/10.1073/pnas.1608193113.

122 Hu, W.Q., Lum, G.Z., Mastrangeli, M. et al. (2018). Small-scale soft-bodied robot with multimodal locomotion. *Nature* 554 (7690): 81–85. https://doi.org/10.1038/nature25443.

123 Kim, Y., Yuk, H., Zhao, R. K. et al. (2018). Printing ferromagnetic domains for untethered fast-transforming soft materials. *Nature* 558(7709): 274-+. https://doi.org/10.1038/s41586-018-0185-0

124 Cui, J.Z., Huang, T.Y., Luo, Z.C. et al. (2020). Nanomagnetic encoding of shape-morphing micromachines. *Nature* 578 (7793): E10–E10. https://doi.org/10.1038/s41586-019-1888-6.

125 Park, O.K., Tiwary, C.S., Yang, Y. et al. (2017). Magnetic field controlled graphene oxide-based origami with enhanced surface area and mechanical properties. *Nanoscale* 9 (21): 6991–6997. https://doi.org/10.1039/c7nr01054k.

126 Jamin, T., Py, C., and Falcon, E. (2011). Instability of the origami of a ferrofluid drop in a magnetic field. *Phys. Rev. Lett.* 107 (20): 204503. https://doi.org/ARTN 20450310.1103/PhysRevLett.107.204503.

127 Tang, J.D., Yin, Q.F., Qiao, Y.C. et al. (2019). Shape morphing of hydrogels in alternating magnetic field. *ACS Appl. Mater. Interfaces* 11 (23): 21194–21200. https://doi.org/10.1021/acsami.9b05742.

128 Patel, A. and Mequanint, K. (2011). Hydrogel biomaterials. In: *Biomedical Engineering-Frontiers and Challenges*. IntechOpen.

129 Randhawa, J.S., Leong, T.G., Bassik, N. et al. (2008). Pick-and-place using chemically actuated microgrippers. *J. Am. Chem. Soc.* 130 (51): 17238–17239. https://doi.org/10.1021/ja806961p.

130 Wu, Z.L., Moshe, M., Greener, J. et al. (2013). Three-dimensional shape transformations of hydrogel sheets induced by small-scale modulation of internal stresses. *Nat. Commun.* 4 (1): 1586. https://doi.org/ARTN 158610.1038/ncomms2549.

131 Karnaushenko, D., Munzenrieder, N., Karnaushenko, D. D. et al. (2015). Biomimetic microelectronics for regenerative neuronal cuff implants. *Adv. Mater.* 27(43): 6797-+. https://doi.org/10.1002/adma.201503696

132 Zheng, Z.Q., Wang, H.P., Dong, L.X. et al. (2021). Ionic shape-morphing microrobotic end-effectors for environmentally adaptive targeting, releasing, and sampling. *Nat. Commun.* 12 (1): 411. https://doi.org/ARTN 41110.1038/s41467-020-20697-w.

133 Jamal, M., Zarafshar, A.M., and Gracias, D.H. (2011). Differentially photo-crosslinked polymers enable self-assembling microfluidics. *Nat. Commun.* 2 (1): 527. https://doi.org/ARTN52710.1038/ncomms1531.

134 Li, W.M., Huang, G.S., Yan, H. et al. (2012). Fabrication and stimuli-responsive behavior of flexible micro-scrolls. *Soft Matter* 8 (27): 7103–7107. https://doi.org/10.1039/c2sm25366f.

135 Abdullah, A.M., Li, X.L., Braun, P.V. et al. (2018). Self-folded gripper-like architectures from stimuli-responsive bilayers. *Adv. Mater.* 30 (31): https://doi.org/ARTN180166910.1002/adma.201801669.

136 Lou, J.M., Liu, Z.H., Yang, L. et al. (2021). A new strategy of discretionarily reconfigurable actuators based on self-healing elastomers for diverse soft robots. *Adv. Funct. Mater.* 2008328. https://doi.org/ARTN200832810.1002/adfm.202008328.

137 Randhawa, J.S., Keung, M.D., Tyagi, P. et al. (2010). Reversible actuation of microstructures by surface-chemical modification of thin-film bilayers. *Adv. Mater.* 22 (3): 407–410. https://doi.org/10.1002/adma.200902337.

138 Xu, B.R., Tian, Z., Wang, J. et al. (2018). Stimuli-responsive and on-chip nanomembrane micro-rolls for enhanced macroscopic visual hydrogen detection. *Sci. Adv.* 4 (4): https://doi.org/ARTN eaap820310.1126/sciadv.aap8203.

139 Magdanz, V., Stoychev, G., Ionov, L. et al. (2014). Stimuli-responsive microjets with reconfigurable shape. *Angew. Chem. Int. Ed.* 53 (10): 2673–2677. https://doi.org/10.1002/anie.201308610.

140 Fernandez-Pacheco, A., Streubel, R., Fruchart, O. et al. (2017). Three-dimensional nanomagnetism. *Nat. Commun.* 8: https://doi.org/ARTN1575610.1038/ncomms15756.

141 Kim, J., Chung, S.E., Choi, S.E. et al. (2011). Programming magnetic anisotropy in polymeric microactuators. *Nat. Mater.* 10 (10): 747–752. https://doi.org/10.1038/Nmat3090.

142 Rus, D. and Tolley, M.T. (2018). Design, fabrication and control of origami robots. *Nat. Rev. Mater.* 3 (6): 101–112. https://doi.org/10.1038/s41578-018-0009-8.

143 Bandari, V.K., Nan, Y., Karnaushenko, D. et al. (2020). A flexible microsystem capable of controlled motion and actuation by wireless power transfer. *Nat. Electron.* 3 (3): 172–180. https://doi.org/10.1038/s41928-020-0384-1.

11

Rolled-up Electronics and Origami

Wen Huang[1] *and Xiuling Li*[2]

[1] *Hefei University of Technology, School of Microelectronics, Department of Electronic Science, 485 Danxia Road, Hefei, Anhui 230601, People's Republic of China*
[2] *University of Illinois at Urbana-Champaign, School of Engineering, Department of Electrical and Computer Engineering, 208 North Wright Street, Urbana, Illinois 61801, USA*

11.1 Introduction

Rolled-up architectures, either cylindrical or helical, are prevalent in the three-dimensional (3D) world that we live in, providing many critical mechanical functions. Prior to the invention of integrated circuits in the 1950s, electronics employing 3D structural designs emerged from common sense engineering. Fundamental electronic devices have been designed and fabricated in the fashion of rolled-up architectures, such as solenoid inductors and helical antennas. However, rolled-up design thinking has been suppressed by layer-by-layer planar semiconductor processing, along with the imaginations of generations of engineers and scientists, due to the difficulty of constructing 3D architectures from 2D processing. A nanotechnology fabrication platform that overcomes these fundamental challenges in mechanics and form will enable high-performance on-chip electronic systems that are otherwise impossible to achieve with existing planar-integrated circuit technology.

Fabricating rolled-up architectures off-chip at a macroscale, based on self-assembling mechanics, is typically facilitated by a stress mismatch between components in the structure. Thin-film stresses and stress mismatches are not new phenomena in semiconductor processing, being inevitable and potentially useful at many processing steps such as epitaxial growth and physical and chemical deposition. In the year 2000, Prinz et al. demonstrated a rolled-up architecture utilizing stress mismatch between epitaxial layers for the first time [1]. Since then, rolled-up nanotechnology has become a hot field of research and has been developed as a platform technology for interdisciplinary engineering, including the improvement of monolithic electronic devices [2–11]. Compared with planar electronic on-chip devices, self-rolled-up structures offer a higher degree of freedom to electronic device designers. In essence, electronic devices are engineered to guide the direction

Nanomembranes: Materials, Properties, and Applications, First Edition.
Edited by Yongfeng Mei, Gaoshan Huang, and Xiuling Li.
© 2022 WILEY-VCH GmbH. Published 2022 by WILEY-VCH GmbH.

of flow, velocity distribution, concentration, and inertia of charge carriers under different amplitudes and frequencies of excitation signals. For most electronic devices, self-rolled-up structures can be used to flexibly guide the direction of flow and velocity distributions of carriers in 3D spaces to facilitate new electrical characteristics or improve traditionally defined performance. More importantly, the natural requirement of self-rolling technology for constituent film thicknesses to be thin (10s of nm) makes it very easy for low-dimensional materials to be integrated into self-rolling platforms, taking advantage of physical effects such as carrier velocity and concentration through a material phase transformation, and adjusting inertia by increasing carrier effective mass in electronic devices [12–14]. The potential to realize new electrical characteristics breaks the conventional cognitive framework of device design.

Constructing 3D rolled-up architectures on-chip using conventional planar semiconductor processing can be considered as an art of origami enabled by advanced nanotechnology, which involves an interdisciplinary knowledge base, including material science, mechanical engineering, electronics engineering, and thermal engineering, among other disciplines. To present a systematic illustration, this chapter focuses first on representative works that have built the theoretical basis and yield the most sophisticated rolled-up electronics and origami, followed by a discussion of the potential directions for development in this field.

11.1.1 Rolled-up Origami for Electronics

Structurally, self-rolling electronic devices generally consist of four parts: a substrate, a sacrificial layer, a strain layer, and a functional layer. To be compatible with semiconductor or MEMS-integrated circuit processes, the substrates of self-rolling electronic devices are usually hard semiconductor or insulator materials, such as silicon, quartz or fused silica, sapphire, and gallium arsenide. Sapphire or glass are the best choices for substrate materials to eliminate parasitic effects when self-rolled electronic devices are used as independent components. The material selection of the remaining three parts is not independent from the choice of the first. It is necessary to select appropriate materials and manufacturing processes according to the specific device performance requirements, whether required to operate at high frequencies or with high current densities. According to the requirements of active and passive devices, the material systems used in self-rolled-up technology can vary widely.

It has been proven that the materials for self-rolling active device structures, such as field-effect transistors, can be grown as single-crystal inorganic thin films, a representative case being the single-crystal III–V semiconductor thin-film pair InGaAs/GaAs [15]. Molecular-beam epitaxy (MBE) has been used to grow a 200 nm thick GaAs buffer layer on a semi-insulating GaAs (001) wafer and regenerate a 40 nm thick AlAs film as the sacrificial layer, which can be etched by a solution of hydrochloric acid. Above the sacrificial layer, 20 nm $In_{0.2}Ga_{0.8}As$ and 30 nm doped GaAs thin-film channel layers were grown by lattice mismatch. The doping concentration was 3.5×10^{18} cm^{-3}. The surface of the GaAs thin films needs to be passivated, which was achieved by depositing a layer of 11 nm thick alumina as a passivating dielectric layer. Titanium and chromium films of 16 and 22 nm, respectively, were deposited as metal electrodes sequentially. Finally, a layer of alumina was redeposited as capping layer. The total thickness of film on the sacrificial layer was about 110 nm. After the sacrificial layer was removed, the films roll together into a multilayer hollow cylinder with an inner diameter of only 5.3 μm, as shown in Figure 11.1.

Compared to the implementation of active devices, self-rolled-up nanotechnology provides a more advantageous platform to fabricate passive devices. Similar to the architecture of active self-rolled-up devices, passive self-rolled-up devices require a sacrificial layer, strain layer, and functional layer. Firstly, the strain layer needs to be composed of an insulating material to prevent leakage current; secondly, Young's modulus of the sacrificial layer material should be maximized to minimize the strain energy absorption of strain layer; thirdly, the surface smoothness of sacrificial layer material should be maximized to reduce the accumulation effect of unevenness, thereby improving the yield of the devices; lastly, the rolling resistance introduced by the functional layer should be minimized to control the structure's inner diameter, and enable more complex 2D structures to self-rolled up into the desired 3D structure. The two-dimensional thin films satisfying the above structures can be self-assembled into hollow multi-circle cylindrical or spiral structures. These two structures can be regarded as the most basic structural forms based on self-rolled-up nanotechnology. By choosing different sacrificial layer materials, controlling local stress, pretreating the thin films, and post-rolling heat treatment, a variety of self-rolled-up structures can be constructed, which may be used in passive device design, as shown in Figure 11.2 [16].

Silicon nitride thin films are often selected as the strain layer material for self-rolled-up passive electronic devices because plasma-enhanced chemical vapor deposition (PECVD) is fully compatible with conventional semiconductor technology. This insulating material is also nontoxic, transparent under visible light, and can therefore be used in biological and optical applications. Silicon nitride thin films deposited by PECVD are amorphous thin films, whose mechanical properties can be regarded as isotropic. Therefore, the rolling up process of silicon nitride films can be controlled by the etching parameters of the sacrificial layers. For example, choosing a crystalline or amorphous sacrificial layer affects the rolling direction, resulting in different final structural forms, as shown in Figure 11.3a,b [16].

The strain layer films in Figure 11.3a are deposited on polycrystalline germanium sacrificial layers. These rectangular strain layer films are arranged in a radial pattern. From the experimental results, it can be seen that the rolling direction of the film is only related to the position where the sacrificial layer begins to be etched, which is related to the position of the etching window outlined by the dotted line. However, as shown in Figure 11.3b, if a single-crystal material is chosen as the sacrificial layer, such as on single-crystal silicon (111) substrate, the rolling direction of the silicon nitride film will tend to follow the ⟨110⟩ crystal direction, which has nothing to do with the orientation of the film. Moreover, the chirality of the rolling structure can be controlled by adjusting the angle between the direction of rolling up of the film and the (110) substrate direction.

As previously described, the thickness of the strain layer and internal residual stress is generally uniform. In theory, the inner diameter after rolling should be

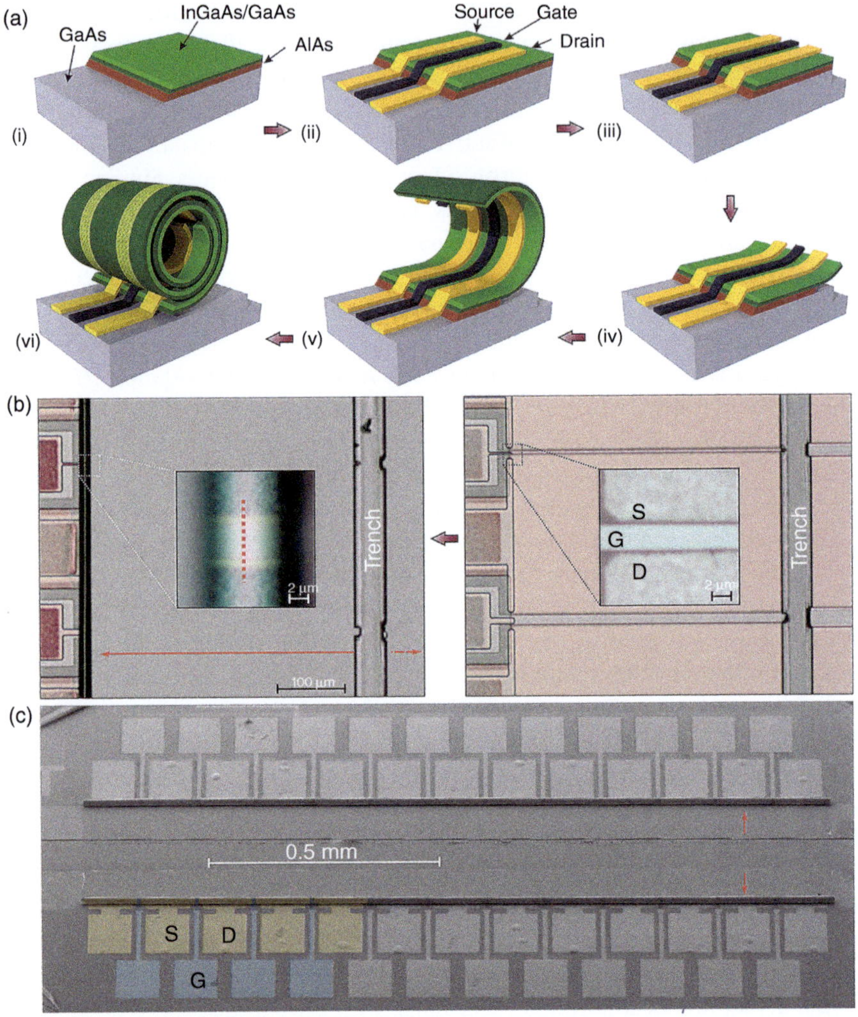

Figure 11.1 Process flow for rolled-up TFTs. (a) Schematic fabrication of RUFETs by the roll up of strained semiconducting nanomembranes. Strained single-crystalline GaAs and $In_{0.2}Ga_{0.8}As$ (green) on top of AlAs layers (red) are patterned by standard 2D techniques: (i) Device isolation by shallow wet etching. (ii) Structuring planar TFT with ohmic contacts (yellow) and gate electrodes (blue). (iii) Deep etching allows access to the sacrificial AlAs layer. (iv)–(vi) Selective wet etching of the underlying AlAs layer releases the strained layer stack into a 3D tubular shape. (b) Optical image of unrolled planar (right) and RUFET devices (left). Insets: close-ups of the channel region after rapid thermal annealing. (c) SEM overview of millimeter-long tubes, partly color coded. The red arrows indicate the rolling directions. Source: Adapted with permission from Grimm and Bof Bufon [15]. Copyright (2012) by the American Chemical Society.

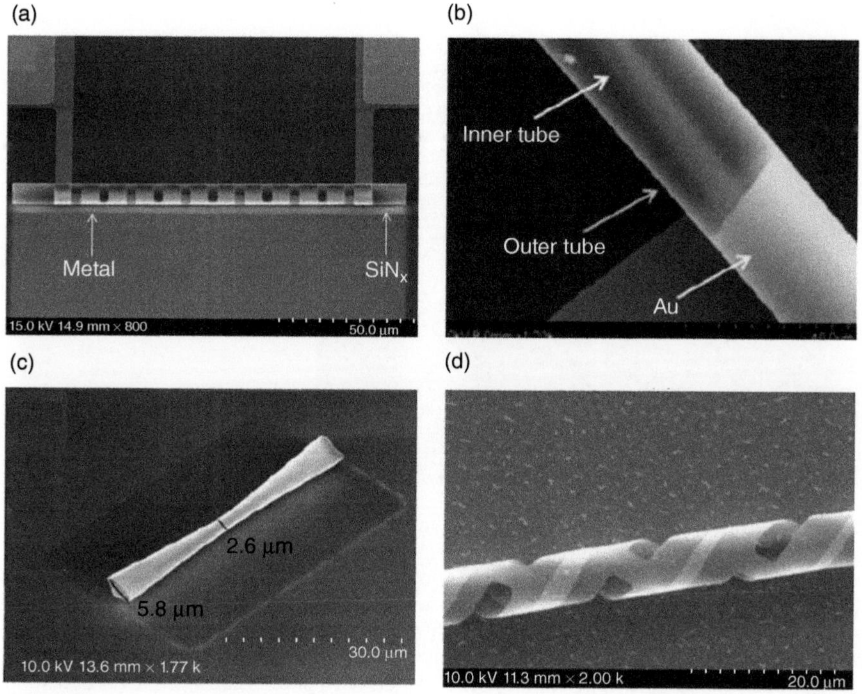

Figure 11.2 SEM images of a variety of self-rolled-up structures. (a) The multiple turn 3D self-rolled-up coils made of metal and SiN_x nanomembrane connected in series. (b) Coaxial tube due to addition of compressively strained Au (70 nm) onto a 16 nm LF/20 nm HF SiN_x bilayer. (c) Local diameter variations (5.8–2.6 μm) in 50 μm long single-roll microtubes subjected to a post-rolling anneal at 600 °C for 60 s. (d) Local thickness variation in chiral tubes induced by angular variations relative to the (110) etch plane on a Si (111) substrate. Source: Froeter et al. [16]. Adapted with permission from IOP Publishing.

uniform everywhere, and the gap between adjacent layers (turns) should be zero. In fact, the distribution of rolling stress and resistance on two-dimensional thin films can be quantitatively changed by the design of the functional layer, thus forming a controllable structure with nonuniform inner diameter and layer-to-layer spacing. As shown in Figure 11.4, when the functional layer is selected as a conventional metal, the local rolling resistance can be changed by depositing metal of different thicknesses in different regions. With the scheme shown in the figure, it is easy to get structures such as a roll-in-roll, which has practical applications as coaxial transmission lines [8].

Functionalized silicon nitride films can be pretreated before rolling to form a special two-dimensional structure. After rolling, it can be self-assembled into a more complex three-dimensional structure. For example, through the combination of microfabrication techniques, such as lithography and plasma etching of thin films, a through-hole structure with controllable size, density, and periodic array alignment can be formed, as shown in Figure 11.5 [17]. After rolling, compared with the previous structure, a hollow multilayer cylindrical structure wall will appear

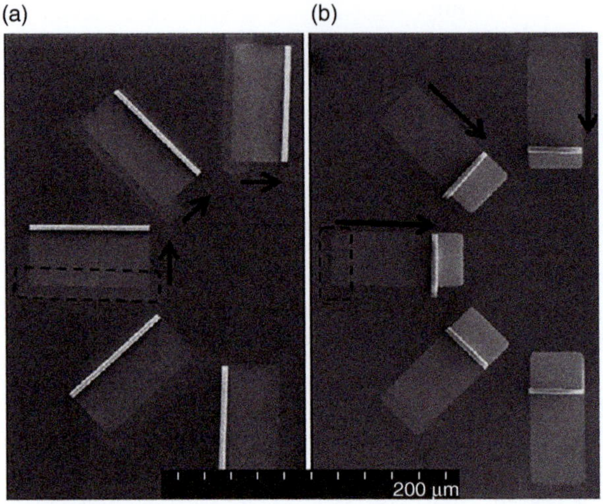

Figure 11.3 Control of rolling direction: SEM images of microtubes rolled-up rectangular-shaped SiN_x thin-film membranes (16 nm LF/20 nm HF) oriented in a radially distributed pattern (only half the wheels are shown). The etching window (outlined by dashed lines) is opened by optical lithography and dry etched to expose the sacrificial layer sidewall along either the long edge (a) or short edge (b) of each rectangular pad. The arrows indicate the pads' rolling directions, which originates from the etch window and proceeds forward. Source: Froeter et al. [16]. Adapted with permission from IOP Publishing.

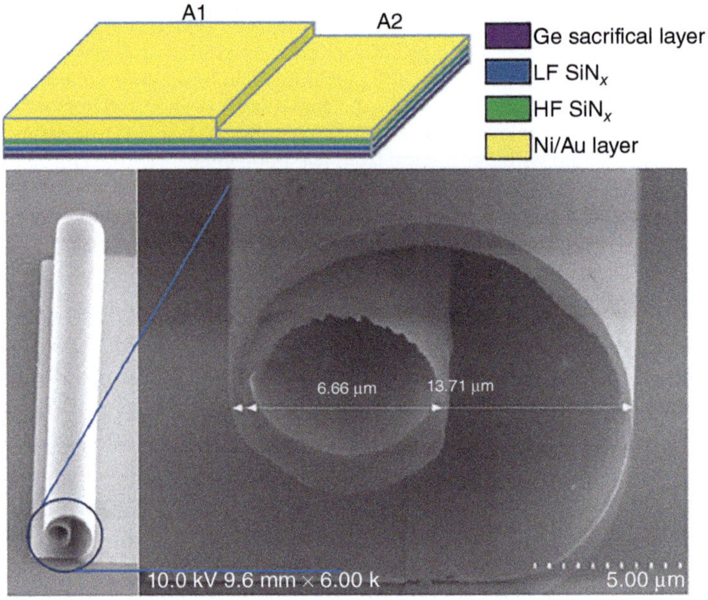

Figure 11.4 A coaxial structure rolled up by local stress control. Above schematic view is the design of the planar structure required to form a coaxial rolled-up structure with two different inner diameters. The SEM image shows the fabricated structure with measured inner diameter. The inner diameter is 6.66 μm for the first turn and 13.71 μm for the second turn. The thickness of LF SiN_x is 20 nm and the thickness of HF SiN_x is 20 nm. The thickness of metal layer in area A1 is 5 nm Ni/20 nm Au. The thickness of metal layer in area A2 is 5 nm Ni/100 nm Au. The thickness of Ge sacrificial layer is 10 nm. Source: Huang et al. [8]. Adapted with permission from American Chemical Society.

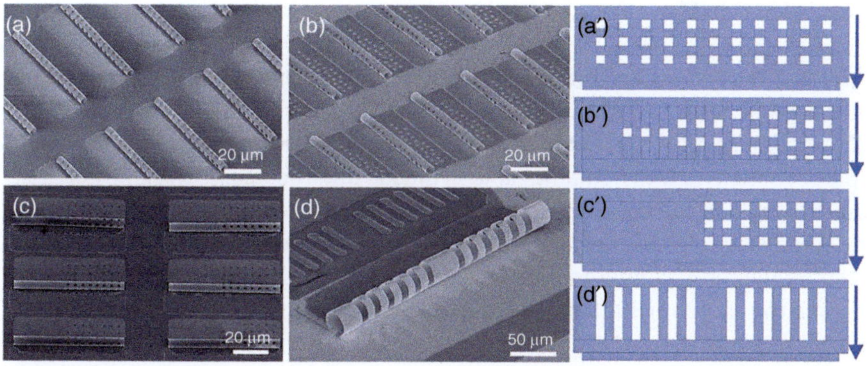

Figure 11.5 Arrays of SiN_x microtubes self-rolled up from 2D sheets with different perforation patterns. (a) uniform square lattice perforations; (b) gradient perforations; (c) abrupt one-sided perforations; (d) rectangular perforations; the corresponding 2D patterns are shown on the right as (a′), (b′), (c′), and (d′), respectively, with the arrow indicating rolling direction. Source: Michaels et al. [17]. Adapted with permission from John Wiley & Sons, Inc.

periodically arranged through holes, which may overlap completely, forming a direct contact structure from the outer wall to the inner wall, or may alternately cover.

These diverse construction methods provide a choice for passive devices with different functions. In addition, the thermal resistance of the film will also change by removing mass, which provides an effective way to improve the thermal stability of self-rolling passive devices based on the need to work under higher power conditions. The thermal tolerance of self-rolling passive devices based on silicon nitride thin films should be considered because the components of silicon nitride thin films deposited in an ammonia/silane/nitrogen environment by PECVD do not contain only Si–N bonds. In fact, the chemical reactions occur as follows [18]:

$$SiH_4(g) + 4NH_3(g) \xrightarrow{\text{Plasma}} Si(NH_2)_4(s) + 4H_2(g) \tag{11.1}$$

In addition to the Si—N bond, there are many N—H bonds in the films. It is noted that the film deposition temperature is 300°C, so when the curled structure is heated again, and the temperature is higher than the deposition temperature, the following reactions occur:

$$3Si(NH_2)_4(s) \xrightarrow{T_a>300\,°C} Si_3N_4(s) + 8NH_3(g) \tag{11.2}$$

The N—H bonds in the films will be released mainly in the form of ammonia, resulting in a smaller density of the films and more chemical bonds between atoms in the form of Si—N bonds. This phenomenon has different effects on the residual stress of silicon nitride films deposited at different frequencies. The mean free path of free radicals in plasma deposited at low frequencies, such as 380 kHz, is larger than that at high frequencies, such as 13.56 MHz. Therefore, at low frequencies, it is easier to deposit thin films on the substrate surface, which makes the density of low-frequency silicon nitride thin films much higher than that of high-frequency silicon nitride thin films. The chemical bonds are arranged tightly in low-frequency silicon nitride, and the compressive stress exhibits a tendency to expand. On the contrary, the atomic chemical bonds of high-frequency silicon nitride films mainly depend on the attraction of Si—N bonds, and the tensile stress tends to induce shrinkage forces. After annealing at temperatures greater than 300°, the compressive stress of low-frequency silicon nitride films decreases with the decrease of film density. On the contrary, more chemical bonds in the form of Si—N bonds in the films increase the tensile stress of the high-frequency silicon nitride films. For low-frequency and high-frequency silicon hydride thin films with thicknesses of 20 nm, the experimental results show that rapid annealing can reduce the compressive stress of low-frequency silicon nitride thin films from -1195 ± 44.7 to -1109 ± 33.3 MPa, a small change of only 7.2%. For high-frequency silicon nitride thin films, the tensile stress changes from 415 ± 29.2 to 873 ± 29.4 MPa, and the change is as high as 110%. As shown in Figure 11.2c, the middle inner diameter of the double-film structure formed by depositing high-frequency silicon nitride film on low-frequency silicon nitride film is much smaller than its initial value after rolling and rapid annealing. This characteristic can be used in the design of reconfigurable passive electronic devices and as a reference standard for the maximum operating temperature of self-rolling electronic devices.

11.2 Rolled-up Origami Modeling

In addition to active devices, the electrical performance of most passive devices depends primarily on their structural dimensions. Especially for passive devices operating at higher frequencies, a minute change in a structural parameter can influence performance tremendously. For example, the working wavelength of passive electronic devices operating in the millimeter-wave band is in millimeter scale. This scale is comparable to the inner diameter of most self-rolling structures reported, about 10s of microns to 100s of microns. It shows that self-rolled-up devices operating at this frequency cannot be regarded as completely electrically lumped. The appearance of distributed effects makes the electrical performance of these devices more sensitive to changes in their physical size and configuration. Therefore, accurate predictions of the physical sizes of two-dimensional films after self-rolling into three-dimensional structures are very important for the self-rolling platform to become a practical passive device design platform.

For the conventional uniform self-rolled-up structure, the most important predictive parameter is the inner diameter after rolling. There are two methods for calculating this parameter: analytical methods and numerical methods. As early as 1925, Timoshenko had proposed an analytical formula for calculating the curvature of bimetallic thermostats. The expression is as follows [19]:

$$\frac{1}{\rho} = \frac{6(a_2 - a_1)(t - t_0)(1 + m)^2}{h\left[3(1 + m)^2 + (1 + mn)\left(m^2 + \frac{1}{mn}\right)\right]} \tag{11.3}$$

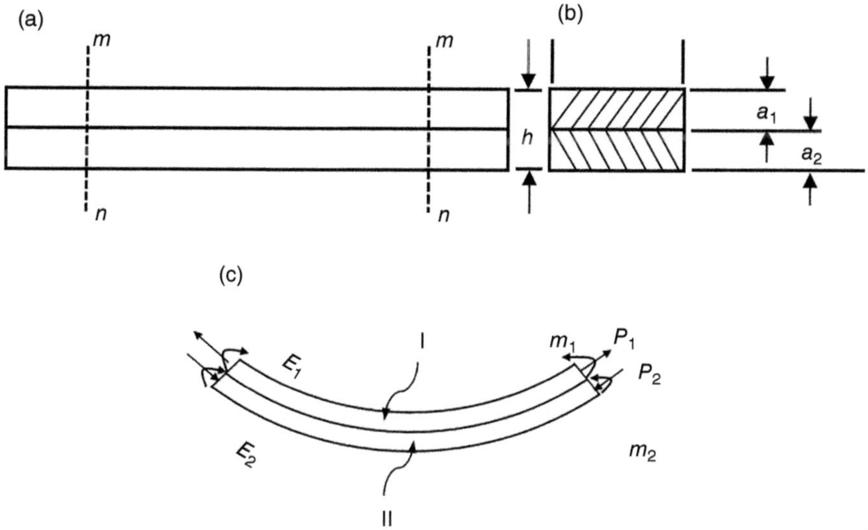

Figure 11.6 Deflection of a bi-metal strip. (a) The side view of a bi-metal strip with cross section line labeled as $m-n$. (b) The corresponding cross section in (a) with the two metal strip thicknesses labeled. (c) Deflection of the bi-metal strip due to different strains of the two metal strips. E_1 and E_2 are Young's modulus of the two metal strips. Source: Adapted with permission from Timoshenko [19]. Copyright (1925) by Journal of the Optical Society of America.

The structural parameters correspond to those shown in Figure 11.6. $M = a_1/a_2$, $n = E_1/E_2$, and subscripts 1 and 2 denote two different films that are in contact. This method gives a more accurate analytical algorithm for curvature prediction in the simplest case of a two-film system but suffers clear shortcomings. When the film is singular or a stack greater than two layers, or when the stress distribution of the film is not uniform, this method poorly predicts the size of the curled structure. These problems are addressed by establishing a mechanical finite element model and by using numerical methods for simulation calculations.

The most significant geometrical parameter that determines the final size of the device is the diameter of the innermost tube. The electrical performance of the high-frequency electronics and optics is extremely sensitive to the dimensional parameters meaning that it is critical to control the inner diameter precisely. For the structure comprising more than two sublayers with the top strained layer patterned, it is impossible to determine the inner diameter of the rolled-up structure by analytical methods discussed above. Instead, a transient quasi-static FEM modeling method for the rolling process of multilayer membrane structures that is given in Ref. [8] is able to predict the rolled-up dimension.

Assuming that we are going to simulate the rolled inner diameter of a bilayer SiN_x-based S-RuM device, the top strained bilayer is held in place by a Ge sacrificial layer and then released. Therefore, a fixed boundary condition is applied to all nodes at the bottom of the LF SiN_x thin film to simulate this effect. The reported FEM method assumes that all materials are isotropic and linearly elastic. According to

Mindlin–Reissner shell theory, shell elements can be used to simulate multilayer structures with controlled precision. Depending on the design of the structure, different layer thicknesses and material properties can be assigned to the respective layers. For example, Young's modulus E of the PECVD LF SiN_x and HF SiN_x thin films could be set to 210 GPa. Poisson's coefficient of the double layer could be set to 0.28 according to the literature. For both LF SiN_x and HF SiN_x membranes, their residual stress is modeled by the coefficient of thermal expansion in the FEM. For temperature increments, all nodes in the simulation have the same setup value. Different coefficients of thermal expansion are assigned to different layers for simulating compression and tensile stresses. The values of compression and tensile stress could be measured by the FSM 500TC metrology tool. The exact value of the stress can be determined by applying an appropriate temperature increment. The thermal coefficient of LF SiN_x is taken from the literature; its temperature increment is determined to be 1450 °C to reach the measured value. When the temperature increment is fixed at 1450 °C, the thermal coefficients of other materials can be determined to achieve their respective measured residual stresses. Table 11.1 summarizes all the material properties used in this example.

To model the dynamic etching progress of the Ge sacrificial layer, a moving boundary condition is used in FEM method. The rolling process is a nonlinear large deformation transient quasi-dynamic process, and it is modeled by a series of FEM simulations of static deformation by releasing the constraints on the bottom segments in sequence. In simulation example, the length of each segment is set to be less than 1/200 of the estimated circumference of the first turn. A simulation loop, as shown in Figure 11.7, is implemented as the moving boundary condition. The simulation loop starts by applying a fixed boundary condition at the bottom of the bilayer, as shown in Figure 11.7a. In Figure 11.7b, the constraint on the first segment ΔX is released, and all the nodes in the released part are given a temperature increment ΔT = 1450 °C. Then, an updated geometry is obtained, as shown in Figure 11.7c, after static simulation is performed. By repeating the first three steps, the next segment is released and the same temperature increment is

Table 11.1 Material properties set in FEM simulation.

Sub-layer	Residual stress (MPa)	Young's modulus (GPa)	Poisson coefficient	Thermal expansion coefficient (1/°C)	Temperature increment (°C)
LF SiN_x	−1168	210	0.28	2.75×10^{-6}	1450
HF SiN_x	+406.95	210	0.28	-9.61×10^{-7}	1450
Ni	+798.4	200	0.31	-1.9×10^{-6}	1450
Au	+379	79	0.44	-1.85×10^{-6}	1450

Note: Signs – and + for the residual stress denote the compressive and tensile stresses, respectively.

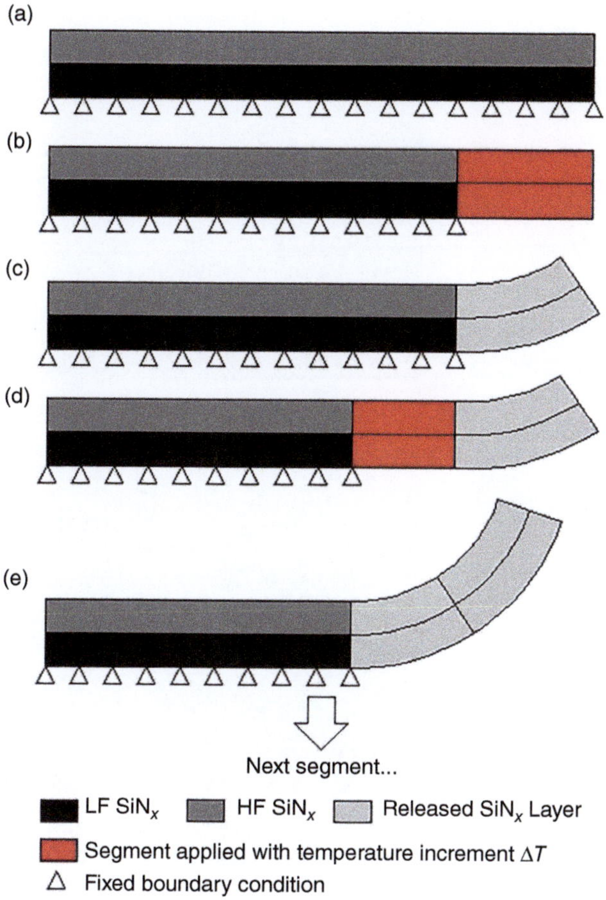

Figure 11.7 Transient quasi-static FEM modeling of the rolling process of bilayer membrane. (a) Initial structure with applied fixed boundary condition to all element nodes. (b) Unfix the first segment and apply a temperature increment ΔT to it. (c) Update the structure after static simulation. (d) Unfix the next segment and apply a temperature increment ΔT to it. (e) Update the structure after static simulation and continue the loop until the last segment is simulated. Source: Adapted with permission from Huang et al. [8]. Copyright (2014) by Nano Letters.

applied to obtain the next updated geometry, as shown in Figure 11.7d, until the last segment is simulated.

The FEM method has been demonstrated to have extremely high precision to predict the rolled-up geometry of complex layered membranes, especially the inner diameter. Figure 11.8 shows the scanned electronic microscope (SEM) pictures of computationally designed SiN_x bilayer rolled-up with 1/4, 2/4, 3/4, or a full turn with a 101 μm inner diameter. With precisely simulated value of the inner diameter, a "closed" tube with nearly no seam is fabricated on silicon wafer shown in Figure 11.8d.

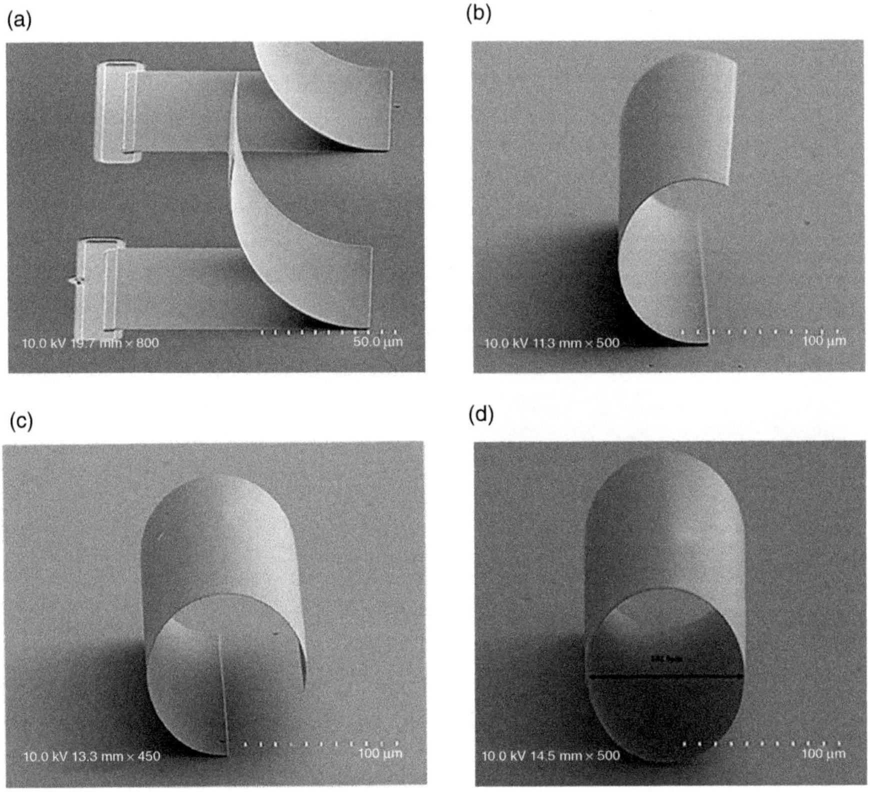

Figure 11.8 SEM images of precisely fabricated 100 μm inner diameter rolled-up SiN_x bilayer structures with designed fraction of turns. (a) 1/4 turn. (b) 2/4 turn. (c) 3/4 turn. (d) Full turn with inner diameter measured as indicated. Source: Huang et al. [8]. Adapted with permission from American Chemical Society.

11.3 Rolled-up Radio-Frequency Electronics

11.3.1 Rolled-up RF Inductors

Inductors are extensively used for impedance matching, signal filtering, and magnetic energy storage in almost all high-frequency electronic circuits, especially in the radio-frequency (RF) domain [20, 21]. For over 200 years, since the first inductor was invented by Michael Faraday, the utilization of coiled structures has been the only way to obtain the desired inductance (L), which is determined by Faraday's law of induction. Therefore, inductance becomes a function of the induced magnetic field and the architecture of the inductor itself. As applications operating in the RF band naturally maintain relatively large values of angular frequency ω, the reactance $j\omega L$ is suitable in most scenarios with a value of inductance L less than 100s of nH, which is achievable until the inductors are integrated on-chip monolithically with tight requirements for geometrical miniaturization. On-chip design and fabrication of RF inductors must be compliant with layer-by-layer planar semiconductor processing.

To be compatible with CMOS technology, inductor conduction materials are limited to copper and aluminum. Since the 1960s, the architecture of on-chip inductors has been largely unchanged, with the spiral structure–based construction dominating the RF-integrated circuit industry. However, two-dimensional (2D) structures are far from the ideal choice for constructing inductors, a conclusion manifested by comparing the design of on- and off-chip inductor archetypes, such as the conventional solenoid inductor. Intrinsic issues such as relatively weak magnetic coupling between turns, significant parasitic substrate interactions, and the difficulty of integrating a magnetic core are associated with planar spiral structures. Spiral on-chip inductors have survived until now because of their relative simplicity of design and fabrication within the requirement of inductances than 100s of nH. Under this condition, the cost and the on-chip area occupation of spiral on-chip inductors have been acceptable, especially in the early stages of IC industry. Nowadays, the leading transistor node is approaching 2 nm, and advanced 3D IC integration technologies have significantly improved the integration density of everything on-chip except for passive devices, and most significantly lack progress for the number one bottleneck for miniaturization – inductors. For RFICs, RF inductors have become the barrier for further improvement, and many strategies have been demonstrated to overcome the intrinsic issues of spiral inductors, such as suspending the spiral coils and stacking vertically connected (by TSVs) coils, etc. [22–25]. Bulk micromachining and surface micromachining processes were developed to obtain such structures, but may introduce more complicated processing, incompatibility with preexisting CMOS technology, and prohibitive cost. There is a common theme in the above methods that all of them consider solutions in the 2D design framework, meaning that trade-offs between the size and overall inductor electrical performance are inevitable. A new design and fabrication method that is not only compatible with current semiconductor processing but also able to achieve 3D architectures from a 2D layout and fabrication process is desired to solve these intrinsic issues, and bring a revolution to the current 2D design and fabrication framework.

Rolled-up origami provides a practical way to achieve 3D architectures from 2D processing, which is the platform similarly needed for the revolution of on-chip inductors. SiN_x rolled-up membrane (S-RuM) nanotechnology has been demonstrated as an excellent method to precisely obtain high-density 3D coils [26–28]. Since 2012, after the design and fabrication concept of the S-RuM RF inductor was proposed, there have been several generations of improvements on this platform to increase fabrication yield and improve electrical performance across a wide frequency spectrum. The quality factor (Q factor) has been of forerunning interest. The most recent and advanced design and fabrication process of S-RuM RF inductors is reported in reference [29] and is summarized in Figure 11.9.

The largest improvement of the latest S-RuM RF inductor platform compared to many of the rolled-up origami technologies is the gas vapor–based lateral etching of the sacrificial layer. A wet etching (e.g. H_2O_2 for etching Ge) technique was widely used before that, and the common release speed of the sacrificial layer was ~10 μm/h. For rolling paths with lengths of 100s μm, the time required to fully etch the sacrificial layer is far beyond an acceptable amount of time. Furthermore,

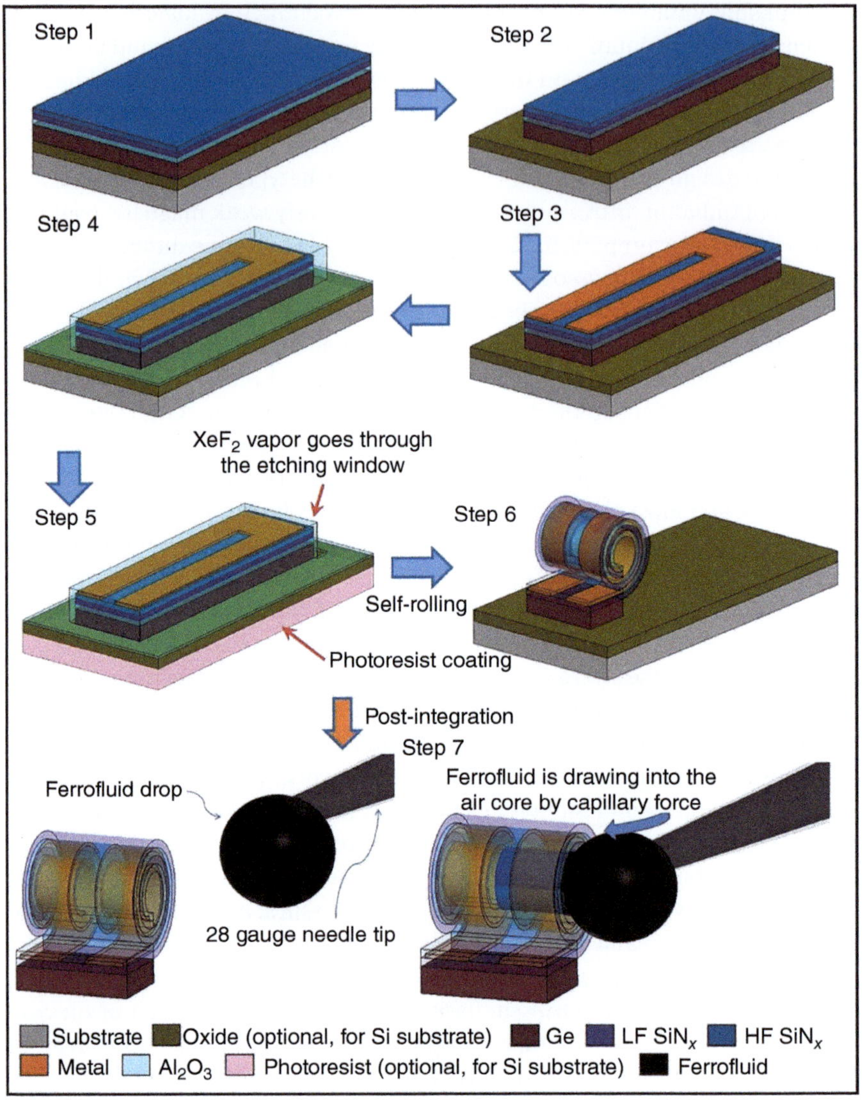

Figure 11.9 Fabrication process of S-RuM RF/power inductors. Illustration of the fabrication process flow for an air-core S-RuM inductor, including the vapor phase releasing and the post-fab capillary core-filling approach for strong magnetic induction. Source: Adapted from Huang et al. [29].

the extended submersion of the device in the liquid environment introduces a possibility for the etchant molecules to go through SiN_x membranes and etch the sacrificial layer at unwanted locations, which must be prevented to achieve high fabrication yield. Step 1 in Figure 11.8 shows the general stack of films before the first lithography step is performed, in which, other than the substrate, the above films are deposited in sequence for different functions. If only passive devices are integrated on a chip, the function of the substrate is for mechanical support only.

So, if the substrate material is a doped semiconductor material, such as p-type or n-type silicon, a lightly doped high resistivity silicon substrate is preferred to reduce the impacts of parasitic effects on device electrical performance. Technically, there is no strict requirement on the substrate material as long as it absorbs little strain energy from the SiN_x bilayer. Soft materials such as Kapton film or PDMS are theoretically suitable to be used as the substrate for the S-RuM inductors. The oxide layer is for electrical isolation and decoupling, and it is not necessary unless a doped silicon substrate is used. Germanium is chosen as the sacrificial layer due to its smooth surface and relatively large Young's modulus, which is important for reducing the surface roughness accumulation and for avoiding strain energy absorption. Between the sacrificial layer and the SiN_x bilayer for providing strain energy, an Al_2O_3 thin film deposited by ALD is inserted as a protection layer to avoid uncontrolled sacrificial layer etching issues caused by PECVD SiN_x pin holes. Generally, the residual stress of the LF and HF SiN_x thin films is maximized to generate the largest torque moment to roll up a metal thin film on top, desired to be thick and wide. Then, step 2 defines the mesa by using Freon RIE. Following, in step 3, the functional layer (metal layer) is deposited using E-beam evaporation and patterning using i-line lithography. Square wave-like metal patterns are then deposited and patterned on top of the HF SiN_x membrane at step 3, which forms 3D coils after rolling up into the EM functional architecture. Usually, Cu is used as the metal material due to its compatibility with conventional semiconductor processing. However, another metal thin film such as nickel must be used underneath the Cu to enhance the adhesion between the metal layer and the SiN_x membrane. Notice that the cross-sectional area of the metal layer of on-chip RF inductor is usually as large as 10's of μm^2 to reduce the ohmic loss. Therefore, for S-RuM inductors, either a similarly large cross-sectional area of rollable Cu thin film or a much larger stored magnetic energy density, which could be quantified as inductance/ohm, must be achieved to obtain an acceptable Q factor. We will analyze these two conditions in detail below to clearly show the critical considerations in the design of S-RuM RF inductors.

It is important to understand the operating mechanism to estimate the electrical performance of S-RuM RF inductors. Compared to numerical methods such as the finite element method, e.g. the high-frequency structure simulator (HFSS), which can simulate the S-parameters of S-RuM RF inductors given reasonable structural and material models, analytical methods provide an understanding of the underlying device physics and elucidate the complex relationship between the electrical performance and structural and material parameters through mathematical equations. Furthermore, if analytical equations without integral or differential operations could be derived, the calculation speed of the analytical methods will be significantly faster than that of numerical methods. According to reference [26], by implementing the following assumptions, relatively simple expressions for calculating the inductance L_{e_total} and the Q factor Q_{total} versus the frequency can be derived, shown from Eqs. (11.4) and (11.5).

$$L_{e\_total} = \frac{\mathrm{Im}\left(-\frac{1}{Y_{12}}\right)}{\omega} = \frac{\mathrm{Im}\left(\frac{1}{Y_s}\right)}{\omega} = \frac{N_c\{L' - C_c[R^2 + \omega^2(L')^2]\}}{1 + \omega^2 C_c^2[R^2 + \omega^2(L')^2] - 2\omega^2 C_c L'} \tag{11.4}$$

$$Q_{\text{total}} = \left|\frac{\text{Im}(Y_{11})}{\text{Re}(Y_{11})}\right| = \left|\frac{\text{Im}(Y_p + Y_s)}{\text{Re}(Y_p + Y_s)}\right| = \frac{\omega\left|L' - (C_c + N_c^2 C_s)\,[R^2 + \omega^2 (L')^2]\right|}{R} \tag{11.5}$$

where $L'=L-2(1-N_c-1)M$, N_c is the number of cells, L is the self-inductance of each cell (see Ref. [26] for detailed derivations).

The first assumption is that the magnetic induction density inside any cross section of the tubular structure is uniform. The second assumption is that the magnetic coupling between nonadjacent 3D coils is negligible. The precision of this method is relatively high when the width of metal strip and the inner diameter is within 10s of μm. Notice that the direction of current flow in adjacent 3D coils is opposite, meaning a minor cancelation of magnetic coupling exists. However, by calculation, a relatively small gap between adjacent 3D coils is able to reduce the canceling magnetic coupling coefficient to be less than 5% of the uncanceled magnitude. Compared to planar spiral RF inductors, the advantage of S-RuM RF inductors is clear. As the total thickness of all the rolled-up membranes (10s of nm) is negligible compared to the inner diameter (10s of μm), the increase of the total projected on-chip area of the S-RuM RF inductor, which is the summation of the side wall thickness and the inner diameter, is nearly unchanged despite a potentially large number of turns having been rolled up. Therefore, the magnetic coupling between turns is very strong due to the almost identical shared magnetic flux. As a result of the majority of the EM field distribution being above the substrate, the inductor electrical performance is nearly immune to the resistance level of the substrate, implying that the maximum operating frequency of the S-RuM RF inductor could be much higher than that of the planar spiral RF inductor with the same inductance. Therefore, the inductor behaves as a lumped device and is able to be used at microwave frequencies, which is critical for the miniaturization of monolithic microwave-integrated circuits. Although there is a significant reduction of the substrate parasitic capacitance, the electrical performance of the S-RuM RF inductors is not immune to other unique device-embedded capacitances. As the length of one metal strip is 100s of μm, phase differences in the signal are introduced between adjacent turns. Even if the phase difference is relatively small, cross-talk capacitive effects exist and are proportional to the length of the tubular structure. It should be noted that the capacitances among all of the adjacent turns in one 3D coil are connected in series, implying that a larger number of turns reduce the total cross-talk capacitance. Therefore, to achieve good performance at high operating frequencies, short tubular structures with a large number of turns are preferred. A device with many turns also benefits from an enhancement of the inductance density but compromises ohmic losses. A thick metal layer is required to achieve a large cross-sectional area when the width of the metal strip is limited to a short tubular structure. According to reference [29], a 180 nm Cu strip on a 5 nm nickel adhesion layer is able to be rolled up without any noticeable rolling issues except for a μm scale air gap between adjacent turns. Other than performance tolerance deterioration, air gaps between turns actually help reduce the cross-talk capacitances and provide space for post-fabrication electroplating of the metal layer to reduce the net series resistance.

11.3.2 Rolled-up RF Capacitors

In addition to RF inductors, RF capacitors are also critical devices in RFIC design to perform functions such as DC signal isolation, impedance matching, and signal filtering [30–32]. Conventional on-chip capacitors can be categorized by several archetypes according to their operating mechanism and architecture, including the metal–insulator–metal (MIM) RF capacitor, the interdigital RF capacitor, the p–n junction capacitor, and the MOS capacitor. Among them, MIM and the interdigital capacitors are commonly used in RFIC design due to their simplicity and high-frequency operation capabilities. The design and fabrication of these two types of RF capacitors are also limited by planar semiconductor processing, so, to obtain large capacitance, the MIM capacitors have to "go inside" the substrate to form a 3D multilayer architecture, while the interdigital capacitors have to increase the number of fingers contained by occupying more on-chip area. These two strategies either require more complicated fabrication processing or an enlargement of the footprint of the device. Furthermore, more serious substrate parasitic effects will be introduced, lowering the maximum operating frequency of the RF capacitors. Similarly, as RF inductors can be developed on the S-RuM platform, constructing RF capacitors in a 3D fashion adds one more degree of freedom in design to obtain better electrical performance, most importantly in capacitance density.

In Ref. [33], 3D interdigital RF capacitors were successfully developed on the SiN_x S-RuM platform. The rolled-up platform structural design is the same as the S-RuM platform used for RF inductor design except for the pattern of the metal layer. The metal pattern is shown in Figure 11.10, with primary dimensional parameters labeled, which is exactly the same as the electrode arrangement of planar interdigital RF capacitors. In fact, the electric field energy is indeed stored in between the electrode fingers before rolling, and the capacitance depends on the gap between the two adjacent electrodes and the length of electrodes. An analytical

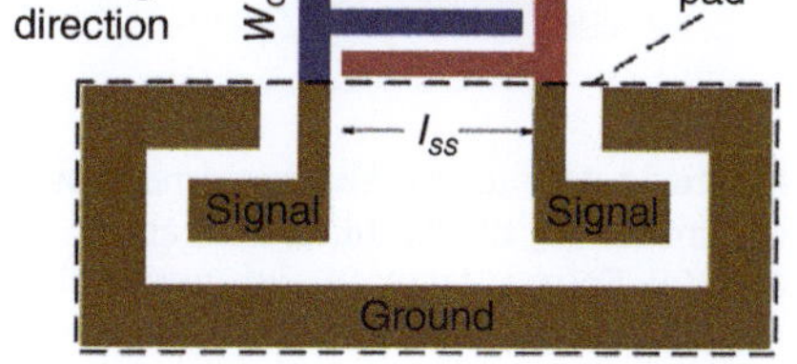

Figure 11.10 The planar layout design of S-RuM interdigital RF capacitors with measurement fixture and primary dimensional parameters labeled. Source: Adapted with permission from Sang et al. [33]. Copyright (2019) by the Nanotechnology.

expression is difficult to derive even considering the simplest structure in which the width of the finger w_{cp} equals the gaps g_l and w_{cs}, and the critical dimension of the device is much smaller than the operating wavelength. According to Ref [34], the following expression could be used to estimate the value of the capacitance in the above case.

$$C_p = \left(\frac{\varepsilon_r + 1}{w}\right) l_{ss}[(N-3)A_1 + A_2]\ \text{pF/unit length} \tag{11.6}$$

where N is the number of fingers, w is the base width of the interdigital capacitor, l_{ss} is the length of the fingers, ε_r is the relative dielectric constant of the substrate, and the values of A_1 and A_2 depend on the ratio of T/g_l (T is the thickness of the substrate). It is clear that, after rolling up, the substrate is "gone," and replaced by the very thin dielectric layers underneath the metal pattern. Therefore, most of the electric field penetrates the dielectric layers and extends into the air surrounding the fingers, meaning a significant decrease of the capacitance obtained by the electric coupling between electrode fingers. Furthermore, the rolled-up feedlines form two multiple turn coils, introducing parasitic inductance connected in series with the capacitance, which decreases the maximum operating frequency of the capacitor. Although the above two changes of the traditional planar interdigital RF capacitors have negative impacts on the overall performance, the 3D configuration of the electrodes leads to an enhancement of the working mechanism. There are infinite configuration possibilities between opposite polarity electrodes, dependent on the inner diameter. As shown in Figure 11.11, two extreme status cases are illustrated – status A shows the overlapping electrodes on adjacent turns having the same polarity, while status B shows just an opposite situation. If we define a unit length of l_u as shown in Figure 11.10, then, without considering the increment of side wall thickness due to the negligible single-turn side wall thickness compared to the inner diameter, the quotient of the circumference of one turn divided by the unit length of l_u determines the electrode configuration status quantitatively. In other words, odd and even quotients represent the status A and B, respectively. In between status

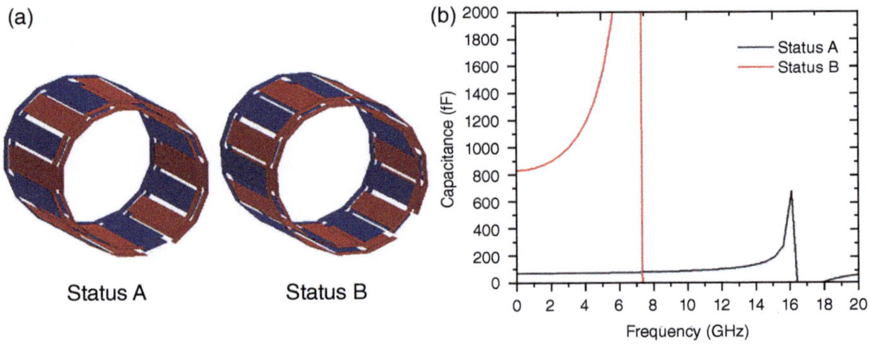

Figure 11.11 (a) the two extreme cases of the overlapping configuration of adjacent electrodes. (b) the simulated capacitance versus frequency of the two extreme cases status A and B. Source: Adapted with permission from Sang et al. [33]. Copyright (2019) by the Nanotechnology.

A and B, the adjacent face-to-face electrodes are partially overlapped with each other, implying that the electrical performance should also be in between that of status A and B. It is obvious that status B constructs parallel-plate capacitors between the adjacent face-to-face electrodes, and all these capacitors are connected in parallel, meaning a significant enhancement of the capacitance after the planar interdigital capacitor is rolled up into a 3D interdigital capacitor. Oppositely, status A cancels out all the parallel-plate capacitors leading to the minimum capacitance value obtained from the 3D electrode configuration.

In Ref. [33], demonstrative samples were fabricated on a sapphire substrate, as shown in Figure 11.12. The reported capacitance of a rolled-up device of status between A and B shows a capacitance of 21.5 pF after rolling up, which is 17.2× larger than that of the same planar interdigital capacitor before rolling up. The corresponding capacitance density is as large as 371 pF/mm^2 that has been improved by ~115× after rolling. Figure 11.13 shows the side-view SEM image of the

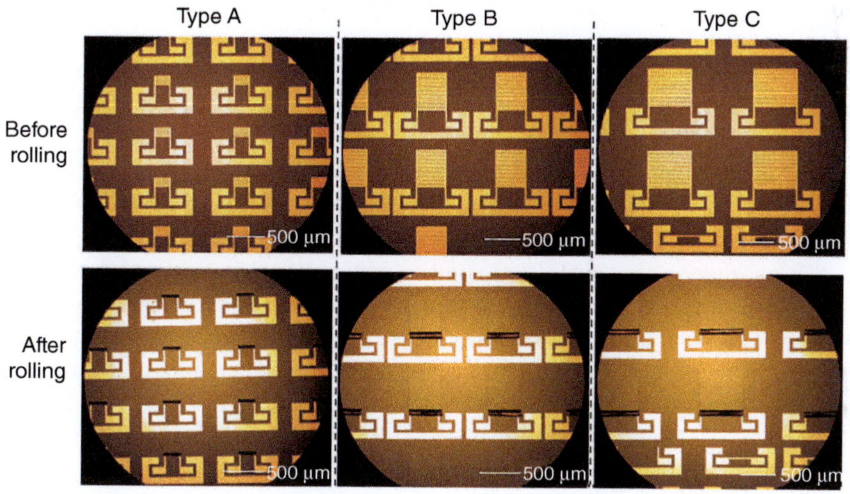

Figure 11.12 A set of photo images showing S-RuM capacitors with different designs (Types A, B, and C) before and after rolling. Source: Adapted with permission from Sang et al. [33]. Copyright (2019) by the Nanotechnology.

Figure 11.13 Comparison of measured inner diameter from the SEM image and simulated structure. Source: Sang et al. [33] Adapted with permission from IOP Publishing.

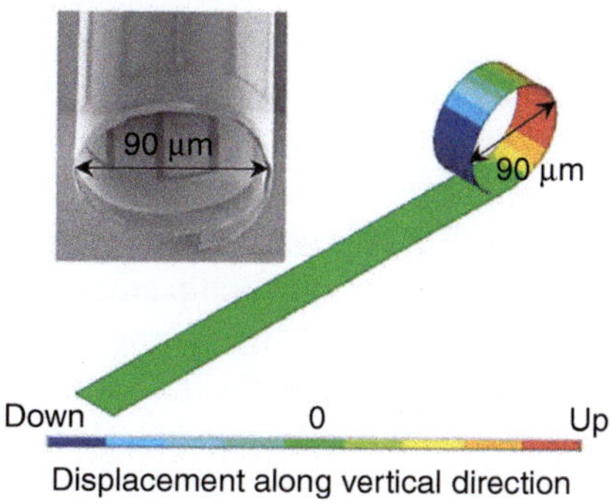

device in which very small air gaps could be found between turns. The inner diameter of 90 μm is also labeled and compared to that of the simulated result indicating a precise design is obtained by using FEM modeling. For obtaining excellent performance tolerance, air gaps should be eliminated. However, if the air gaps could be filled with high relative dielectric constant materials, as reported in reference [35], the capacitance could be further enhanced to reach the level of supercapacitors.

11.3.3 Rolled-up Antennas

Since the 1980s, mobile communication has developed from low-speed and low-frequency analog communication in the 1G era to high-speed and high-frequency communication utilizing large-scale antenna arrays (massive MIMO) as a standard technology in current, inchoate 5G era [36–38]. In the 5G millimeter-wave band, specifically, antenna design has evolved from a single antenna with a fixed beam to a beamforming antenna array [39–41]. Although the size of the 5G antenna for millimeter-wave band operation has been reduced when compared with the size of antennas working at lower frequencies in previous generations, due to the increasing requirements for mobile terminals for front-end chip integration, the millimeter-wave antenna and antenna array still urgently require innovation to further miniaturize the system.

The working wavelength of a millimeter-wave antenna is a centimeter or less, which reduces the cost to monolithically integrate an antenna and front-end communication circuit to an acceptable level. Antennas integrated as such are called AOCs (antenna-on-chip). Compared to conventional in-package antennas, on-chip antennas reduce the interconnection loss and delay from the antenna to the front-end communication circuit, and are simple to design, with high-performance consistency through mature semiconductor fabrication technology. Therefore, AOCs are an effective solution for the design and manufacture of 5G and future mobile terminal HF antenna arrays. Presently, on-chip antennas are usually fabricated on the top metal layer of the chip during wafer fabrication, which belongs to the back-end technology of front-end communication chip processing. As a result of semiconductor fabrication technology being restricted to planar processing, the structure of on-chip antennas is also limited to the two-dimensional plane of a wafer. As a result, the planar radiation elements are inevitably close to the substrate. Although the substrate material plays the role of mechanically supporting the antenna, it also introduces serious parasitic effects. Low resistivity substrates, obtained by pre-doping the wafer, suffer the greatest parasitic losses. The electromagnetic wave radiated from the antenna will electromagnetically couple to a substrate with low resistivity, which greatly reduces the radiation efficiency of the antenna. Parasitic coupling has become one of the primary bottlenecks restricting the improvement of on-chip antennas.

Over the years, domestic and foreign research teams have tried to propose new platform solutions for the design and manufacture of on-chip antennas to eliminate undesired substrate coupling. The most representative research results can be divided into solution spaces involving micromachining, substrate proton injection,

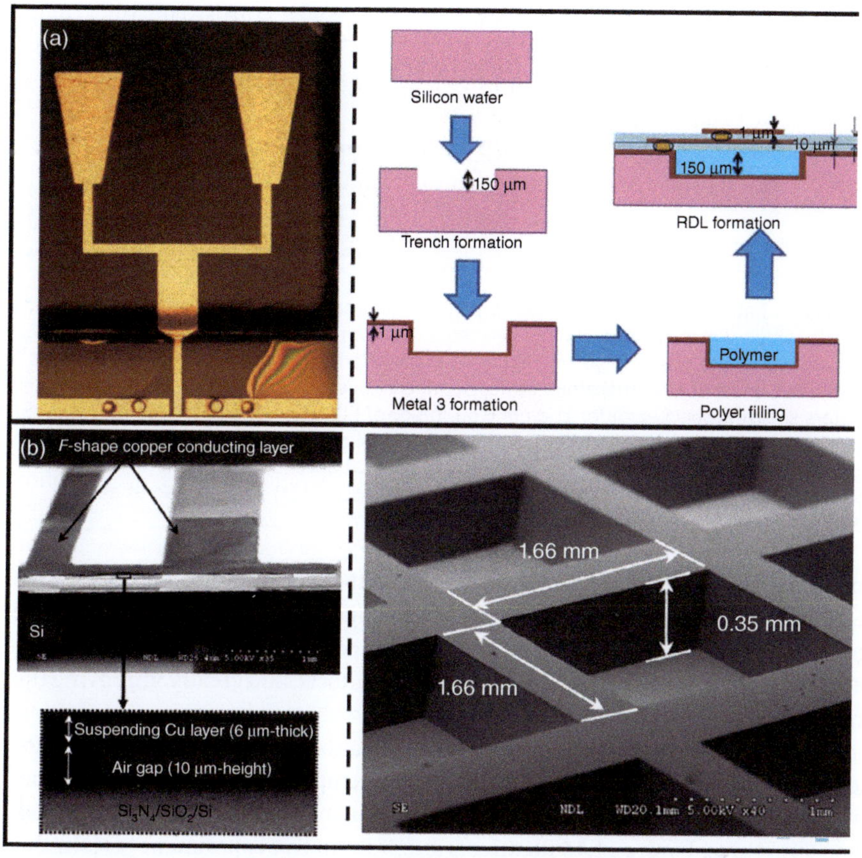

Figure 11.14 On-chip antennas designed and manufactured with micromachining technology. (a) 2-units array antenna on silicon substrate operating at 135 GHz. Left side on the dotted line is the top view of the antenna, and the right side is the fabrication processing flow to form cavity under the antenna. (b) An SEM image shows a planar antenna on silicon substrate operating at a few GHz. Left side on the dotted line is the cross-section view showing the suspended Cu conductive layer with micromachining technology, and the right side is the SEM image of the backside showing the periodical silicon cavity structure with micromachining technology. Source: Modified from Huang et al. [42]. Adapted with permission from Elsevier.

substrate thinning, and radiation shielding, among which the most commonly reported design is based on improvements from micromachining. Figure 11.14 shows two on-chip antennas designed and manufactured with micromachining technology [42, 43]. These technologies improve the radiation efficiency of the antenna to varying degrees and obtain high gain, but all of them drastically increase the manufacturing complexity and cost of the on-chip antenna without exception, which presents prohibitive obstacles to high integration of the antenna with the front-end communication circuit. In essence, there is no simple and effective design and manufacturing method that is simultaneously compatible with conventional semiconductor technology, breaks through the limitations of

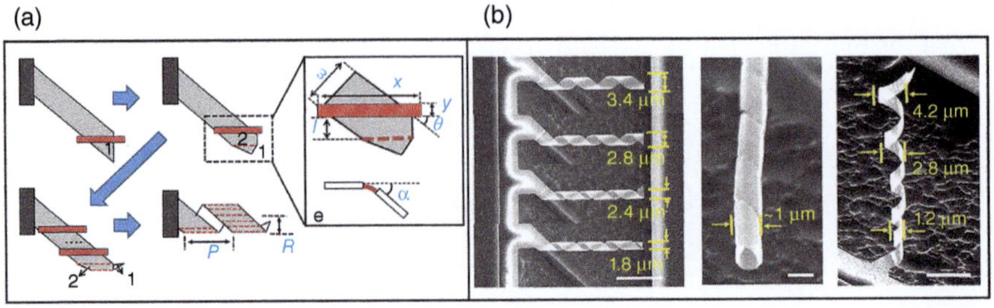

Figure 11.15 Fabrication process of an optical helical antenna. (a) The process was based on FIB-SID, which was conducted on predefined aluminum strips from its free to fixed ends sequentially, to realize the transformation from a 2D strip into 3D helix. The inset figure is the details of the fabrication parameters. (b) Sample helices with different structural parameters showing the flexibility and controllability of FIB-SID. Left: four helices with radii tuned from 1.8 to 3.4 μm, respectively. Scale bar: 5 μm. Middle: a metallic "nanotube" with dimensions less than 1 μm. Scale bar: 1 μm. Right: vertical antenna with dimensions that gradually changed from 1.2 to 4.2 μm. Scale bar: 5 μm. Source: (a) and (b) Mao et al. [44] Adapted with permission from John Wiley & Sons, Inc.

two-dimensional radiator design, and expands the design dimensionality into three-dimensional space. With an on-chip three-dimensional structure, an antenna design could be mechanically self-supporting and could be elevated from the substrate, reducing the influence of parasitic substrate effects, and greatly improving the antenna gain.

In recent years, based on advanced micro- and nanoelectronics processing methods, domestic and foreign research teams have attempted to design and manufacture real antennas to make a worthy attempt at realizing on-chip three-dimensional antenna structures. Figure 11.15 illustrates a representative experiment [44]. In this work, FIB-SID technology is used to bombard an aluminum metal film in different regions of the film in a sequential manner to change its local stress and produce directionally controlled bending, so as to obtain a 3D spiral structure on the chip with a controllable inner diameter. As a result of the inner diameter of the spiral structure being significantly smaller than typical RF wavelengths, it is only used in the design of an optical antenna. Although this work successfully breaks the 2D design wall of on-chip antennas, its disadvantages are also numerous. One disadvantage is that the deformation method is inefficient and hard to scale toward mass production; the other is that the antenna operating frequency is too high to be practical.

In a parallel paradigm, the development of self-rolling film nanotechnology provides a new way to realize 3D micro/nanostructures on-chip. Among them, self-rolling nanotechnology based on SiN_x thin films is the most widely used in the fields of electrical, optical, and biological research due to its material compatibility with conventional semiconductor fabrication technology, nontoxicity to biological systems, and optical transparency. By cutting one of the right angles of the rectangular diamond film, the mechanical stress of the film will become anisotropic. When hydrogen fluoride (HF) releases the SiO_2 sacrificial layer, the rolling axis of the film will deviate from the geometric axis of the film, thus realizing a directional spiral curl. The primary disadvantage of this method is that the rolling angle is difficult to control accurately, so it cannot be used in the design of an on-chip antenna working in the millimeter-wave band. Figure 11.2d shows (111) monocrystalline silicon (Si), anisotropically etched in a potassium hydroxide solution (KOH), as the sacrificial layer. The etching speed along the (110) surface is much faster than (111) surface, creating a helical structure with chirality that can be controlled by adjusting the angle between the stepped SiN_x bilayer film and the (110) surface. This fabrication method provides a way to precisely control the helix rolling angle, but when it is necessary to roll a large SiN_x film to manufacture an antenna, issues caused by pinholes in the PECVD SiN_x film prevent practical yield. Subsequently, these pinhole problems prevent the fabrication of an array of antennas.

Inspired by the structural design of S-RuM RF inductors with fabrication flow shown in Figure 11.9, the following fabrication method could be considered as shown in Figure 11.16 Combining both "bottom-up" and "top-down," preparation methods, the main body of the spiral self-rolling structure could be constructed on a (111) single-crystal silicon (Si) substrate, using a single-crystal silicon sacrificial layer and a SiN_x bilayer film deposited by PECVD as the strain layer. To reduce the detrimental influence of pinholes in the SiN_x bilayer film of the radiation unit, atomic layer deposition (ALD) is used to deposit an alumina film (Al_2O_3) between the (111) single-crystal silicon (Si) and the SiN_x bilayer film. The SiN_x deposition environment of the PECVD was composed of silane (SiH_4)/nitrogen (N_2)/ammonia (NH_3). By optimizing the RF power supplied to the process plasma, gas flow rates, and other process parameters, a 20 nm thick LF SiN_x thin film was deposited at a

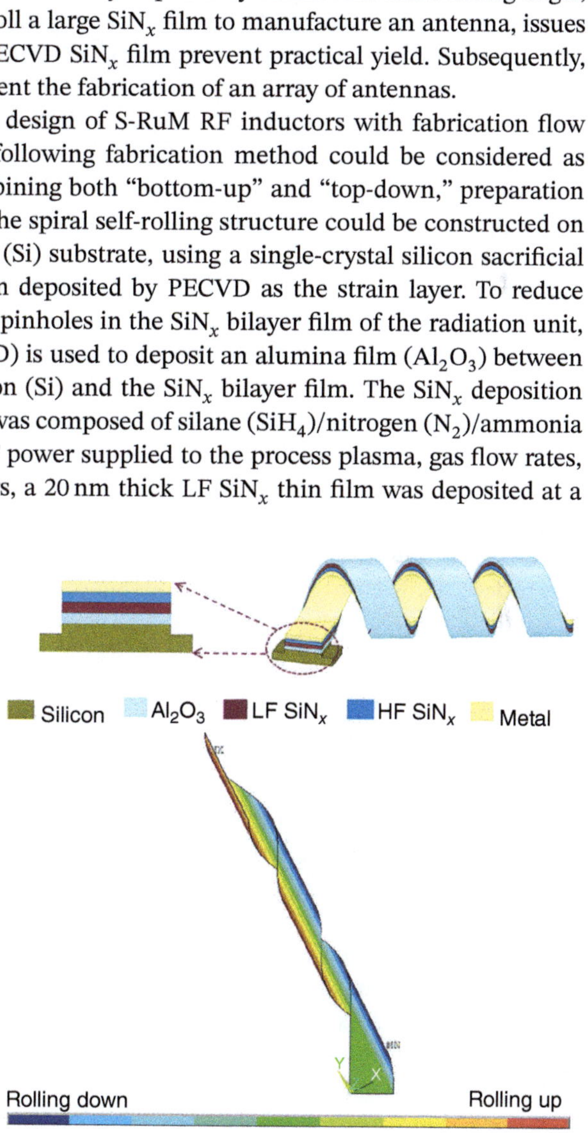

Figure 11.16 Potential S-RuM helical antenna structural design (above) and the schematic view of the resulting spiral structure prediction obtained through FEM simulation (below).

lower RF frequency (LF) (380 kHz), engineered to contain a maximized value of residual compressive stress. Directly following the deposition of the LF SiN_x layer, a 20 nm thick HF SiN_x film was deposited at a high RF frequency (13.56 MHz) to obtain a film with maximized tensile stress. Due to the stress mismatch between the LF SiN_x and HF SiN_x film layers, the composite structure of the films tends to roll upwards when released from the substrate. The conductive layer will be deposited on top of the SiN_x bilayer film. After the mesa pattern is defined by photolithography, ICP-RIE and Freon RIE processes will be used to etch the multilayer graphene, SiN_x bilayer film, Al_2O_3 film, and the (111) single-crystal silicon (Si) to form a mesa. Finally, the (111) Si sacrificial layer will be etched by KOH along the (110) surface. To fix the rolled-up structure on the substrate surface and prevent detachment, and also to link the antenna to the feed network, a metal anchor will be deposited on the mesa according to the electrical design needs and forming the corresponding spiral structure on the rolled-up SiN_x bilayer.

Incidental to the precise control of the size of the planar figure defined by the lithography process, the internal diameter and the rolling angle of the spiral structure after rolling can be accurately calculated, so that the characteristic size of the spiral antenna can be precisely controlled. The internal diameter of spiral structure can be obtained by quasi-static finite element mechanical simulation. Based on Mindlin–Reissner shell theory, the integrated structure of heterogenous films is regarded as a finite element shell element without thickness. The material properties (Young's modulus, residual stress, and Poisson's coefficient) and thicknesses of each layer are assigned to the corresponding layer, or shell element in a mathematical sense, according to the manufacturing sequence. The lateral etching process of sacrificial layer is simulated by dynamic boundary conditions. The tensile stress and compressive stress are characterized by the assumed positive and negative thermal expansion coefficients of the HF and LF SiN_x films, respectively, and all layers are given the same temperature increase, so the specific value of the thermal expansion coefficient is fit by comparing the residual stress test data. This method has been successfully applied to the mechanical simulation of other self-rolling film structures, and can accurately predict the size of the structure after rolling. Assuming a certain rolling direction, and provided that the material properties of each layer of the film stack are known, Figure 11.16 shows a schematic view of a resulting spiral structure prediction obtained through FEM simulation.

Knowing the inner diameter, we can design the circumference of the helix (C), the number of turns of the helix, the spacing between each turn (S), and the pitch angle (α), as shown in Figure 11.17a. The corresponding schematic view of rolled-up on-chip helical antenna with a coplanar transmission feed line is shown in Figure 11.7b. Following the design rule of helical antennas [45], this helical antenna has 10 turns with inner diameter of 44 μm, pitch 39.3 μm, and α equals 14°. Figure 11.17c,d shows the simulation results of all the important antenna characteristics. For the axial-mode operation, the maximum gain can be as high as 10.5 dB@2.15 THz and the HPBW is 30°@$\theta = 90°$.

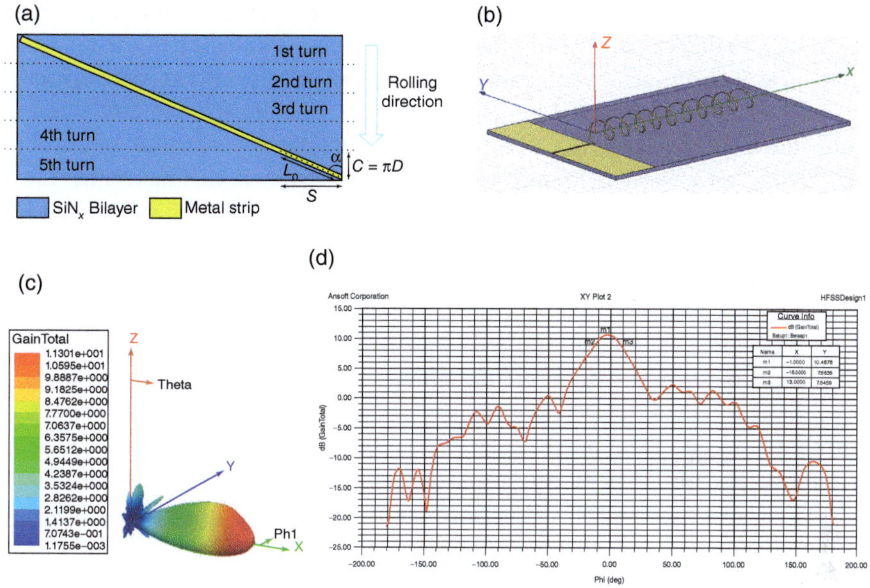

Figure 11.17 Design of microtube helical antenna and the simulated important performance metrics. (a) The planar pattern before being rolled up with design parameters labeled. (b) A 3D rolled-up schematic view of microtube helical antenna with a coplanar transmission feed line. (c) 3D gain pattern with maximum gain 11.3 (10.5 dB) @ 2.15 THz. (d) HPBW is 30° found from the gain rectangular plot at $\theta = 90°$.

11.4 S-RuM Power Passive Electronics

11.4.1 S-RuM Power Inductors

Compared to on-chip RF inductors, on-chip power inductors are more difficult to design and fabricate. Usually, power inductors are made off-chip due to the requirement for at least 100s of nH and up to Henry-level inductances, 100s of mA up to ampere-level current ratings, and minimized direct current (DC) resistances, typically in the mΩ range. Historically, the power density of power inductors was not a critical parameter until the integration density of integrated circuits became one of the primary design considerations in addition to performance tolerance. The best way to maximize the above two metrics is to design and fabricate power inductors monolithically. A typical application of power inductors is in DC/DC power converters for magnetic energy storage, which is widely used in electronic devices such as Internet of Things (IoT) units, personal computers, and mobile phones to regulate the input level of the voltage supply from a few volts to tens of volts. This major application of power inductors introduces another critical parameter: the switching frequency, which has been pushed as high as 10s to 100s of MHz, far beyond the frequency range of current devices operating at 1 or 2 MHz. If we compare the size of each component inside the DC/DC power converter, we will find that bulky power inductors dominate the overall size of power converters due to the extreme difficulty in scaling inductors compared to transistors, demonstrated in the historical scaling of highly integrated logic control circuits. Therefore, monolithic design and fabrication of high-performance on-chip power inductors for enabling compact, highly efficient, and fast switching power converters become more important. However, no matter which popular technology platform is used, the power system in package (PSiP) or the power system on chip (PSoC), no platform is able to avoid using planar spiral structures as the integrated power inductor. Identical to the case involving planar spiral RF inductors, low inductance density, and serious substrate parasitic effects, such as eddy currents and parasitic capacitance effects, inherently limit the inductance density and the switching frequency of planar power inductors. A promising route to solve these issues is to improve the S-RuM RF inductor platform for power applications.

To improve the S-RuM platform for power applications, there are several major factors that need to be considered. Primarily, the 3D coil density must be significantly increased, meaning that a greater number of turns must be achieved. Magnetic materials with good high-frequency performance must be also integrated. S-RuM inductors must handle ampere-level currents, meaning that a much larger metal layer cross section is imperative to lower ohmic heat generation. Finally, an efficient way to increase the overall heat conductivity to lower the operating temperature is desired. As mentioned in Section 11.3.1, a XeF_2 vapor-based release method provides the ability to roll up centimeter-long membranes with a fast-rolling speed under a heavy rolling load. Following the rolling process, the hollow core can be filled with a magnetic material, which should be a simple method compatible with planar semiconductor processing. In reference [29], a clever method has been demonstrated to effectively fill the hollow core by utilizing the capillary force between the tubular structure and a small ferrofluid drop. A ferrofluid is a colloidal solution of magnetic nanoparticles. Figure 11.18 shows the three-step post-fabrication method to integrate the ferrofluid inside the hollow core after rolling up the inductor.

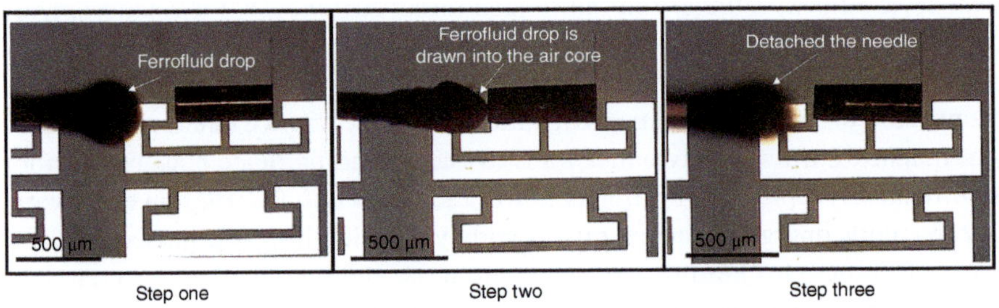

Figure 11.18 Three-step core-filling process utilizing capillary force. Step one: ferromagnetic fluid drawn into a micropipette by capillary action with a droplet hanging at the tip; Step two: the pipette tip makes contact with the S-RuM air-core inductor tube, and capillary action forces the ferrofluid into the core of the inductor tube; Step three: then the pipette withdrawn and detached from the core-filled S-RuM inductor tube. Source: Adapted from Huang et al. [29].

The first step, "alignment," is to align the tip of the micropipette with one end of the tubular structure. The second step, "filling," is to deliver the ferrofluid drop hanging on the tip of the micropipette by bringing the droplet into contact with the end of the tubular structure. The capillary force between the large surface tension of the liquid droplet and the microscale diameter of the air-core tubular structure is large enough to pull the ferrofluid into the core in under a few seconds. The last step, "detachment," is to retract the micropipette from the end of the tubular structure. It can be inferred that this capillary core-filling method could be reversible, which implies a reconfigurable on-chip power inductor could be realized without reconstructing the coiled structure, using a simple, fast, clean, and controllable method. Now, the material properties of the nanoparticles dispersed in the ferrofluid become the critical parameters to determine the electrical performance of the on-chip S-RuM power inductors. Typically, spherical Fe_3O_4 nanoparticles with a mean diameter of ~10 nm are dispersed in an organic carrier fluid containing surfactants, which demonstrate superparamagnetic behavior and a larger ferromagnetic resonance frequency compared to that of Fe_3O_4 macroscale solid materials due to the low thermal energy barrier to spontaneously reorient the magnetic moment of a single domain nanoparticle on a nanosecond time scale dictated by the Néel relaxation time and a Brownian motion component. Experimentally, ferrofluid-filled S-RuM power inductors with 1000s of nH are able to work at frequencies larger than 100 MHz, with roadmaps to improvement indicating that the highest performance that the S-RuM platform can deliver is yet to be achieved. Regarding the electro-thermal co-optimization of S-RuM power inductors, work is ongoing to systematically investigate this issue and a corresponding solution. Some experimental data and intuitive explanations were reported, which shows that air-core S-RuM power inductors with 10s of nH are able to handle DC current as large as 250 mA without irreversible damage, but the measured maximum operating temperature is as high as 350 °C.

11.4.2 Rolled-up Microsupercapacitors

Recent developments in portable and intelligent electronics require high-energy-density power storage components, such as microsupercapacitors (MSCs). Differing by their electrode design configuration, there are three common MSC structural categories: sandwich, interdigital, and fibrous types. Standard semiconductor layer-by-layer planar processing is capable of realizing high capacitance density on-chip capacitors, but, if the capacitor is manufactured individually, the fabrication flow is relatively complex and is largely incompatible with post-processing. The S-RuM platform has been demonstrated to provide a 3D configuration of electrodes to greatly enhance the capacitance density of MSCs. Two types of sandwich MSCs can be realized by using the self-rolled-up membrane-based structural design shown in Figure 11.19 [46]. Case I uses inorganic materials while case II introduces an organic self-assembled material (SAM) in between the metal and insulating layers to reduce leakage current. Ge with a thickness of 10 nm in case I and 20 nm in case II were used as the sacrificial layers. The surface of the Ge layer could be intentionally oxidized by thermal treatment in an oxygen-rich environment, and

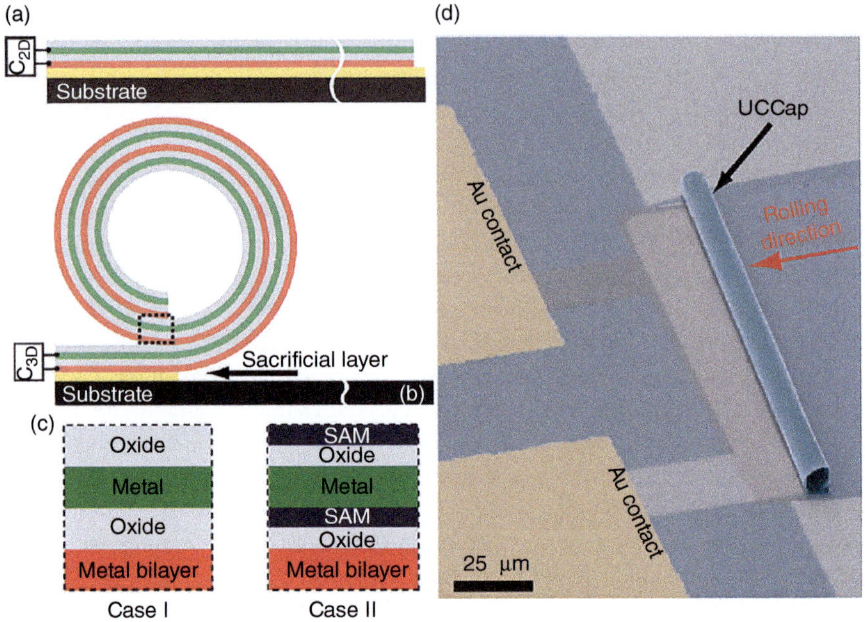

Figure 11.19 (a) Layer sequence used to create the nanomembrane-based rolled-up capacitors, which in (b) is shown in their rolled state. In (c) the layer sequence obtained by the self-winding process for inorganic (Case I) and hybrid organic/inorganic (Case II) capacitors is depicted. Source: Adapted with permission from Bof Bufon et al. [46]. Copyright (2010) by American Chemical Society.

the wet release of the membrane above the sacrificial layer could occur in H_2O_2 diluted with H_2O or in DI H_2O alone. If a faster release speed is desired, a higher concentration of H_2O_2/H_2O is recommended. The bottom capacitor electrode, also functioning as the strained layer, is a Ti/Cr metal layer with a total thickness of 15–35 nm, deposited by e-beam evaporation. The capacitor dielectric material is an Al_2O_3 thin film with a thickness of a few nanometers, deposited by ALD. The top capacitor electrode is a Cr thin film with a thickness of a few nanometers, deposited by indirect thermal deposition.

By theoretical analysis, the capacitance of the rolled-up capacitor C_{3D} can be estimated by the following equation:

$$C_{3D} = C_{2D}\left(2 - \frac{1}{N}\right) \tag{11.7}$$

where C_{2D} is the capacitance of device before rolling, and N is the number of turns. The equation implies that a sandwich S-RuM MSC with a high number of turns has a maximum capacitance two times larger than that of the same capacitor structure prior to self-rolling. Compared to the capacitance enhancement capability of S-RuM interdigital RF capacitors, sandwich S-RuM MSCs clearly doesn't utilize the advantage of the 3D electrode configuration. However, as the electrodes of sandwich S-RuM MSCs are very close to each other, with a nearly uniform gap, defined by the thickness of the ALD Al_2O_3 layer, the value of C_{2D} could be very large, potentially

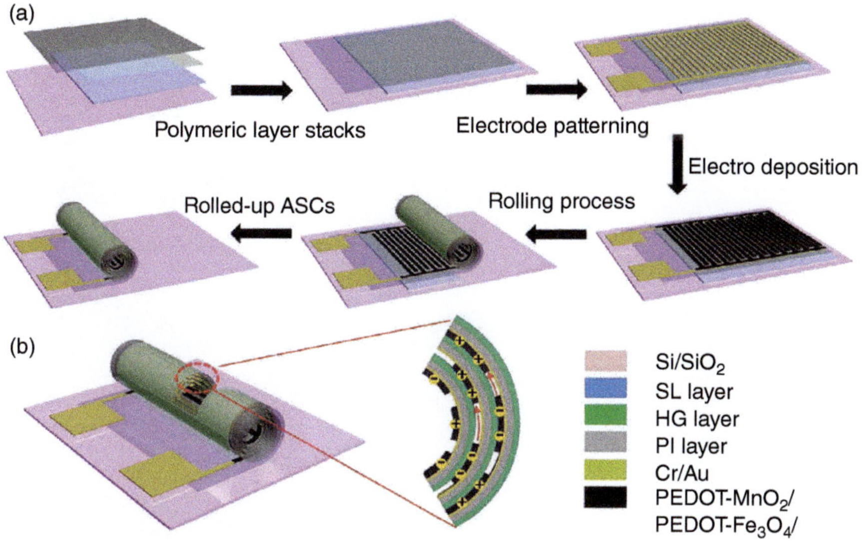

Figure 11.20 (a) Schematic illustration of the design and fabrication of rolled-up asymmetric MSC. (b) Conceptual scheme of a rolled-up tubular MSC. Negative ion transport is indicated by red arrow when charging process takes place inside the tubular device. Source: Adapted with permission from Li et al. [47]. Copyright (2019) by Advanced Science.

in the nF range and leading to a rolled-up capacitance density of 100s of $\mu F/cm^2$. As a compromise, a small gap between electrodes lowers the maximum sustainable applied voltage, inhibiting high power operation.

S-RuM asymmetric MSCs using an interdigital electrode configuration are demonstrated in reference [47], which achieve an ultrahigh capacitance density of 88.6 mF/cm^2 and energy density of 28.69 mWh/cm^2. A schematic illustration of the design and fabrication of this device is shown in Figure 11.20 In this design, a 400 nm thick lanthanum acrylate-based sacrificial layer is patterned on a Si/SiO_2 substrate, which is etched by an acidic potassium diethylenetriaminepentaacetic solution with a pH of 10. The rolling vehicle is a 900 nm thick photocrosslinked hydrogel-based swelling layer, and the self-rolling is triggered after submersion into the etchant. A reinforced layer composed of a 1700 nm thick polyimide is fabricated above, and, on top of it, 10 nm Cr and 50 nm Au thin films are deposited in sequence by e-beam evaporation and patterned to form interdigital electrodes to collect charge. The majority of the capacitance is attributed to the pseudocapacitive cathode and anode materials, MnO_2 and Fe_3O_4, dispersed in poly(3,4-ethylenedioxythiophene) ($PEDOT/MnO_2$ and $PEDOT/Fe_3O_4$), as shown in Figure 11.21. Optical and SEM images of the rolled-up device before and after rolling are shown in Figure 11.22 with a high magnification image of an open window into the device cut by FIB. Although the gap between turns is not uniform, indicating a low-performance tolerance, the reported device array connected in series or parallel was able to power a timer and an LED. Furthermore, these devices demonstrate remarkable cycling stability, retaining up to 91.8% of their original capacitance after 12 000 cycles.

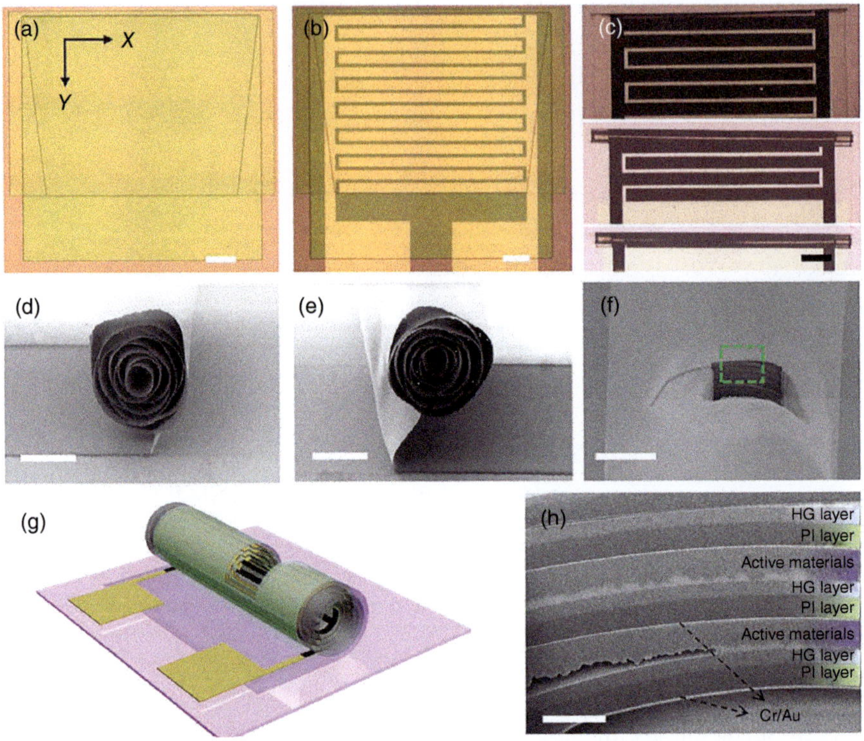

Figure 11.21 (a)–(c) Optical microscope images of the three layers, Cr/Au current collectors, electrodeposited materials, and the rolling process (scale bar (a)–(c): 500 μm). (d), (e) SEM images of tube openings, which were taken from two ends of the tube (scale bar (d), (e): 100 μm). (f)–(h) Schematic illustration and SEM images of the tube cross section prepared by FIB-cutting (scale bar (f): 50 μm, scale bar (h): 5 μm). Source: (a)–(h) Li et al. [47]. Copyright (2019) by Advanced Science.

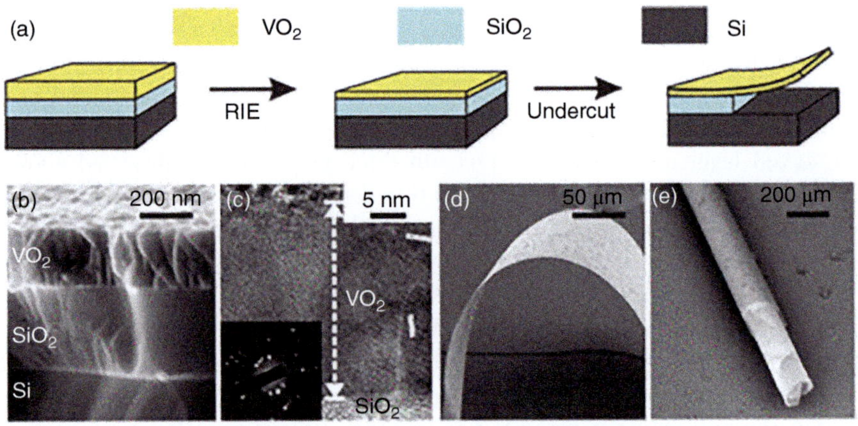

Figure 11.22 Fabrication and characteristics of an ultrathin VO_2 nanomembrane. (a) Schematic of the microfabrication process using rolling-up nanotechnology. (b) Cross-sectional SEM image of the VO_2 nanomembrane before the thinning process. (c) Cross-sectional TEM image of the thinned VO_2 nanomembrane. (d) SEM image of the ultrathin VO_2 nanomembrane. (e) SEM image of the VO_2 microtube. Source: (a)–(e) Tian et al. [48] Adapted with permission from American Chemical Society.

11.5 Reconfigurable Rolled-up Electronics

Wearable electronic devices, miniature and lightweight, have gradually become mainstream with diverse functionality, challenging the traditional electronic system design paradigm. Presently, the integrated circuit design remains in the two-dimensional plane. This traditional design constraint limits the integration complexity and diversity of functions that a single chip can perform. Reconfigurable self-rolled-up technology is proposed to address this problem. Compared with planar devices, 3D devices manufactured by self-rolled-up technology have better electrical performance and boast significantly reduced on-chip area footprints. The performance of the device often depends on its materials and structure. Self-rolled-up technology provides a good three-dimensional structure and material composition to the designer, making the design of reconfigurable devices possible.

The reconfigurability of self-rolled-up technology can be realized in different ways. It has been proven that reconfigurability can be realized by forming three-dimensional devices through self-rolled-up technology containing phase transition materials. The phase transition characteristics of vanadium dioxide have been under extensive investigation by researchers. Y. F. Mei et al. demonstrated the use of vanadium dioxide (VO_2) to achieve reconfigurable device functionality, as shown in Figure 11.22 [48]. A VO_2 thin film with a thickness of 200 nm was deposited onto a silicon dioxide insulation (SiO_2) layer of 300 nm on a silicon substrate. The VO_2 thin film was then thinned by fluoride plasma etching to achieve the desired thickness. Then, the underlying SiO_2 was removed by diluted hydrofluoric acid. Driven by the internal stress of the film, the VO_2 thin film curled along the stress gradient. During the process of continuous thinning, by measuring and comparing the resistance of the VO_2 film at different thicknesses, it was concluded that the thickness of the film does not affect the transmission characteristics. On the contrary, a thinner film thickness allows the film to bend more easily, and by adding a Cr layer on the VO_2 film, the VO_2 film is induced to curl and form a self-rolled-up microtube. In this work, a new theoretical model is proposed: the redistribution of the VO_2 strain force is related to the thickness and initial strain of the deposited Cr thin film, which makes the self-rolled-up microtube structurally adjustable. At the same time, the stress redistribution of the film bilayer is adjusted by changing the diameter of the microtubule, so that the phase transformation temperature is reduced from 341 to 329 °K. The threshold voltage of the transformation is also reduced. It can be imagined that when the tubular structure is not pinned down, the Cr/VO_2 bilayer strain film can be made to roll up and lay down (relax) periodically by changing the environmental temperature. Using this periodically changing structure, the electrical performance of the device can be reconfigured.

Periodically reconfiguring the VO_2 thin-film phase change-based device through the film's thermal characteristics mentioned above is primarily achieved by triggering a change of internal stress by changing the phase of the material itself. In the process of advanced electronic system design, it is often necessary to sense parameters of the ambient environment, which inevitably require a variety of sensors. As shown in Figure 11.23, the exploration of palladium-based materials by

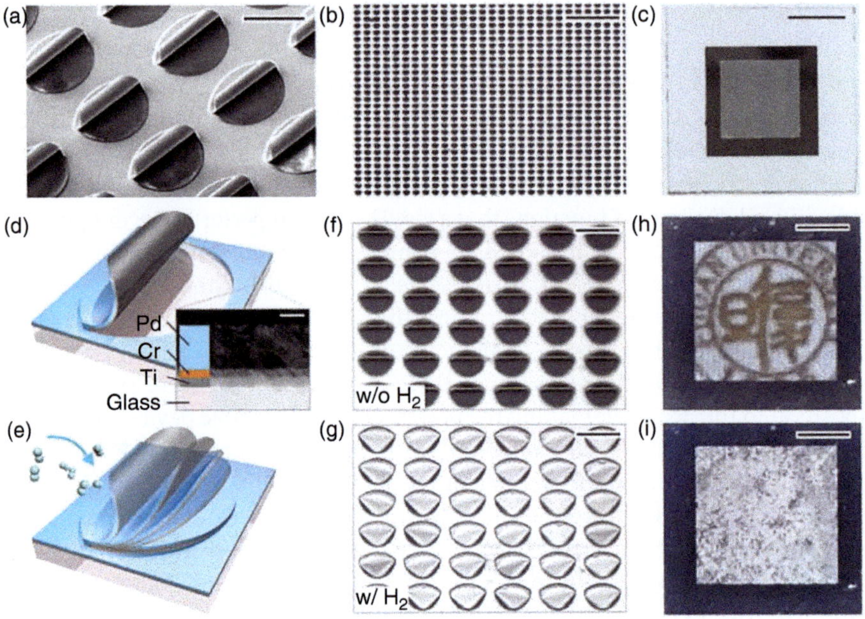

Figure 11.23 Configuration and stimuli-responsive performance of high-density stimuli-responsive NRs. (a) SEM images depicting the structure of NRs. Scale bar, 50 μm. (b) Optical image depicting the successful fabrication of NRs with high density. Scale bar, 500 μm. (c) Photograph of 5.5-mm × 5.5-mm array of NRs on a glass substrate. The light gray part is NRs, whereas the dark gray part is the planar nanomembrane as a contrast. Scale bar, 4 mm. (d) Scheme depicting the configuration of NR. The inset is the cross-section TEM image of the deposited trilayer nanomembrane system consisting of Ti, Cr, and Pd from the bottom up. Scale bar, 25 nm. (e) Scheme depicting stimuli-responsive behavior of NR. The NRs change from tubular structure into planar status with hydrogen stimuli. (f) and (g) Optical images depicting the NR array responding to hydrogen stimuli. Scale bars, 100 μm. (h) and (i) Visual images depicting the NR array responding to hydrogen stimuli. Scale bars, 2 mm. Source: Xu et al. [49]. Adapted with permission from AAAS.

Y. F. Mei and others provides a mechanism to design complex hydrogen sensors [49]. They deposited titanium (Ti), chromium (Cr), and palladium (Pd) with thicknesses of 5, 5, and 20–50 nm, respectively, to construct a prestressed film stack. Through conventional photolithography and wet stripping technology, compatible with planar semiconductor manufacturing technology, they successfully fabricated on-chip stimuli-responsive self-rolled-up nanomembrane arrays. The palladium film undergoes a phase transformation when the self-rolled-up nanomembrane arrays are placed in a hydrogen environment, which induces the volume expansion of the Pd nanomembrane. One end of the film is fixed to the substrate, so the top layer of the three-dimensional structure will expand, becoming planar; when the hydrogen is removed, the nanofilm array will roll up again. When the hydrogen concentration and Pd film thicknesses are optimized, the response time and recovery time for a complete cycle are 3.4 and 7.6 s, respectively, which is quite fast for hydrogen detection. After comparing the diameter change of the nanofilm over one cycle, it is found that the diameter of the rolled-up nanofilm is inversely proportional to the hydrogen concentration, and the thickness of the Pd film is inversely proportional to the response time to hydrogen. In the limiting case, the 20 nm thick Pd film can detect 1% hydrogen, which exceeds the limitations of traditional mechanical hydrogen detectors based on phase transitions. This work shows that the large area nanomembrane array has excellent nonelectric hydrogen detection performance, and the stress response of Pd to different concentrations of hydrogen provides a basis for the design of miniaturized sensors. At the same time, it can also use the phase transition characteristics of Pd to enable three-dimensional devices with variable structures.

11.6 Conclusion and Outlook

The application of self-rolled-up film nanotechnology to electronics is a young research field. The most important characteristic of this platform is that it can translate 2D film stacks into 3D structures, utilizing the preexisting suite of manufacturing techniques available for wafer fabrication. This method of spatial mapping converts a two-dimensional figure into a three-dimensional figure, thereby changing the conduction path of carriers. Alternatively, by engineering a device's induced structural strain, the constituent material properties can be controlled. These techniques add at least one dimension to the design flexibility and fabrication of devices. Traditional electronic devices therefore have opportunities to be broadly enhanced by the self-rolled-up platform. It is exciting that most self-rolled-up electronic devices that have been previously demonstrated can be manufactured with conventional semiconductor technology, meaning that it is possible to integrate these three-dimensional devices on-chip with active digital, analog, or photonic-integrated circuits. There are very few examples of self-rolled-up active electronic devices that have been realized. Active devices are challenging to fabricate utilizing the self-rolled-up platform because further development of active devices on this platform involves rigorous processing consistency improvements, high precision nanoscopic device layout techniques, understanding of physical models, and accurate simulation. Passive electronic devices dominate the self-rolled-up platform, with inductors, capacitors, and antennas at the forefront. These self-rolled-up devices greatly improve energy density but face a common bottleneck, that is, they can't curl high conductivity materials with micron thicknesses, preventing a significant reduction of ohmic losses associated with high-resistance thin-film conductive layers. From the perspective of technology development, how to curl or otherwise integrate thicker conductive materials, or to alternatively find high conductivity two-dimensional materials, is a primary motive in the development of self-rolled-up passive electronic technology. Additionally, reconfigurable self-rolled-up structures based on phase change materials and secondary deformation caused by physical and chemical stimulation receive continuous attention, which is of great significance to the realization of both high-power and high-frequency microsystems.

References

1 Prinz, V.Y., Seleznev, V.A., Gutakovsky, A.K. et al. (2000). Free-standing and overgrown InGaAs/GaAs nanotubes, nanohelices and their arrays. *Physica E* 6 (1–4): 828–831.

2 Li, X. (2008). Strain induced semiconductor nanotubes: from formation process to device applications. *J. Phys. D: Appl. Phys.* 41 (19): 193001.

3 Froeter, P., Huang, Y., Cangellaris, O.V. et al. (2014). Toward intelligent synthetic neural circuits: directing and accelerating neuron cell growth by self-rolled-up silicon nitride microtube array. *ACS Nano* 8 (11): 11108–11117.

4 Li, X. (2011). Self-rolled-up microtube ring resonators: a review of geometrical and resonant properties. *Adv. Opt. Photon.* 3 (4): 366.

5 Huang, W., Zhou, J., Froeter, P.J. et al. (2018). Three-dimensional radio-frequency transformers based on a self-rolled-up membrane platform. *Nat. Electron.* 1 (5): 305–313.

6 Chun, I.S., Challa, A., Derickson, B. et al. (2010). Geometry effect on the strain-induced self-rolling of semiconductor membranes. *Nano Lett.* 10 (10): 3927–3932.

7 Chun, I.S. and Li, X. (2008). Controlled assembly and dispersion of strain-induced InGaAs/GaAs nanotubes. *IEEE Trans. Nanotechnol.* 7 (4): 493–495.

8 Huang, W., Koric, S., Yu, X. et al. (2014). Precision structural engineering of self-rolled-up 3D nanomembranes guided by transient quasi-static FEM modeling. *Nano Lett.* 14 (11): 6293–6297.

9 Huang, G., Mei, Y., Thurmer, D.J. et al. (2009). Rolled-up transparent microtubes as two-dimensionally confined culture scaffolds of individual yeast cells. *Lab Chip* 9 (2): 263–268.

10 Mei, Y., Solovev, A.A., Sanchez, S., and Schmidt, O.G. (2011). Rolled-up nanotech on polymers: from basic perception to self-propelled catalytic microengines. *Chem. Soc. Rev.* 40 (5): 2109.

11 Macrelli, E. (2018). Roll up for three-dimensional transformers. *Nat. Electron.* 1 (5): 270–271.

12 Huang, W. and Li, X. (2018). Downscaling inductors with graphene. *Nat. Electron.* 1 (1): 6–7.

13 Baselmans, J. (2012). Kinetic inductance detectors. *J. Low Temp. Phys.* 167 (3–4): 292–304.

14 Luomahaara, J., Vesterinen, V., Grönberg, L., and Hassel, J. (2014). Kinetic inductance magnetometer. *Nat. Commun.* 5 (1): 4872.

15 Grimm, D., Bof Bufon, C.C., Deneke, C. et al. (2013). Rolled-up nanomembranes as compact 3D architectures for field effect transistors and fluidic sensing applications. *Nano Lett.* 13 (1): 213–218.

16 Froeter, P., Yu, X., Huang, W. et al. (2013). 3D hierarchical architectures based on self-rolled-up silicon nitride membranes. *Nanotechnology* 24 (47): 475301.

17 Michaels, J.A., Wood, D.R., Froeter, P.J. et al. (2019). Effect of perforation on the thermal and electrical breakdown of self-rolled-up nanomembrane structures. *Adv. Mater. Interfaces* 6: 1901022.

18 Smith, D.L., Alimonda, A.S., Chen, C.-C. et al. (1990). Mechanism of SiN_xH_y Deposition from NH -SiH_4 Plasma. *J. Electrochem. Soc.* 137 (2): 10.

19 Timoshenko, S. (1925). Analysis of bi-metal thermostats. *J. Opt. Soc. Am. A* 11 (3): 233.

20 Park, M., Lee, S., Kim, C.S. et al. (1998). The detailed analysis of high Q CMOS-compatible microwave spiral inductors in silicon technology. *IEEE Trans. Electron Devices* 45 (9): 1953–1959.

21 Yue, C.P. and Wong, S.S. (2000). Physical modeling of spiral inductors on silicon. *IEEE Trans. Electron Devices* 47 (3): 560–568.

22 Tang, C.-C., Wu, C.-H., and Liu, S.-I. (2002). Miniature 3-D inductors in standard CMOS process. *IEEE J. Solid-State Circuits* 37 (4): 471–480.

23 Jiang, H., Wang, Y., Yeh, J.-L.A., and Tien, N.C. (2000). On-chip spiral inductors suspended over deep copper-lined cavities. *IEEE Trans. Microwave Theory Tech.* 48 (12): 2415–2423.

24 Zou, J., Liu, C., Trainor, D.R. et al. (2003). Development of three-dimensional inductors using plastic deformation magnetic assembly (PDMA). *IEEE Trans. Microwave Theory Tech.* 51 (4): 1067–1075.

25 Yoon, J.-B., Choi, Y.-S., Kim, B.-I. et al. (2002). CMOS-compatible surface-micromachined suspended-spiral inductors for multi-GHz silicon RF ICs. *IEEE Electron Device Lett.* 23 (10): 591–593.

26 Huang, W., Yu, X., Froeter, P. et al. (2012). On-chip inductors with self-rolled-up SiN_x nanomembrane tubes: a novel design platform for extreme miniaturization. *Nano Lett.* 12 (12): 6283–6288.

27 Yu, X., Huang, W., Li, M. et al. (2015). Ultra-small, high-frequency and substrate-immune microtube inductors transformed from 2D to 3D. *Sci. Rep.* 5 (1): 9661.

28 Huang, W., Zhou, J., Froeter, P. et al. (2017). CMOS-compatible on-chip self-rolled-up inductors for RF/mm-wave applications. In: *2017 IEEE MTT-S International Microwave Symposium (IMS)*, 1645–1648. Institute of Electrical and Electronics Engineers.

29 Huang, W., Yang, Z., Kraman, M.D. et al. (2019). Monolithic mTesla Level Magnetic Induction by Self-Rolled-up Membrane Technology. *Sci. Adv.* 6 (3): 1–9.

30 Chen, W., McCarthy, K.G., Mathewson, A. et al. (2010). High-performance MIM capacitors using novel PMNT thin films. *IEEE Electron Device Lett.* 31 (9): 996–998.

31 Choi, J.-Y., Kang, L.-S., Cho, K.-H. et al. (2009). Structural and electrical properties of Mn-doped $Bi_4Ti_3O_{12}$ thin film grown on $TiNSiO_2Si$ substrate for RF MIM capacitors. *IEEE Trans. Electron Devices* 56 (8): 1631–1636.

32 Sul, W.S. and Pyo, S.G. (2014). RF characteristic analysis model extraction on the stacked metal–insulator–metal capacitors for radio frequency applications. *IEEE Trans. Electron Devices* 61 (8): 3011–3013.

33 Sang, L., Zhou, H., Yang, Z. et al. (2019). Monolithic radio frequency SiN_x self-rolled-up nanomembrane interdigital capacitor modeling and fabrication. *Nanotechnology* 30 (36): 364001.

34 Alley, G.D. (1970). Interdigital capacitors and their application to lumped-element microwave integrated circuits. *IEEE Trans. Microwave Theory Tech.* 18 (12): 1028–1033.

35 Wang, J., Bandari, V.K., Karnaushenko, D. et al. (2019). Self-assembly of integrated tubular microsupercapacitors with improved electrochemical performance and self-protective function. *ACS Nano* 13 (7): 8067–8075.

36 Bjornson, E., Hoydis, J., and Sanguinetti, L. (2018). Massive MIMO has unlimited capacity. *IEEE Trans. Wireless Commun.* 17 (1): 574–590.

37 Busari, S.A., Huq, K.M.S., Mumtaz, S. et al. (2018). Millimeter-wave massive MIMO communication for future wireless systems: a survey. *IEEE Commun. Surv. Tutorials* 20 (2): 836–869.

38 Qiao, D., Qian, H., and Li, G.Y. (2016). Broadbeam for massive MIMO systems. *IEEE Trans. Signal Process.* 64 (9): 2365–2374.

39 Li, L., Wang, D., Niu, X. et al. (2018). mmWave communications for 5G: implementation challenges and advances. *Sci. China Inf. Sci.* 61 (2): 021301.

40 Shafi, M., Molisch, A.F., Smith, P.J. et al. (2017). 5G: a tutorial overview of standards, trials, challenges, deployment, and practice. *IEEE J. Select. Areas Commun.* 35 (6): 1201–1221.

41 Shahmansoori, A., Garcia, G.E., Destino, G. et al. (2018). Position and orientation estimation through millimeter-wave MIMO in 5G systems. *IEEE Trans. Wireless Commun.* 17 (3): 1822–1835.

42 Huang, I.-Y., Sun, C.-H., and Hsu, K.-Y. (2012). Improving bandwidth and return loss of Si-based MEMS antenna using suspending and electromagnetic band-gap structures. *Sens. Actuators, A* 174: 33–42.

43 Chu, H., Guo, Y.-X., Lim, T.-G. et al. (2012). 135-GHz micromachined on-chip antenna and antenna array. *IEEE Trans. Antennas Propagat.* 60 (10): 4582–4588.

44 Mao, Y., Zheng, Y., Li, C. et al. (2017). Programmable bidirectional folding of metallic thin films for 3D chiral optical antennas. *Adv. Mater.* 29 (19): 1606482.

45 Milligan, T.A. (2005). *Modern Antenna Design*. Hoboken, N.J.: IEEE Press, Wiley-Interscience.

46 Bof Bufon, C.C., Cojal González, J.D., Thurmer, D.J. et al. (2010). Self-assembled ultra-compact energy storage elements based on hybrid nanomembranes. *Nano Lett.* 10 (7): 2506–2510.

47 Li, F., Wang, J., Liu, L. et al. (2019). Self-assembled flexible and integratable 3D microtubular asymmetric supercapacitors. *Adv. Sci.* 60 (20): 1901051.

48 Tian, Z., Xu, B., Hsu, B. et al. (2018). Reconfigurable vanadium dioxide nanomembranes and microtubes with controllable phase transition temperatures. *Nano Lett.* 18: 3017–3023.

49 Xu, B., Tian, Z., Wang, J. et al. (2018). Stimuli-responsive and on-chip nanomembrane micro-rolls for enhanced macroscopic visual hydrogen detection. *Sci. Adv.* 4: eaap8203.

12

Rolled-Up Whispering-Gallery Mode Optical Microcavities

Yunqi Wang, Shuo Yang, Gaoshan Huang, Borui Xu, and Yongfeng Mei

Fudan University, Department of Materials Sciences, 220 Handan Road, Shanghai 200433, People's Republic of China

12.1 Introduction

Inspired by origami and kirigami, novel three-dimensional structures can be built up by engineering the strain gradient of the nanomembranes [1, 2]. Many types of microstructures can be built by rolling, folding, and bending patterned nanomembrane layers. Among them, if the gradient of strain distribution is large enough, the nanomembrane rolls up into a curved structure to build a tubular geometry [3]. Since more than 10 years ago, tubular microcavities have been fabricated in a precise way by combining top-down and bottom-up approaches [4]. Nanomembrane with predesigned strain gradient is supposed to be deposited on the top of a patterned sacrificial layer, and then through selective etching, the strained layer can be separated from the sacrificial layer. As a consequence, the intrinsic strain gradient causes the bilayer to curve into a tubular microcavity spontaneously [5, 6]. Through tunable strain engineering of the nanomembranes, the diameter, length, and winding number can be precisely controlled, which can be convenient in fundamental studies and practical applications [7, 8].

The microtubes fabricated by rolled-up technique have been widely used in optical microcavities, which play a significant part in light–matter interaction by effectively confining light through resonant circulating [9]. According to different light confinement strategies, optical microcavities can be divided into three main types, including Fabry–Perot microcavities [10, 11], photonic crystal microcavities [12, 13], and whispering-gallery mode (WGM) microcavities [14, 15]. Researches on WGM resonance can be traced back to more than 100 years ago. Whispering-gallery type resonance was first observed and investigated in the acoustic mode of St Paul's Cathedral, then this type of resonance was named WGM and researches on this topic extended to the radiofrequency and optical domains gradually. In a WGM optical microcavity, light travels along the surface of the microcavity by total internal reflection and is confined in a small volume, achieving high light intensity and energy density [16]. Owing to the desired characteristics of ultrahigh quality factor (Q factor) and small

Nanomembranes: Materials, Properties, and Applications, First Edition.
Edited by Yongfeng Mei, Gaoshan Huang, and Xiuling Li.
© 2022 WILEY-VCH GmbH. Published 2022 by WILEY-VCH GmbH.

mode volume, WGM microcavities are attracting increasing interest in various fields, including sensors [17, 18], filters [19], and lasers [20, 21].

According to the geometry shape, WGM microcavities can be divided into microspheres [22–24], microdisks [25], microtoroids [19], and microtubes [26–28]. Among those, rolled-up microtubes possess some unique characteristics owing to their unique geometry. Firstly, with the cylindrical channel, properties of fluids can be introduced into the system, such as carrying nanoparticles (NPs), cells, and molecules. Secondly, typical subwavelength-thickness walls lead to evanescent wave with severe leakage, which is of great importance in sensing applications. Thirdly, the asymmetric tube structure can be precisely designed in axial direction or through the lobe shape, adding extra degrees of freedom for mode selection, mode tuning, and directional emission. Lastly, rolled-up technique can be utilized among a variety of materials, including metals [29], semiconductors [30, 31], insulators [32, 33], and 2D materials [34, 35], and the geometry characteristics of the tubes can be tuned precisely.

In this chapter, we first theoretically analyze the WGM in rolled-up microtubes with wave equation and ray optics model. Then move to discuss light propagation in the structure, including the optical loss, evanescent field coupling, and optical characterization. Structural asymmetry induced by rolled-up technique and designing in axial direction is also introduced in this section. In Section 12.4, we continue to investigate different materials in rolled-up optical microcavities, such as semiconductors, oxide, and two-dimensional materials. Finally, some applications based on rolled-up microtubes are discussed.

12.2 Theoretical Analysis

Theoretical analysis is an important part of the designing process. From the theoretical calculation and numerical simulation, we can obtain the resonant wavelength, Q factor, electric field distribution, and other characteristics of WGM microcavities that can be beneficial for further investigation. In this section, analysis based on wave equation and ray optics model is discussed, some tools including numerical simulation methods and Husumi functions are also briefly introduced, which may be useful in practical applications.

12.2.1 Wave Equation

In this section, theoretical analysis based on wave equation for tubular microcavity is provided. Through solving Maxwell's equation, the distribution of electric field in the rolled-up microcavities can be obtained. Besides, some numerical simulation methods are also introduced in this section, which can be useful in complicated situations.

As for a dielectric cavity, the geometry feature can be described by the spatial distribution of the refractive index n. To get the eigenmodes of the microtube, we use cylindrical coordinates to calculate the electric fields $\vec{E}(r, \varphi, z)$, as shown in Figure 12.1, then the wave equation becomes

$$-\frac{1}{n^2(r,\varphi,z)}\nabla^2\vec{E}(r,\varphi,z) = k^2\vec{E}(r,\varphi,z) \tag{12.1}$$

In Eq. (12.1), k and n represent the vacuum wave vector and the refractive index of the material, respectively. As seen from the equation, a WGM can be defined as a time-harmonic solution of Maxwell's equations with k. Since tubular microcavities are open systems with light leaking out of the cavity, a WGM can be treated as a quasibound state or quasinormal mode decaying exponentially in time [36]. As for rolled-up tubular microcavities, the thickness is usually much smaller than the resonant wavelength of WGMs. In this case, WGMs with transverse-magnetic (TM)-polarized are preferred, where TM-polarized is defined as that the electric field vector is polarized perpendicular to the microtube cross section. The reason why TM-polarized WGMs are preferred while transverse-electric (TE)-polarized WGMs are suppressed in rolled-up microcavities with ultrathin wall thickness may be ascribed to the electric field distribution discontinuities at the interface, which makes TE-polarized modes more sensitive to the notches induced by the rolling edges [37]. Moreover, as the experimental results indicate, TE-polarized modes only appear in microcavities with thickness larger than a critical value [38]. Consequently, here we mainly discuss TM-polarized WGMs in rolled-up tubular microcavities. Considering TM-polarized modes, Eq. (12.1) can be derived into Helmholtz equation for z component $E_Z(r, \varphi, z)$ of the electric field:

$$-\frac{1}{n^2(r,\varphi,z)}\nabla^2 E_Z(r,\varphi,z) = k^2 E_Z(r,\varphi,z) \tag{12.2}$$

Due to the cylindrical symmetry of the microtube and the circular nature of the WGMs, with little discrimination of the tube diameter along z direction, the WGMs can be treated with adiabatic approximation (Born–Oppenheimer approximation) to separate the azimuthal from the axial propagation [39, 40]. Thus, the solution of Eq. (12.2) can be written as:

$$E_Z(r,\varphi,z) = \Phi(r,\varphi,z)\Psi(z) \tag{12.3}$$

In Eq. (12.3), $\Phi(r, \varphi, z)$ is the wave function for light traveling at a fixed z plane, $\Psi(z)$ represents the envelope of the solution along z direction. Through the adiabatic approximation, the Helmholtz equation can be divided into two separate equations. The electric field distribution $\Phi(r, \varphi, z)$ has to satisfy the two-dimensional wave equation at every axial position

$$-\frac{1}{n^2(r,\varphi,z)}\nabla^2_{r,\varphi}\,\Phi(r,\varphi,z) = k^2_{\mathrm{circ}}(z)\Phi(r,\varphi,z) \tag{12.4}$$

where $k_{\mathrm{circ}}(z)$ represents the wave vector in the circular plane at a fixed z position, and then $\Psi(z)$ satisfy the axial propagation:

$$-\frac{1}{n^2}\frac{\partial^2}{\partial z^2}\Psi(z) + k^2_{\mathrm{circ}}(z)\Psi(z) = k^2\Psi(z) \tag{12.5}$$

As Eq. (12.5) indicates, diversity of n is included in $k_{\mathrm{circ}}(z)$, thus the mentioned n refers to the refractive index of the material. Similar to a Schrodinger equation for

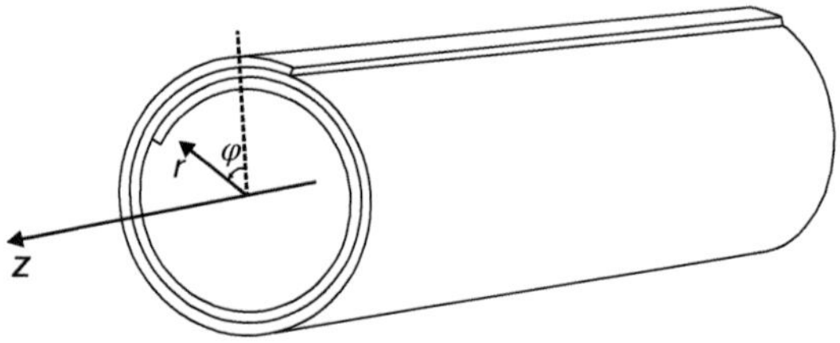

Figure 12.1 Schematic of a roll-up tubular microcavity in cylindrical coordinate system.

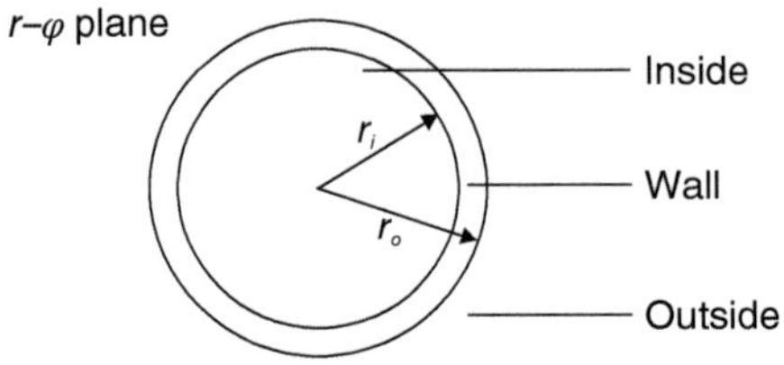

Figure 12.2 Schematic of the simplified ring section in the $r-\varphi$ plane.

a particle wave in one-dimensional potential, Eq. (12.5) can be called as photonic quasi-Schrodinger equation [7, 41]. Though the variety in $k_{\text{circ}}(z)$, the kinetics in $r-\varphi$ plane can lead to a quasipotential $V_{\text{eff}}(z)$ for light propagating in z direction. Generally, the desired axial modes can be obtained by numerically solving Eq. (12.5).

As for light propagation in the circular plane, if the rolling edges with asymmetric features of microcavities are neglected, the $r-\varphi$ plane can be treated as a perfect ring with rotational symmetry, as shown in Figure 12.2. Usually, the microcavity is placed in air, in this case, refractive index jumps from 1 (n_{air}) to n at the inner radius r_{i} and returns to 1 at the outer radius r_{o} of the microtube, thus the area is consequently divided into three regions. The solutions of Eq. (12.4) can be written as linear combination of functions $J_m(nk_{\text{circ}}r)\text{e}^{im\varphi}$ and $N_m(nk_{\text{circ}}r)\text{e}^{\text{i}m\varphi}$, which refers to Bessel functions and von Neumann functions, respectively. For the clearance of expression, Hankel functions are introduced as $H_m^1(nk_{\text{circ}}r) = J_m(nk_{\text{circ}}r) + iN_m(nk_{\text{circ}}r)$ represents radially outgoing waves and $H_m^2(nk_{\text{circ}}r) = J_m(nk_{\text{circ}}r) - iN_m(nk_{\text{circ}}r)$ represents radially incoming waves. Then the electric field distributions of WGMs in a perfect ring section can be expressed as:

$$\begin{aligned} &A_m J_m(k_{\text{circ}}r) \quad r < r_{\text{i}} \\ &B_m H_m^1(nk_{\text{circ}}r) + C_m H_m^2(nk_{\text{circ}}r) \quad r_{\text{i}} < r < r_{\text{o}} \\ &D_m H_m^1(k_{\text{circ}}r) \quad r_{\text{o}} < r \end{aligned} \tag{12.6}$$

where, m is the azimuthal number and A_m, B_m, C_m, and D_m are coefficients that need to be calculated according to certain boundary conditions, e.g. the electric field and its derivative value have to be continuous in TM-polarized WGMs. The obtained k_{circ} are a series of complex values, where the real parts represent the energy of WGMs and the imaginary parts describe the loss [7, 41]. As has been mentioned above, the tubular microcavities are open systems, the curved boundary allows light to leak out of the cavity, this theoretical loss leads to quality factor between 10^9 and 10^{11} [41].

However, due to complexity of the practical problems, analytical treatment may be hard to practice in some cases. Several strategies such as finite-difference time-domain (FDTD) method, finite-element method (FEM), and finite-difference frequency-domain (FDFD) method have been used for numerical simulation of WGM microcavities. Among these methods, FDTD method is the most prominent one [42, 43]. Through discretizing time and space into tens of thousands of grids, Maxwell's equation can be converted into finite-difference equations. Owing to its flexibility, FDTD simulation is widely used in complicated structures with different materials [7]. However, for short-wavelength regime, in other words, when the resonant wavelength is small compared to the characteristic length scales of the structure, demanding computational time and memory are required. Similar to FDTD method, FEM also divides the entire domain into finite grids, moreover, nonuniform spatial and nonorthogonal grids can be used to discretize the simulation domain, e.g. triangle grids in two-dimensional simulation. This feature is of great benefit while calculating structures with curved boundaries, which is the main advantage over the FDTD method [7]. Besides, compared to FDTD method, FEM can be used in both time domain and frequency domain. Dealing with the long-lived modes in time domain is usually too time-consuming, it is more convenient to simulate in the frequency domain, then the frequency-domain FEM and FDTD method can be taken into consideration [36]. Every method has its own advantages and shortcomings, so simulation method must be chosen based on the actual situation, e.g. structural features and computational memory, in some cases, a few methods can also be used in combination to figure out one individual problem.

12.2.2 Ray Optics

Ray (geometrical) optics is the short-wavelength limit of wave optics, through tracing rays guided in the given structure, the problem can be analyzed in a more direct way. Firstly, the rolling edges of rolled-up microtubes are neglected for simplification, then the cross section of rolled-up microtube becomes a perfect ring. Considering WGMs in the structure, as shown in Figure 12.2, light travels along the outer surface of the ring through multiples of total internal reflections and returns in phase. Obviously, the approximate condition for WGM resonance is

$$2\pi R_0 n_{\text{eff}} = m\lambda \tag{12.7}$$

In Eq. (12.7), R_0 is the outer radius of the ring section, n_{eff} and λ refer to the effective refractive index and the vacuum wavelength of light, respectively. m respects the number of total internal reflections in a circle, and it is supposed to be an integer. Besides, m in Eq. (12.7) also equals the azimuthal number m in the solution to the wave equation. The above equation is a simplified model to help us get a total picture of WGMs in ray optics. However, the real trajectories can be quite complicated, each light line reflects along the outer surface of the microtube, the reflected wave experiences a phase delay, which is a function of the wavelength λ, incident angle χ, and the dielectric-to-surrounding refractive index ratio. As a consequence, the optical distance is different from $2\pi R_0 n_{\text{eff}}$ in the simplified model [23].

Since tracing trajectories in real space can be very complicated, it is more appropriate and convenient to study the trajectories in phase space. Poincaré surface of section (SOS), consisting of the intersection points of a set of trajectories with the

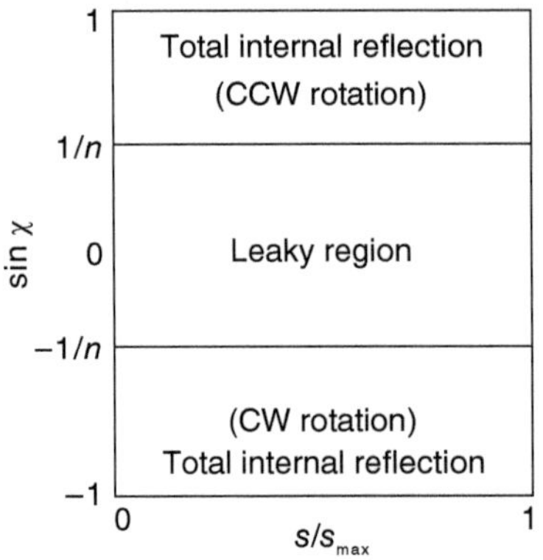

Figure 12.3 Typical SOS of a microcavity (open system).

boundary, can be employed to analyze the dynamics in the microcavities in phase space, as shown in Figure 12.3. As for each ray reflected at the boundary of the microcavity, the position in terms of arclength coordinate along the boundary s and another parameter defined as $p = \sin\chi$ (χ refers to the incident angle) are recorded on the SOS. Here, $\sin\chi > 0$ represents counterclockwise (CCW) propagation and $\sin\chi < 0$ means clockwise (CW) propagation. It is worth noting that the microcavities are open systems, and rays can escape from the cavity. According to Snell's law, if the incident angle χ is smaller than the critical angle χ_c for total internal reflection, rays leave the cavity through refraction process, e.g. $\sin\chi_c = 1/n$ when the cavity is placed in air. Therefore, in the SOS of a microcavity, the region between lines that satisfy $\sin\chi = \pm 1/n$ is defined as "leaky region," as shown in Figure 12.3. With the refractive index n increasing, the area of the leaky region decreases correspondingly.

Finally, Husimi function is employed to study waves in dynamical systems. By calculating the overlap integral of the wave function with a coherent state representing a minimal-uncertainty wave packet, the desired Husimi function can be obtained [44]. Among four Husimi functions, the function for internal waves is employed here:

$$H^{\text{inc(em)}}(s,p) = \frac{nk}{2\pi}\left|\mathcal{F}h_{\psi}(s,p) + \frac{i}{k\mathcal{F}}h_{\partial\psi}(s,p)\right|^2 \tag{12.8}$$

with weighting factor $\mathcal{F} = \sqrt{n\sqrt{1-p^2}}$. The function

$$h_{\psi}(s,p) = \int_0^{s_{\max}} ds' g(s')\xi(s';s,p) \tag{12.9}$$

refers to the overlap of the wave function ($g = \psi$) or its normal derivative ($g = \partial\psi$) on the boundary of the microcavity with the minimal-uncertainty wave packet

$$\xi(s';s,p) = (\sigma\pi)^{-1/4}\sum_{l=-\infty}^{\infty}\exp\left[-\frac{(s' + s_{\max}l - s)^2}{2\sigma} - inkp(s' + s_{\max}l)\right] \tag{12.10}$$

The wave packet $\xi(s';s,p)$ is centered around (s,p) and the wave packet uncertainty is determined by the parameter σ. Using Eq. (12.8), Fang et al. [45] investigated the mode chirality of a rolled-up tubular microcavity, as shown in Figure 12.4. The CW mode and CCW mode in rolled-up microcavity can be easily divided in the phase

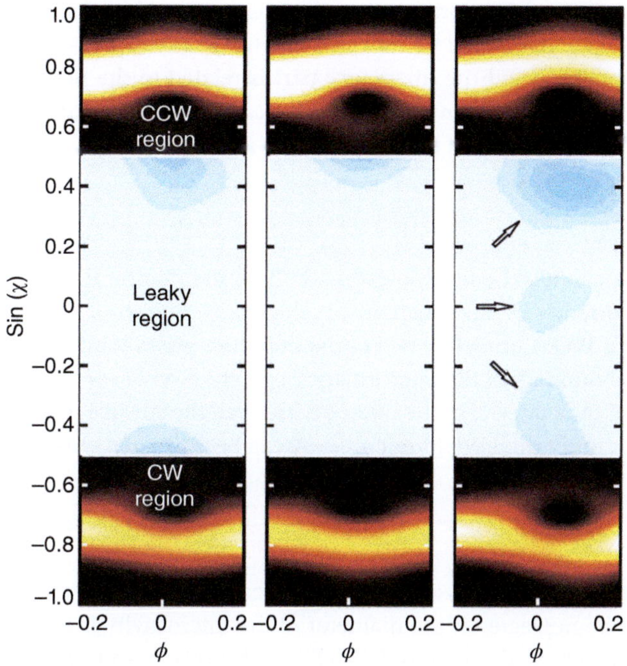

Figure 12.4 Derived CCW and CCW components of WGM at the outer notch of the rolled-up microtube using Husimi function. Source: Adapted from Fang et al. [45]. Copyright 2016, American Physical Society.

space using this method. Besides, the intensity in the leaky region can reflect the evolution of Q factor, indicating that the local maximum Q factors are induced by the high mode chirality of WGMs in rolled-up microtubes with particular geometry shapes. More details are discussed in Section 12.3.

12.3 Light Propagation in Tubular WGM Microcavities

12.3.1 *Q* Factor and Optical Loss

As one of the most significant parameters of WGM microcavities, Q factor is a characteristic to value the light storage ability and optical loss. It can be defined as the value of time-averaged energy divided by the power loss per cycle [46]:

$$Q = \omega \frac{\text{Stored energy}}{\text{Power loss}} \tag{12.11}$$

where ω refers to the angular resonance frequency. Q factor can also be calculated by the quotient of the resonant frequency ω and the considered full width at half maximum (FWHM) bandwidth $\Delta\omega$ of the Lorentz-type peak [46]:

$$Q = \frac{\omega}{\Delta\omega} = \omega\tau \tag{12.12}$$

where τ refers to the decay time of the stored energy. Besides, as mentioned in the wave equation section, the optical loss of a WGM can be conveniently expressed by a complex-valued frequency ω_{com}, and the imaginary part is related to the photon lifetime via $\tau = -1/(2\,\mathrm{Im}(\omega_{com}))$ with $\mathrm{Im}(\omega_{com}) < 0$. In this case, Q factor compares the lifetime τ with the oscillation period of the light $T = 2\pi/\,\mathrm{Re}(\omega)$ [36]:

$$Q = 2\pi\frac{\tau}{T} = -\frac{\mathrm{Re}(\omega_{com})}{\mathrm{Im}(\omega_{com})} \tag{12.13}$$

As a consequence, Q factor can also be used to estimate the finite photon lifetime. It can also be determined directly by the mechanisms that cause light loss. In the ideal situation, Q factor of a WGM only depends on its radiation losses while light circulates along the curved boundary of the microcavity. In practice, however, due to absorption and scattering in the material of the microcavities and the surface roughness, the total Q factor of an individual WGM microcavity can be estimated with the formula as followed [23]:

$$Q_0^{-1} = Q_{rad}^{-1} + Q_{mat}^{-1} + Q_{s.s}^{-1} + Q_{cont}^{-1} \tag{12.14}$$

In Eq. (12.14), Q_{rad} represents radiative losses. Since Q_{rad} decays exponentially with increasing diameter, it can be neglected if the diameter of the microcavity is larger than 10 μm [23, 47]. The ability of the materials (such as absorption and scattering) is reflected in Q_{mat}, and the surface inhomogeneity-caused scattering losses are described in $Q_{s.s}$. Q_{cont} stands for the losses induced by surface contaminates during the fabrication process. As for tubular WGM microcavities, Eq. (12.14) can be simplified as [48, 49]:

$$Q_0^{-1} \approx Q_{wall}^{-1} + Q_{s.s}^{-1} \tag{12.15}$$

where Q_{wall} and $Q_{s.s}$ refer to the partial Q factors influenced by loss of the microcavity wall, and loss due to the surface scattering, respectively [7, 48]. According to Mie scattering theory, as for a microtube resonator, Q factor increases with increasing the wall thickness and effective index, and decreases with longer resonant wavelength [49, 50].

As for rolled-up tubular microcavities, their Q factors (<5000) are usually lower than that of other types of optical microcavities of a tubular shape ($>10^4$) owing to their asymmetry features (e.g. notches and cone effects) [7, 51]. To improve the Q factor and optical performance of rolled-up microcavities, several approaches can be adopted. Firstly, using materials with high refractive index. For instance, Y_2O_3/ZrO_2 rolled-up tubular microcavities have higher Q factors than that of SiO/SiO_2 [49]. Secondly, coating layers of high refractive index material on the tube wall. It was previously reported that coating HfO_2 layers can improve the optical performance more effectively than Al_2O_3 coating since HfO_2 material has a higher refractive index [52]. Thirdly, a U-shape mesa can be employed to suspend the middle part of the rolled-up microcavities from the substrate to reduce radiative loss [53]. Finally, special lobe structures can be designed to improve light confinement, which is discussed later [54].

12.3.2 Evanescent-Field Coupling and Optical Characterization

As mentioned in Section 12.2.2, every time light reflects along the outer surface of the microtube, the reflected wave experiences a phase delay. At the same time, the reflected light shifts a little, in other words, the points that light incidents and reflects do not coincide perfectly. This phenomenon can be attributed to the Goos–Hanchen effect, and the shift of totally reflected light along the boundary is called the Goos–Hanchen shift (GHS) [55]. That is to say, part of the light turns to travel a short distance in the medium with lower refractive index in the form of “evanescent wave” [52]. Larger GHS leads to a stronger evanescent wave intensity as it seeps into the lower-index medium near the surface. Evanescent wave tightly influences optical loss of the microcavities, thus affecting the Q factor, and larger evanescent wave (i.e. larger GHS) corresponds to smaller Q factors [56].

For different polarization states, GHS in the incident plane can be different, which can be expressed as followed:

$$d_{\mathrm{TM}} = \frac{\lambda}{\pi} \frac{\sin(\theta)}{[n^2 \sin^2(\theta) - 1]^{1/2}} \tag{12.16}$$

$$d_{\mathrm{TE}} = \frac{d_{\mathrm{TM}}}{[(1 + n^2)\sin^2(\theta) - 1]^{1/2}} \tag{12.17}$$

Equations (12.16, 12.17) describe the relationships between the GHS versus the incident angle, which are numerically deduced from Artmann’s formulas [57]. $d_{\mathrm{TM(TE)}}$ stands for the GHS for the TM (TE) mode. θ refers to the incident angle and n is the refractive index. GHS in TE mode is larger than that in the TM mode, thus TE mode is proved to have a severe optical loss. In this way, the phenomenon that TM mode is preferred in rolled-up microcavity can be further explained.

Furthermore, the evanescent waves can be used to fabricate new optical devices and study the mode coupling phenomenon. Through the evanescent field, WGM microcavities can be coupled to become a system called a “photonic molecule,” which is an analog to the electron states in a chemical molecule [58]. As an effective strategy to manipulate and modulate resonant light, photonic molecule systems are of high interest for both fundamental research and practical applications. With a concentric microcavities structure fabricated by depositing high-index material onto both the outer and inner side of a low-index rolled-up tubular microcavity, Wang et al. [59] realized mode coupling of two WGMs in the two high-index layers. The coupled modes in the concentric structure broke up into two supermodes, and the two supermodes were tuned through the continuous variation of cavity thickness along the tube axis. Wang et al. [60] also achieved mode coupling between WGMs in a rolled-up microtube and a microsphere by manipulating a microsphere inside the tube cavity, as shown in Figure 12.5a. Due to the significant difference between the sizes of the two coupled microcavities, through regulating the excitation position on and off the contact point of the microcavities, the coupling strength can be modulated by moving the excitation position off the tangent point of the system along the axis. Except for coupling with WGM resonance in another microcavity, WGMs in a rolled-up microtube can also interact with localized surface plasmon resonances. In a photonic–plasmonic system (i.e. microcavities with metal on the

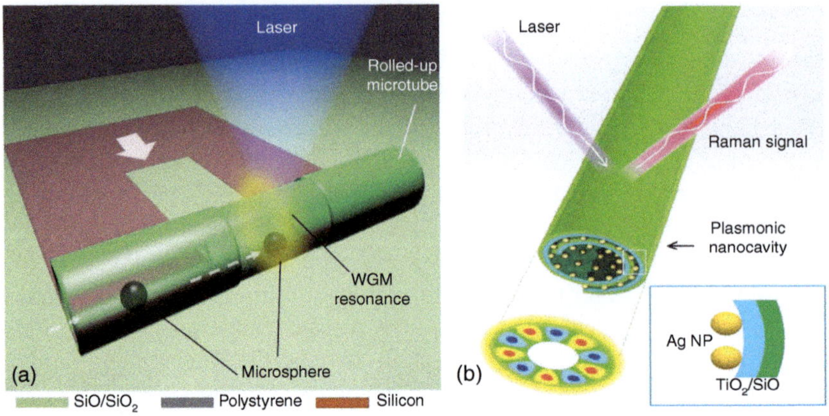

Figure 12.5 Mode coupling via the evanescent field. (a) Schematic of a coupled photonic molecule formed by trapping a microsphere in a tubular microcavity. Source: Adapted from Wang et al. [60]. Copyright 2018, Wiley-VCH. (b) Schematic configuration of a rolled-up WGM microcavity with silver particles on it for enhanced Raman signal. Source: Adapted from Zhang et al. [61]. Copyright 2015, Springer Nature.

surface), the plasmon modes at the surface of the photonic–plasmonic system can enhance light–matter interactions and improve the Q factor of the structure [62, 63]. Zhang et al. [61] fabricated rolled-up microtubes with silver nanoparticles on the surface, as shown in Figure 12.5b. With the coupling of the plasmonic and WGM resonance, the optical fields were highly concentrated and the Raman signals were greatly enhanced. Yin et al. [64] achieved *in situ* production of silver nanoparticles in a photonic–plasmonic system. By precisely changing the power and irradiation time of the laser source, the size and shape of the silver nanoparticle can be changed, thus the coupling strength can be tuned conveniently.

Besides, the evanescent waves can also be employed to characterize the resonant properties of rolled-up microtubes. To couple light into and out a WGM microcavity, an evanescent-field-based coupler is often used. For instance, researchers place a tapered optical fiber perpendicular to the axis (length) direction of the microtube. When light propagates in the fiber, some of the light can escape from the waist of the fiber, and then enter into the microtube via evanescent field [65]. At the same time, the residual light in the fiber continues to propagate forwards and some of the light can be reflected into the fiber to propagate backward, thus the desired resonances in the microtube can be obtained and further analyzed by the transmission (reflection) spectra from the fiber [65]. Since only light at certain wavelength can be coupled into the test sample, transmission spectra with discrete Lorentzian dips (peaks in reflection spectra) can be obtained by sweeping the lasing wavelength transmitting in the coupled fiber [27].

12.3.3 Structural Asymmetry Induced by Rolled-Up Technique

Fabricated by the self-rolled-up technique, rolled-up tubular microcavities have a unique geometry feature. In this section, some characteristics induced by the structural asymmetry of the rolled-up microcavities are introduced.

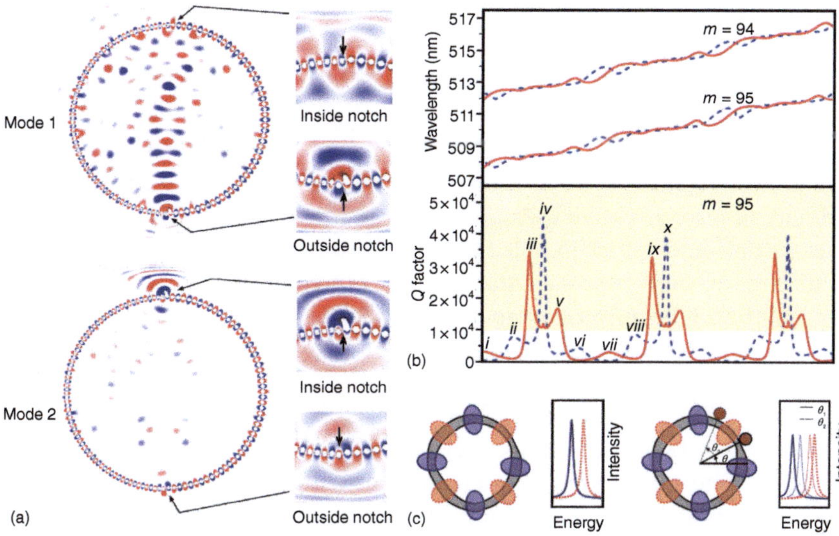

Figure 12.6 Structural asymmetry induced mode charity in the spiral-like cross section. (a) Electric field distributions of the two modes in a rolled-up microtube. The local maximum of the WGM peaks is transformed to a white level to show the electric field distribution near the boundary more clearly. Source: Adapted from Hosoda and Shigaki [66]. Copyright 2007, American Institute of Physics. (b) Evolution of resonant wavelength and Q factor of nearly degenerate mode pairs as the overlap length L increases. Source: Adapted from Fang et al. [45]. Copyright 2016, American Physical Society. (c) Schematic of detection mechanism with a rolled-up microtube. Due to different scattering levels of the CW and CCW components of the WGM in a rolled-up microcavity, mode degeneracy vanishes and mode splitting emerges, which can be perturbed by nearby nanoparticle, as illustrated in the right part of (c). Source: Adapted from Li et al. [67]. Copyright 2013, American Physical Society.

The cross section of a rolled-up microcavity features a spiral shape, as shown in Figure 12.6, notches on the inner and outer surface of the microtube ruin the rotational symmetry. Hosoda and Shigaki [66] first studied the mode splitting in the rolled-up microcavities using FDTD. They found that a pair of degenerate WGMs in completely cylindrical microcavities would break up into two modes and appear as a doublet in the simulation spectra. The energy of mode splitting is about a few meV. However, the splitting reported in the experimental results is not very pronounced due to the relatively low Q factors in rolled-up microtubes [26, 68]. Simulation results indicate that the degenerate mode breaks up into two modes with different field distributions (as shown in Figure 12.6a) and optical lengths. As for one mode, scattering caused by the outside edge is dominant toward the inside, while for the other mode, scattering caused by the outside edge is weaker than the inside [66]. In other words, the wavelength difference (i.e. mode splitting) in rolled-up microcavities is caused by the edge-induced field distribution difference.

To further go into the problem, the chirality of the rolled-up microtube has to be figured out. The structure has a deterministic structural chirality, which is determined by the rolling direction of the strained nanomembrane, and the rolling direction can be controlled by engineering the strain gradient. Here we discuss the upward-rolled microtube for example. The boundary of the structure at a fixed plane can be defined as

$$\rho(\varphi) = R\left(1 - \frac{t}{2\pi}\varphi\right) \tag{12.18}$$

where R is the largest radius according to $\varphi = 0$ and t refers to the thickness of the nanomembrane (single layer). The deterministic structural chirality of the upward-rolled microtube is CCW, while this structural chirality feature determines the chirality of resonant modes. Fang et al. [45] further discussed the modulation of Q factors and wavelengths in rolled-up microcavities with FEM simulation. They found that the resonant wavelengths of the two modes both increased with the overlap length (the length that nanomembrane overlaps along the circumstance), and Q factors presented a periodic and obvious modulation at the same time. Furthermore, the local maximum of Q factors can be achieved when the resonant wavelength splitting of the modes is minimized, which indicates an open or dissipative system [36, 41]. Besides, they also found that resonant modes with a high Q factor were supposed to have a high mode chirality, and the mode chirality was also in accordance with the chirality of the structure at the same time. Another important characteristic of modes with high Q factors in rolled-up microcavities is that the electric field mostly concentrates in the thicker part of the microtube to avoid possible scatter loss near the notches [45, 66].

The structural asymmetry and mode chirality in rolled-up microcavities are expected to have a wide prospect in both fundamental research and practical applications. Li et al. [67] achieved angular position detection of a single nanoparticle using rolled-up tubular microcavities. Since one of the degenerate WGMs is in a certain phase that minimizes the energy mode of the system under a constant structure, and the other is orthogonal to it, the pair of modes are thus determined by the rolled-up geometry, which can be served as a reference pattern for the electric field distribution near the microcavity for sensing applications [69]. Besides, the modes split depended on the relative angular position of the nanoparticle and the rolled-up structure, which is superior to symmetry structure, thus the responses of neighboring WGMs caused by a nearby disturbance can achieve angular position detection of single nanoparticles, as explained in Figure 12.6c.

However, since a rolled-up microtube is a three-dimensional structure, a thorough understanding cannot be achieved with the two-dimensional system mentioned above. The structural features in the axial direction also have an effect on the resonance. Firstly, rolling is a self-assembly process that cannot fabricate a constant spiral shape in the axial direction. Fluctuation in shape, diameter, and winding number will have influences on the WGM resonance in the rolled-up microtubes. Secondly, considering the two-dimensional model, tube length in the axial direction is supposed to be infinite, while in a three-dimensional model, this length is limited, thus the confinement in this direction also influences the resonance at the same time. Finally, since the self-rolled-up technique serves as a convenient and tunable method to fabricate microtubes, winding number can be designed to vary along the tube axis through the lithography process. With a special-designed lobe structure (i.e. rolling edges), some unique characteristics can be added to the rolled-up microtubes. Strelow et al. [41] proposed a rolled-up microtube with a parabolic lobe, which converts the structure into a bottle-like microcavity, as shown in Figure 12.7. In this structure, spatially integrated photoluminescence (PL) spectrum shows that WGMs present as a series of discrete sharp peaks, and each eigenmode peak is accompanied by a series of higher-order peaks with equal energy spacing. All modes in one group can be classified by the same azimuthal mode number m [49, 70]. Besides, in the spatially resolved spectrum, modes are observed to localize near the lobe center, and the modes in every group show increasing axial antinodes with increasing energy. This phenomenon indicates that the modes are confined to the lobe position and emerge a series of higher-order axial modes simultaneously. The physical mechanism of the axial mode eigenenergies and field distributions can be illustrated in analogy to particle waves by a quasi-Schrödinger equation with a quasipotential [41, 71]. With a parabolic lobe, the winding number of a microtube changes quadratically in a finite region in the axial direction. $k_{\text{circ}}(z)$, i.e. the wave vector in the circular cross section, is dependent on the winding number linearly. Thus, the relation between the quasipotential and the axial coordinate z can be described as

$$V^*_{\text{eff}}(z) = \frac{\hbar c}{e} k_{\text{circ}}(z) = az^2 + b \tag{12.19}$$

where the parameter b refers to the energy of propagation in the circular plane at the center cross section ($z = 0$) and a describes the curvature of the quasipotential. The solution to the Eq. (12.19) suggests that the parabolic potential leads to a series of axial modes with equidistant frequencies (eigenenergies at the same time) and confinement of the mode distribution in the axial direction. Besides, the axial modes can also be modulated by changing the coefficient a and b in $V^*_{\text{eff}}(z) = az^2 + b$. As for lobes with the same curvature (a), the energy spacings between the axial modes remain unchanged, but fewer higher harmonic modes are formed for the smaller lobe owing to its shallower quasipotential (smaller b). While with the decreasing of the lobe curvature, the energy spacings of the axial modes decrease correspondingly. In other words, by designing the shape of the parabolic lobes, the mode energies can be precisely modulated.

Except for parabolic shape, the lobes can also be designed to have a triangular or rectangular shape. Strelow et al. [41] also calculated rolled-up microcavities with triangular and rectangular lobes and studied the corresponding spatially integrated spectra. Different from the parabolic lobes, the axial energy spacings decrease with higher energies in the triangular scheme and increase in the rectangular one. To conclude, different-shaped lobes lead to corresponding quasipotential and the energy spacings between axial modes can be modulated: parabolic lobes lead to a constant spacing, triangular to a decrease, and the rectangular to an increase, respectively. With a triangular lobe structure, Wang et al. [72] realized direction light emission of axial modes with polarization selectivity, which is discussed later.

The axial modes can also be tuned through an external-assistant method after the designing process. Li et al. [73] demonstrated a dynamic axial mode turning strategy using a near-field probe in a tubular rolled-up microcavity. As shown in Figure 12.8a, the axial modes are located within the lobe and are location-dependent, thus the modes can be tuned by changing the positions of the probes. Due to the perturbation theory, a redshift of the resonance peaks can be observed as the probe

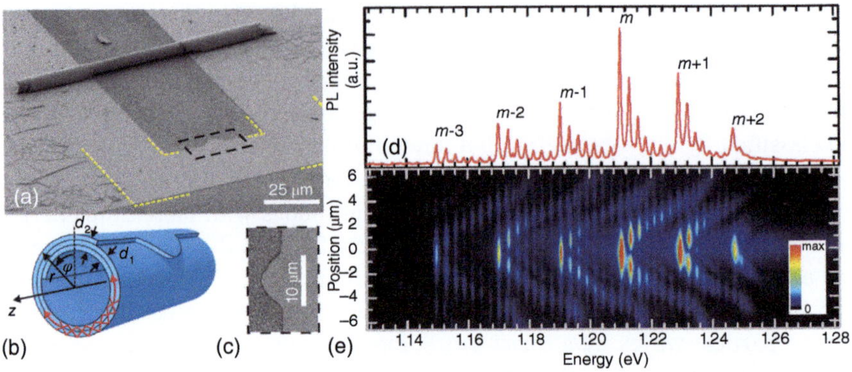

Figure 12.7 Axial confinement of WGMs in rolled-up microtubes. (a) Scanning electron micrography (SEM) image of a rolled-up microtube. (b) Schematic of a rolled-up tubular microcavity with parabolic lobe. (c) Enlarged SEM image of the parabolic lobe in (a). (d) Spatially integrated photoluminescence (PL) spectrum of the tubular microcavity in (a). (e) Spatially and energetically resolved PL image of (a). Source: Strelow et al. [41]. Adapted with permission from American Physical Society.

slightly touches the surface of the microcavity, and the level of the redshift is related to the spatial overlap between the probe and evanescent field, as explained in Figure 12.8b,c [75]. The sensitivity of different axial modes is different owing to the corresponding resonant field distributions. Among all kinds of WGMs, the fundamental mode shifts the most since it has the largest overlap when the probe is in the middle of the structure. Wang et al. [74] proposed an external-strain-based method to modulate the axial modes and to modify the tube shape, as shown in Figure 12.8d. By using polydimethylsiloxane (PDMS) as the substrate material, a controllable mechanical stretching can be added to the microtube. While increasing the PDMS strain along the axial direction, mode energy and axial mode spacing decrease correspondingly, which can be attributed to the deformation of the rolled-up microtubes. Besides, the strain-caused deformation is worth noticing, which is confirmed to modify the cross section of the structure into a more "round" geometry and eliminate interlayer voids at the same time. This shape modification can be served as an effective way for enhancing Q factors [54, 76].

Besides the lobes, some extra designs can be added along the axis of the rolled-up microcavities. Kietzmann et al. [77] incorporated striped luminescent organic (naphthyl end-capped bithiophene molecules) films into inorganic (AlInP) microtubes. These organic stripes acted as light-emitting sources, at the same time, they also induced the extra light confinement along the axial direction. The combination of organic layers also exhibited strong light–matter interaction and high fluorescence yield. Tian et al. [78] realized single WGM emission with a mesostructured diamond microcavity, in which the rolled-up microtube was fabricated with a designed unique geometry along the axial direction. The microtube with subwavelength-thin wall was further designed along the axis to achieve a single-mode selection, as explained in Figure 12.9: a wavelength-scale axial barrier was added to restrict the optical field with the fundamental axial mode and periodic

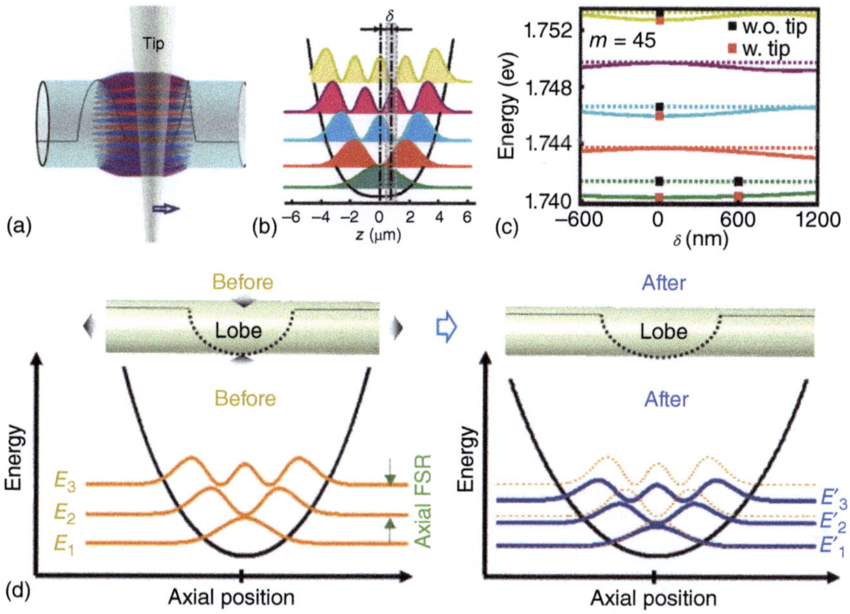

Figure 12.8 Axial modes tuning by external-assistant methods. (a) Schematic of the dynamic tuning with tip probing. (b) The first five axial modes as a function of the axial position and the predicted alternating overlap of the mode energies with the probe. (c) Predicted mode shift with different axial positions. Dashed (solid) lines describe the predicted (calculated) energy with (without) the probe, respectively. Source: (a–c) Adapted from Li et al. [73]. Copyright 2012, American Institute of Physics. (d) Schematic of axial mode distribution changes before and after modification with PDMS substrate. Dashed (solid) lines describe the axial potential well before and after modification, predicted (calculated) energy with (without) the probe, respectively. Source: (d) Adapted from Wang et al. [74]. Copyright 2018, American Chemical Society.

hole arrays were designed on both sides of the cavity with mirror symmetry to select the single azimuthal mode. The hole array structure induces distributed feedback effect in the microcavity, which can select the WGM that satisfies the phase-matching condition at the Bragg gap. As shown in Figure 12.9b, the resonant wavelength of selected single WGM can be tuned by changing the geometry of the periodic structure. As we can see, axial direction design gives an extra degree of freedom to modulate the WGMs in the rolled-up microtubes, which is a remarkable advantage of tubular geometry.

Despite the designs mentioned above, axial confinement and modulation can also be realized by asymmetrically shaping the rolled-up microcavities intentionally. Bolaños Quiñones et al. [51] fabricated asymmetric cone-like microtubes by unevenly rolling up strained nanomembranes, as shown in Figure 12.10a. In usual cases with uniform photoresist thickness, microtubes of symmetric geometry are supposed to be obtained, in which the winding number of the nanomembranes varies from the middle to the end symmetrically [80]. However, with a sloped photoresist layer, the circular-shaped nanomembrane rolls up unevenly, resulting in a

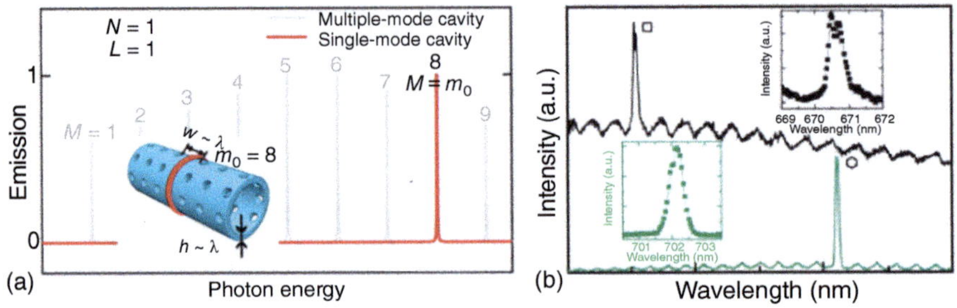

Figure 12.9 Single-mode WGM emission in a mesostructured diamond microcavity. (a) Emission spectra for the multimode (gray line) and single-mode (red line) microcavities. Inset shows schematic of a single-mode mesostructured microtube. (b) PL spectra of WGM emission at different wavelengths. Source: Adapted from Tian et al. [78]. Copyright 2018, American Chemical Society.

cone-like asymmetric microtube with winding numbers and radius varying along the tube axis. In the axial direction, WGMs continuously blueshift from the top end to the bottom end, as shown in Figure 12.10b. With a photonic quasi-Schrödinger equation, enhanced axial confinement is confirmed with a larger spacing of axial potential. Besides, the polarization state of the WGM in such an asymmetric microtube also differs from that of a symmetric microtube. Ma et al. [79] studied the spin–orbit coupling of the modes in rolled-up asymmetric microtubes, as shown in Figure 12.10c. In this structure, the right and left circular components acquired a geometric phase with opposite signs: $a(0) = a_{+}e^{-i\varphi} + a_{-}e^{-i\varphi}$, where φ refers to the geometric phase, $|a_{+}|^2$ and $|a_{-}|^2$ are redistributed amplitudes for each component owing to the mode conversion. Thus, through the mode conversion, polarization changed from linear to elliptical. The orientation of polarization major axis was determined by the geometry phase angle φ, which can be obtained from corresponding experimental results. The polarization tuning phenomenon confirms the rolled-up asymmetry microtube as an ideal system to realize spin–orbit coupling, which serves as a significant essential physical process of a non-Abelian evolution [79].

12.4 Materials and Techniques in Rolled-Up Tubular Optical Microcavities

Over the past few decades, materials of rolled-up tubular optical microcavities have been developed from a limited range to a variety of materials, including oxide, semiconductor, 2D materials, and others. Besides, owing to discrepancy in the chemical properties of the desired materials, different techniques are selected to produce the nanomembranes, such as molecular-beam epitaxy (MBE), electron-beam evaporation, and spin coating. In this section, typical materials and techniques used in fabricating rolled-up optical microcavities are summarized. For the sake of integrality, application of the microcavities is briefly introduced in Section 12.4, while more

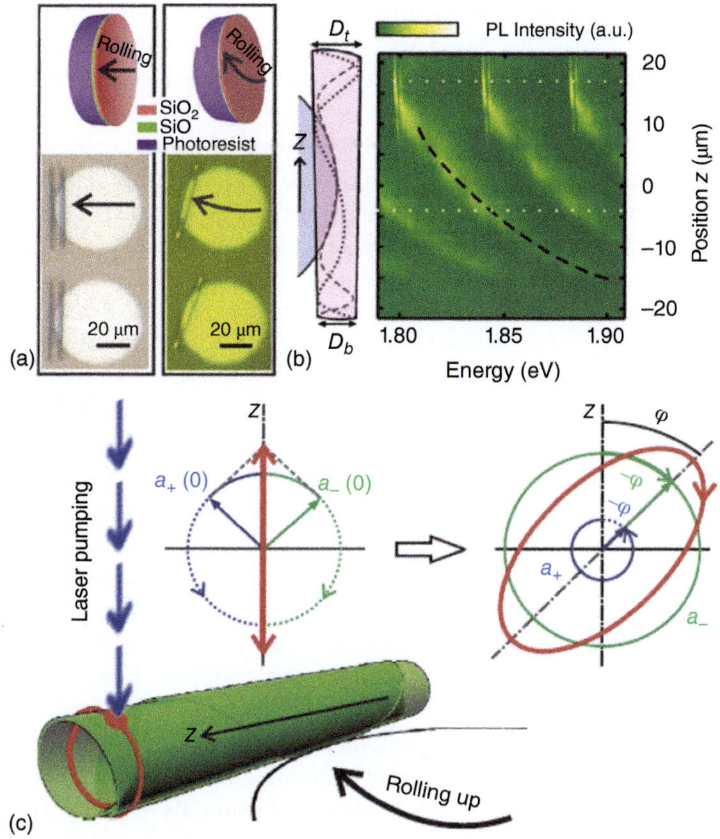

Figure 12.10 Asymmetrically shaping procedure of the rolled-up microcavities. (a) Schematic and optical images of rolled-up microtubes with symmetric (left) and asymmetric (right) geometry. The asymmetric microtubes were obtained by designing uniformly thick and sloped photoresist patterns. (b) PL intensity mapping in the axial direction of an asymmetric microtube. The schematic on the left shows the corresponding axial positions. Source: (a, b) Adapted from Quiñones et al. [51]. Copyright 2012, Optical Society of America. (c) In a designed asymmetric microtube, the right and left components of light redistribute through coupling and change the polarization state from linear to elliptical. Source: Adapted from Ma et al. [79]. Copyright 2012, Springer Nature.

details about the concerning application of rolled-up tubular microcavities are presented in Section 12.5.

12.4.1 Semiconductor

Rolled-up semiconductor microcavities exhibit numerous characteristics, including tunable size, ultrahigh Q-factors, and directional emission. Optical microcavities rolled up from semiconductor nanomembrane have gained considerable interest for their outstanding light confining ability. As for semiconductor materials, MBE technique is often adopted to fabricate rolled-up microtubes. MBE technique allows the formation of monocrystalline nanomembranes in high quality. In this case, the strain gradient and corresponding rolling process can be precisely controlled by engineering the component of each layer. Besides, extra designs such as lobes and U-shape to add axial confinement and prevent radiation loss can be fabricated through the etching process.

In 2009, Li and Mi [81] fabricated InGaAs/GaAs quantum dot microtubes by selectively etching AlAs as sacrificial layer and releasing the coherently strained epitaxial InGaAs/GaAs layer from GaAs substrate. Optical microcavities with high Q factor can be pumped optically at room temperature with an ultralow threshold of ~4.0 μW and a minimum intrinsic linewidth of ~0.2–0.3 nm. Besides, InGaAsP can also be utilized in rolled-up tubes. Coherently strained epitaxial bilayers InGaAsP that consist of ~36 nm $In_{0.81}Ga_{0.19}As_{0.41}P_{0.59}$ and ~15 nm $In_{0.81}Ga_{0.19}As_{0.41}P_{0.59}$ were grown on InP substrate and self-organized InAs quantum dots/dashes were incorporated in $In_{0.81}Ga_{0.19}As_{0.41}P_{0.59}$ layer. Due to the minimization of strain, coherently strained bilayer rolled into tubular microcavities with subwavelength wall thickness when selectively released from InP substrate. Lasing at room temperature was achieved under continuous-wave optical pumping from roll-up InAs/InGaAsP quantum decorated microtubes with a measured threshold ~6 μW [82]. Despite III–V group materials, group IV semiconductors such as GeSn can also be fabricated into microtubes with rolling process. Wu et al. achieved GeSn rolled-up microtubes with a diameter of 7.3 μm through MBE, and the Q factor was characterized to be around 800 [83].

Usually, to achieve coupling of light into optical modes, quantum dots or quantum wells can be combined in rolled-up microcavities as gain materials inside the strained layers, which give rise to internal emitters inside the microcavities. However, epitaxial growth severely limits the variety of materials and shapes available for rolled-up microcavities. To solve this problem, luminescent materials were deposited on the strained layers before releasing from substrate. A hybrid inorganic/organic optical microcavity was fabricated by combining AlInP and stripes naphthyl end-capped bithiophene molecules films. As organic material acted as visible-light emitters, an efficient rolled-up AlInP semiconductor microtube resonator was formed. The hybrid inorganic/organic optical microcavity system provided opportunities for novel research on light–molecule interaction and application [77].

12.4.2 2D Materials

Graphene has attracted great interest for its superior optical and electronic properties. Graphene has great potential in optical and optoelectronic devices by integrating with 3D tubular geometry [35]. In 2019, Yin et al. [84] reported a rolled-up graphene-activated optoplasmonic microcavity, which was employed for *in situ* monitoring the photodegradation dynamics of organic dye molecules (R6G) in real time. The degradation of R6G molecules was triggered by laser irradiation and monitored by resonance mode peak position shift, as shown in Figure 12.11. Graphene layer could significantly strengthen the electric field at the surface of Au nanomembrane, which greatly improved the surface detection sensitivity.

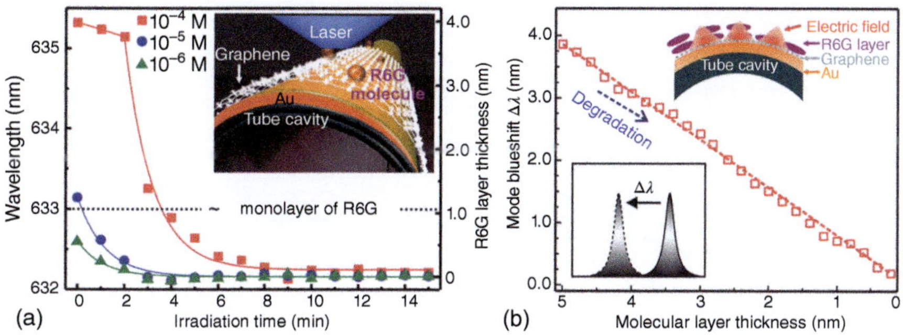

Figure 12.11 Optical characterization of graphene-activated optoplasmonic cavity. (a) Measured resonant mode shift upon different R6G concentrations. Measurement data (scattered symbols) are fitted by exponential decay curves (solid lines). (b) Calculated mode shift in response to the change of R6G molecule layer thickness upon degradation on the graphene-activated optoplasmonic cavity. Source: Adapted from Yin et al. [84]. Copyright 2019, American Chemical Society.

12.4.3 Oxide

In 2008, Mei et al. [6] applied photoresist as sacrificial layer in the rolling process, which permits a broad range of materials and their combination as functional layer. The rolled-up microcavities were fabricated by selectively dissolving the underlying sacrificial photoresist layer with acetone. It is worth noticing that in this photoresist-assisted rolling process, the patterning process is directly completed by photolithography of the sacrificial layer. Compared with traditional techniques used in rolled-up microcavities, the methodology of depositing and releasing pre-stressed nanomembranes on polymers provided an economical and feasible option to fabricate tubular microcavities with designed geometries from a variety of materials, whilst avoiding the use of corrosive acid and expensive epitaxial membranes.

Using the photoresist-assisted rolling process for oxide materials, various applications in photonics have been investigated. Huang and coworkers [85] reported ring resonators that could be operated as optofluidic components in 2010, as shown in Figure 12.12. By selectively etching photoresist as sacrificial layer and releasing prestressed SiO/SiO_2 bilayer nanomembranes, tubular optical microresonators from rolled-up nanomembranes with subwavelength wall thickness were constructed. The SiO/SiO_2 bilayer nanomembrane thickness is less than 40 nm with a thickness ratio of ~1 : 4, and the corresponding microtube diameter is less than 10 μm. Atomic layer deposition (ALD) is employed to increase tube wall thickness and refractive index. 30 nm HfO_2 layer was coated by ALD and significantly increased mode intensities, detection limit, and optical sensitivities. The combination of rolled-up WGM microcavities and ALD posttreatment demonstrated a method to coordinate sensitivity and detection limitations for sensing application [52].

Harazim et al. [86] reported the fabrication of high *Q*-factor tubular rolled-up optofluidic ring resonators (RU-OFRRs) in 2012, and the fabrication process is

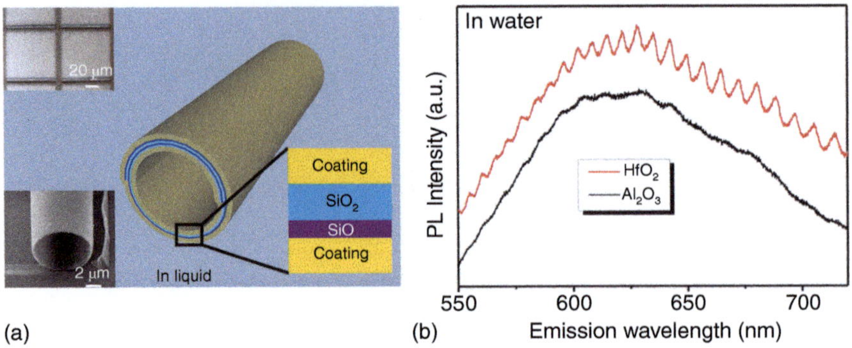

Figure 12.12 Schematic diagram and PL spectra of a rolled-up microcavity. (a) Three-dimensional schematic diagram of a rolled-up nanomembrane in liquid. Inset on the top is SEM image of the array of microcavities. Inset on the bottom is magnified microcavity. (b) PL spectra of an optical microtube cavity with a diameter of ~9 μm in air, deionized water, ethanol, and water/ethanol mixture (1 : 1 in volume). Source: Huang et al. [52]. Adapted with permission from American Chemical Society.

illustrated in Figure 12.13. Photoresist pattern of 2.4 μm thickness was spinned on Si substrate as sacrificial layer. Two-layer SiO_2 nanomembranes were deposited on sacrificial layer to induce strained SiO/SiO_2 nanomembrane. By selectively resolving sacrificial layer with acetone, SiO/SiO_2 nanomembrane rolled up into three-dimensional microtubes. The fabricated microtubes were then picked up from the mother substrate and transferred to designed microchips for subsequent integration. The transferred microtubes were fixed to designed position on microchips by atomic layer depositing 10 nm Al_2O_3. Optical signals were characterized by PL spectroscopy, detecting a Q factor ~2900. The RU-OFRRs' minimum detection limit is 3.4×10^{-4} per refractive index unit (RIU) and a maximum sensitivity of 880 nm/RIU.

Ultrathin wall and low refractive indices of rolled-up oxide nanomembranes are an inherent restriction for good light confinement in microcavity walls, which significantly reduces Q factors [87]. Owing to the great compatibility of the photoresist-assisted rolling process, oxides with high refractive indices can be introduced to the structure. In 2012, Wang et al. [49] introduced materials with high refractive indices for rolled-up microcavities to improve light confinement in the wall of microcavities. High-index contrast between microcavity wall and the surrounding medium could reduce light loss in the wall of tubular cavity and enhance Q factor, as shown in Figure 12.14. They fabricated ~30 nm Y_2O_3/ZrO_2 bilayer high-index-contrast nanomembranes on polymer sacrificial layers and rolled up into ultrathin wall microcavities with Q factor of ~1600. This work demonstrated a cheap and practicable way to fabricate three-dimensional optical confinement oxide WGM microtubular cavities with high Q factor [49].

Meanwhile, oxide rolled-up microtubes can be further fabricated with functionalized layers for sensing applications. Ma et al. [88] rolled strained Si/SiO_2 bilayer nanomembranes into asymmetric microtube cavities for molecular detection. Different from previous work, the microtube resonators were fabricated by releasing a Si/SiO_2 bilayer nanomembranes from a sloped photoresist pattern and a 30 nm HfO_2

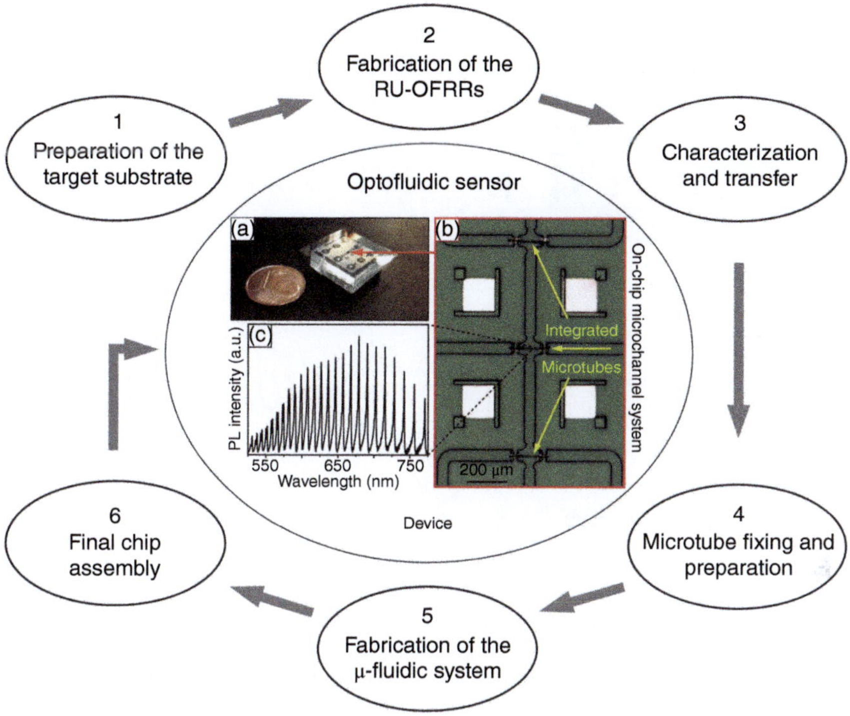

Figure 12.13 Fabrication process of high-quality rolled-up optofluidic resonators. Source: Harazim et al. [86]. Adapted with permission from Royal Society of Chemistry.

layer was deposited on the microtube surface by using ALD. The high electronegativity difference of Hf-O leads to high affinity for polar molecule adsorption, as shown in Figure 12.15. In 2019, Yin et al. [89] reported an *in situ* detection of the dynamic evolution of water layer on an amorphous oxide tubular microcavity's surface probed by optical resonances.

Besides, nanoparticles can also be employed in the microtubes for further research. Surface-enhanced Raman scattering (SERS) can be highly strengthened with cooperation of WGMs based microcavities. Zhang et al. [61] realized WGM rolled-up TiO_2/SiO tubular microcavities decorated with silver nanoparticles. The interaction between WGM-based tubular microcavities and surface plasmon contributed to an extra enhancement at the order of 10^5 compared with flat SERS substrates, as shown in Figure 12.16. Madani et al. [90] deposited luminescent cadmium phosphide (Cd_6P_7) nanoparticles (NPs) on two-dimensional pre-strained TiO_2 nanomembranes. By rolling up the strained TiO_2 nanomembranes into three-dimensional tubular microcavity, the NPs are embedded in tube wall's windings and serve as light source for WGM tubular microcavities' light source under laser excitation. A proper choice of the emitting NPs allows tuning of the resonant frequency from the visible to the IR spectral range. They experimentally demonstrated a monolithic integration of two vertically rolled-up microtube

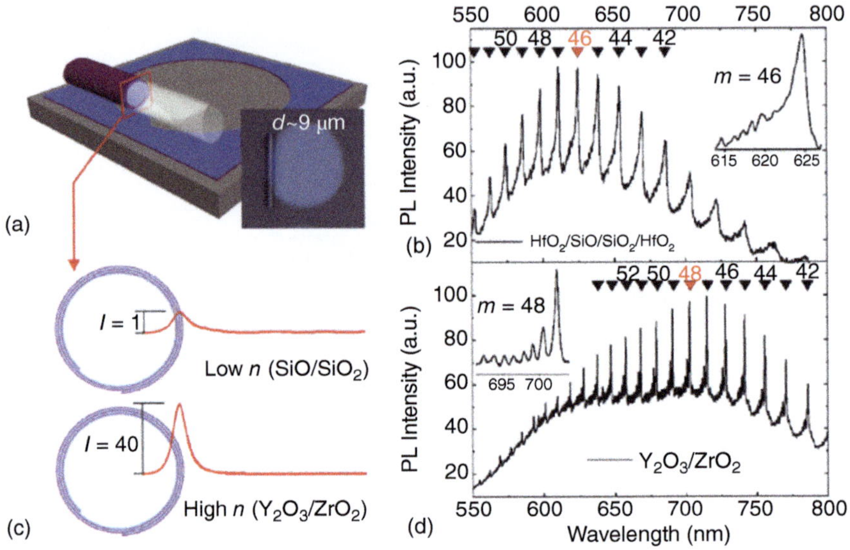

Figure 12.14 Schematic diagram and optical characterizations of a rolled-up microcavity. (a) Schematic diagram illustrating the fabrication process of a rolled-up Y_2O_3/ZrO_2 tubular microcavity. (b) The upper part shows the radial field intensity distributions of the TE modes for the tubes with a low refractive index SiO/SiO_2 and the lower part with a high refractive index Y_2O_3/ZrO_2. (c) The PL spectrum from the middle of a microtubular cavity rolled from a circular SiO/SiO_2 nanomembrane coated with 30 nm HfO_2. (d) The PL spectrum from the middle of a middle of a microtubular cavity rolled from a circular Y_2O_3/ZrO_2 nanomembrane. Source: Adapted from Wang et al. [49]. Copyright 2012, The Optical Society.

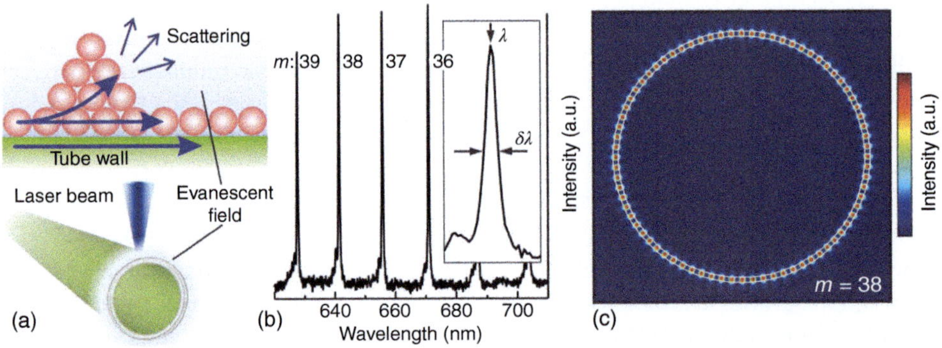

Figure 12.15 Schematic diagram and optical characterizations of a rolled-up microcavity with a layer of water. (a) The bottom panel shows schematically a microtube cavity being excited by a laser beam. The top panel shows a schematic of resonant light propagating in the tube wall and the surface molecular layer, being scattered by molecular clusters. (b) Measured optical resonant mode spectrum labeled by mode numbers $m = 36–39$. The inset shows an individual mode where the mode shift characterizes the thickness of the molecular layer, and the variation of the mode width characterizes the light scattering by the surface clusters. (c) Electric field profile of mode $m = 38$ taken to quantify the resonant mode shift induced by adsorbed molecular layers. Source: Adapted from Yin et al. [89]. Copyright 2019, American Association for the Advancement of Science.

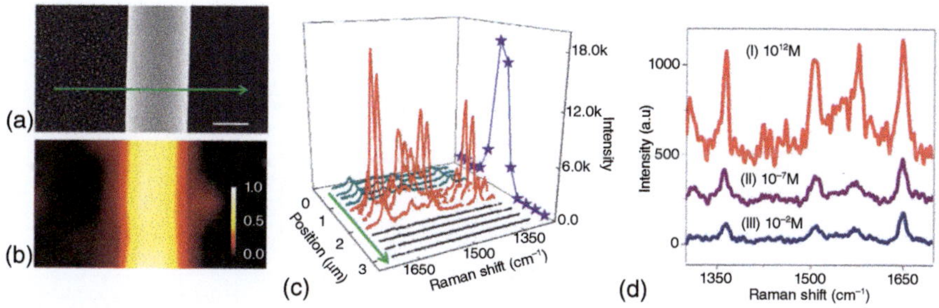

Figure 12.16 Raman enhancement in a whispering-gallery plasmon nanocavity. SEM image (a) and Raman intensity mapping (b) of the R6G signal on a rolled-up plasmon nanocavity. (c) Raman spectra of the line scan along the green arrow marked in (a). The stars refer to the intensity of 1650 cm^{-1} band extracted from the spectra. (d) A comparison of SERS detection limits on Ag NP-decorated plasmon nanocavities (I), undecorated nanotubes (II), and flat silver-NP-decorated nanomembrane (III). Source: Zhang et al. [61]. Adapted with permission from Springer Nature.

resonators on polymer-based 1×5 multimode interference waveguides to achieve three-dimensional multichannel coupling [91].

12.5 Applications

With the advantages of tunable structure parameters and unique shape features, rolled-up tubular microcavities have a promising future in a broad range of applications. Generally, the applications based on rolled-up tubular microcavities are related to the evanescent wave and the specially designed geometry. With the cylindrical channel, properties of fluids can be introduced, such as carrying nanoparticles, cells, and molecules. In this section, several applications of rolled-up tubular microcavities are introduced, including sensing [92], lasing [78, 93, 94], and microfluidics [52, 95, 96].

12.5.1 Sensing

The typical thin tube wall of the rolled-up microcavity ensures a strong interaction between the evanescent field and the analytes, and the tubular geometry is of great advantage for liquids used as carriers of analytes [8]. Therefore, rolled-up tubular microcavities can be utilized as effective sensors for refractive index, single particles, and real-time detection of the dynamic process of surface molecular [52, 76, 88].

As for fluidic sensing, microtubes are supposed to be immersed into the test solution or coupled with a tapered glass capillary full of liquid solutions during the sensing process [96–98]. Bernardi et al. [99] first used Si/SiO_x microtube for refractive index sensing as a refractometer. Sugar solution was inserted into the tubular microcavity, which led to a change in refractive index and a shift of WGMs correspondingly. The sensitivity of this refractometer was measured to be ~62 nm/RIU. Song et al. [97] further used SiO_x/SiN_x rolled-up microcavities for

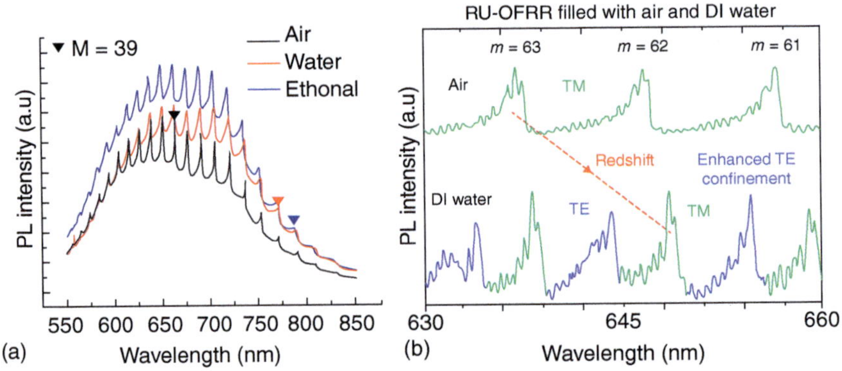

Figure 12.17 Fluidic sensing. (a) PL spectra of rolled-up tubular microcavities in different solvents. Source: Adapted from Song et al. [97]. Copyright 2018, IOP Publishing. (b) PL spectra of the rolled-up cavity ring resonator with air (black line) and DI water (green line) as the sensing sample. Source: Adapted from Harazim et al. [86]. Copyright 2012, Royal Society of Chemistry.

solvent sensing. The microtubes were fabricated by rolling-up plasma-enhanced chemical vapor deposition-synthesized SiO_x/SiN_x nanomembranes, which were confirmed to have better optical properties and sensing abilities at the same time. A sensitivity of 510 nm/RIU and a detection limitation of 10^{-4} RIU was obtained in this system, as shown in Figure 12.17a. However, sophisticated setup procedures are required in the liquid sensing processes mentioned above, which may be less feasible for further integration in practical applications [8]. Harazim et al. [86] reported high Q rolled-up optofluidic microcavities, the microcavities were well integrated into a lab-on-chip platform. The on-chip optofluidic system could detect changes in the refractive index of the liquids in the channels, and a maximum sensitivity of 880 nm/RIU and a minimum detection limit of 3.4×10^{-4} RIU was obtained, as shown in Figure 12.17b. This method provides convenience for practical applications based on integrating rolled-up tubular microcavities into lab-on-chip systems, which largely simplifies the sample delivery process and minimizes the required sample volume [8].

Besides, sensors based on rolled-up tubular microcavities can also be used to monitor the kinetics of physical processes and chemical reactions. Owing to the typical subwavelength wall of the rolled-up microcavity, the evanescent field of WGMs is confirmed to be sensitive to tiny changes and reactions near the inner and outer surface of the tube walls. Miao et al. [98] achieved the evaporation kinetics detection with an optofluidic rolled-up microcavity system. In the optofluidic microtube system, colloidal quantum dots of Cd_3P_2 were coated on the SiO/SiO_2 microtubes. The solvent evaporation can be monitored through the behavior of the WGM peaks, including the wavelengths and Q factors. In this case, WGMs peaks shifted following a logarithmic function and the Q factors enhanced linearly with the evaporation time, as shown in Figure 12.18a,b. This phenomenon can be explained considering the concentration change during the evaporation process, which leads to changes in refractive index correspondingly [100]. The optofluidic system offers a strategy

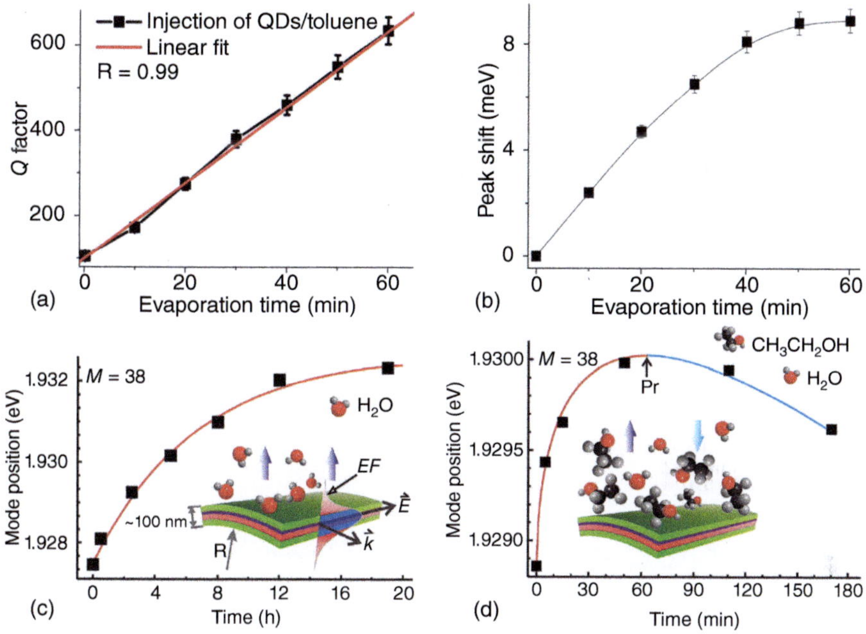

Figure 12.18 Kinetics process monitoring. Changes of *Q*-factor (a) and mode shift (b) of the WGMs with increasing evaporation time of QDs/toluene solution. Source: Adapted from Miao et al. [98]. Copyright 2015, Wiley-VCH. (c) H_2O molecule desorption monitoring by analyzing the WGM wavelength shift during the reaction process. The data is fitted by the linear-driving-force model (red line). The inset is an illustration of the desorption process. (d) WGM mode position during the bidirectional dynamic process of ethanol molecule desorption and H_2O molecule adsorption. The inset shows a schematic of the desorption/adsorption processes. Source: Adapted from Ma et al. [88]. Copyright 2013, Wiley-VCH.

to calculate the kinetics of evaporation process with reproducible WGMs. Ma et al. [88] realized dynamic process detection with rolled-up $SiO/SiO_2/HfO_2$ microtubes, in this system, dynamic molecular processes of H_2O and C_2H_5OH could be detected efficiently on the surface of the microtube wall. With a HfO_2 coating layer, the structural stability and surface polarity can be enhanced, thus providing convenience for polar molecular adsorption [101]. According to the linear-driving-force model, the H_2O molecules in air can be adsorbed onto the HfO_2 surface and build a natural wetting layer, and the wetting layer maintains a stable surface after reaching a dynamic equilibrium [102]. Therefore, when the microcavity is transferred into a dry environment, the equilibrium will be broken and the H_2O molecules turn to desorb from the surface. This dynamic process can be monitored by the continued blueshift of the WGM peaks in the spectrum, as shown in Figure 12.18c,d. It is confirmed that the dynamic detection of C_2H_5OH desorption can also be achieved using this mechanism [88]. This sensitive sensing ability of system can be used to probe molecular level changes on the surface, which can be served as a convenient system to detect surface reaction in a label-free mode.

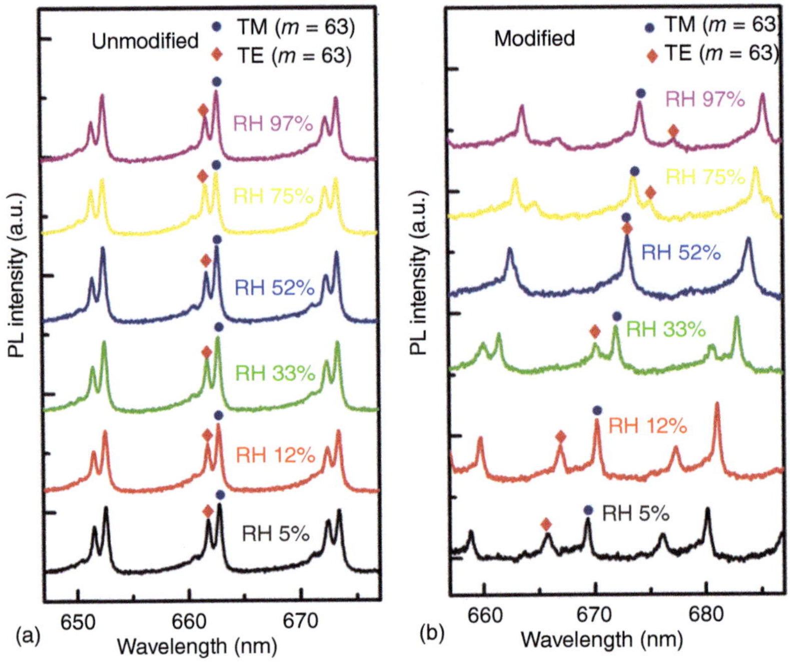

Figure 12.19 WGM shift before (a) and after (b) polymer layer modification. Circle (diamond) symbols refer to TM (TE) mode WGM peaks, respectively. Source: Adapted from Zhang et al. [103]. Copyright 2014, The Royal Society of Chemistry.

Zhang et al. [103] also realized humidity sensing with a rolled-up polymer/oxide/polymer microcavity. Through coating poly(acrylic acid)/poly (ethylenimine) (PAA/PEI) polymers layer, the humidity detection sensitivity was greatly enhanced. Owing to the interaction between PAA and PEI through electrostatic and van der Waals force, the H_2O-sensitive functional layer turns out to be uniform and thickness-changeable [104]. With a gradually increased relative humidity (RH), H_2O molecules get diffused into the polymer layer, leading to the expansion of the polymer layer and corresponding redshift of WGMs, as shown in Figure 12.19. Sensitivity of ~130 pm per RH% in sandwiched polymer/oxide/polymer microcavities is ~10 times larger than WGM-based tubular microcavities.

As mentioned above, plasmonic nanostructures can be coupled to the rolled-up microtubes with the evanescent field, which can lead to high Q factor, small mode volume, and enhanced electric field intensity [105]. These unique advantages are crucial for sensing activities. Coupling with metal nanostructures, hybrid photon–plasmon modes with enhanced light–matter interaction can also be used for angular sensing. Yin et al. [106] designed a rolled-up tubular microcavity with graded silver nanogaps coating on it. As shown in Figure 12.20, the coupling strength and mode polarization were dependent on the azimuthal angle, and the polarization state with higher intensity gradually transferred from TE mode (at area with metal cap) to TM mode (at the sides of the microcavity). Besides, the intensity

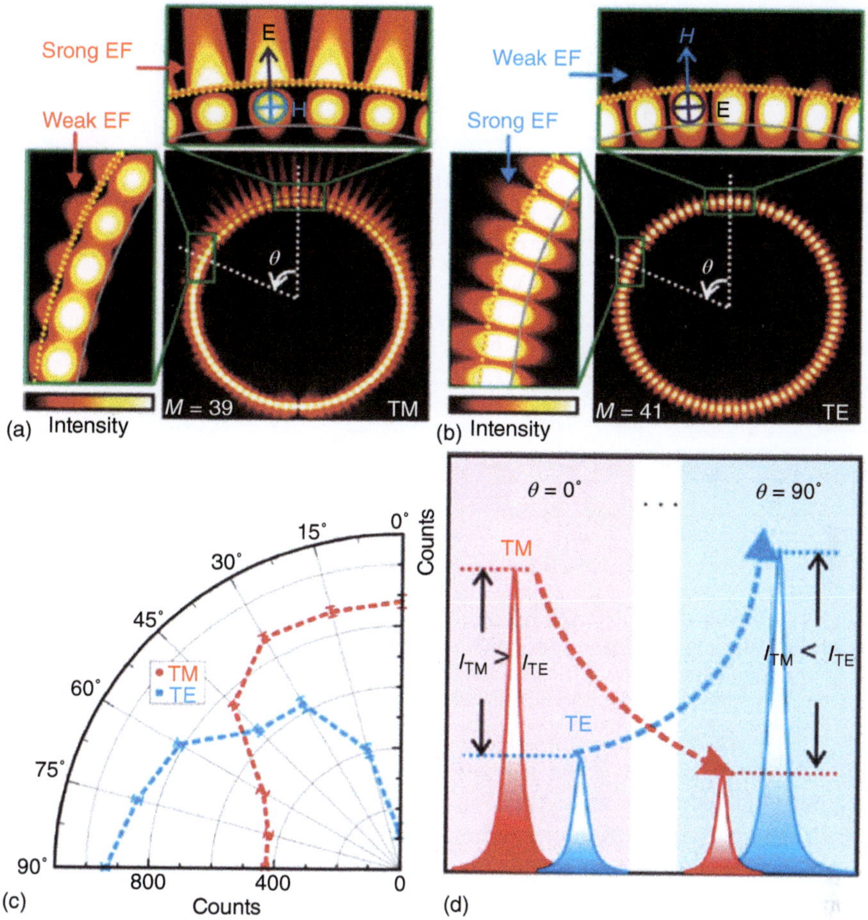

Figure 12.20 Photodegradation dynamics monitoring with graphene-activated optoplasmonic microcavities. (a, b) Optical field distributions of the hybrid WGM for the hybridized TM and TE modes, respectively. (c) Intensity variations between TM mode and TE mode are plotted in the polar diagram as a function of the azimuthal angle θ. (d) Illustration of the far-field mode intensity changes with different azimuthal angles. According to different definitions, the polarization of TM mode and TE mode in this figure are diametrically opposite from the definition in Section 12.2. Source: (a, b, c, d) Adapted from Yin et al. [106]. Copyright 2017, American Chemical Society.

difference of the hybridized TM and TE modes exhibits enhanced sensitivity to surface perturbations at the nanoscale, which provides a novel method for angular sensing.

Furthermore, rolled-up microtubes can also be used as biosensors. Smith et al. [76] captured embryonic fibroblast mouse cells successfully with rolled-up microtubes, and the condition of the cells was monitored by analyzing peak sharpening and mode shifts of WGMs in the microtubes. As mentioned earlier, an intrinsic void exists in the rolled-up structure due to the loose rolling process. When a cell enters the microtube system, the nanogap (void) turns to decrease owing to the cell being

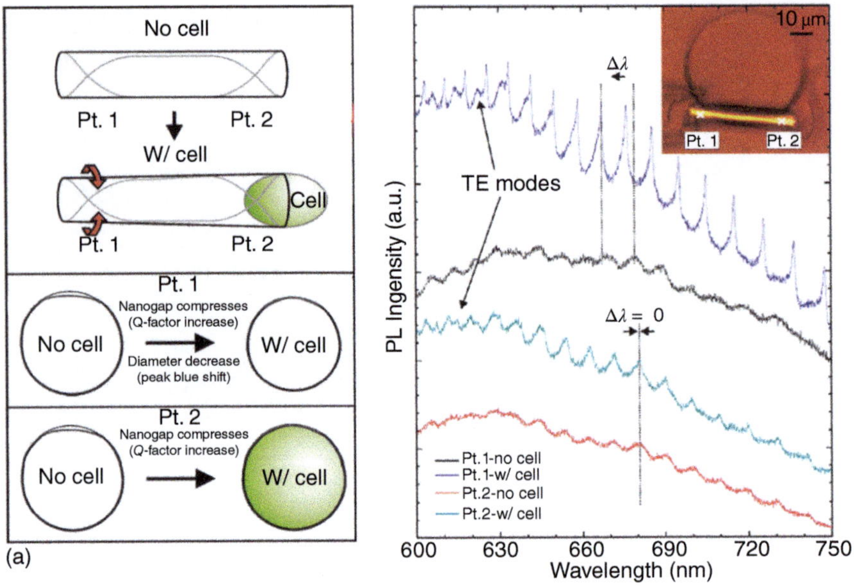

Figure 12.21 Individual mouse cells detection. (a) Schematic diagrams of the process that can be further analyzed for sensing an individual mouse cell. (b) PL spectra of the two points on the tube with (without) the cell, respectively. Source: Smith et al. [76]. Adapted with permission from American Chemical Society.

squeezed inside of the tube. The closing of the nanogap leads to a lower optical loss and higher Q factor at the same time, as demonstrated in Figure 12.21a. Besides, Figure 12.21b indicates the tighter winding structure corresponds to a smaller diameter, which causes the blueshift of the resonant modes. The detection of individual cells paves the way for further investigation of single animal cells.

12.5.2 Lasing

WGM microcavities have been widely used in lasing applications due to the advantages of high Q factor and small mode volume (V). Large Q/V ratio correspondingly leads to high energy density, which is confirmed to provide convenience for enhanced light amplification and strong light–matter interaction between the gain media and the microcavity interior, offering the possibility of obtaining lasers with low threshold and narrow linewidths [107]. As a kind of unique WGM microcavity, the rolled-up tubular microcavities exhibit several distinct characteristics, including direction emission and controlled polarization, which are of great benefit in practical applications [82].

In 2009, Strelow et al. [93] presented lasing behavior in rolled-up microtubes, where GaAs quantum wells were used as the lasing gain material. Through time-resolved studies, fast turn-on times were observed. Dastjerdi et al. [82] investigated the lasing behavior of rolled-up InGaAsP microtubes, wherein self-organized InAs quantum dots or dashes were employed as the gain material. In the rolled-up

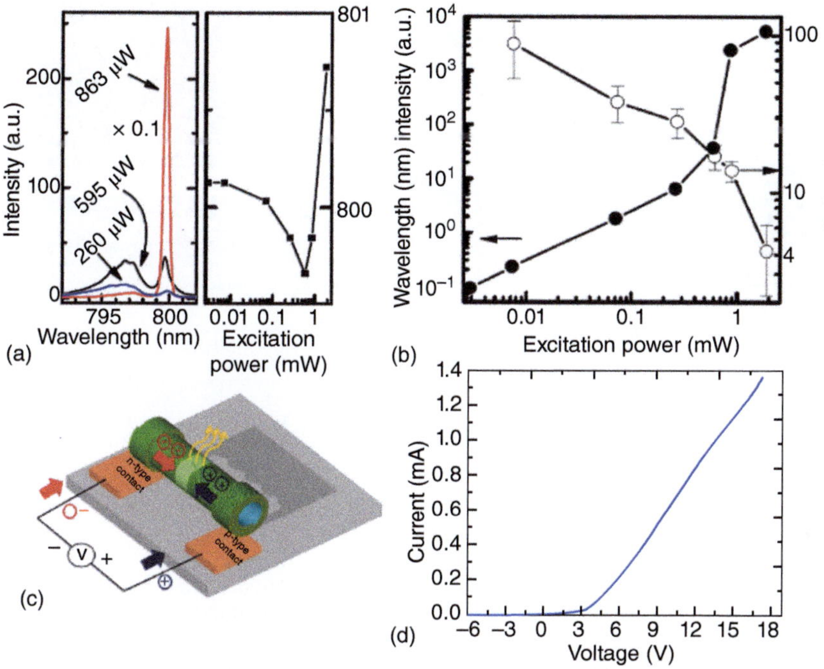

Figure 12.22 Lasing behavior of rolled-up microcavities. (a) PL spectra, wavelength of the lasing mode in the time-integrated spectra, and (b) relationship between the lasing intensity (left), delay time (right), and excitation power of the microlaser. Source: Adapted from Strelow et al. [93]. Copyright 2009, American Institute of Physics. (c) Schematic of the electrically pumping rolled-up tubular microlaser. (d) Current–voltage relationship of the microlaser at room temperature. Source: Adapted from Dastjerdi et al. [20]. Copyright 2013, IOP Publishing.

InAs/InGaAsP structure, lasing at room temperature under continuous optical pumping was observed, and the threshold was measured to be ~6 μW. Due to the comparatively large surface-to-volume ratio of tubular geometry, the threshold power of rolled-up tubular microlasers may be higher than that of microdisk lasers, and therefore, replacing optical pumping with electrical pumping can be more efficient and convenient. Strelow et al. [93] studied the lasing behaviors of electrical-pumped rolled-up microtube lasers. Figure 12.22a,b shows the excitation power dependent measurements, with increasing excitation power, higher lasing intensity and shortening lifetime can be obtained. Dastjerdi et al. [20] further demonstrated rolled-up InAs/InGaAsP quantum well heterostructure with electrical injection. The device showed strong emission in the communication wavelength (~1.5 μm), and a lasing threshold of ~1.05 mA was measured at 80 K, as shown in Figure 12.22c,d. The designed microlaser paves the way for on-chip optical communications as an electric-injected integrated microdevice.

Compared to other kinds of WGM microcavities, such as microdisks, microspheres, and microtoroids, rolled-up tubular microcavities have extra freedom in the axial direction, which can be further designed and utilized. As mentioned earlier

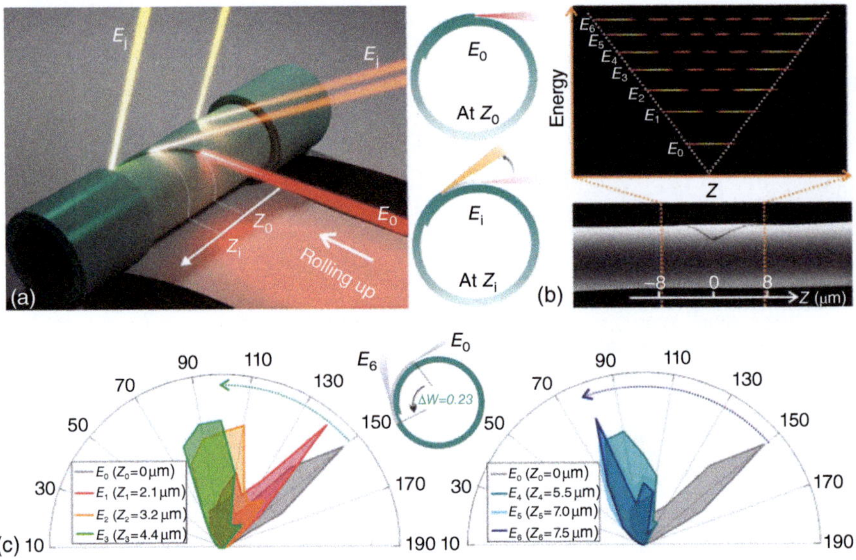

Figure 12.23 Directional emission with a special-designed rolled-up microcavity. (a) Schematic of the multidirectional light emission from a rolled-up microtube with triangle lobe. (b) Spatially and energetically resolved optical field distributions of an as-designed triangular lobe (top). SEM image of a three-dimensional microcavity with a triangular lobe (bottom). (c) Polar plots of the far-field intensity distribution. Inset: Schematic showing the cross-sectional view of the microtube with corresponding emission angle for E_0 and E_6. Source: Wang et al. [72]. Adapted with permission from American Chemical Society.

in Section 12.3, with an axial-designed mesostructured diamond microcavity, Tian et al. [78] realized tunable single-mode WGM emission. The resonant wavelength of selected single WGM can be tuned by precisely designing the geometry of the periodic structure. Wang et al. [72] designed and investigated the triangle lobes of rolled-up microtubes. The unique "exserted" triangle lobe structure serves as a new degree of freedom in the axial direction, and therefore, the higher-order axial modes are not only supported in this structure but also guided into different directions, as demonstrated in Figure 12.23. A maximum emission angle change up to 70° was experimentally demonstrated in high-order axial modes. Besides, the mode chirality can be manipulated by designing the relative distance between the two edges of the tube structure through tuning the conversion between CW and CCW components. This structure opens up new possibilities for 3D tunable emission directionality, which may be of great benefit in fundamental studies and applications.

12.6 Summary and Outlook

In this chapter, we have summarized recent progress in rolled-up WGM optical microcavities. From the perspective of theoretical analysis, both wave equation and ray optics model can be taken into consideration. Through solving Maxwell's equation, the resonant wavelength, *Q* factor, and electric field distribution can be calculated and further investigated. Besides, numerical simulations such as FDTD method and FEM can also be employed to solve the problem in complicated situations. As for ray optics model, problems can be transferred into the phase space. In this case, the CW and CCW weight of WGM can be separated and the mode charity is further investigated. Considering the rolled-up microtubes with unique geometry characteristics, the typical subwavelength-thickness wall should be noticed. Owing to the ultrathin tube wall, evanescent field with severe leakage and the corresponding optical loss is of great significance in both fundamental studies and practical applications. Furthermore, fabricated by rolled-up technique, the microtubes can be precisely designed in the axial direction or through the lobe, which is beneficial for mode tuning and directional emission.

Due to the wide-range compatibility of rolled-up technique, a variety of materials, including oxide, semiconductor, and 2D materials can be fabricated into different kinds of microdevices. Besides, with the cylindrical channel, properties of fluids can be added into the system, which is beneficial in sensing applications, including refractive index sensing, reaction monitoring, and individual cell detection. Furthermore, the enhanced energy density inside the microcavity is favored for lasing applications. Axial direction adds new degrees of freedom to the structure for further investigation.

As for the future development for rolled-up WGM microcavities, new materials, high precision fabrication methods, and functionalized integration should be taken into consideration. With the further exploration of materials, new materials may be potential to be employed for next-generation microdevices, e.g. transition metal dichalcogenide for ultrasensitive gas detection [108], PVA-PAA hydrogel with environmental sensitivity [109]. Novel materials with special electronic, electromechanical, thermoelectric, optoelectronic, optomechanical, and photonic properties can also be introduced in rolled-up microtube systems, which have a wide applicate prospect in smarter microdevices. Moreover, different kinds of materials can be combined in an individual structure, e.g. with multilayers of polymers coating onto a single microcavity, multicomponent gas sensing can be achieved.

Considering the optical loss due to the imperfection of the structure, fabrication methods with higher precision are supposed to be utilized to prevent the scattering loss. With less optical loss and corresponding enhanced Q factor, better light confinement and higher energy density can be obtained inside the cavity, which is of great potential in lasing applications. High Q factor also leads to narrow linewidth of WGM peak in the spectra, and mode splitting and mode broaden mechanisms can be selected for ultrahigh sensing activities, such as single virus detection and DNA detection [110–113]. Finally, there is an increased need for the integration and miniaturization of micro/nano-devices. Thus, developing rolled-up microcavities with different functionalities into a highly integrated system is of great importance in future applications. The system is supposed to involve rolled-up microcavity units for different purposes, including microelectronic, microfluidic, optoelectronic, etc. Such a system could be highly desired for both fundamental studies and further applications. Continuous efforts are expected to enable broader applications of rolled-up optical microcavities and better micro-/nano-systems.

References

1 Rogers, J., Huang, Y., Schmidt, O.G. et al. (2016). Origami MEMS and NEMS. *MRS Bull.* 41 (2): 123–129.

2 Chen, Z., Huang, G., Trase, I. et al. (2016). Mechanical self-assembly of a strain-engineered flexible layer: wrinkling, rolling, and twisting. *Phys. Rev. Appl.* 5 (1): 017001.

3 Huang, G. and Mei, Y. (2012). Thinning and shaping solid films into functional and integrative nanomembranes. *Adv. Mater.* 24 (19): 2517–2546.

4 Schmidt, O.G. and Eberl, K. (2001). Nanotechnology – thin solid films roll up into nanotubes. *Nature* 410 (6825): 168.

5 Huang, G. and Mei, Y. (2018). Assembly and self-assembly of nanomembrane materials – from 2D to 3D. *Small* 14 (14): 1703665.

6 Mei, Y., Huang, G., Solovev, A.A. et al. (2008). Versatile approach for integrative and functionalized tubes by strain engineering of nanomembranes on polymers. *Adv. Mater.* 20 (21): 4085–4090.

7 Wang, J., Zhan, T., Huang, G. et al. (2014). Optical microcavities with tubular geometry: properties and applications. *Laser Photonics Rev.* 8 (4): 521–547.

8 Wang, J.W., Karnaushenko, D., Medina-Sanchez, M. et al. (2019). Three-dimensional microtubular devices for lab-on-a-chip sensing applications. *ACS Sensors* 4 (6): 1476–1496.

9 Vahala, K.J. (2003). Optical microcavities. *Nature* 424 (6950): 839–846.

10 Pelton, M., Santori, C., Vuckovic, J. et al. (2002). Efficient source of single photons: a single quantum dot in a micropost microcavity. *Phys. Rev. Lett.* 89 (23): 233602.

11 Wu, X.Q., Chen, Q.S., Wang, Y.P. et al. (2019). Stable high-Q bouncing ball modes inside a Fabry–Perot cavity. *ACS Photonics* 6 (10): 2470–2478.

12 Painter, O., Lee, R.K., Scherer, A. et al. (1999). Two-dimensional photonic band-gap defect mode laser. *Science* 284 (5421): 1819–1821.

13 Bekele, D., Yu, Y., Yvind, K. et al. (2019). In-plane photonic crystal devices using Fano resonances. *Laser Photonics Rev.* 13 (12): 1900054.

14 McCall, S.L., Levi, A.F.J., Slusher, R.E. et al. (1992). Whispering-gallery mode microdisk lasers. *Appl. Phys. Lett.* 60 (3): 289–291.

15 Sak, M., Taghipour, N., Delikanli, S. et al. (2020). Coreless fiber-based whispering-gallery-mode assisted lasing from colloidal quantum well solids. *Adv. Funct. Mater.* 30 (1): 1900054.

16 Armani, D.K., Kippenberg, T.J., Spillane, S.M. et al. (2003). Ultra-high-Q toroid microcavity on a chip. *Nature* 421 (6926): 925–928.

17 Vollmer, F. and Arnold, S. (2008). Whispering-gallery-mode biosensing: label-free detection down to single molecules. *Nat. Methods* 5 (7): 591–596.

18 Zhang, Y.N., Zhu, N.S., Zhou, T.M. et al. (2020). Research on fabrication and sensing properties of fiber-coupled whispering gallery mode microsphere resonator. *IEEE Sens. J.* 20 (2): 833–841.

19 Monifi, F., Friedlein, J., Oezdemir, S.K. et al. (2012). A robust and tunable add-drop filter using whispering gallery mode microtoroid resonator. *J. Lightwave Technol.* 30 (21): 3306–3315.

20 Dastjerdi, M.H.T., Djavid, M., and Mi, Z. (2015). An electrically injected rolled-up semiconductor tube laser. *Appl. Phys. Lett.* 106 (2): 021114.

21 Kippenberg, T.J., Kalkman, J., Polman, A. et al. (2006). Demonstration of an erbium-doped microdisk laser on a silicon chip. *Phys. Rev. A* 74 (5): 051802.

22 Bae, K., Zhu, J., Wolenski, C. et al. (2020). Indium tin oxide nanoparticle-coated silica microsphere with large optical nonlinearity and high quality factor. *ACS Photonics* 7 (11): 3042–3048.

23 Chiasera, A., Dumeige, Y., Feron, P. et al. (2010). Spherical whispering-gallery-mode microresonators. *Laser Photonics Rev.* 4 (3): 457–482.

24 Hanumegowda, N.M., Stica, C.J., Patel, B.C. et al. (2005). Refractometric sensors based on microsphere resonators. *Appl. Phys. Lett.* 87 (20): 201107.

25 Guo, Z., Qin, Y., Chen, P. et al. (2020). Hyperboloid-drum microdisk laser biosensors for ultrasensitive detection of human IgG. *Small* 16 (26): 2000239.

26 Lin, X., Fang, Y., Zhu, L. et al. (2016). Self-rolling of oxide nanomembranes and resonance coupling in tubular optical microcavity. *Adv. Opt. Mater.* 4 (6): 936–942.

27 Li, Y.L., Fang, Y., Wang, J. et al. (2016). Integrative optofluidic microcavity with tubular channels and coupled waveguides via two-photon polymerization. *Lab Chip* 16 (22): 4406–4414.

28 Wang, L., Li, S., Zhang, B. et al. (2018). Asymmetrically curved hyperbolic metamaterial structure with gradient thicknesses for enhanced directional spontaneous emission. *ACS Appl. Mater. Interfaces* 10 (9): 7704–7708.

29 Yin, Y., Qiu, T., Ma, L. et al. (2012). Exploring rolled-up Au–Ag bimetallic microtubes for surface-enhanced Raman scattering sensor. *J. Phys. Chem. C* 116 (48): 25504–25508.

30 Wang, J., Song, E., Yang, C. et al. (2017). Fabrication and whispering gallery resonance of self-rolled up gallium nitride microcavities. *Thin Solid Films* 627: 77–81.

31 Xu, C., Pan, R., Guo, Q. et al. (2019). Ultrathin silicon nanomembrane in a tubular geometry for enhanced photodetection. *Adv. Opt. Mater.* 7 (21): 1900823.

32 Tian, Z., Zhang, L., Fang, Y. et al. (2017). Deterministic self-rolling of ultrathin nanocrystalline diamond nanomembranes for 3D tubular/helical architecture. *Adv. Mater.* 29 (13): 1604572.

33 Yang, J., Wang, Y., Wang, L. et al. (2020). Tubular/helical architecture construction based on rolled-up AlN nanomembranes and resonance as optical microcavity. *J. Semicond.* 41 (4): 042601.

34 Zhou, X., Tian, Z., Kim, H.J. et al. (2019). Rolling up $MoSe_2$ nanomembranes as a sensitive tubular photodetector. *Small* 15 (42): 1902528.

35 Wang, L., Tian, Z., Zhang, B. et al. (2019). On-chip rolling design for controllable strain engineering and enhanced photon–phonon interaction in graphene. *Small* 15 (23): 1805477.

36 Cao, H. and Wiersig, J. (2015). Dielectric microcavities: model systems for wave chaos and non-Hermitian physics. *Rev. Mod. Phys.* 87 (1): 61–111.

37 Cai, M., Painter, O., and Vahala, K.J. (2000). Observation of critical coupling in a fiber taper to a silica-microsphere whispering-gallery mode system. *Phys. Rev. Lett.* 85 (1): 74–77.

38 Poellinger, M., O'Shea, D., Warken, F. et al. (2009). Ultrahigh-Q tunable whispering-gallery-mode microresonator. *Phys. Rev. Lett.* 103 (5): 053901.

39 Matsko, A.B. and Ilchenko, V.S. (2006). Optical resonators with whispering-gallery modes – Part I: basics. *IEEE J. Sel. Top. Quantum Electron.* 12 (1): 3–14.

40 Lan, Y.W., Li, S.L., Cai, Z.Y. et al. (2017). Semi-analytical calculation of resonant modes in axially asymmetric microtube resonators. *Opt. Commun.* 386: 72–76.

41 Strelow, C., Schultz, C.M., Rehberg, H. et al. (2012). Light confinement and mode splitting in rolled-up semiconductor microtube bottle resonators. *Phys. Rev. B* 85 (15): 155329.

42 Kim, S.K., Kim, S.H., Kim, G.H. et al. (2004). Highly directional emission from few-micron-size elliptical microdisks. *Appl. Phys. Lett.* 84 (6): 861–863.

43 Fang, W., Cao, H., and Solomon, G.S. (2007). Control of lasing in fully chaotic open microcavities by tailoring the shape factor. *Appl. Phys. Lett.* 90 (8): 081108.

44 Hentschel, M., Schomerus, H., and Schubert, R. (2003). Husimi functions at dielectric interfaces: inside–outside duality for optical systems and beyond. *Europhys. Lett.* 62 (5): 636–642.

45 Fang, Y., Li, S., and Mei, Y. (2016). Modulation of high quality factors in rolled-up microcavities. *Phys. Rev. A* 94 (3): 033804.

46 Ward, J. and Benson, O. (2011). WGM microresonators: sensing, lasing and fundamental optics with microspheres. *Laser Photonics Rev.* 5 (4): 553–570.

47 Schwesyg, J.R., Beckmann, T., Zimmermann, A.S. et al. (2009). Fabrication and characterization of whispering-gallery-mode resonators made of polymers. *Opt. Express* 17 (4): 2573–2578.

48 Lacey, S., White, I.M., Sun, Y. et al. (2007). Versatile opto-fluidic ring resonator lasers with ultra-low threshold. *Opt. Express* 15 (23): 15523–15530.

49 Wang, J., Zhan, T., Huang, G. et al. (2012). Tubular oxide microcavity with high-index-contrast walls: Mie scattering theory and 3D confinement of resonant modes. *Opt. Express* 20 (17): 18555–18567.

50 Zhan, T., Xu, C., Zhao, F. et al. (2011). Optical resonances in tubular microcavities with subwavelength wall thicknesses. *Appl. Phys. Lett.* 99 (21): 211104.

51 Quiñones, V.A.B., Ma, L., Li, S. et al. (2012). Enhanced optical axial confinement in asymmetric microtube cavities rolled up from circular-shaped nanomembranes. *Opt. Lett.* 37 (20): 4284–4286.

52 Huang, G., Quiñones, V.A.B., Ding, F. et al. (2010). Rolled-up optical microcavities with subwavelength wall thicknesses for enhanced liquid sensing applications. *ACS Nano* 4 (6): 3123–3130.

53 Kipp, T., Welsch, H., Strelow, C. et al. (2006). Optical modes in semiconductor microtube ring resonators. *Phys. Rev. Lett.* 96 (7): 077403.

54 Boettner, S., Li, S., Trommer, J. et al. (2012). Sharp whispering-gallery modes in rolled-up vertical SiO_2 microcavities with quality factors exceeding 5000. *Opt. Lett.* 37 (24): 5136–5138.

55 Goos, F. and Hanchen, H. (1947). *Ein neuer und fundamentaler versuch zur totalreflexion. *Ann. Phys.* 1 (7–8): 333–346.

56 Saito, M. and Koyama, K. (2012). Spectral and polarization characteristics of a deformed droplet laser. *J. Opt.* 14 (6): 065002.

57 Artmann, K. (1948). *Berechnung der seitenversetzung des totalreflektierten strahles. *Ann. Phys.* 2 (1–2): 87–102.

58 Rakovich, Y.P. and Donegan, J.F. (2010). Photonic atoms and molecules. *Laser Photonics Rev.* 4 (2): 179–191.

59 Wang, J., Yin, Y., Hao, Q. et al. (2018). Curved nanomembrane-based concentric ring cavities for supermode hybridization. *Nano Lett.* 18 (11): 7261–7267.

60 Wang, J.W., Yin, Y., Hao, Q. et al. (2018). Strong coupling in a photonic molecule formed by trapping a microsphere in a microtube cavity. *Adv. Opt. Mater.* 6 (1): 1700842.

61 Zhang, J., Li, J., Tang, S. et al. (2015). Whispering-gallery nanocavity plasmon-enhanced Raman spectroscopy. *Sci. Rep.* 5: 15012.

62 Rottler, A., Broell, M., Schwaiger, S. et al. (2011). Tailoring of high-Q-factor surface plasmon modes on silver microtubes. *Opt. Lett.* 36 (7): 1240–1242.

63 Yin, Y., Li, S.L., Giudicatti, S. et al. (2015). Strongly hybridized plasmon-photon modes in optoplasmonic microtubular cavities. *Phys. Rev. B* 92 (24): 241403.

64 Yin, Y., Wang, J.W., Lu, X.Y. et al. (2018). In situ generation of plasmonic nanoparticles for manipulating photon-plasmon coupling in microtube cavities. *ACS Nano* 12 (4): 3726–3732.

65 Boettner, S., Li, S., Jorgensen, M.R. et al. (2013). Vertically aligned rolled-up SiO_2 optical microcavities in add-drop configuration. *Appl. Phys. Lett.* 102 (25): 251119.

66 Hosoda, M. and Shigaki, T. (2007). Degeneracy breaking of optical resonance modes in rolled-up spiral microtubes. *Appl. Phys. Lett.* 90 (18): 181107.

67 Li, S., Ma, L., Boettner, S. et al. (2013). Angular position detection of single nanoparticles on rolled-up optical microcavities with lifted degeneracy. *Phys. Rev. A* 88 (3): 033833.

68 Huang, G.S., Kiravittaya, S., Quiñones, V.A.B. et al. (2009). Optical properties of rolled-up tubular microcavities from shaped nanomembranes. *Appl. Phys. Lett.* 94 (14): 141901.

69 Hardy, A., Belanger, P.A., and Siegman, A.E. (1982). Orthogonality properties of phase conjugate optical resonators. *Appl. Opt.* 21 (6): 1122–1124.

70 Zhong, Q.H., Tian, Z.B., Dastjerdi, M.H.T. et al. (2013). Characterization of azimuthal and longitudinal modes in rolled-up InGaAs/GaAs microtubes at telecom wavelengths. *Opt. Express* 21 (16): 18909–18918.

71 Strelow, C., Rehberg, H., Schultz, C.M. et al. (2008). Optical microcavities formed by semiconductor microtubes using a bottlelike geometry. *Phys. Rev. Lett.* 101 (12): 127403.

72 Wang, J., Yin, Y., Yang, Y.-D. et al. (2019). Deterministic yet flexible directional light emission from spiral nanomembrane cavities. *ACS Photonics* 6 (10): 2537–2544.

73 Li, S., Ma, L., Zhen, H. et al. (2012). Dynamic axial mode tuning in a rolled-up optical microcavity. *Appl. Phys. Lett.* 101 (23): 231106.

74 Wang, J.W., Yin, Y., Hao, Q. et al. (2018). External strain enabled post-modification of nanomembrane-based optical microtube cavities. *ACS Photonics* 5 (5): 2060–2067.

75 Johnson, S.G., Ibanescu, M., Skorobogatiy, M.A. et al. (2002). Perturbation theory for Maxwell's equations with shifting material boundaries. *Phys. Rev. E* 65 (6): 054101.

76 Smith, E.J., Schulze, S., Kiravittaya, S. et al. (2011). Lab-in-a-tube: detection of individual mouse cells for analysis in flexible split-wall microtube resonator sensors. *Nano Lett.* 11 (10): 4037–4042.

77 Kietzmann, S., Strelow, C., Tavares, L. et al. (2015). Organic molecular films as light-emitting and light-confining material in rolled-up AlInP semiconductor microtube resonators. *ACS Photonics* 2 (11): 1532–1538.

78 Tian, Z., Li, S., Kiravittaya, S. et al. (2018). Selected and enhanced single whispering-gallery mode emission from a mesostructured nanomembrane microcavity. *Nano Lett.* 18 (12): 8035–8040.

79 Ma, L.B., Li, S.L., Fomin, V.M. et al. (2016). Spin-orbit coupling of light in asymmetric microcavities. *Nat. Commun.* 7: 10983.

80 Vicknesh, S., Li, F., and Mi, Z. (2009). Optical microcavities on Si formed by self-assembled InGaAs/GaAs quantum dot microtubes. *Appl. Phys. Lett.* 94 (8): 081101.

81 Li, F. and Mi, Z.T. (2009). Optically pumped rolled-up InGaAs/GaAs quantum dot microtube lasers. *Opt. Express* 17 (22): 19933–19939.

82 Dastjerdi, M.H.T., Djavid, M., Arafin, S. et al. (2013). Optically pumped rolled-up InAs/InGaAsP quantum dash lasers at room temperature. *Semicond. Sci. Technol.* 28 (9): 094007.

83 Wu, X., Tian, Z., Cong, H. et al. (2018). Infrared tubular microcavity based on rolled-up GeSn/Ge nanomembranes. *Nanotechnology* 29 (42): 42LT02.

84 Yin, Y., Pang, J., Wang, J. et al. (2019). Graphene-activated optoplasmonic nanomembrane cavities for photodegradation detection. *ACS Appl. Mater. Interfaces* 11 (17): 15891–15897.

85 Wang, J., Huang, G.S., and Mei, Y.F. (2014). Modification and resonance tuning of optical microcavities by atomic layer deposition. *Chem. Vap. Deposition* 20 (4–6): 103–111.

86 Harazim, S.M., Quiñones, V.A.B., Kiravittaya, S. et al. (2012). Lab-in-a-tube: on-chip integration of glass optofluidic ring resonators for label-free sensing applications. *Lab Chip* 12 (15): 2649–2655.

87 Fang, Y., Lin, X., Tang, S. et al. (2017). Temperature-dependent optical resonance in a thin-walled tubular oxide microcavity. *Prog. Nat. Sci.: Mater. Int.* 27 (4): 498–502.

88 Ma, L., Li, S., Quiñones, V.A.B. et al. (2013). Dynamic molecular processes detected by microtubular opto-chemical sensors self-assembled from prestrained nanomembranes. *Adv. Mater.* 25 (16): 2357–2361.

89 Yin, Y., Wang, J., Wang, X. et al. (2019). Water nanostructure formation on oxide probed in situ by optical resonances. *Sci. Adv.* 5 (10): eaax6973.

90 Madani, A., Ma, L., Miao, S. et al. (2016). Luminescent nanoparticles embedded in TiO_2 microtube cavities for the activation of whispering-gallery-modes extending from the visible to the near infrared. *Nanoscale* 8 (18): 9498–9503.

91 Madani, A., Naz, E.S.G., Harazim, S. et al. (2018). Multiplexing and tuning of a double set of resonant modes in optical microtube cavities monolithically integrated on a photonic chip. *Opt. Lett.* 43 (19): 4703–4706.

92 Yu, X., Arbabi, E., Goddard, L.L. et al. (2015). Monolithically integrated self-rolled-up microtube-based vertical coupler for three-dimensional photonic integration. *Appl. Phys. Lett.* 107 (3): 031102.

93 Strelow, C., Sauer, M., Fehringer, S. et al. (2009). Time-resolved studies of a rolled-up semiconductor microtube laser. *Appl. Phys. Lett.* 95 (22): 221115.

94 Heo, J., Bhowmick, S., and Bhattacharya, P. (2012). Threshold characteristics of quantum dot rolled-up microtube lasers. *IEEE J. Quantum Electron.* 48 (7): 927–933.

95 Madani, A., Harazim, S.M., Bolanos Quiñones, V.A. et al. (2017). Optical microtube cavities monolithically integrated on photonic chips for optofluidic sensing. *Opt. Lett.* 42 (3): 486–489.

96 Quiñones, V.A.B., Ma, L., Li, S. et al. (2012). Localized optical resonances in low refractive index rolled-up microtube cavity for liquid-core optofluidic detection. *Appl. Phys. Lett.* 101 (15): 151107.

97 Song, P., Chen, C., Qu, J. et al. (2018). Rolled-up SiO_x/SiN_x microtubes with an enhanced quality factor for sensitive solvent sensing. *Nanotechnology* 29 (41): 415501.

98 Miao, S., Chen, D., Madani, A. et al. (2015). Optofluidic sensor: evaporation kinetics detection of solvents dissolved with Cd_3P_2 colloidal quantum dots in a rolled-up microtube. *Adv. Opt. Mater.* 3 (2): 187–193.

99 Bernardi, A., Kiravittaya, S., Rastelli, A. et al. (2008). On-chip Si/SiO_x microtube refractometer. *Appl. Phys. Lett.* 93 (9): 415501.

100 Szewczuk, A. and Blahut, M. (2011). Applications of gradient index multimode interference structures in the technology of optical sensor. *Acta Phys. Pol. A* 120 (4): 740–743.

101 Ma, L., Kiravittaya, S., Quiñones, V.A.B. et al. (2011). Tuning of optical resonances in asymmetric microtube cavities. *Opt. Lett.* 36 (19): 3840–3842.

102 Reid, C.R. and Thomas, K.M. (1999). Adsorption of gases on a carbon molecular sieve used for air separation: linear adsorptives as probes for kinetic selectivity. *Langmuir* 15 (9): 3206–3218.

103 Zhang, J., Zhong, J., Fang, Y.F. et al. (2014). Roll up polymer/oxide/polymer nanomembranes as a hybrid optical microcavity for humidity sensing. *Nanoscale* 6 (22): 13646–13650.

104 Shen, L., Fu, J., Fu, K. et al. (2010). Humidity responsive asymmetric free-standing multilayered film. *Langmuir* 26 (22): 16634–16637.

105 Xavier, J., Vincent, S., Meder, F. et al. (2018). Advances in optoplasmonic sensors – combining optical nano/microcavities and photonic crystals with plasmonic nanostructures and nanoparticles. *Nanophotonics* 7 (1): 1–38.

106 Yin, Y., Chen, Y., Naz, E.S.G. et al. (2017). Silver nanocap enabled conversion and tuning of hybrid photon-plasmon modes in microtubular cavities. *ACS Photonics* 4 (4): 736–740.

107 Pelton, M., Vuckovic, J., Solomon, G.S. et al. (2002). Three-dimensionally confined modes in micropost microcavities: quality factors and Purcell factors. *IEEE J. Quantum Electron.* 38 (2): 170–177.

108 Yang, S., Jiang, C., and Wei, S.-H. (2017). Gas sensing in 2D materials. *Appl. Phys. Rev.* 4 (2): 021304.

109 Li, W., Huang, G., Yan, H. et al. (2012). Fabrication and stimuli-responsive behavior of flexible micro-scrolls. *Soft Matter* 8 (27): 7103–7107.

110 He, L., Oezdemir, S.K., Zhu, J. et al. (2011). Detecting single viruses and nanoparticles using whispering gallery microlasers. *Nat. Nanotechnol.* 6 (7): 428–432.

111 Jiang, X., Qavi, A.J., Huang, S.H. et al. (2020). Whispering-gallery sensors. *Matter* 3 (2): 371–392.

112 Shao, L., Jiang, X.-F., Yu, X.-C. et al. (2013). Detection of single nanoparticles and lentiviruses using microcavity resonance broadening. *Adv. Mater.* 25 (39): 5616–5620.

113 Zhi, Y., Yu, X.-C., Gong, Q. et al. (2017). Single nanoparticle detection using optical microcavities. *Adv. Mater.* 29 (12): 1604920.

13

2D Materials Nanomembrane

Yang Zong, Binmin Wu, Xinyi Ke, Yongfeng Mei, and Jizhai Cui

Fudan University, Department of Materials Science, No. 220 Handan Road, Shanghai 200433, People's Republic of China

13.1 The Development History of 2D Materials

2D layered materials exist in nature and have been pointed a long time ago, but monolayer materials had not been discovered until twenty-first century. For a long time, scientists believed that thermodynamic fluctuations do not allow any 2D crystals to exist at a finite temperature. In 2004, physicists Andre Geim and Konstantin Novoselov of the University of Manchester in the United Kingdom successfully isolated graphene from graphite using the mechanical cleavage method [1], thus confirming that graphene can exist alone at room temperature. Meanwhile, this outstanding work also shows the unusual electrical transmission characteristics in graphene. Novoselov et al. also won the 2010 Nobel Prize in Physics for their pioneering experiments in graphene materials, in recognition of their outstanding research in graphene materials.

Graphene is known as a new material with unique optical, electrical, and mechanical properties. It has important application prospects in materials science, micro/nanoprocessing, energy, biomedicine, and drug delivery. It is considered a revolutionary material in the future. The enthusiasm for graphene materials has also triggered interest in other 2D materials, such as hexagonal boron nitride, silylene [2], germanene [3], transition metal oxides, and 2D transition metal dichalcogenides (TMDCs). The research fields include solid-state physics, materials science and engineering, and diverse applications. According to theoretical calculations, 826 kinds of materials can be converted into stable 2D materials. The rich physical and chemical connotations give 2D materials the intended membrane properties, making them shine in the fields of micromechanics, microstructures, and sensors.

Nanomembranes: Materials, Properties, and Applications, First Edition.
Edited by Yongfeng Mei, Gaoshan Huang, and Xiuling Li.
© 2022 WILEY-VCH GmbH. Published 2022 by WILEY-VCH GmbH.

13.2 Characteristics of 2D Materials Nanomembrane

Generally, a 2D material refers to a thin-membrane material in which layers composed of one or several atoms are stacked by van der Waals forces [4]. For a perfect 2D material, there are no dangling bonds in the plane, which can provide an atomically flat surface. 2D materials are usually combined through van der Waals forces when assembled with other materials, and the van der Waals forces between 2D material layers are much weaker than the covalent bonds and ionic bonds in the layers. Due to the special assembly method of 2D materials, it is possible to use 2D materials and other 2D materials or other dimensional materials (such as quantum dots, nanowires, and bulk materials) to be assembled into heterojunctions conveniently and efficiently, and problems such as lattice mismatching and process incompatibility can be ignored [5]. The construction of high-quality, low-cost van der Waals heterojunctions provides an extremely valuable platform for electronics, optoelectronics, and chemical catalysis. It also promotes research and development in important fields, such as materials science, condensed matter physics, and engineering.

Although the thickness of 2D materials is only a few atomic layers thick, they still have the properties of high tensile strength and excellent toughness [6, 7], which gives them the universal characteristics of ultra-high mechanical flexibility. Experimentally, the morphology, position, and stacking method of 2D materials can be manipulated by micro/nanomanipulation methods, such as needle tips and capillaries. This is impossible for bulk materials. Based on the 2D material nanomembrane, a series of constructed three-dimensional (3D) structures are shown in Figure 13.3: nano-scrolls (NSs) [8], origami [9], and kirigami [10]. The assembly of 2D material nanomembrane into 3D polyhedral structures while retaining the excellent inherent properties of 2D materials has aroused great interest in the development of new device applications.

Achieving a fully flexible electronic device usually requires the ability to fold, bend, and stretch, but still maintain its original performance. Conventional electrical systems and related technologies based on hard silicon and brittle silicon cannot meet this requirement, which has prompted scientists to explore other types of highly elastic materials. Through the assembly process of folding and rolling, a relatively simple structure can be converted into a complex topology with unique and original outstanding characteristics, such as NSs. Theoretical calculations have predicted the unique topological structure of NS based on 2D material nanomembrane, resulting in abnormal electronic and optical properties. Therefore, these NSs are expected to be used to construct flexible electronics, microfluidics, energy storage, automatic micromechanical, and optical resonator modules. In graphene, boron nitride, and transition metal sulfides, such high-quality NSs have been experimentally achieved, and they exhibit high strength and chemical inertness.

Figure 13.1a illustrates a very simple method for manufacturing high-quality TMD-NSs [8]. This method requires only a drop of ethanol solution to roll *in situ* chemical vapor deposition (CVD) of a single layer of transition metal sulfur within five seconds. Due to the high field effect transistor (FET) mobility, TMD-NSs are expected to exhibit potential applications in optoelectronic devices. Due to

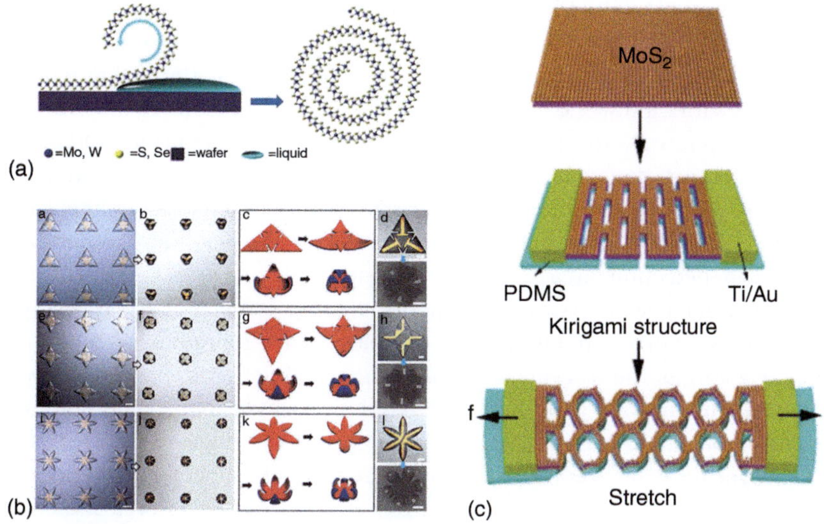

Figure 13.1 3D structures based on 2D material nanomembranes. (a) Rolling up nanoscrolls. Source: Cui et al. [8]. Copyright 2018, Springer Nature, (b) Origami. Source: Xu et al. [9]. Copyright 2019, American Chemical Society, and (c) Kirigami. Source: Zheng et al. [10]. Copyright 2018, American Chemical Society.

the self-encapsulated structure, the optical and electrical properties of the NSs are more insensitive to external factors, while the electrical properties of the 2D material sheet vary with the underlying substrate and environment. In addition, due to the internal open topological structure, the interlayer spacing of TMD-NS can be easily expanded to adapt to various functional materials, including small organic molecules, polymers, nanoparticles, and 2D materials, as well as biological substances. These functions are very attractive for solar cells, photodetectors, flexible logic circuits, energy storage, and sensor applications.

2D layered materials have been widely used in electronics, sensing, and energy applications. Single-layer 2D materials have unique characteristics, including atomic-scale thickness, excellent electrical and thermal characteristics, high stability, and mechanical strength. Although significant progress has been made, most of the single-layer functional devices based on 2D materials are still configured on rigid planar substrates, which will limit device functions and increase overall device size. Because the single-layer 2D material is very thin, the bending stiffness is low, which will benefit researchers to bend or fold them and create a 3D flexible reconfigurable device with a small size. The 3D structure provides significant advantages for optoelectronic applications. At the same time, 2D materials have unique advantages over conventional semiconductors in the optoelectronic field, including light absorption over a wide energy spectrum, ultrafast carrier dynamics, adjustable optical properties through doping, and low dissipation rate.

In Figure 13.1b, the reversible conversion of a single layer of molybdenum disulfide (MoS_2) into a complex 3D shape is shown by attaching it to a differentially cross-linked and stimulus-responsive polymer nanomembrane with integrated gold electrodes [9]. First, based on the adjustable light crosslinking of the stimulus-responsive polymer nanomembrane, we can control the degree and direction of MoS_2 folding. This control can fold complex Miura-ori patterns that require two-way hinges with rigid sections. The realization of the Miura folding mode is also an important step toward the realization of more complex origami/origami styles and deployable functional devices. Secondly, the arrangement of MoS_2 and the reconfiguration of the pre-designed 3D pattern of photoelectric-based photoelectric devices have been realized, resulting in spatially or angularly resolved photoelectric detection and a detection area with a large adjustable range between the flat state and the folded state. Stimulus-response folding is completely reversible, no tether or metal wire is required, and compared with inorganic or metal 3D structures, it also has the advantages of flexibility and softness, paving the way for the next generation of adaptives and bionics. A reversible, reconfigurable, and stimulus-response 3D photoelectric device based on single-layer 2D materials has not yet been realized. This is an important step in the development of bionic, adaptive, intelligent, wearable, and robotic devices. 2D layered materials (such as MoS_2) are very attractive for flexible devices due to their unique layered structure, novel physical and electronic properties, and high mechanical strength. However, their limited mechanical strain (<2%) cannot meet the load condition requirements of most flexible and stretchable equipment applications. Inspired by the ancient Japanese paper-cutting art of kirigami, the scientists built a nanoscale kirigami architecture based on a 2D MoS_2 nanomembrane on a soft substrate [10], as shown in Figure 13.1c.

13.3 Structure and Design of 2D Material Nanomembrane

13.3.1 Suspended Structure

In addition to the conventional film structure attached to the substrate, the 2D materials can also form a single nanomembrane-suspended structure. This kind of suspended structure removes the effect of substrate, which makes the 2D materials better display their own superior performance.

The existence of substrate will inevitably introduce unnecessary impurities into the 2D materials devices. The stress between the substrate and the 2D materials will also produce a certain intrinsic strain on the 2D material lattice. These impurities and strains will affect the electrical properties of 2D materials. The graphene on SiO_2-insulating substrates obtained 5000 $cm^2/V/s$ carrier mobilities in the previous research [11]. While the researchers removed the substrate to gain a suspending graphene nanomembrane, the result in mobilities achieved 200 000 $cm^2/V/s$ at low temperatures [12]. Using these changes in electrical properties caused by strain, suspended 2D nanomembrane can be used as sensors. For example, for the suspended graphene sensor on the top of the microfluidic channel, the resistivity of the graphene nanomembrane decreases rapidly by about 25% after water is injected into the channel. The decrease of resistivity is caused by the graphene nanomembrane filling and spreading through the channel. When the water flows into the channel, the suspended graphene nanomembrane above will deform under the pressure in the pipeline, resulting in the change of electron mobility and the decrease of resistivity. The change of resistivity can be used to detect the change of water flow in the channel [13]. At the same time, the existence of the substrate will also affect the overall mechanical properties, Young's modulus E of single graphene can reach 0.8–1.0 TPa, which is the "hardest" material on earth. However, after bonding with the substrate, the mechanical properties of the composite structure will inevitably be affected by the substrate with low elastic modulus. Therefore, due to the excellent mechanical properties and adjustable excellent electrical properties of the suspended 2D material film, suspended 2D materials have become an ideal choice for nanomembrane materials in microelectromechanical systems/nanoelectromechanical systems (MEMS and NEMS). Suspended graphene has been widely used in resonators [14–18], mechanical sensors [19–23], strain sensors [24, 25], photodetectors [26], and other NEMS devices [27–37].

Apart from the excellent mechanical and electrical properties, suspended 2D materials also have unique optical properties. Take graphene as an example; graphene bubble is a special structure in suspended graphene. This kind of structure is similar to the bulge in the macroworld. It is connected with the substrate on all sides and bulges upward in the middle. This kind of bubbles has many different geometries, for example, circle, triangle, and diamond. A graphene-based lens with adaptive focus by applying an electric field to control the curvature of a spherical graphene bubble was reported by Georgiou et al [38]. When combined with some transparent liquid, graphene optical lens with adjustable focal length can be made. This kind of graphene optical lens mainly utilizes the flexible control space of suspended graphene film and the excellent optical properties of graphene film itself (Figure 13.2).

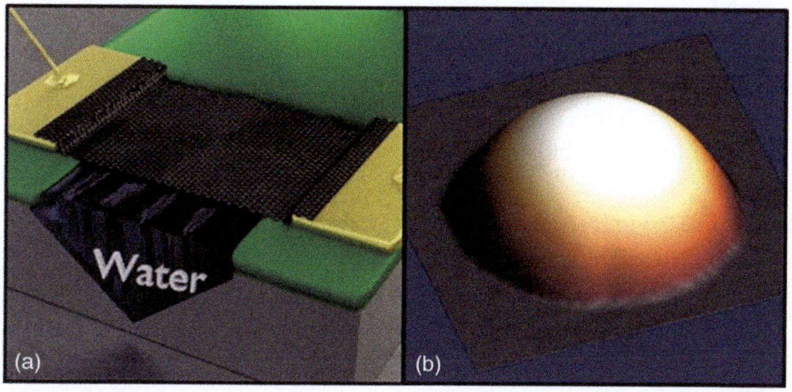

Figure 13.2 Suspended graphene in different application situations. (a) Sensor on the top of the microfluidic channel. Source: Meireles et al. [13]. Copyright 2020, Wiley-VCH). (b) Bubble optical lens. Source: Georgiou et al. [38]. Copyright 2011, American Institute of Physics)..

13.3.2 Kirigami of 2D Material Nanomembrane

2D material nanomembranes are similar to paper in the macroworld. In real life, we can build beautiful and complex structures through kirigami. Similarly, we can also use the "kirigami" method to construct complex microscale structures by 2D material nanomembrane [39].

The kirigami principle of 2D material nanomembrane is divided into two parts. The first part is the patterning of continuous 2D material nanomembrane. Like paper in the macroworld, continuous 2D material nanomembrane can also be cut to form various patterns. Different from ordinary paper, 2D material nanomembrane cutting needs lithography and etching technology. Lithography is used to design the shape of 2D material film. Then, we can cut 2D material nanomembrane by etching. The second part is to release the trimmed 2D material nanomembrane. The common release method is to immerse the trimmed 2D material nanomembrane and substrate in the proportioned developer to release the 2D material nanomembrane. This technique requires the addition of a sacrificial layer between the 2D material nanomembrane and the substrate. The sacrificial layer can be dissolved in the solution to separate the 2D material nanomembrane from the substrate. For example, an aluminum sacrificial layer is added to the standard graphene production process. After patterning and cutting the film, sinking the sample in a deionized water/HCl solution to remove the aluminum release layer is necessary. Then, we can get the freestanding graphene released in the solution.

2D material nanomembranes are widely used in flexible electronic devices due to their low out-of-plane stiffness and high in-plane stiffness. The low out-of-plane stiffness makes the 2D material nanomembrane very flexible, but the high in-plane stiffness makes it unable to adapt to the large deformation along the in-plane direction, which limits its application in flexible devices to a certain extent. After kirigami, the 2D material nanomembrane can produce greater deformation in the plane, which overcomes this problem. Like ordinary A4 paper, it is very difficult for A4 paper to have very large deformation in the plane without kirigami. However, after a certain design of kirigami, A4 paper can produce large deformation in the in-plane direction. The same is true for 2D material nanomembranes. The 2D material nanomembrane after kirigami operation can not only have large in-plane tensile strength, but also have very low in-plane stiffness. These characteristics can be very well applied to flexible electronic devices (Figure 13.3).

13.3.3 Origami of 2D Material Nanomembrane

The above 2D material nanomembrane is in the inherent plane geometry. However, some applications, such as wearable electronic devices, biological or dispersible sensors, and actuators, require bending and folding structures with small-size features. Therefore, we can also use "origami" method to build a 3D structure using a 2D material nanomembrane.

Mechanical stimulation can induce the folding of 2D materials [40]. Take graphene as an example; graphene has three different stable states: planar, "armchair," and "zigzag." The latter two are in the self-folding state of graphene.

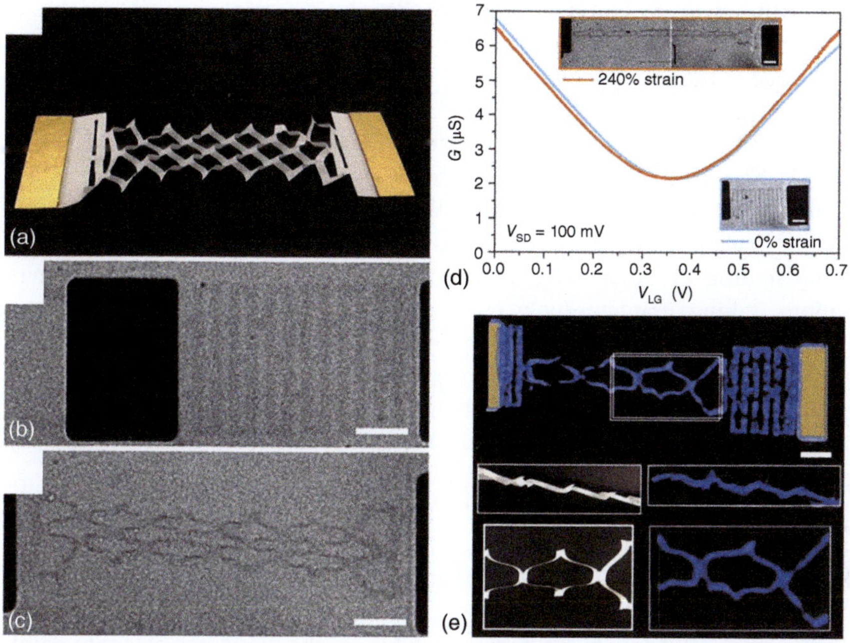

Figure 13.3 Graphene kirigami. (a), (b), (c) Paper and graphene in-plane kirigami springs, respectively. (d) Electrical properties of stretched graphene springs. (e) Three-dimensional reconstruction from a z-scan focal series of a graphene spring. Source: Blees et al. [39]. Copyright 2015, Springer Nature.

Mechanical stimulation can transform graphene from planar to folded. When the planar graphene absorbs the energy of external mechanical stimulation, such as ultrasound, the system will cross the energy barrier and change into the other two folded states.

Using interface force is another method of 2D material nanomembrane origami [41]. The folding of graphene can also be realized by using the interface force. Bin Wang et al. obtained the folded graphene structure by folding polymethyl methacrylate (PMMA)/graphene with water between PMMA/graphene and substrate. They first constructed hydrophilic and hydrophobic regions on the same side of the substrate, and then transferred PMMA with graphene to the substrate. When the substrate and PMMA were immersed in water, the water entered between the substrate and PMMA and slowly separated. The PMMA on the hydrophobic side was still firmly bonded to the substrate. After controlling the direction of substrate removal from water, graphene with folded structure can be obtained. The graphene was folded by the interfacial force between water, PMMA, and substrate.

Prestressing can also be used for 2D material nanomembrane origami [42]. Some metal layers with internal stress can be deposited on the 2D material nanomembrane as hinges. When the whole 2D material film is released, the film will fold along the direction of the internal stress of the hinge under the action of these metal hinges with internal stress. Taking graphene as an example, Daeha Joung et al. designed a cube box of graphene. Metal hinges made of Cr are deposited on the 12 edges of the cube. When the graphene box is released into the solution, graphene will turn into a square structure from the original two-dimensional cross structure. This is because Cr on the edge of the square body is released from the substrate, and the adhesion graphene is bent and folded together.

Finally, we can also use interface force to achieve 2D material nanomembrane origami [43]. This kind of origami method often needs to combine some substrates, which are responsive to external stimuli. When the 2D material nanomembrane is combined with the corresponding exciting substrate, the nanomembrane can follow the folding or other behaviors of the substrate in response to external stimulation. Taking graphene/poly (*N*-isopropyl acrylamide) (PNIPAM) as an example, PNIPAM is a thermal-responsive polymer. When the external temperature changes, the internal phase of polymer will change, resulting in macrodeformation. Graphene/PNIPAM nanomembrane system can be designed into various shapes. This petal-shaped structure is shown in the picture. When the solution was heated, the petals changed from unfolded state to folded state.

13.4 2D Material Nanomembrane Sensors

With the advent of the new technological revolution, the world has begun to enter the information age. In the process of using information, the first thing to be solved is to obtain accurate and reliable information, and sensors are the main way and means to obtain information in the fields of nature and production. A sensor is a detection device that can receive the information to be measured, and can transform the information into electrical signals or other required forms of information output according to a certain rule to meet the requirements of information transmission, processing, storage, display, and recording. It is the first link to realize automatic detection and automatic control. Although traditional materials still occupy the sensor market, it is gradually difficult to meet the new requirements of current sensors: miniaturization, digitization, intelligence, multifunction, systemization, and networking. Here, we will review new sensors based on 2D material nanomembrane: gas sensors, force sensors, etc.

13.4.1 2D Material Gas Sensors

Gas sensors are used to detect gas concentration and composition. They convert the type of gas and its concentration-related information into electrical signals. According to the strength of these electrical signals, information related to the presence of the gas to be measured in the environment can be obtained, so that it can be detected, monitored, and alarmed. It can also form an automatic detection, control, and alarm system through an interface circuit and a computer. Gas sensors play an extremely important role in environmental protection, emission control, industrial sectors, and medical diagnosis. Due to the wide variety of gases and their different properties, it is impossible to use one sensor to detect all types of gases. Therefore, there are many types of sensors that can realize gas–electric conversion. According to the composition of the gas sensor materials, it can be divided into two categories: non-semiconductors (such as surface acoustic wave devices) and semiconductors (such as metal oxides and 2D van der Waals layered material).

For surface acoustic wave devices, the surface wave speed and frequency will drift with changes in the external environment [44]. The gas sensor uses this performance to coat a layer of gas-sensitive film (Pb phthalocyanine, Pd, Pt, etc.) [45] that selectively adsorbs a certain gas on the surface of the piezoelectric crystal. When the corresponding gas-sensitive film interacts with the gas to be measured [46] (NO_2, H_2, NH_3, etc.), it causes the film quality and conductance of the gas-sensitive film to change, causing the surface acoustic wave frequency of the piezoelectric crystal to drift. The gas concentration affects the quality of the film and the degree of change in conductivity, which can cause different changes in the frequency of surface acoustic waves. By measuring the change in the frequency of the surface acoustic wave, an accurate change in the concentration of the reaction gas can be obtained.

In 1970, Bergveld developed an ion-sensitive field-effect transistor that responded to hydrogen ions [47], marking the birth of ion-sensitive semiconductor sensors. Since then, semiconductor sensors have attracted the attention and interest of scientists from all over the world because of their easy integration, miniaturization, and high sensitivity. The semiconductor gas sensor uses the change in physical properties such as electrical conductivity when the gas to be measured is in contact with the semiconductor surface to detect the gas. According to the interaction between the semiconductor and the gas, the changes that occur are limited to the semiconductor surface or deep into the semiconductor. It can be divided into two types: the gas adsorbed on the surface of the semiconductor and the semiconductor is absorbed by electrons, resulting in changes in physical properties (such as the conductivity of the semiconductor), but the internal chemical composition has not changed [48, 49]. The other is that the reaction between the semiconductor and the gas changes the internal composition of the semiconductor [50], thereby changing the conductivity of the semiconductor.

After years of development, although great progress has been made in the research of surface acoustic wave devices, it has also been applied in practice. However, surface acoustic wave devices are still difficult to meet the requirements of the new era for gas sensors: simple structure, miniaturization, high sensitivity, and good dynamic performance. Therefore, the most practically used semiconductor gas sensor is the current one. Because metal oxides exhibit the best characteristics in terms of responsiveness and sensitivity, they stand out from conductive polymers, carbon nanotubes, nanorods, and nanowires, and are widely used. However, metal oxide--based resistive gas sensors usually work at high temperatures, which can lead to high power consumption and response drift [51]. Therefore, it is urgent to find new materials to supplement or replace the shortcomings of traditional materials. Since the first discovery, due to its peculiar structural characteristics, the 2D thin-film material with an atomic layer thickness is considered to be the most promising material.

2D materials usually only have a thickness of single or a few atomic layers, so they also have an ultra-high surface-to-volume ratio compared to bulk materials, which leads to high atom utilization and easy chemical modification. All the carbon atoms of the graphene atomic film are exposed to the external environment, which is very conducive to physical or chemical modification, thus showing great potential in gas detection. In the gas detection process, the physical or chemical interaction between the detected molecule and the graphene can affect the conductivity of the graphene to the greatest extent. Schedin et al. used graphene to detect the adsorption of individual gas molecules [52]. The breakthrough in the resolution of the gas sensor is mainly due to the extremely low noise (Johnson noise and flicker noise) in the graphene device that cannot drown the signal of a single molecule. O. Leenaerts et al. used the density functional theory (DFT) system to study the adsorption of H_2O, NH_3, CO, NO_2, and NO on graphene [53]. For each adsorbed gas, consider three adsorption sites at the top of the carbon atom, the center of the carbon hexagon, and the center of the carbon—carbon bond. For these positions, the different orientations of the molecules were checked and the adsorption energy of all molecules was calculated to determine their exact orientation on the surface and their preferential binding sites. According to the state density and the highest occupied molecular orbital and the lowest unoccupied molecular orbital of the adsorbate, the charge transfer between different molecules and the graphene surface is discussed. The characteristics of the donor (NH_3, CO, NO) or acceptor (H_2O, NO_2) of different molecular dopants have been determined, which are consistent with the experimental results.

Beyond graphene, graphene derivatives, TMDs, and assembled heterostructures have gradually emerged in the field of gas detection [54–59]. In addition to continuously improving the crystal quality of 2D materials, the gas detection performance of 2D material films can also be further improved through structural design. As shown in Figure 13.4 [60], a device with a suspended structure can increase the contact area between the gas and the 2D material. At the same time, the structure can also avoid the scattering of charge carriers in the 2D material due to the surface roughness of the substrate, charged impurities, and remote phonons.

13.4.2 2D Material Mechanical Sensors

Mechanical sensors have attracted much attention due to their wide application in wearable devices, electronic skins, and artificial intelligence robots. For monitoring human health or environmental changes, there is a great need for flexible sensors with high sensitivity, high performance, lightweight, convenience in carrying, sturdiness and durability, and low energy consumption. Traditional semiconductor materials are almost generally fragile and are not suitable for flexible mechanical sensors. The outstanding mechanical and electrical properties of 2D materials make them the best candidates for the next generation of flexible electronic products. Here, we emphasize 2D flexible mechanical sensors, including flexural strain sensors (bending and stretching) and pressure sensors.

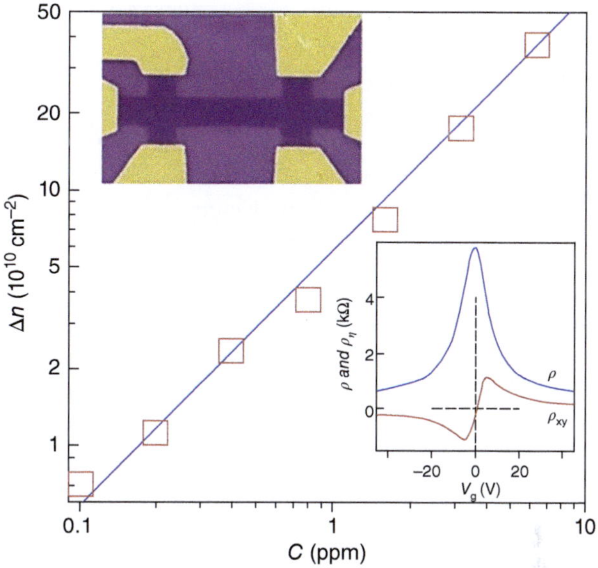

Figure 13.4 Graphene gas sensors. Graphene oxide acts as a conduction channel for FET devices by bridging the source and drain. Source: Ganhua et al. [60]. Copyright 2011, American Chemical Society.

Strain sensors are usually used to prepare sensing materials and flexible electrodes with piezoelectric properties on a flexible substrate. The deformation caused by mechanical strain can affect the lattice–electron interaction, thereby effectively changing the energy band structure of 2D semiconductor crystals. The conductivity of the sensing material is changed by applying an external force to the flexible substrate. Due to the structural characteristics of 2D semiconductors, there must be anisotropy in the plane and the direction perpendicular to the plane. Depending on whether there is anisotropy in the plane, 2D materials can be divided into two categories: in-plane anisotropic (ReS_2, BP, etc.) and in-plane isotropic (graphene, MoS_2, etc.) materials. 2D materials have symmetric/asymmetric in-plane piezoresistive effects, which are very suitable for direction-sensitive strain sensing. ReS_2 is a typical 2D material with in-plane anisotropy. ReS_2 has two anisotropic axes with an angle of about 60°, which has completely opposite piezoresistive effects. As shown in Figure 13.5a, the ReS_2-bending strain sensor [61] built on flexible polyimide can sensitively sense bending from different directions. In addition to the bending strain sensor, the more challenging one is the tensile strain sensor. Although it has a larger strain limit than semiconductor materials, such as silicon, the original form of 2D materials still cannot meet current needs. Zheng et al. adopted the strain engineering concept inspired by the ancient paper-cutting art kirigami, and developed a reversible stretchable electronic device of flexible MoS_2 [10], as shown in Figure 13.1c. The kirigami structure has flexible deformability, which significantly improves the reversible stretchability of flexible MoS_2, and is used on the surface of the elbow joint of the robot. Sensing materials based on the kirigami structure have sensitive strain sensing. However, the electrodes under the same structure need to avoid strain disturbance. As shown in Figure 13.5b, Emmanuel Okogbue et al. built a $PtSe_2$ kirigami conductor [62] with a stretchability of 2000% while keeping the conductivity constant.

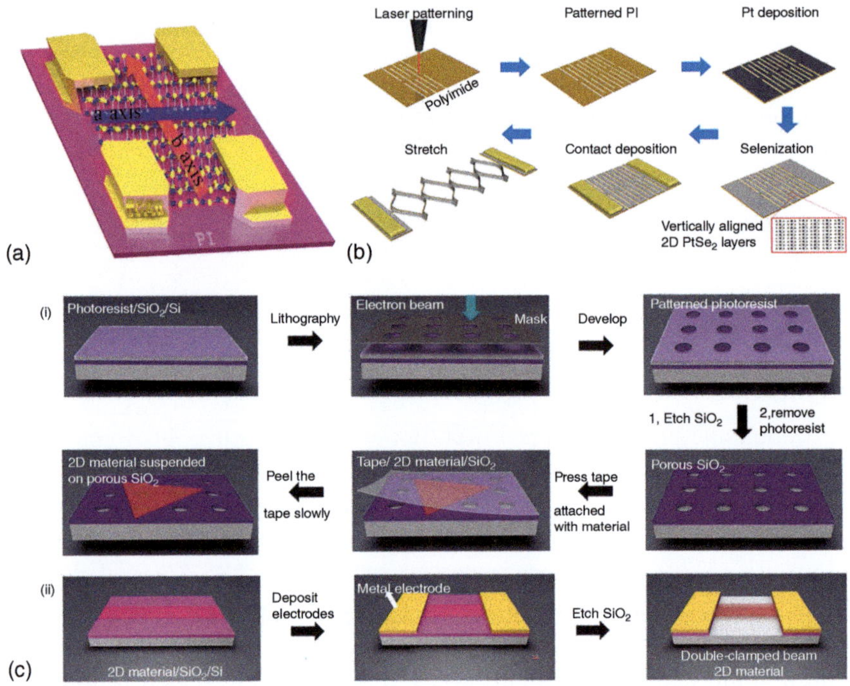

Figure 13.5 2D material mechanical sensors. (a) ReS_2-bending strain sensor with anisotropy. Source: An et al. [61]. Copyright 2019, American Chemical Society. (b) Stretchable $PtSe_2$ metal wire. Source: Okogbue et al. [62]. Copyright 2019, American Chemical Society. (c) Preparation process of 2D material-suspended platform. Source: Jiang et al. [63]. Copyright 2019, Wiley-VCH.

Pressure sensors is one of the most commonly used sensors in industrial practice. They are widely used in various industrial automation environments, involving many industries such as aerospace, road transportation, artificial intelligence, and electronic skin. Recently, pressure sensors based on 2D materials such as graphene have attracted attention due to their outstanding performance. 2D materials have been proven to have piezoelectric properties. Because 2D materials have only an atomic level of thickness, they are susceptible to interference from external physical factors, among which the influence of the substrate (such as doping and parasitic charges) is the most serious. Therefore, suspended 2D materials are used to characterize the intrinsic properties of 2D crystals. As shown in Figure 13.5c, there are two general processes for preparing freely suspended 2D materials [63]: (i) Use dry transfer to transfer the 2D material to the substrate with voids; (ii) Use an acid solution to transfer the 2D material underneath. Part of the sacrificial layer is etched away. Measurement by nanoindentation method proves that graphene is the strongest material ever. Wenjun Chen et al. reported a resistive pressure sensor based on two-layer wrinkle graphene with nanoscale wrinkles [64], which has an ultra-high working sensitivity (6.92 kPa^{-1}). This highly sensitive, low-power pressure sensor is expected to be integrated into flexible electronic devices to achieve tactile sensing.

13.5 2D Material Nanomembrane Robots

Graphene, TMDs, and other 2D layered materials can be used to fabricate microrobots for their competitive properties [65]. For example, 2D layered materials have superior photoelectric performance [66, 67], high in-plane strength, low bending stiffness [32], and magnetic and piezoelectric properties, which have been used in designing ultrathin actuators [68, 69]. These novel researches and excellent characteristics make 2D layered materials striking in microrobots.

13.5.1 Fabrication of 2D Material Nanomembrane Actuators and Robots

The fabrication and processing of 2D thin-film microrobot is mainly composed of the growth (CVD) and the 3D fabrication and the composite (separated in polymer).

Graphene and other 2D layered materials have been heated for their special properties. CVD has been widely used to prepare and produce graphene and other 2D layered materials. These 2D layered materials firstly grow on Ni, Cu, and SiO_2/Si substrates [70, 71]. Take graphene CVD on Ni film for an example, firstly, to enlarge the size of grain, the nickel film is annealed in an Ar/H_2 atmosphere, and then the H_2/CH_4 atmosphere is used to contact the film. During this step, the carbon atoms, obtained by decomposed hydrocarbons, are dissolved into the nickel film to form a solid solution. At last, the chip is cooled in argon. Carbon atoms diffuse out of the Ni–C solid solution and precipitate on the Ni surface to form a graphene film [72]. During the precipitation, Ni film can provide growth sites on its surface because of its similar lattice constants and densely packed hexagonal lattice of graphene [73]. For further fabrication, we should remove the original substrate and transfer the graphene to another specific substrate. Firstly, the graphene was coated by PMMA with a small thickness and then underwent a baking process. The etchant was used to remove the metal sacrificial layer, and the PMMA/graphene film was obtained. Then a transferred process was used to transfer the film onto a new substrate. At last, acetone was used to remove the PMMA. A graphene film on a new substrate can be fabricated [72]. Then the graphene for the main component to fabricate robots was fabricated.

Recently, the fabrication of microrobots with main component of graphene is mainly developed for the progress of micro/nanofabrication and origami/kirigami folding (Figure 13.6). For instance, Zhu et al. approached a graphene/epoxy hybrid bimorph actuator [74]. Graphene films were firstly grown on a SiO_2/Si substrate coated with Ni catalyst layer by CVD. Then the substrate was cut to pieces as a handling substrate. Gold contact pads and a serpentine microheater pattern were formed by photolithography and O_2 plasma reactive ion etching (RIE). At last, a thin epoxy (SU-8) cantilever is used as a support body. During this fabrication process, the graphene was patterned by micro/nanofabrication to meet the demand of actuator. In addition to micro/nanofabrication, origami folding is another way to

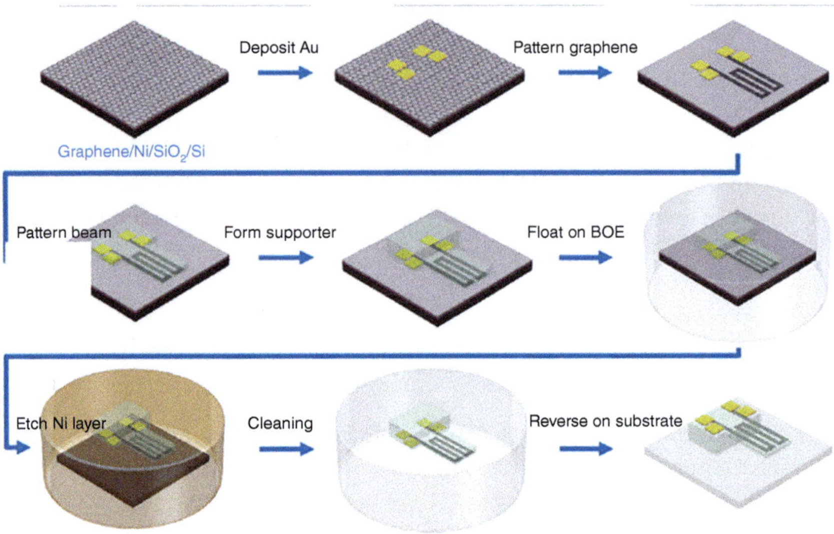

Figure 13.6 Schematic diagram of the fabrication process. Graphene and epoxy hybrid cantilever system. Source: Zhu et al. [74]. Copyright 2011, American Chemical Society.

transform two-dimensional shapes into 3D structures. A three-dimensional layered materials-based robot is constructed by self-folding of 2D materials.

2D layered materials can serve as polymer nanocomposites for their atomically thin structure and abundant surface area, which will largely enhance interactions between the two-dimensional layered materials and the polymer matrix. Zhao et al. realized the controllable distribution of graphene oxide (GO) particles in a hydrogel matrix to design different bionic hydrogel actuators [75]. They developed entirety and locality infiltration to produce hydrogels with GO. For entirety infiltration, GO aqueous solution is used to infiltrate the hydrogels because of the porous microstructure of the hydrogels. For locality infiltration, the GO aqueous solution is injected into the designed hydrogel bodies without water inside to realize local mixing.

Based on these 2D layered materials polymer nanocomposites, people develop the two-dimensional layered materials 3D printing [76]. For instance, 3D printing requires the printed fluid to have good rheological properties. Responsive GO sheets can form a non-covalent network. This network has shear thinning behavior and a sufficiently high elastic shear modulus (G'), which is very suitable for building self-supporting 3D structures. This structure has proven to have rheological properties very suitable for 3D printing. Eduardo Saiz et al. used this construction idea to create a pH-responsive 3D printing polymer with 2D layered materials inside.

13.5.2 Motion and Deformation of 2D Material Nanomembrane Actuators and Robots

The mechanism of motion and deformation of 2D material nanomembrane micro/nanorobot or actuators can be divided into two kinds. The first one is to drive and deform the two-dimensional material film by using its piezoelectric, magnetic, thermal expansion, and other physical properties [68, 69, 76]. The second is to drive and deform under the stimulation of external conditions by using the characteristics of the substrate combined with the 2D material film.

2D materials themselves have many unique properties, which can be utilized in microactuators. For instance, MXenes have been used to act as electrical-responsive, carbon dioxide- and temperature-responsive, relative humidity-responsive, and light-responsive actuators [77]. For example, MXenes can be used in photothermal actuators because of their good thermal conductivity and electromagnetic wave absorption. In these light-heat actuators, MXenes act as a light-to-heat converter, converting light energy into heat energy. Therefore, they have been successfully applied to photothermal actuators. In addition to photothermal actuators, MXenes can serve as humidity-responsive actuators. MXenes, like other 2D materials, show high sensitivity toward moisture, and hence could be good candidates for humidity-responsive actuators [78]. MXenes can also be used in humidity-responsive actuators because of their diverse polar functional groups and freely changeable interlayer spacing. There are many hydrogen bonds on the surface functional groups of MXenes, which will adsorb water molecules, and the relative humidity will affect the number of water molecules adsorbed. Therefore, changing the humidity will cause the number of adsorbed water molecules to change, leading to a change in the interplanar distance, resulting in a macroscopic larger bending displacement. Applied in the field of soft robotics, MXenes can serve as electroactive ionic soft actuators. The introduction of MXenes to ionic actuators significantly enhanced their performance because MXenes show best blocking forces and bending strains. Two-dimensional materials have large areas between layers, which can absorb electrochemical reaction ions reversibly. Based on this characteristic, many two-dimensional materials, including graphene, MoS_2, CNT, and MXenes, have been widely used in electrochemical actuators. An electrochemical actuator based on $Ti_3C_2T_x$ with two-layer and three-layer structures was reported by Di Pang et al. [79]. $Ti_3C_2T_x$ has good stability. In the double-layer structure, the $Ti_3C_2T_x$ MXene film is adhered to the transparent tape and placed in the H_2SO_4 solution (Figure 13.7). At this time, reversible ion intercalation/deintercalation will occur in the middle of the MXene sheet, thereby changing the interlayer spacing. In this case, the H_2SO_4 electrolyte readily produces proton H^+ for reversible redox reaction during the charge and discharge cycles of electrochemical processes, which cause the volume of $Ti_3C_2T_x$ to expand or contract. Another mechanism is to utilize the thermal expansion of 2D materials, Zhu et al. reported a graphene-based bimorph microactuator, which is a graphene/epoxy hybrid bilayer structure [74]. There is a heating circuit connected to the graphene/epoxy hybrid bilayer structure. When current flows through the graphene film, the graphene will generate resistive heat. At the same time, the organic epoxy film will absorb heat. There is a strong π—π bond between the two layers of graphene and epoxy. This allows the two layers to produce 0.4% strain without delamination.

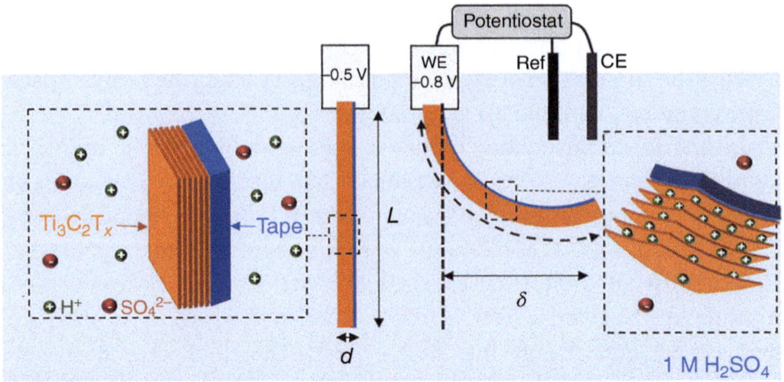

Figure 13.7 Deformation of 2D material nanomembrane actuators. The change of the gap between MXene sheets according to stimulus. Source: Pang et al. [79]. Copyright 2019, American Chemical Society.

Some 2D materials-based microactuators may not deform by the properties of two-dimensional materials, but by the characters of matrix. Some matrices, such as a responsive polymer (a branched copolymer surfactant, BCS), can drive and deform under the stimulation of external conditions. The existence of 2D materials makes up for some defects of these responsive matrices. For example, liquid crystal elastomers (LCEs) are ordered polymers that can exhibit a reversible shape change in response to heat, light, or solvent exchange. The inner part of LCE is many single domains with orientation. When there is no external stimulation, the single domain is in the same row. When stimulated by external heat, light, or solvent exchange, the ordered single domain will turn into disordered one, resulting in macroscopic deformation. LCE actuators generally respond slowly to external stimuli because of the low thermal conductivity of their materials. Yang et al. prepared a high response rate near infrared (NIR)-responsive composite material by adding uniformly arranged graphene sheets to the LCE matrix [80]. This material has a reversible mechanical drive.

In addition to near-infrared light and heat, chemical stimuli can also be used to drive graphene-based soft actuators/robots. For example, Leeladhar and Singh prepared a graphene-PDMS nanocomposite/gold bilayer-based multi-response soft actuator with a dual-drive mode [81]. Using optomechanical drive, the 3D tubular structure can be transformed into a 2D film under NIR irradiation. Using chemical mechanical drive, the fully internally rotated actuator rolls in the opposite direction after being exposed to acetone or aldehyde. In this case, graphene acts as a nanoscale heater, absorbing infrared light, causing the temperature to rise. At this time, the material's coefficient of thermal expansion (CTE) mismatch results in a large mismatch strain and deformation occurs. This achieves optomechanical drive. The graphene nanoplatelets (GNPs)-PDMS layer here is an active layer to achieve chemical driving. As an inert metal, gold is not affected by any chemical solution. Therefore, when the actuator contacts with chemicals, such as acetone and aldehyde, the GNP-PDMS layer will be affected by the chemicals and expand, while the gold layer will not be affected. As a result, the strain adaptation at the double-layer interface leads to the deformation of the macroscopic shape.

In addition to the above methods, electrical stimulation can also drive graphene nanocomposites. Graphene has excellent electrical conductivity, which allows it to be used in enhanced electrical response rGO/poly(2-acrylamide-2-methylpropanesulfonic acid-*co*-acrylamide) nanocomposites [82]. The addition of rGO nanosheets improves the speed of electrical response. When an electric field is applied, the hydrogel will shrink and deform and also produce ion movement. Graphene promotes ion transport, leading to an increase in osmotic pressure, and exerting force on the negatively charged sulfonic acid groups in the gel. This stress gradient causes the hydrogel to shrink. Compared with pure gel, the curvature of the gel with rGO nanosheets is larger and the speed of deformation is faster.

13.6 Summary

2D material nanomembrane is different from other films fixed on the substrate. It has unique electrical, optical, and mechanical characteristics after freestanding. The 2D material can be separated from the substrate as a freestanding nanomembrane, and can also be used as a layer or doping components to form a composite nanomembrane structure with other polymer films. 2D materials nanomembrane has three structures: suspended structure, kirigami structure, and origami structure. The suspended structure is a semi-released state, similar to a drum surface, which will produce some changes in physical properties when deformation occurs. Kirigami and origami are two ways to construct 2D material nanomembrane. Through kirigami and origami, 2D material nanomembrane with special geometric can be constructed to realize some specific functions. 2D material nanomembrane can be used in a variety of sensors and robots systems with these characteristics. Different from other 2D material film sensors, freestanding 2D material nanomembrane sensors can make more flexible use of their own deformation and change some physical characteristics to respond to external stimuli. This is the flexibility of 2D material nanomembrane sensor. For the field of robotics, 2D material nanomembrane provides some characteristics that can be used for actuation, such as magnetism, interlayer expansion, and piezoelectricity. In addition, the peculiar mechanical properties of the 2D material itself are also very conducive to the deformation and support of the robot. With the continuous exploration of nanofabrication technology, 2D material nanomembrane will further expand opportunities for applications in sensors and robotic fields.

References

1 Novoselov, K.S., Geim, A.K., Morozov, S.V. et al. (2004). Electric field effect in atomically thin carbon films. *Science* 306: 666–669.

2 Lalmi, B., Oughaddou, H., Enriquez, H. et al. (2010). Epitaxial growth of a silicene sheet. *Appl. Phys. Lett.* 97 (22): 223109.

3 Dávila, M.E., Xian, L., Cahangirov, S. et al. (2014). Germanene: a novel two-dimensional germanium allotrope akin to graphene and silicene. *New J. Phys.* 16: 095002.
4 Binmin Wu, X.W., Tang, H., Jiang, W. et al. (2020). Multifunctional MoS_2 transistors with electrolyte gel gating. *Small* 6: 2000420.
5 Liu, Y., Huang, Y., and Duan, X. (2019). Van der Waals integration before and beyond two-dimensional materials. *Nature* 567 (7748): 323–333.
6 Lee, C., Wei, X., Kysar, J.W. et al. (2008). Measurement of the elastic properties and intrinsic strength of monolayer graphene. *Science* 321: 385–388.
7 Zhu, H., Wang, Y., Xiao, J. et al. (2015). Observation of piezoelectricity in free-standing monolayer MoS(2). *Nat. Nanotechnol.* 10 (2): 151–155.
8 Cui, X., Kong, Z., Gao, E. et al. (2018). Rolling up transition metal dichalcogenide nanoscrolls via one drop of ethanol. *Nat. Commun.* 9 (1): 1301.
9 Xu, W., Li, T., Qin, Z. et al. (2019). Reversible MoS_2 origami with spatially resolved and reconfigurable photosensitivity. *Nano Lett.* 19 (11): 7941–7949.
10 Zheng, W., Huang, W., Gao, F. et al. (2018). Kirigami-inspired highly stretchable nanoscale devices using multidimensional deformation of monolayer MoS2. *Chem. Mater.* 30 (17): 6063–6070.
11 Gibertini, M., Koperski, M., Morpurgo, A.F. et al. (2019). Magnetic 2D materials and heterostructures. *Nat. Nanotechnol.* 14 (5): 408–419.
12 Liu, J., Xia, F., Xiao, D. et al. (2020). Semimetals for high-performance photodetection. *Nat. Mater.* 19 (8): 830–837.
13 Meireles, L.M., Neto, E.G.S., Ferrari, G.A. et al. (2020). Graphene electromechanical water sensor: the wetristor. *Adv. Electron. Mater.* 6 (2): 1901167.
14 Bunch, J.S., van der Zande, A.M., Verbridge, S.S. et al. (2007). Electromechanical resonators from graphene sheets. *Science* 315 (5811): 490–493.
15 Miao, T., Yeom, S., Wang, P. et al. (2014). Graphene nanoelectromechanical systems as stochastic-frequency oscillators. *Nano Lett.* 14 (6): 2982–2987.
16 Will, M., Hamer, M., Mueller, M. et al. (2017). High quality factor graphene-based two-dimensional heterostructure mechanical resonator. *Nano Lett.* 17 (10): 5950–5955.
17 van der Zande, A.M., Barton, R.A., Alden, J.S. et al. (2010). Large-scale arrays of single-layer graphene resonators. *Nano Lett.* 10 (12): 4869–4873.
18 Chen, C., Rosenblatt, S., Bolotin, K.I. et al. (2009). Performance of monolayer graphene nanomechanical resonators with electrical readout. *Nat. Nanotechnol.* 4 (12): 861–867.
19 Bhaskaran, M.K.H. (2015). Ultrasensitive room-temperature piezoresistive transduction in graphene-based nanoelectromechanical systems. *Nano Lett.* 15 (4): 2562–2567.
20 Smith, A.D., Niklaus, F., Paussa, A. et al. (2013). Electromechanical piezoresistive sensing in suspended graphene membranes. *Nano Lett.* 13 (7): 3237–3242.
21 Dolleman, R.J., Davidovikj, D., Cartamil-Bueno, S.J. et al. (2016). Graphene squeeze-film pressure sensors. *Nano Lett.* 16 (1): 568–571.
22 Davidovikj, D., Scheepers, P.H., van der Zant, H.S.J. et al. (2017). Static capacitive pressure sensing using a single graphene drum. *ACS Appl. Mater. Interfaces* 9 (49): 43205–43210.

23 Tao, L.-Q., Zhang, K.-N., Tian, H. et al. (2017). Graphene-paper pressure sensor for detecting human motions. *ACS Nano* 11 (9): 8790–8795.

24 Zhao, J., Zhang, G.-Y., and Shi, D.-X. (2013). Review of graphene-based strain sensors. *Chin. Phys. B* 22 (5): 57701.

25 Tian, H., Shu, Y., Cui, Y.-L. et al. (2014). Scalable fabrication of high-performance and flexible graphene strain sensors. *Nanoscale* 6 (2): 699–705.

26 Liu, C.-H., Chang, Y.-C., Norris, T.B. et al. (2014). Graphene photodetectors with ultra-broadband and high responsivity at room temperature. *Nat. Nanotechnol.* 9 (4): 273–278.

27 Sun, J., Schmidt, M.E., Muruganathan, M. et al. (2016). Large-scale nanoelectromechanical switches based on directly deposited nanocrystalline graphene on insulating substrates. *Nanoscale* 8 (12): 6659–6665.

28 Tian, H., Li, C., Mohammad, M.A. et al. (2014). Graphene earphones: entertainment for both humans and animals. *ACS Nano* 8 (6): 5883–5890.

29 Zhou, Q. and Zettl, A. (2013). Electrostatic graphene loudspeaker. *Appl. Phys. Lett.* 102 (22): 223109.

30 Zhou, Q., Zheng, J., Onishi, S. et al. (2015). Graphene electrostatic microphone and ultrasonic radio. *Proc. Natl. Acad. Sci. U.S.A.* 112 (29): 8942–8946.

31 Todorovic, D., Matkovic, A., Milicevic, M. et al. (2015). Multilayer graphene condenser microphone. *2D Mater.* 2 (4): 045013.

32 Akinwande, D., Brennan, C.J., Bunch, J.S. et al. (2017). A review on mechanics and mechanical properties of 2D materials-graphene and beyond. *Extreme Mech. Lett.* 13: 42–77.

33 Davidovikj, D., Bouwmeester, D., van der Zant, H.S.J. et al. (2018). Graphene gas pumps. *2D Mater.* 5 (3): 031009.

34 Davidovikj, D., Alijani, F., Cartamil-Bueno, S.J. et al. (2017). Nonlinear dynamic characterization of two-dimensional materials. *Nat. Commun.* 8: 1253.

35 Davidovikj, D., Poot, M., Cartamil-Bueno, S.J. et al. (2018). On-chip heaters for tension tuning of graphene nanodrums. *Nano Lett.* 18 (5): 2852–2858.

36 Dolleman, R.J., Houri, S., Davidovikj, D. et al. (2017). Optomechanics for thermal characterization of suspended graphene. *Phys. Rev. B* 96 (16): 165421.

37 Goldsche, M., Sonntag, J., Khodkov, T. et al. (2018). Tailoring mechanically tunable strain fields in graphene. *Nano Lett.* 18 (3): 1707–1713.

38 Georgiou, T., Britnell, L., Blake, P. et al. (2011). Graphene bubbles with controllable curvature. *Appl. Phys. Lett.* 99 (9): 093103.

39 Blees, M.K., Barnard, A.W., Rose, P.A. et al. (2015). Graphene kirigami. *Nature* 524 (7564): 204.

40 Zhang, J., Xiao, J., Meng, X. et al. (2010). Free folding of suspended graphene sheets by random mechanical stimulation. *Phys. Rev. Lett.* 104 (16): 166805.

41 Wang, B., Huang, M., Kim, N.Y. et al. (2017). Controlled folding of single crystal graphene. *Nano Lett.* 17 (3): 1467–1473.

42 Joung, D., Nemilentsau, A., Agarwal, K. et al. (2017). Self-assembled three-dimensional graphene-based polyhedrons inducing volumetric light confinement. *Nano Lett.* 17 (3): 1987–1994.

43 Xu, W., Qin, Z., Chen, C.-T. et al. (2017). Ultrathin thermoresponsive self-folding 3D graphene. *Sci. Adv.* 3 (10): e1701084.

44 Caliendo, C., D'amico, A., Varadi, P. et al. (1988). Surface acoustic wave H_2 sensor on silicon substrate. In: *IEEE 1988 Ultrasonics Symposium Proceedings*, 569–574.

45 Jr, B., David, S., and Wohltjen, H. (1989). Surface acoustic wave devices for chemical analysis. *Anal. Chem.* 61: 704A–715A.

46 RAPP, M., BINZ, D., KABBE, I. et al. (1991). A new high-frequency high-sensitivity SAW device for NO_2 gas detection in the sub-ppm range. *Sens. Actuators, B* 4: 103–108.

47 Bergveld, P. (1970). Development of an ion-sensitive solid-state device for neurophysiological measurements. *IEEE Trans. Biomed. Eng.* 1: 70–71.

48 Kong, J., Franklin, N.R., Zhou, C. et al. (2000). Nanotube molecular wires as chemical sensors. *Science* 287: 622–625.

49 Collins, P.G., Bradley, K., Ishigami, M. et al. (2000). Extreme oxygen sensitivity of electronic properties of carbon nanotubes. *Science* 287: 1801–1804.

50 Crowther, A.C., Ghassaei, A., Jung, N. et al. (2012). Strong charge-transfer doping of 1 to 10 layer graphene by NO_2. *ACS Nano* 6: 1865–1875.

51 Yang, S., Jiang, C., and Wei, S.-h. (2017). Gas sensing in 2D materials. *Appl. Phys. Rev.* 4 (2): 021304.

52 Schedin, F., Geim, A.K., Morozov, S.V. et al. (2007). Detection of individual gas molecules adsorbed on graphene. *Nat. Mater.* 6 (9): 652–655.

53 Leenaerts, O., Partoens, B., and Peeters, F.M. (2008). Adsorption of H_2O,NH_3, CO,NO_2, and NO on graphene: a first-principles study. *Phys. Rev. B* 77 (12): 125416.

54 He, Q., Zeng, Z., Yin, Z. et al. (2012). Fabrication of flexible MoS_2 thin-film transistor arrays for practical gas-sensing applications. *Small* 8 (19): 2994–2999.

55 Late, D.J., Doneux, T., and Bougouma, M. (2014). Single-layer $MoSe_2$ based NH_3 gas sensor. *Appl. Phys. Lett.* 105 (23): 233103.

56 Li, H., Yin, Z., He, Q. et al. (2012). Fabrication of single- and multilayer MoS_2 film-based field-effect transistors for sensing NO at room temperature. *Small* 8 (1): 63–67.

57 Niu, Y., Wang, R., Jiao, W. et al. (2015). MoS_2 graphene fiber based gas sensing devices. *Carbon* 95: 34–41.

58 Feng, Z., Xie, Y., Chen, J. et al. (2017). Highly sensitive $MoTe_2$ chemical sensor with fast recovery rate through gate biasing. *2D Mater.* 4 (2): 025018.

59 Donarelli, M. and Ottaviano, L. (2018). 2D materials for gas sensing applications: a review on graphene oxide, MoS(2), WS(2) and phosphorene. *Sensors (Basel)* 18 (11): 3638.

60 Ganhua, L., Park, S., Yu, K. et al. (2011). Toward practical gas sensing with highly reduced graphene oxide – a new signal processing method to circumvent run-to-run and device-to-device variations. *ACS Nano* 5: 1154–1164.

61 An, C., Xu, Z., Shen, W. et al. (2019). The opposite anisotropic piezoresistive effect of ReS_2. *ACS Nano* 13 (3): 3310–3319.

62 Okogbue, E., Han, S.S., Ko, T.J. et al. (2019). Multifunctional two-dimensional $PtSe_2$-layer kirigami conductors with 2000% stretchability and metallic-to-semiconducting tunability. *Nano Lett.* 19 (11): 7598–7607.

63 Jiang, H., Zheng, L., Liu, Z. et al. (2019). Two-dimensional materials: from mechanical properties to flexible mechanical sensors. *InfoMat* 2 (6): 1077–1094.

64 Chen, W.J., Gui, X.C., Liang, B.H. et al. (2017). Structural engineering for high sensitivity, ultrathin pressure sensors based on wrinkled graphene and anodic aluminum oxide membrane. *ACS Appl. Mater. Interfaces* 9 (28): 24111–24117.

65 Novoselov, K.S., Fal'ko, V.I., Colombo, L. et al. (2012). A roadmap for graphene. *Nature* 490 (7419): 192–200.

66 Yin, Z., Zhu, J., He, Q. et al. (2014). Graphene-based materials for solar cell applications. *Adv. Energy Mater.* 4 (1): 1300574.

67 Lopez-Sanchez, O., Lembke, D., Kayci, M. et al. (2013). Ultrasensitive photodetectors based on monolayer MoS_2. *Nat. Nanotechnol.* 8 (7): 497–501.

68 Wu, W., Wang, L., Li, Y. et al. (2014). Piezoelectricity of single-atomic-layer MoS_2 for energy conversion and piezotronics. *Nature* 514 (7523): 470.

69 Cheon, G., Duerloo, K.-A.N., Sendek, A.D. et al. (2017). Data mining for new two- and one-dimensional weakly bonded solids and lattice-commensurate heterostructures. *Nano Lett.* 17 (3): 1915–1923.

70 De Arco, L.G., Zhang, Y., Kumar, A. et al. (2009). Synthesis, transfer, and devices of single- and few-layer graphene by chemical vapor deposition. *IEEE Trans. Nanotechnol.* 8 (2): 135–138.

71 Yu, Q., Lian, J., Siriponglert, S. et al. (2008). Graphene segregated on Ni surfaces and transferred to insulators. *Appl. Phys. Lett.* 93 (11): 113103.

72 Zhang, Y., Zhang, L., and Zhou, C. (2013). Review of chemical vapor deposition of graphene and related applications. *Acc. Chem. Res.* 46 (10): 2329–2339.

73 Eizenberg, M. and Blakely, J.M. (1979). Carbon monolayer phase condensation on Ni(111). *Surf. Sci.* 82 (1): 228–236.

74 Zhu, S.-E., Shabani, R., Rho, J. et al. (2011). Graphene-based bimorph microactuators. *Nano Lett.* 11 (3): 977–981.

75 Zhao, Q., Liang, Y., Ren, L. et al. (2018). Bionic intelligent hydrogel actuators with multimodal deformation and locomotion. *Nano Energy* 51: 621–631.

76 Garcia-Tunon, E., Barg, S., Franco, J. et al. (2015). Printing in three dimensions with graphene. *Adv. Mater.* 27 (10): 1688.

77 Nguyen, V.H., Tabassian, R., Oh, S. et al. (2020). Stimuli-responsive MXene-based actuators. *Adv. Funct. Mater.* 30 (47): 1909504.

78 Park, S., An, J., Suk, J.W. et al. (2010). Graphene-based actuators. *Small* 6 (2): 210–212.

79 Pang, D., Alhabeb, M., Mu, X. et al. *Nano Lett.* 19 (10): 7443–7448.

80 Yang, Y., Zhan, W., Peng, R. et al. (2015). Graphene-enabled superior and tunable photomechanical actuation in liquid crystalline elastomer nanocomposites. *Adv. Mater.* 27 (41): 6376–6381.

81 Leeladhar and Singh, J.P. (2018). Photomechanical and chemomechanical actuation behavior of graphene-poly(dimethylsiloxane)/gold bilayer tube for multimode soft grippers and volatile organic compounds detection applications. *ACS Appl. Mater. Interfaces* 10 (40): 33956–33965.

82 Yang, C., Liu, Z., Chen, C. et al. (2017). Reduced graphene oxide-containing smart hydrogels with excellent electro-response and mechanical properties for soft actuators. *ACS Appl. Mater. Interfaces* 9 (18): 15758–15767.

14

Strain-Tuning of 2D Transition Metal Dichalcogenides

Zhao An, Michael Zopf, and Fei Ding

Leibniz Universität Hannover, Institut für Festkörperphysik, Appelstraße 2, 30167 Hannover, Germany

14.1 Introduction

Since the discovery of graphene, atomically thin two-dimensional nanomaterials have become a highly attractive and widely studied field of research. Their unique electronic, optical, mechanical, and chemical properties offer a great potential for scientific research and industrial application. Graphene is known for its extraordinary mechanical strength and fascinating quantum transport phenomena due to the presence of massless Dirac Fermions [1]. However, the lack of an intrinsic bandgap limits its potential application in (quantum) optics or optoelectronics. Many different types of 2D nanomaterials are being investigated beyond graphene, such as transition metal dichalcogenides (TMDs, e.g. MoS_2, WSe_2), elementary 2D materials (e.g. silicene, germanene), layered ionic solids (e.g. KLn_2TiO_3), or magnetic insulators (e.g. CrI_3), providing new properties and applications. Until now, more than 100 types of 2D materials have been obtained by different fabrication methods such as chemical vapor deposition (CVD), physical vapor deposition (PVD), mechanical exfoliation, or Li intercalation.

2D TMDs are atomically thin semiconductors whose mechanical, electronic, optical, and optoelectronic properties [2–4] have received increasing attention in recent years. They exhibit intrinsic bandgaps (and correspondingly photoluminescence [PL]) in the visible range (around 400–760 nm) and further unique properties such as strong luminescence, high absorption, strong spin–orbit coupling (SOC), and high carrier mobility [5]. The implementation of heterostructures allows for the fabrication of P–N junctions. Potential applications, therefore, include photonic or nanoelectronic devices (e.g. field-effect transistors [6], photodetectors [7], and light-emitting diodes [8]) but are also found in other research fields such as energy and hydrogen storage applications [9] or DNA sequencing [10].

The properties of TMDs can be manipulated by applying strain, magnetic, or electric fields, allowing for a deeper insight into the underlying physics and new possibilities for application. Most 2D nanomaterials have very different mechanical properties compared to their bulk counterparts. Typically, the presence of lattice

Nanomembranes: Materials, Properties, and Applications, First Edition.
Edited by Yongfeng Mei, Gaoshan Huang, and Xiuling Li.
© 2022 WILEY-VCH GmbH. Published 2022 by WILEY-VCH GmbH.

defects leads to low failure strain in three-dimensional solids. In contrast, atomically thin materials like 2D TMDs can withstand large strains (well above 10%) without rupture, permitting for tuning their properties over a large range. This makes strain engineering a particularly suitable tool for the investigation and control of the properties of 2D TMDs.

This chapter starts off by looking at the composition and chemical bonds of 2D TMDs. It then moves on to the structural level and the effects related to symmetry such as optical, electronic and vibrational properties as well as spin–valley physics. After that, the strain-tuning technique for 2D TMDs is reviewed by elaborating on different types of strain fields and tuning techniques as well as their effect on the material properties. As an outlook, representative building blocks of potential optical and optoelectronic applications and the related prospective research are briefly introduced.

14.2 Structure and Properties of 2D Transition Metal Dichalcogenides

14.2.1 Composition and Chemical Bonds

TMDs are layered structures of the type MX_2, where M is the transition metal atom (e.g. Mo, W, Ti, Zr, Re, Pt, Nb, and Ta) and X the chalcogen atom (e.g. S, Se, and Te). Different chemical compositions result in unique properties of 2D TMDs: for instance, $NbSe_2$ exhibits superconductivity, while MoS_2 is a semiconductor and TiS_2 shows metallic properties. TMDs monolayers comprise a X–M–X trilayer, where the M atom layer is sandwiched by two X atom layers. Each M atom is six-coordinated, and the atomic interaction is governed by the chemical bonds between M and X, as shown in Figure 14.1. Interlayer interaction between different TMD monolayers is determined by Van der Waals forces, far weaker than the chemical bonds within each layer. The characteristics of each material are influenced by the thickness, geometry, and electronic configuration [11]. The interplay of different binding mechanisms allows for a certain freedom in designing new materials systems by vertically stacking different types of TMD monolayers. In the following, we focus mostly on the most frequently studied TMDs: MoS_2, $MoSe_2$, WS_2, and WSe_2.

In silicon or graphene, the electronic properties are determined by the hybridization of s and p orbitals. In contrast, the d-orbital filling of the transition metals has a major effect on the properties of 2D TMDs. The transition metal atoms contain partly filled d orbitals in the configuration $5s^1 4d^5$ (Mo) and $6s^2 4f^{14} 5d^4$ (W). In the chalcogen atoms, the orbitals are filled regularly, with $3s^2 3p^4$ (S) and $4s^2 4p^4$ (Se). The difference of electronegativities between M (above 1.36) and X atoms (2.58 for S, 2.55 for Se, and 2.12 for Te) is not large enough (typically <1.7) for developing fully ionic bonds. In the resulting mixture of covalent and ionic bonds, the electron density shifts from the d-orbital of the M atoms closer toward the X atoms. The degree of d-orbital filling influences the atomic structure of TMD layers, resulting in two types of phases: trigonal prismatic (H – hexagonal symmetry) or octahedral (T – tetragonal symmetry). In the trigonal prismatic phase, the X atoms are vertically aligned with

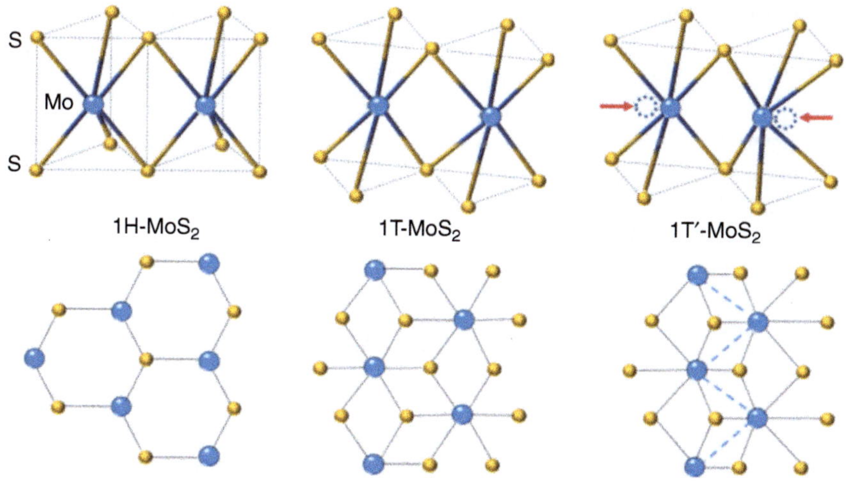

Figure 14.1 Different phase structures of monolayer MoS_2. The top row shows the side view, the bottom row the top view. Source: Adapted from Yin et al. [6]. https://www.nature.com/articles/s41467-017-00640-2. Licensed under CC BY 4.0.

each other, while in the octahedral phase, the top X atom layer is rotated $\pi/3$ around the vertical axis passing through the M atom (see Figure 14.1). In the 1H phase (a single TMD layer) the d orbitals tend to split into three degenerate substates, d_{zx}, d_{yz}, and $d_{x^2-y^2}$, d_{xy} and d_{z^2} with ≈1 eV splitting between the latter two states. In the 1T phase the d orbitals split into two orbitals $d_{x^2-y^2}$, d_{z^2} and d_{zx}, d_{yz}, d_{xy} [12].

T-type monolayers of the group 6 TMDs considered here are metastable and can relax into other phases [12] such as the zigzag phase (T′ – blue, dashed line in Figure 14.1). T (or T′) type monolayers are metallic, while the H phase shows semiconducting properties. The electronic properties are a result of the d-orbital filling since the p orbitals of the X atoms are energetically much lower than the Fermi energy. For example in 1H-WSe_2, there are no electrons left in the d orbital resulting in semiconducting characteristics. In comparison, in 1T-VSe_2 (a group 5 TMD) metallic behavior is observed due to partial filling of the d orbitals. Different phases of TMDs can interconvert, e.g. via strain engineering; the principle is shown in Ref. [13]. The interatomic in-plane distances in the T′ phase are larger than those in the H phase, which can be one reason for a semiconductor to metal transition in response to tensile strain. The lattice mismatch between the H and T′ phase determines the strain required to induce the phase transition [14]. Compared to other 2D TMDs, the phase transition for $MoTe_2$ occurs particularly fast, as the lattice mismatch is merely 3% (see AFM [atomic force microscope]-tuning technique).

The monolayer trigonal phase is also called 1H, as in the left sketch of Figure 14.1. Multilayer stacks of the trigonal phase result in hexagonal and rhombohedral symmetry denoted with 2H and 3R, respectively, featuring different symmetry and interlayer charge transfer dynamics. Naturally occurring crystals are mostly H type and most frequently studied. Below, all the TMDs mentioned are H-type, unless specified.

14.2.2 Structural Symmetry

The symmetry of the crystal lattice is one of fundamental features determining the properties of solids. Here, the symmetry of 2D TMDs and its implications are discussed. We start by looking at the X and M atom layers. In each individual layer, X and M are placed in a honeycomb-type structure, similar to the arrangement of carbon atoms in graphene. However, only half of the sites in the TMD honeycomb structure are occupied, resulting in a C_3 or $2\pi/3$ rotational symmetry which is inherited by the X–M–X trilayer. An additional symmetry in the (H-type) trilayer is the mirror symmetry, i.e., the top and bottom X layers are mirrored with respect to the central M atom layer. Mirror symmetry has great importance: On the one hand, it requires the lifting of the band degeneracy induced by the SOC to be in the out-of-plane (z) direction [15]; on the other hand, the mirror symmetry forces the Bloch functions in the top and bottom layers to be invariant. Lastly, the time-reversal symmetry of **K** and **−K** valleys (Figures 14.3 and 14.4) in reciprocal space is broadly discussed in the literature and is reviewed in Section 14.2.4.

Multiple layers of TMDs can be stacked on top of each other, much like lego bricks. If only one kind of monolayer TMD (i.e. one X–M–X trilayer type) is used, 2H type $\left(D_{4h}^{6}\right)$, 3R type $\left(C_{3v}^{5}\right)$, and further multiple (homo)layer TMDs can be obtained. Symmetries in these assembled systems are determined by the underlying symmetry in the monolayer TMDs. One common feature is that their mirror symmetry is broken.

3R TMDs comprise three X–M–X layers on top of each other. The position of the three layers can be defined by the in-plane translation of each monolayer, with the X atoms in each layer being laterally placed on top of the M atoms in the respective lower layer (see Figure 14.2). The interlayer hopping of charge carriers, including the contribution of valence band and conduction band, vanishes at the **K** and **−K** points in reciprocal space, which is a consequence of the rotational symmetry of the band edge Bloch function. 3R TMDs share many properties with monolayer TMDs.

2H TMDs are composed of two X–M–X layers, with the upper monolayer comprising a rotation of π compared to the bottom monolayer. In contrast to 3R, 2H TMDs are inversion symmetric. Different from monolayer and 3R TMDs, interlayer hopping of charge carriers does not vanish at the **K** and **−K** points in 2H TMDs, because the contribution from the valence band still exists [15].

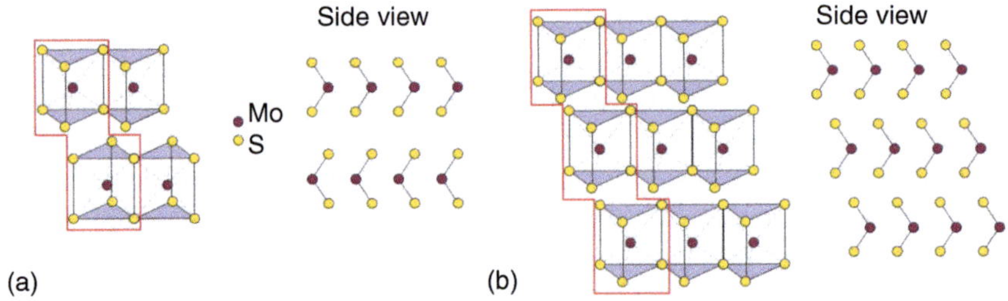

Figure 14.2 Crystal structure of 2H (a) and 3R (b) MoS_2. Source: Adapted from Zhao et al. [11]. https://www.nature.com/articles/lsa2016131. Licensed under CC BY-NC-SA 4.0.

An even number of layers in a TMD result in inversion symmetry, in contrast to the case of an odd number. This symmetry gives rise to many different phenomena: 2D materials with an odd number of layers are piezoelectric while with an even number they are not. Piezoelectricity requires the crystal not to be inversion symmetric, so the positive and negative centers of charge will not coincide when subjected to external strain. Additionally, the breaking of the inversion symmetry allows for studying valleytronics [20].

Materials containing different types of monolayers are called heterostructures or heterojunction samples. We limit our discussion to the simplest case of two different monolayers (called heterobilayers). Due to the comparatively weak out-of-plane Van der Waals interaction, lattice mismatch and a twist angle between the monolayers can exist. While underlying monolayer symmetries can be inherited, these additional degrees of freedom enable the formation of higher-order periodical structures, so-called Moiré patterns. The structure of this superlattice will therefore introduce an additional symmetry.

In heterobilayers, the superlattice constant is $b \approx a/\sqrt{\delta^2 + \theta^2}$, where a is the lattice constant of the monolayer TMD, θ is the twist angle, and δ is the lattice mismatch. The larger the twist angle and lattice mismatch, the smaller the superlattice cell will be. The experimentally observed lattice constant of Moiré patterns is usually in the order of several tens of nm, however can also be as low as 0.5 nm [21].

Heterobilayers have two types of stacking formations as determined by a twist angle of 0 or π, corresponding to H- and R-type stacking (names come from the 2H and 3R stacking in TMDs monolayer). Additionally, different twist angles and lattice mismatches result in distinguished local atomic arrangements or registries. There are six high-symmetry points in Moiré patterns (three in H-type stacking, three in R-type stacking). In these high-symmetry points, the C_3 symmetry is still intact. However, the optical selection rules are different [22] because of the broken mirror symmetry in the heterobilayer. Moreover, the interlayer hopping of charge carriers introduces an extra deviation into the local potential distribution.

After discussing the structural properties in real space, we now move on to the electronic characteristics in reciprocal space.

14.2.3 Band Structure

In the following, we focus our discussion on the four most commonly studied (VIB group) TMDs: MoS_2, $MoSe_2$, WS_2, and WSe_2. Monolayer TMDs can be exfoliated from their bulk counterparts. In bulk TMDs, the conduction and valence band edges are in the $\mathbf{Q}(\mathbf{\Lambda})$ point and $\mathbf{\Gamma}$ point, respectively, therefore, forming an indirect bandgap. As the thickness of TMDs decreases, the energy of the conduction band increases and that of the valence band decreases, resulting in a widened bandgap. Furthermore, the indirect bandgap turns into a direct one due to the shift of the band-edges. The band character at different points in the reciprocal lattice can be assigned to the different contributions of the involved atomic orbitals: At the $\mathbf{K}$ and $-\mathbf{K}$ points the d orbitals from the M atoms contribute most, and their localized nature results in stable bands at the $\mathbf{K}$ and $-\mathbf{K}$ points when the TMD thickness decreases. As a consequence, the band edges transfer to the **K** and **−K** points. The photon emission and light–matter interaction [23] in the monolayer are much stronger than that in the bilayer counterpart.

Multilayer systems exhibit slightly different properties compared to monolayer TMDs due to the additional interlayer interaction. The different orbitals of M and X atoms exhibit distinct contributions to the band edges, causing the band edges to be site-dependent. In multilayer structures, this results in different magnitude of interlayer hopping at the **K** and **Q**(**Λ**) points (Figure 14.3). The work function and bandgap in the four TMDs considered here result in any combination of these TMDs in a heterolayer having a staggered band [24–29] alignment (type II band edges), as sketched in Figure 14.3c. As a result, the photoexcited electrons and holes will relax to the conduction band edge of MoX_2 and the valence band edge of WX_2. When irradiated by laser light, the generated carriers in one layer can transfer to the other layer before recombination. Driven by the large band offset of the conduction (several tens of meV) and valence bands (several hundred meV) of the different layers, the charge transfer is very fast (<1 ps) [30]. In comparison, the decay time of intra-layer electrons is typically in the range of 1–100 ps. The electron and hole residing in the two respective layers will form an exciton because of the strong Coulomb interaction. The spatial separation of the carriers and indirect exciton momentum will lead to long radiative lifetimes that can reach 100 ns [31]. Compared to the in-plane monolayer exciton, the interlayer excitons show new interesting features, revealing the potential for valley-functional optoelectronic devices [32].

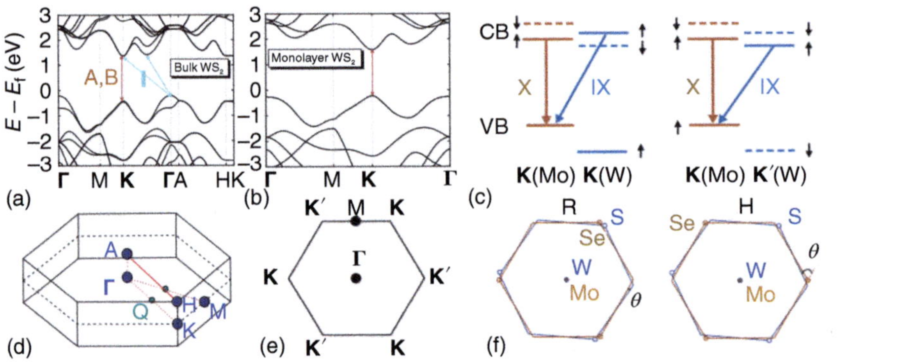

Figure 14.3 Band structures and Brillouin zones for TMDs. (a, b) Calculated band structure of bulk and monolayer WS_2, showing indirect and direct bandgaps, respectively. (d, e) The corresponding three-dimensional and two-dimensional Brillouin zones, highlighting different symmetry points. K′ denotes the −K valley. (c) Schematic band structure of a $WS_2/MoSe_2$ heterobilayer, showing type II (staggered) band alignment. Black arrows denote the spin orientation of the charges in the conduction band (CB) and valence band (VB). The radiative transitions of intra-layer excitons (X) and interlayer excitons (IX) are shown. (f) Two-dimensional Brillouin zones of a $WS_2/MoSe_2$ heterobilayer with the twist angle (θ). Source: (a, b, d) Chaves et al. [16]. https://www.nature.com/articles/s41699-020-00162-4. Licensed under CC BY 4.0; (e) Yoshikawa et al. [17]. https://www.nature.com/articles/s41467-019-11697-6. Licensed under CC BY 4.0; (c, f) Zhang et al. [18]. https://www.nature.com/articles/s41467-020-19466-6. Licensed under CC BY 4.0.

A further change compared to bulk to monolayer TMDs arises in the band curvatures, and therefore, effective masses of the carriers, as shown in Figure 14.3a, b. The change of effective mass at the band edge is twofold: On the one hand, the density of states at the same points [33] (**Γ** in valence band [VB] and **Λ** in conduction band [CB]) is altered. On the other hand, the band edge shifts from the **Γ** point in the VB (**Λ** point in CB) to the **K** (**K**) point, when changing from bilayer to monolayer TMDs. A similar band edge shift can be induced by strain, as illustrated in Section 14.3.3.

14.2.4 Valley-Contrasting Physics and Optical Selection Rules

Monolayer TMDs provide a platform for investigating spintronics and valleytronics, which show strong potential for information storage and processing. Valleys are the extrema of the conduction and valence bands in reciprocal/momentum space. The **K** and **−K** points contain energy degenerate but in-equivalent valleys, see Figure 14.4. Strong SOC not only leads to lifted degeneracy of the bands at **K** and **−K** points but also to a coupling between spins and valleys. The spin polarization in the different valence bands has to be the opposite in the **K** and **−K** points to preserve time-reversal symmetry [4]. The selection rules are additionally governed by the C_3 symmetry. As a result, the spin-conserving transition between valence and conduction band at the **K** valleys will couple with σ^+ and at the **−K** valleys with σ^-, as experimentally observed by Xu and coworkers [34]. The spin-singlet transition dipole polarized perpendicular to the atomic plane is forbidden (dark exciton) and therefore result in significantly weaker emission intensity (several percent) compared to the in-plane polarized signal [35].

Besides the optical selection rules in the monolayer, the valley Hall effect (a type of anomalous Hall effect) also attracts much attention. It is based on the coupling between the orbital motion of electrons with the valley degree of freedom and causes electrons from different valleys to move in opposite directions, perpendicular to the drift current [36, 37]. This is often explained using pseudovectors, such as the Berry curvature and orbital magnetic moment, which show even parity in the **K** and **−K** valleys in the presence of inversion symmetry. In this case, these quantities at **K** and **−K** valleys take the same value under a uniform electric field. The breaking of the inversion symmetry is therefore necessary to obtain a valley contrast. In the case of

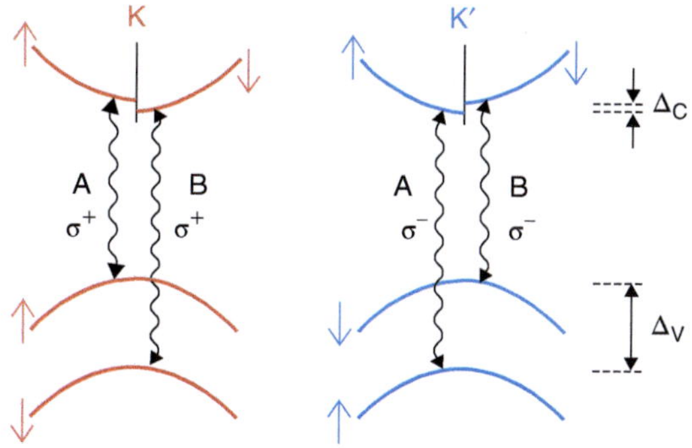

Figure 14.4 Valleys and optical selection rules in monolayer TMDs. The red and blue arrows denote the spin directions, and Δ_c and Δ_v the splitting in the conduction and valence bands, respectively. K′ denotes the −K valley. Source: Stier et al. [19]. https://www.nature.com/articles/ncomms10643. Licensed under CC BY 4.0.

a monolayer, the inversion symmetry is broken and therefore the berry curvatures at **K** and −**K** points give rise to opposite velocities perpendicular to the applied electric field.

The physics in heterobilayers is more complex due to position-dependent atomic registries. Because of the twist angle and lattice mismatch, the overall degree of symmetry is not as high as that in the monolayer [22]. The optical properties have been observed experimentally by Xu and coworkers [25] and other researchers [38, 39].

14.2.5 Excitonic Effects

In this section, excitonic properties in monolayer and heterobilayer TMDs are discussed. Compared with the bulk counterparts, the size and binding energy of excitons in monolayer TMDs are different due to reduced dielectric screening. The heterobilayer furthermore comprises interlayer excitons, in which the electrons and holes are separated in the two layers.

In bulk semiconductors, excitons are mostly of the Wannier type. Their binding energies and exciton radii are proportional to $1/\varepsilon^2$ and ε, respectively, where ε is the dielectric constant. Decreasing the thickness of the semiconductor significantly influences the exciton characteristic: In the extreme case of a single monolayer, the electric field is mainly distributed between the X–M–X layer, leaving the atoms unscreened [40]. This behavior can be understood by calculations [41], showing a reduced dielectric screening effect with nonlocal performance.

In monolayer TMDs the extent of the exciton wavefunction is closer to that of a Wannier–Mott exciton, whereas the binding energy is more of the Frenkel type. The 2D exciton Bohr radius in monolayer TMDs is about 1 nm [40] (several unit cells length), whereas in typical bulk semiconductors such as GaAs it is 13 nm. The exciton binding energies in monolayer TMDs are typically found to be hundreds of meV. In comparison, the binding energy in GaAs is about 5 meV. The electrons and holes of the exciton in monolayer TMDs are usually located around the **K** and −**K** points.

Drawbacks of using 2D TMDs for classical (or quantum) information storage applications are the short radiative lifetimes and valley depolarization times. At cryogenic temperatures, time-resolved PL measurements typically show two or three contributions to the decay [42]: The first one (several picoseconds) is attributed to non-radiative decay by exciton annihilation. Some *ab initio* calculations estimate a time scale of 0.1–1 ps for these processes [43]. The other contributions (e.g. around 33 ps [44]) are explained by radiative decay processes, probably also linked to the emission of excitons localized at crystal potential fluctuations. At the same time, excitons show short valley depolarization time, which is reported to be related to the electron–hole exchange interaction, coupling the valley pseudospin to the exciton center-of-mass motion [45].

Neutral excitons can bind with another electron or hole, forming negative (X^-, T^-) or positive (X^+, T^+) charged excitons (trions). The fluorescence properties of trions are usually characterized by a biexponential decay [44]. Interestingly, optical trion transitions were found to show a splitting of about 7 meV. Yao and coworkers [46] interpreted this splitting as the effect of excess electron–exciton exchange interaction. Besides, the electron-hole exchange interaction in the trion has less influence on the valley depolarization when compared to bright neutral excitons [45]. Further investigations focus on the biexcitons [47] and dark excitons in WSe_2 [42] and dynamics between excitons and trions [48].

The strong excitonic effect gives rise to a significant transfer of oscillator strength from the interband transition to the exciton 1s state [4], leading to strong light–matter interaction. In the visible light range, three kinds of TMDs show strong Van Hove singularities in the density of states (DOS) [23]. The strong light–matter interaction leads to high absorption of light and generation of excitons. For monolayer MoS_2 and WS_2, an absorption of 5–10% was reported [49], which is one order magnitude higher than that in typical bulk semiconductors such as Si and GaAs.

In heterobilayer TMDs, electrons and holes may reside in different layers, forming interlayer excitons. This charge separation greatly extends the lifetime and valley depolarization time. Xu and coworkers [50] reported interlayer exciton lifetimes as long as 1.8 ns in a $WSe_2/MoSe_2$ heterobilayer. Although the Bohr radii and binding energies of interlayer excitons are in the same order of magnitude as for monolayer excitons, the oscillation strength is typically 2–3 orders of magnitude weaker. Another difference is the exciton orientation: In heterobilayers, the bright exciton dipoles have in-plane and out-of-plane components, while in the monolayer they are solely in-plane. The out-of-plane components provide an interface for tuning via the DC Stark effect, which is parity-dependent. The in-plane components lead to a valley-independent repulsive interaction between dipoles that is superimposed by a stronger repulsive intra-valley interaction between the dipoles.

Interlayer charge transfer processes are observed in homo- and heterobilayers. In the absence of an external potential, only Coulomb interaction is present and therefore momentum conservation applies. The optical properties of the imperfectly aligned valleys can be explained by the center-of-mass motion [51]. Compared to the band offset, interlayer hopping is less influential but yet affects the local potential distribution in, e.g., Moiré superlattices.

14.2.6 Vibrational Properties

The vibrational properties and hence the phonon characteristics are typically investigated by means of Raman spectroscopy and charge carrier mobility measurements. With decreasing TMD thickness, the $E^1_{2g}(A_{1g})$ peaks observed in the Raman spectrum, corresponding to the in-plane (out-of-plane) vibrational mode, will show a blueshift (redshift) [52–54]. The anomalous hardening of E^1_{2g} is related to the reduction of dielectric screening. Luo and coworkers [55] suggested that the different behavior of in-plane and out-of-plane vibration is the consequence of the thickness and surface effect: with the decrease of thickness, the frequencies for both modes tend to decrease and modes soften, while the increased surface proximity can harden the in-plane mode. The resulting increased distance between two Raman modes is associated with their different responses.

An additional mode is the B_{2g} mode, which is active in bulk TMDs [56] but becomes Raman inactive in monolayer TMDs. B_{2g} peaks are often used to tell apart monolayer TMDs from multilayer structures, such as the B_{2g} peak of WSe_2 which is observed around 307–310 cm^{-1} according to different reports [57–60].

14.3 Description and Effect of Strain in 2D Transition Metal Dichalcogenides

Deeper insight into the underlying physics in a material can be obtained using external fields or structural variations, often leading to novel applications. Applying electric [50] and magnetic fields [61] or doping [62] is a frequently used tuning technique. In 2D materials, strain engineering is found to be particularly suitable. The enhanced strength and flexibility allow for tuning the strain in a wide parameter space. In 2D materials, the concentration of defects can typically reach low values, such as about 1% [63, 64] in 2D MoS_2. According to Griffith's theory [65], the mechanical strength of a material therefore approaches the theoretical value that is determined by the strength of the chemical bonds. On the other hand, based on continuum theory the reduced dimension gives rise to higher flexibility. Strain can be actively modified in its magnitude and configuration (uniaxial or biaxial, compressive or tensile, discussed in Section 14.3.2) by various methods (discussed in Section 14.3.3), allowing control over structural, electronic, and optical properties.

The research on 2D TMD materials has been greatly influenced by the extensive works graphene. Strain engineering techniques are no exception: e.g. Ding and coworkers [66] reported the reversible tuning of graphene by using piezoelectric actuators. Until now, there are various reports on the strain-tuning of 2D TMDs, covering experimental aspects as well as *ab initio* calculations. After the general discussion of strain, some representative reports are reviewed, starting from simple isotropic strain to the more general anisotropic case.

14.3.1 Description of Strain in 2D Materials

Elastic bodies can move and deform when subjected to an external load. The relative amount of elastic deformation can be described by the dimensionless quantity strain. In a one-dimensional crystal lattice the strain can be described as the relative change of the lattice:

$$\varepsilon = \frac{a' - a}{a}$$

where ε is the strain, a' is the lattice constant after and a before deformation. In a three-dimensional system, the strain tensor is composed of the strain coefficients ε_{ij}:

$$\varepsilon = \begin{pmatrix} \varepsilon_{xx} & \varepsilon_{xy} & \varepsilon_{xz} \\ \varepsilon_{yx} & \varepsilon_{yy} & \varepsilon_{yz} \\ \varepsilon_{zx} & \varepsilon_{zy} & \varepsilon_{zz} \end{pmatrix}$$

Considering the linear-elastic regime and excluding rigid body rotations, ε is symmetric ($\varepsilon_{ij} = \varepsilon_{ji}$) and can be described by only six components. The tensor has the eigenvalues ($\varepsilon_1, \varepsilon_2, \varepsilon_3$), which are the principal strains.

For two-dimensional material systems, a strong focus lies on the in-plane strain [67, 68]. The in-plane strain tensor can be written as

$$\varepsilon = \begin{pmatrix} \varepsilon_{xx} & \varepsilon_{xy} \\ \varepsilon_{yx} & \varepsilon_{yy} \end{pmatrix}$$

Now considering a symmetric tensor again, a rotation of the coordinate system leads to [68]:

$$\varepsilon = \begin{pmatrix} \varepsilon'_{xx} & 0 \\ 0 & \varepsilon'_{yy} \end{pmatrix}$$

With the principal strain components $\varepsilon'_{xx}, \varepsilon'_{yy}$. If the off-diagonal (shear strain) components are not equal, there will be rotation. The strain tensor can then be determined by

$$\varepsilon \cong \begin{pmatrix} \varepsilon''_{xx} & \varepsilon''_{xy} - \theta \\ \varepsilon''_{xy} + \theta & \varepsilon''_{yy} \end{pmatrix} = \mathcal{S}(\varepsilon) + \mathcal{T}(\theta)$$

where $\mathcal{S}(\varepsilon)$ denotes the strain contribution and $\mathcal{T}(\theta)$ the contribution by rotation [67].

Table 14.1 [79] shows a summary of the structural and strain-related properties of four frequently studied 2D TMDs. They exhibit similar lattice constants (from 0.312 to 0.326 nm), and the monolayer thickness is typically in the range of 0.6–0.7 nm. Their Young's modulus is in the range of 200 GPa. The TMDs show tensile strengths in the range of 24–30 GPa and ultimate strain of >0.1. Although their mechanical properties along zigzag and armchair direction are theoretically expected to be anisotropic, this is usually not confirmed experimentally.

Table 14.1 Summary of the structure and mechanical properties of MoS_2, $MoSe_2$, WS_2, and WSe_2.

	MoS_2	$MoSe_2$	WS_2	WSe_2
Lattice constant (nm)	0.312 [69, 70]	0.324 [70, 71]	0.314 [70, 71]	0.326 [70, 71]
Thickness (nm)	0.65 [72, 73]	0.65 [74, 75]	0.65 [73]	0.7 [60, 76, 77]
Young's Modulus (GPa)	222.75(Z) [71] 219.46(A) [71] 270(E) [73, 78]	178.78(Z) [71] 175.97(A) [71] 177(E) [74]	244.18(Z) [71] 240.99(A) [71] 272(E) [73]	196.81(Z) [71] 194.13(A) [71] 167(E) [76]
Tensile strength (GPa)	27.35(Z) [71] 16.90(A) [71]	22.68(Z) [71] 12.86(A) [71] 4.8(E) [74]	29.96(Z) [71] 19.91(A) [71]	24.70(Z) [71] 15.05(A) [71]

(Z) means zigzag direction, (A) armchair direction, and (E) means experimental data.

14.3.2 Types of Strain

Strain can be categorized into various types. Due to the reduced dimension of TMDs, it is natural to first classify between *in-plane and out-of-plane strain*. If the strain is solely directed along the plane of the 2D TMDs and does not comprise additional out-of-plane components, it is called in-plane strain. Experiments with 2D TMDs mostly report on in-plane strain, which has been achieved by different techniques such as substrate bending, substrate thermal expansion, or piezoelectric actuators. Additional out-of-plane strain components can be introduced by applying hydrostatic pressure on the 2D TMDs, such as in a Diamond Avil Cell.

By applying in-plane strain, TMDs can be stretched or compressed, corresponding to *tensile and compressive strain*, respectively. There are only few techniques that reliably allow for applying compressive strain. If compressive strain is applied by techniques based on a deformation of the substrate, the TMD layers are likely to detach or form wrinkles. Therefore, tensile strain is more frequently reported.

Another possible classification is deducted from the uniformity of the strain throughout the TMD layers; *homogeneous and inhomogeneous (local) strain*. A uniform strain distribution leads to a homogeneous change of material properties. Localized strain gives rise to inhomogeneous strain fields, leading to, e.g., modified bandgaps in different positions of the 2D TMD, as shown in Figure 14.5. The resulting local potential minima are energetically favorable for the excitons compared to the surrounding area. Therefore, the excitons drift to the strain center, where they radiatively decay. This is also referred to as the funnel effect due to the structure of the electronic potential and is attractive for PL and electroluminescence applications.

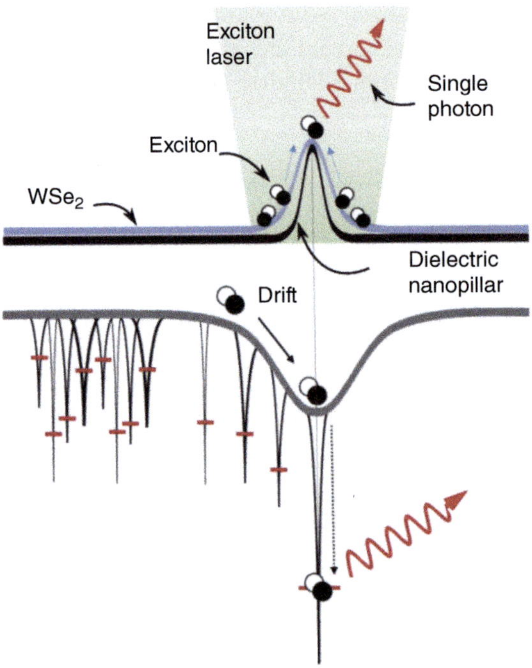

Figure 14.5 Atomically thin WSe_2 is deformed by localized strain that locally modulates the bandgap. Source: Branny et al. [80]. https://www.nature.com/articles/ncomms15053. Licensed under CC BY 4.0.

Inhomogeneous strain can originate from local slippage of TMD flakes induced by in-plane tensile strain, or from wrinkles induced by in-plane compressive strain. Experiementally, AFMtips, patterned substrates, and the pre-stretched substrate method can induce such inhomogeneous strain.

Without considering rotations or shear strain, the in-plane strain can be classified according to the number of orthogonal axes it is applied to. It is denoted by *unaixal* (either ε'_{xx} or ε'_{yy} is 0) and *biaxial strain* (none of ε'_{xx} and ε'_{yy} is 0). Uniaxial strain has been applied to 2D TMDs e.g. by elastic substrate bending, and biaxial strain by elongating the substrate in two orthogonal directions.

Strain can be applied in a *static* or *dynamically* controllable fashion. By patterned substrates, static strain is induced in 2D TMDs, since the substrate pattern cannot be changed once fabricated. In contrast, strain applied to TMDs by piezoelectric actuators can be controlled by setting a voltage dynamically.

14.3.3 Effect of Strain on TMDs

Strain can exert its influence in two fundamental aspects, the distance of atoms and symmetry. Changing the spatial extent and angles of the interatomic bonds leads to changes in the band structure (effective masses, direct to indirect bandgap transitions, etc.) and vibrational modes. Furthermore, the symmetry can be changed, e.g. by breaking the C_3 symmetry when applying uniaxial strain [46, 81]. Considering Moiré superlattices in twisted bilayers, the application of strain has an additional effect, since the material properties are now atom registry-dependent and strain can change atom coordination/registry and lattice mismatch.

Changes in the properties of 2D TMDs by applying strain were reported by experiments and *ab initio* calculations. These changes can be observed using various strain-tuning techniques that allow for the application of different types of strain. While the details about these techniques are discussed in Section 14.3.3, below the dependence of the 2D TMD properties on the different types of strain is discussed.

What all the *ab initio* calculations [33, 82, 83] suggest is that tensile strain leads to a shrinkage of the bandgap while (small) compressive strain gives rise to increased separation of the conduction band and valence band, which has been confirmed in various experiments, see Section 14.4. However, it has been reported that beyond a critical limit, compressive strain results in a shrinking bandgap, see Figure 14.6f. The change of the bandgap is slightly different in each type of 2D TMD monolayer, a qualitative explanation is given in Ref. [88].

Strain can modify the band structure. Once the compressive strain is beyond a critical point, band edges of monolayer TMDs change to **K** and **Λ** points [86], by which the direct bandgap transforms into an indirect one, as shown in Figure 14.6d–f. Sufficient in-plane tensile strain can "pull" the valence band maximum (VBM) from the **K** point to the **Γ** point. Adding to that, tensile strain can also induce phase transitions. Calculations [86, 89] revealed that MoS_2 will undergo a semiconductor–metal transition when the in-plane isotropic tensile strain exceeds 11%. Experimentally, the semiconductor–metal transition has been observed by applying strain via AFM tips or a diamond anvil cell (DAC).

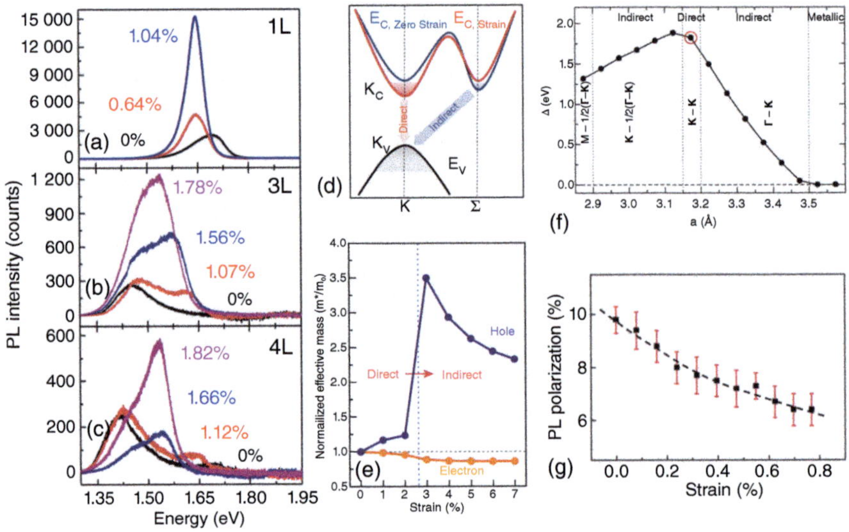

Figure 14.6 Bandgap and PL in response to external strain. (a–c) PL intensity change of WSe_2 (1 layer, 3 layers, and 4 layers) with tensile strain. (d) Changes in the band structure with in-plane strain. Source: (a–d) Adapted from Desai et al. [84]. Copyright (2014) American Chemical Society. (e) Effective mass change with tensile strain. Source: Adapted from Horzum et al. [85]. Copyright (2013) American Physical Society. (f) *Ab initio* calculation result of band structure under compressive and tensile strain, where the red circle denotes the unstrained state. Band edges (VBM-conductance band minimum (CBM) in the figure) and phase transitions are also shown. Source: Adapted from Ghorbani-Asl et al. [86]. Copyright (2013) American Physical Society. (g) Change in the degree of circular polarization by in-plane tensile strain in a monolayer of MoS_2. Source: Adapted from Zhu et al. [87]. Copyright (2013) American Physical Society.

The change in the effective masses of electrons and holes are complex in monolayer TMDs, sine they are influenced by positions (e.g., different electron effective mass in **K** or **Λ** point), directions, and strain configurations [90]. Electron effective masses decrease slightly with in-plane tensile strain [33, 85, 90–93]. It is worth noting that the hole effective mass shows a discontinuous shift [85, 92] when the band edge transition between a direct and indirect bandgap occurs, see Figure 14.6.

Strain can change the intensity of the PL signal. The response to strain at different sites in the reciprocal lattice (e.g. **K** or **Γ** points) is different, which is why a competition of the band exists, as shown in Figure 14.6d. Under tensile strain, the conduction band around the **K** point is lowered, enhancing the direct bandgap contribution and therefore the PL intensity due to the stronger radiative exciton decay. In contrast, MoX_2 and WX_2 show the opposite change. Experimentally, an increasing PL intensity of WSe_2 [84] and WS_2 [94] was observed by applying strain, and a decreasing intensity for MoS_2 [94].

The optical polarization of the emission is determined by the valley-dependent optical selection rules, as discussed in Section 14.2.4. The degree of polarization (DoP) [87] is defined by $\text{DoP} = (I_{co} - I_{cr})/(I_{co} + I_{cr})$, where I_{co} is the intensity of the co-polarized signal and I_{cr} is the intensity of the cross-polarized signal. The DoP is influenced by the quality of the sample, the type of excitons (neutral or charged), and the type and magnitude of strain. Figure 14.6g shows a decrease of the neutral exciton DoP in MoS_2 with tensile strain at room temperature. Recently, in the research area of heterobilayers and Moiré patterns, a circular-to-linear polarization transition was reported [95] due to the strain that occurred in the fabrication process. Changes in the polarization were also reported in 2D TMDs using magnetic field tuning and the Zeeman effect [96]. Fabian and coworkers [97] pointed out in their calculations that the band splittng in VB and CB can be influenced by strain.

The second harmonic generation (SHG), schematically shown in Figure 14.7a, is also influenced by strain. Monolayer TMDs typically show a SHG intensity change dependent on the laser polarization. As shown in Figure 14.7b, a pattern of hexagonal symmetry is observed, due to the broken inversion symmetry and nonzero second-order nonlinear susceptibility [68, 99, 100]. The SHG signal has a clear response to external strain with regard to its shape and signal intensity. Studying the SHG is very helpful in the assembly of heterobilayers, since it is twist-angle dependent.

The change in vibrational properties can be observed using Raman spectroscopy. In-plane tensile strain will induce phonon softening [33]. Furthermore, the in-plane mode E^1_{2g} will split into two discrete peaks E^{1-}_{2g} ($E^{'-}$ in Figure 14.7c) and E^{1+}_{2g} ($E^{'+}$ in Figure 14.7c). Redshifts of the E^{1-}_{2g} peak have also been reported. The distance between the in-plane and out-of-plane modes increases with tensile strain. Changes in the out-of-plane mode have been observed in 2D TMDs subjected to strain by piezoelectric actuators or DACs. Meanwhile, lineshapes of PL emissions can change with strain, which is often phonon associated.

Tensile strain can furthermore increase mobility, because of the reduction of scattering in the valley.

The effect of strain in 2D TMDs is summarized in Table 14.2.

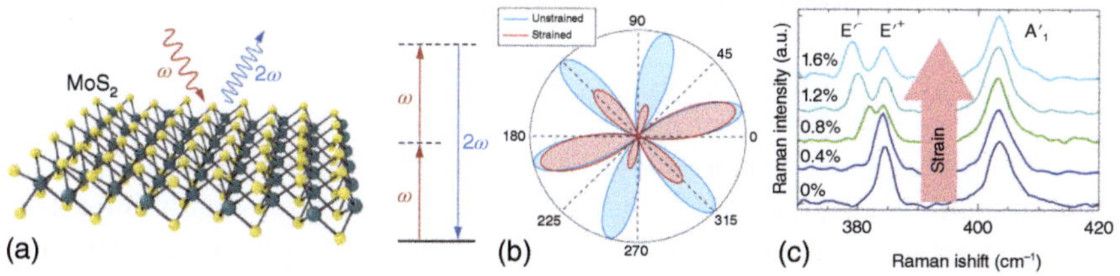

Figure 14.7 (a) Schematic of SHG in MoS_2. (b) Polarization resolved SHG signal intensity. In the unstrained state, the pattern has a hexagonal symmetry. When strained, the shape and intensity change. Source: Mennel et al. [68]. https://www.nature.com/articles/s41467-018-02830-y. Licensed under CC BY 4.0. (c) Raman spectra of MoS_2 for different magnitudes of strain. Source: Adapted from Conley et al. [98]. Copyright (2013) American Chemical Society.

Table 14.2 Summary of experimental reports for different types of strain in 2D TMDs.

Strain configuration	Observed changes in TMD properties
In-plane, homogeneous, uniaxial/biaxial, tensile	Redshift of PL peaks (X, T) [87, 98, 101–105] Redshift of Raman E^1_{2g} peak [98, 104, 106] Splitting of E^1_{2g} peak [106] Raman E^1_{2g} splitting [84, 87] Direct to indirect band transition 2.5% strain [106] Intensity and shape change of SHG [99]
Inhomogeneous, tensile	Redshift of PL [107, 108] Funnel effect [107, 108] Phase transition [109] New Raman peak [109]
Hydrostatic, homogeneous, compressive	Blueshift [110–113] Blueshift of E^1_{2g}, A_{1g} [110–116] Phase transition WS_2 [111–113, 117, 118]
In-plane, tensile/compressive	PL blueshift [119] Raman peaks blueshift [119]

14.4 Strain-Tuning Techniques

Strain can be applied to TMDs with various experimental techniques, which we summarize here. TMDs are usually already subject to strain due to the nature of the fabrication processes such as mechanical exfoliation or material growth using molecular-beam epitaxy or CVD. In the following, the (mostly active) strain-tuning techniques are discussed, which allow for deterministic control of the TMD properties. Typically applied techniques are discussed and classified, starting from AFM tips and concluding with piezoelectric actuators.

14.4.1 Atomic Force Microscopy Tips

AFMs are commonly used in labs across the world and therefore convenient for strain-tuning of 2D TMDs, see Figure 14.8. Dynamic and localized strain fields can be achieved by pressing the tips on the material surface. It is possible to combine this technique with dynamic *in situ* characterization. Researchers utilized AFM tips to measure the mechanical properties of MoS_2 and observe a very high Young's modulus [120]. The generated strain fields can be large enough to induce phase transitions. Lee and coworkers [109] investigated the influence of strain on the structural phase by pressing AFM tips with constant load onto $MoTe_2$. The energetic difference between 2H and 1T′ phase is smaller in this material compared to WSe_2, $MoSe_2$, WS_2, and MoS_2, and the required strain for interconverting the phases is also

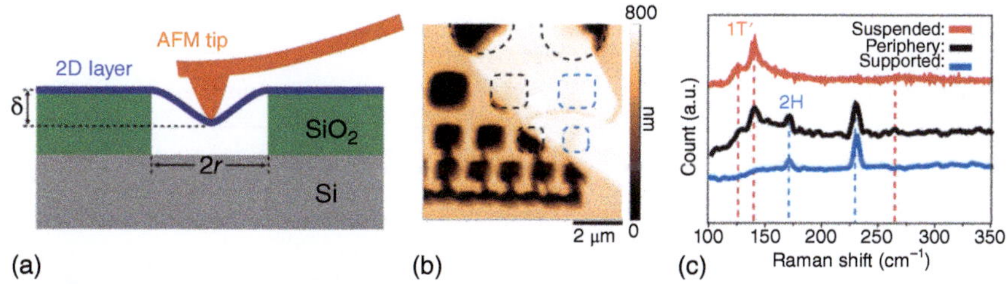

Figure 14.8 (a) Schematic of strain induced by an AFM tip. Source: Adapted with permission from Liu et al. [73]. Copyright (2014) American Chemical Society. (b) AFM topography image of a suspended 2H $MoTe_2$ flake. The suspended area (partially) covered by the TMD flake corresponds to the zones enclosed by dashed lines. (c) Raman spectra of $MoTe_2$ when pressed by an AFM tip. At different positions (suspended, periphery, and supported) the Raman signals are different, revealing a phase transition. Source: (b, c) Adapted with permission from Song et al. [109]. Copyright (2016) American Chemical Society.

lower [121]. Sufficient uniaxial tensile strain for the transition is as low as 3%, referring to the lattice constant difference in the two phases (2H 0.61 nm, 1T′ 0.63 nm).

When using AFM tips to generate strain fields, the response of 2D materials is substrate and position-dependent. Investigations on graphene showed that the strain on suspended parts of the material can be four to five times larger compared to parts connected to the substrate [122].

The magnitude of strain can be extracted as suggested by Lee and coworkers [109] using $\varepsilon = 2\delta^2/3r^2$, where δ is the indentation depth, r the radius, and ε the radial strain. Wu and coworkers used $F = \left(\sigma_0^{2D}\pi\right)\delta + (E^{2D}q^3/r^2)\delta^3$, where F is the applied load, δ is the indentation depth at the center, σ_0^{2D} is the pretension, r is the radius of the hole, $q = 1/(1.05 - 0.15\nu - 0.16\nu^2)$ with ν denoting the Poisson's ratio. One drawback of using AFM tips to generate strain may be that it is not very practical in low-temperature measurements or for the development of scalable devices.

14.4.2 Substrate Deformation

Strain can be induced by the deformation of the substrate. Bending elastic substrates with 2D materials on top has proven to be an effective way to apply uniaxial or biaxial in-plane tensile strain. The 2D TMDs can be transferred onto substrates such as terephthalate (PET) [87], polydimethylsiloxane (PDMS) [105], polycarbonate [98], polyethylene naphthalate (PEN), or polyethylene terephthalate glycol–modified (PETG) [123] films. After the transfer, the 2D materials will stick to the top surface of the elastic substrates via Van der Waals interaction. When the flexible substrate is bent, the strain will be transferred to the 2D TMDs. According to continuum mechanics, the generated strain follows $\varepsilon = d/2R$, where d is the thickness of the substrate and R is the bending radius, as shown in Figure 14.9. For cantilever-type bending, the strain level is evaluated by $\varepsilon = (3t\delta/2L^2)(1 - (x/L))$ [124, 125]. Here, L is the length of the substrate, t the thickness of the substrate, x the distance from the TMDs flake to the fixed root, and δ the deflection of the free edge.

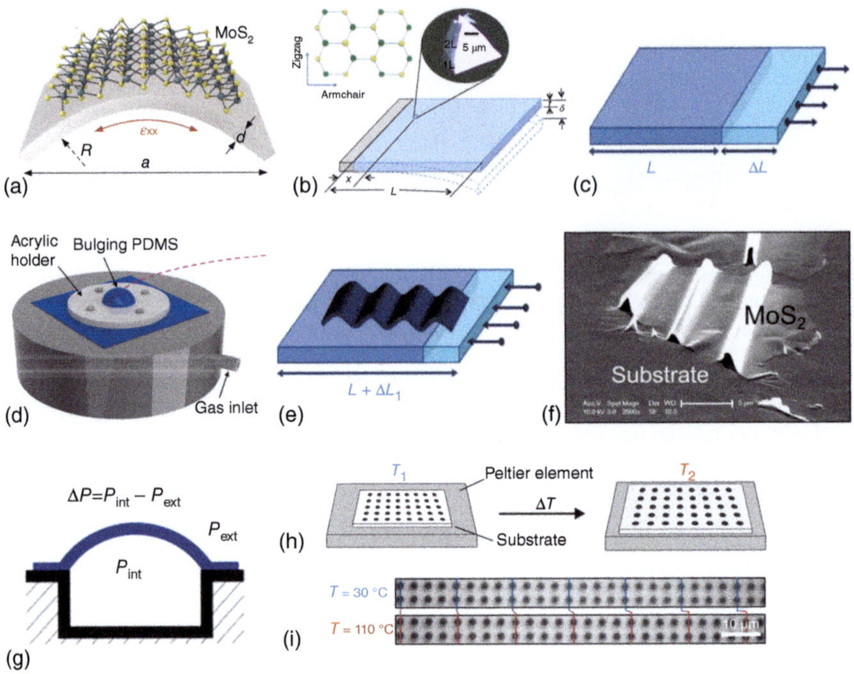

Figure 14.9 Strain induced by elastic substrate deformation. (a) Schematic of substrate bending for inducing tensile strain. R is the curvature radius and d is the thickness of the substrate. Source: Adapted from Mennel et al. [68]. https://www.nature.com/articles/s41467-018-02830-y. Licensed under CC BY 4.0. (b) Schematic of cantilever bending, where one end (grey) is fixed, the other is deflected. x denotes the distance of the MoS_2 flake to the fixed end. L is the length of the whole cantilever, and δ is the vertical deflection of the free end. Source: Adapted with permission from He et al. [124]. Copyright (2013) American Chemical Society. (c) Schematic of substrate elongation. L is the length, and ΔL is the extension. Source: Adapted with permission from Sichert et al. [107]. Copyright (2015) American Chemical Society. (d) and (g) show a schematic of a blown bubble. When the inner pressure (p_{int}) is higher than the external pressure (p_{ext}), the elastic substrate will be blown like a bubble. Gas can be led via the inlet (not shown) into the chamber below the PDMS film. Source: (d) Adapted with permission from Yang et al. [105]. Copyright (2017) American Chemical Society; (g) Adapted with permission from Lloyd et al. [104]. Copyright (2016) American Chemical Society. (e) Schematic of the prestretch method. L is the length in the idle state, and ΔL_1 is the extension in response to external pressure. Releasing the prestress leads to the formation of wavy wrinkles as seen in the SEM image in (f). Source: (e) Adapted from Sichert et al. [107]. Copyright (2015) American Chemical Society; (f) Castellanos-Gomez et al. [108]/with permission of American Chemical Society. (h) and (i) show schematics of strain induced by thermal expansion. Source: Frisenda et al. [103]. https://www.nature.com/articles/s41699-017-0013-7. Licensed under CC BY 4.0.

Applying tensile in-plane strain with this technique is reported to lead to redshifts in the PL A peaks and splitting and shift of the E^1_{2g} Raman peaks [124]. When the bending-induced strain is large enough, slippage of the layers will occur, such as for more than 1.3% strain [99] in 2D MoS_2. Depositing titanium on the substrate to hold the TMDs in place can effectively prevent slippage [98] and increases the transferrable strain to more than 2.5% for monolayer MoS_2. Next to that, epoxy patches [126], PVA [127], and PDMS encapsulation [128] were used for the same purpose.

A second way to deform the substrate is by stretching it. Elongating a PET substrate with WS_2 led to similar phenomena [106] as discussed for the bending of MoS_2. The application of isotropic strain along the armchair and zigzag direction by aligning the substrate-bending axis with the TMD orientation was also studied both experimentally and using *ab initio* calculations [82]. Biaxial strain can be generated by elongating the substrates in two directions at the same time.

Another possibility is to transfer 2D TMDs on pre-stretched elastic substrates [107, 108]. Strain is then induced after releasing the external pre-stress. This technique gives rise to the formation of flat and wrinkled areas and therefore an inhomogeneous strain field distribution. The localized strain can induce the "funnel effect," i.e. exciton confinement by local strain fields.

2D materials usually have different thermal expansion coefficients compared to the substrate they are placed on [102]. Therefore strain can be generated simply by heating the materials, e.g. by thermal wires or laser radiation. Homogeneous or inhomogeneous strain can be applied by different thermal treatments [102]. Another interesting technique called "blown bubble" (see Figure 14.9g) was investigated on MoS_2. It was first introduced by Bunch and coworkers [129] in their research on graphene and extended to MoS_2 by Yang and coworkers [105]. In the original state, the pressure in the cavity (P_{int}) is equal to the pressure outside ($P_{out} = P_0$, P_0 is the ambient gas pressure). Strain can be induced by a pressure difference, which can be achieved by pumping gas into the cavity beneath the 2D material.

14.4.3 Substrate Patterning

Strain can also be induced in 2D TMDs by patterning the substrate. The artificially increased surface roughness or coarseness leads to local strain fields in the layers transferred on top. The patterning can be performed by argon bombardment [130] or electron-beam lithography [80, 131]. The roughness of the substrate can be characterized by AFM and scanning electron microscopy (SEM). Then, the 2D TMDs can be transferred to the surface by a stamping process [133]. Local, tensile strain can then be observed at the "spiky" parts of the substrate, as illustrated in Figure 14.10. The pillars or thorns on the substrate are small in diameter compared to the size of the TMDs flakes. This leads to inhomogeneous strain and therefore the observation of, e.g. the funnel effect.

Sometimes the transferred 2D TMDs flakes will not perfectly follow the surface morphology of the substrate, resulting in tent-like structures as illustrated in Figure 14.10c, therefore impeding the correct estimation of the strain. The local strain together with the funnel effect allows for an accurate localization of the strain sites using PL mapping. One drawback of the method is that the strain cannot be tuned dynamically.

14.4.4 Diamond Anvil Cell

DACs are powerful devices by which very high hydrostatic pressures up to 39.2 GPa [115] can be applied to 2D materials. This technique allows for conducting *in situ* optical characterization. Nayak et al. [110, 115–117] used this technique

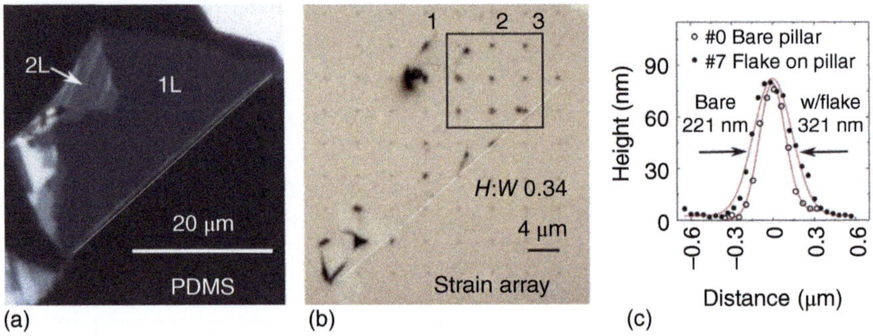

Figure 14.10 Patterned substrate with monolayer WSe_2 flakes. (a) Optical image after transfer of two layers of WSe_2 onto nanopillars (black dotted array). (b) 2D spatial map of integrated PL intensity. (c) AFM profile of a bare nanopillar and a nanopillar that is covered by a WSe_2 flake. Source: Branny et al. [80]/Springer Nature/Licensed under CC BY 4.0.

to investigate MoS_2, WS_2, and the hybrid $Mo_{0.5}W_{0.5}S_2$, resulting in the first experimental evidence of the strain-induced semiconductor to metal transition.

The PL emission peaks of MoS_2 will blueshift with the increase of hydrostatic pressure (compressive strain). The A_{1g} peaks of the Raman spectrum show obvious blueshift rather than slight or no visible shift, because of the additional out-of-plane strain component. With a DAC, compressive, but no tensile strain can be applied. Conveniently, the level of hydrostatic pressure can be measured by a gauge (red Ruby dot in Figure 14.11a) in the chamber.

Similar phase transition (Figure 14.11b), PL peak shift (Figure 14.11c), and band structure change (Figure 14.11d) were also reported in heterobilayer TMDs.

14.4.5 Piezoelectric Actuators

The generation of strain by piezoelectric platforms is based on the inverse piezoelectric effect. Applying an electric field on a piezoelectric material results in its mechanical deformation. Different types of strain (uniaxial, biaxial, or shear strain) can be applied, by altering the crystal orientation and electrical or mechanical boundary conditions. Usually, the electric field is generated by applying a voltage to metal electrodes at the piezoelectric materials. The resulting strain can then be transferred to a 2D material placed on top. This technique is a convenient way to induce strain in 2D TMDs in an independent and dynamically controllable way. For obtaining high electrostriction and therefore higher strain, relaxor ferroelectric crystals are typically used [66]. Lead magnesium niobate-lead titanate $(1-x)[Pb(Mg_{1/3}Nb_{2/3})O_3]\text{-}x[PbTiO_3]$ (PMN-PT) is reported to have the highest piezoelectric charge coefficients [134] for this purpose, d_{33}, and d_{31}.

In 2010, Ding and coworkers [66] applied this technique by investigating the properties of graphene in response to in-plane biaxial strain, as shown in Figure 14.12. Different shifts in the observed Raman modes allowed the extraction of the Grüneisen parameters, in good agreement with theoretical predictions. Hui and coworkers [119] placed a MoS_2 trilayer, sandwiched by a gold and graphene

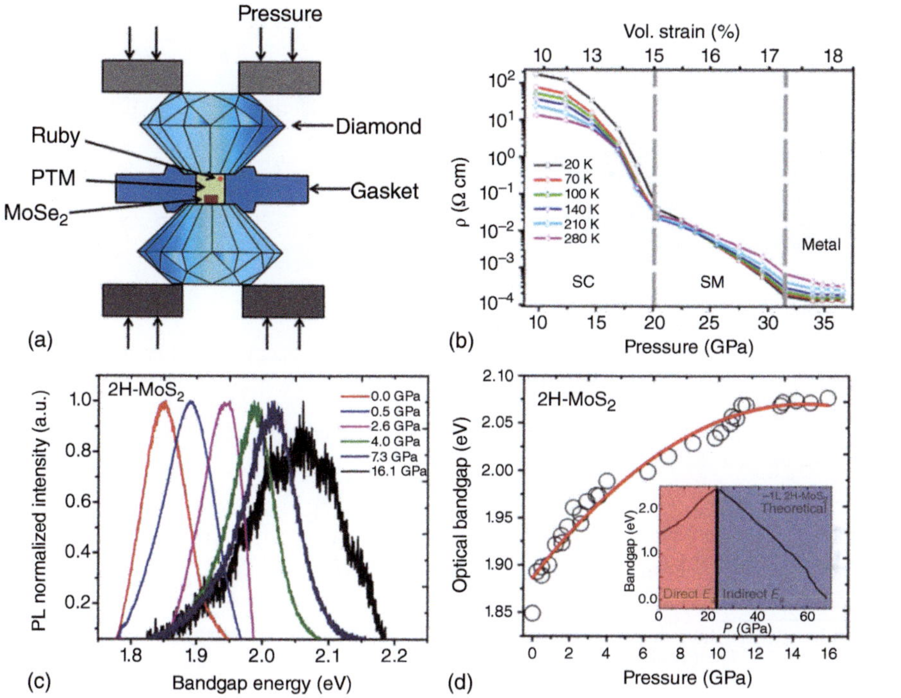

Figure 14.11 Strain induced by DAC. (a) Schematic of the DAC with a MoS_2 flake inside. Ruby dot is the gauge monitoring the pressure. PTM is the pressure transmitting medium. Source: Adapted from Fu et al. [110]. Copyright (2017) American Chemical Society. (b) Phase transition with hydrostatic pressure. The change in resistance (ρ) denotes the phase transition from semiconductor (SC) to semimetal (SM) and metal. Source: Adapted from Nayak et al. [117]. Copyright (2015) American Chemical Society. (c) Spectral PL shift for different pressure. (d) Change of the optical bandgap for different pressure. Source: (c, d) Adapted from Nayak et al. [115]. Copyright (2014) American Chemical Society.

electrode, on a PMN-PT substrate, allowing for the application of biaxial isotropic compressive strain of up to 0.2%. MoS_2 showed a 200% increase in PL intensity, accompanied by about 40% reduction in the linewidth of the emission. A blueshift of the E^1_{2g} and A_{1g} peaks in the Raman spectrum is observed, in contrast to the in-plane strain-tuning techniques mentioned before for which visible shifts of A_{1g} peaks are not detected. A further example of strain-tuning of 2D TMDs with piezoelectric actuators is a report by Vamivakas and coworkers [136], showing the rotation of the polarization direction of defects in monolayer WSe_2 in response to strain.

The magnitude of the strain can be measured in different ways. One possibility is X-ray diffraction (XRD) measurements [119] due to the changes in lattice constant of the material. Hou and coworkers [137] obtained the magnitude of strain by creating micropatterned metallic strain gauges on top of the PMN-PT. By measuring changes in the electric resistance the strain can be obtained by $K\varepsilon = \Delta R/R$, where R is the electric resistance and K is a pattern-specific coefficient.

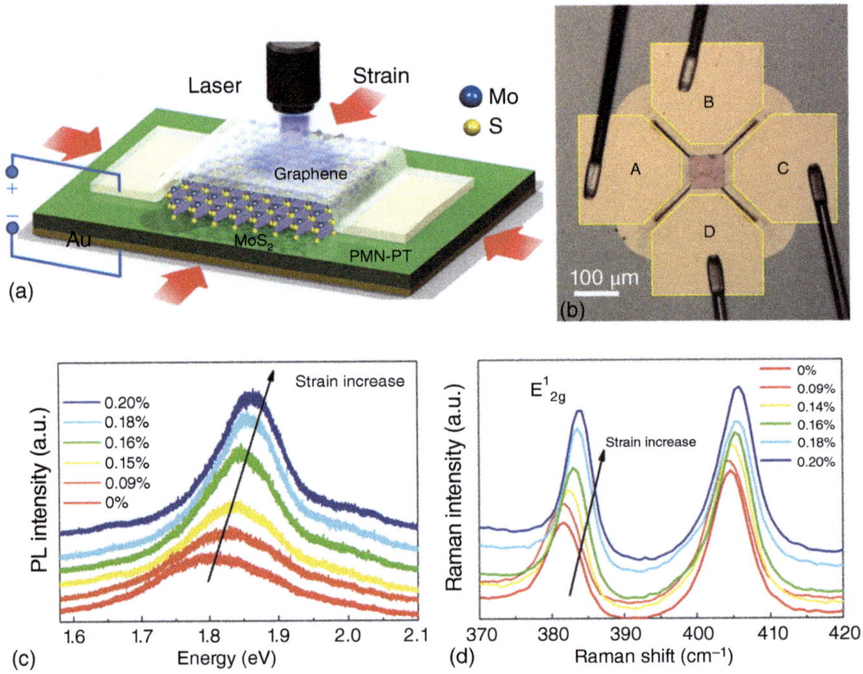

Figure 14.12 Strain induced by piezoelectric actuators. Schematic of bulk (a) and optical image of a patterned (b) piezoelectric strain-tuning platform. Source: (a) Adapted from Hui et al. [119]. Copyright (2013) by American Chemical Society; (b) Chen et al. [135]/Springer Nature/Licensed under CC BY 4.0. (c) PL spectra of trilayer MoS_2 in dependence of compressive strain applied by a piezoelectric actuator. (d) Raman spectra of trilayer MoS_2 in dependence of compressive strain. Source: (c, d) Adapted from Hui et al. [119]. Copyright (2013) by American Chemical Society.

Using piezoelectric strain-tuning platforms allows for much more than the application of in-plane isotropic biaxial strain. Piezoelectric single crystals can be (micro-)fabricated into more complex shapes, which can even allow control over the complete in-plane strain. Figure 14.12b shows an illustration of anisotropic strain-tuning platforms, consisting of four "legs" of the piezoelectric material that can induce biaxial and uniaxial strain simultaneously and independently. In this example, the strain was transferred to a semiconductor quantum dot containing membrane in the center of the device. Uniaxial strain can be obtained by applying a voltage to one pair of opposite contacts (AC or BD), and biaxial strain can be achieved by applying voltages on all contacts at the same time. One can switch between tensile and compressive strain by switching the voltage bias.

In comparison, piezoelectric crystals show some advantages over other techniques. The strain is easy to manipulate, since this platform is controlled by a voltage (or electric field). The small footprint also allows researchers to load it in cryo-chambers for low-temperature measurements. Furthermore, this versatile platform can induce uniaxial/biaxial, compressive/tensile, isotropic, and even

Table 14.3 Comparison of strain-tuning techniques and the types of strain they can apply.

Technique	Type of strain
Substrate bending	In-plane, homogeneous, uniaxial/biaxial, tensile
Substrate elongating	In-plane, homogeneous, uniaxial, tensile
Substrate prestretch	In-plane, tensile, static
Substrate expansion	In-plane, homogeneous, biaxial, tensile
Blown bubble	Homogeneous, biaxial, tensile
AFM tip	Tensile
DAC	In-plane and out-of-plane, homogeneous, compressive
Piezo-platform	In-plane, tensile/compressive, isotropic/anisotropic

anisotropic strain. *In situ* characterization measurements (such as PL or Raman spectroscopy) are combinable with this technique.

14.4.6 Comparison and Summary

A summary of the different techniques that were discussed above and the type of strain that they can induce is shown in Table 14.3.

Apart from the techniques listed above, there are further approaches that have been used in combination with, e.g. nanowires or quantum dots, and may be also applicable for the investigation of TMDs. Goldsche et al. [138], Lu [139] and coworkers transferred graphene between two arms of a device made of Si MEMS (microelectromechanical systems), where one arm is static, another one can move. Fu [140], Chen [141], and their coworkers applied strain on quantum dots/nanowires by motorized stages.

14.5 Applications of Strain-Tuning in 2D Transition Metal Dichalcogenides

Now we turn to the impact of strain-tuning on practical applications of 2D TMDs. Two examples of optoelectronic devices based on piezoelectric/piezophototronic effect are highlighted, followed by a discussion on piezoresistive devices and nanogenerators.

(Opto)electronic devices based on 2D materials can be distinguished into lateral and vertical devices. Figure 14.13a shows an example of a lateral homojunction device. This device was fabricated on a Si/SiO_2 wafer, onto which two separated Pd bottom electrodes were added, allowing to control the voltages V_{G1} and V_{G2}. On top of a monolayer of WSe_2, few-layer graphite flakes, serving as ambipolar

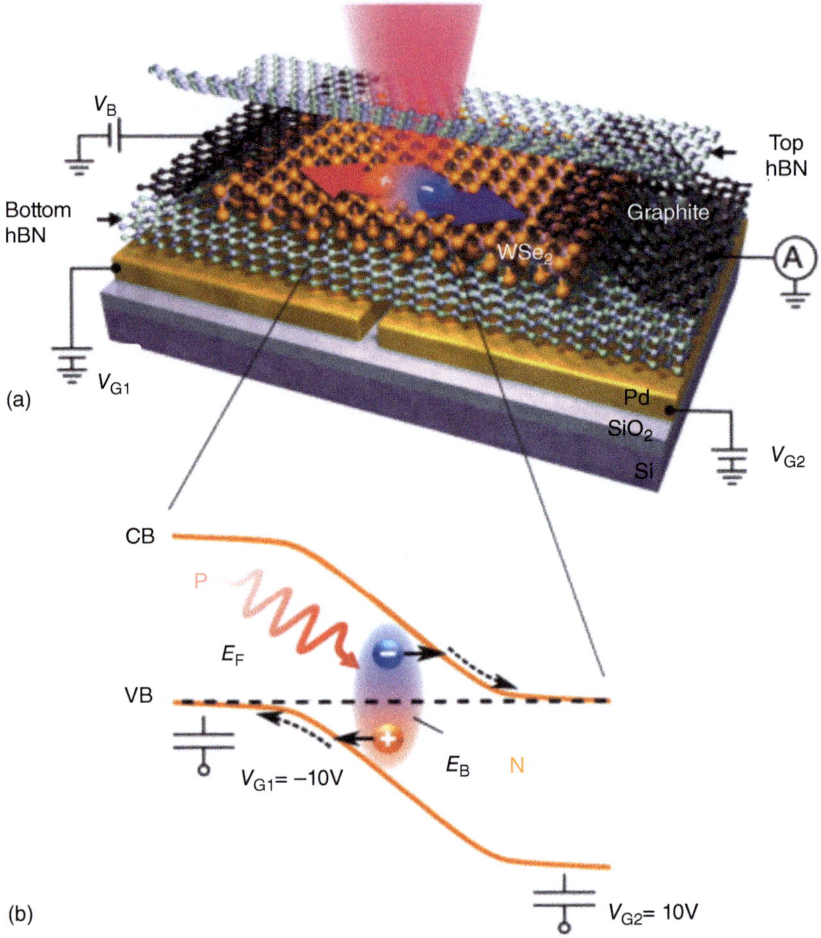

Figure 14.13 (a) Schematic of a monolayer WSe_2 lateral homojunction device, controlled by two local metal gates with voltages V_{G1} and V_{G2}. Two graphite flakes (colored in black) are placed on both sides of the WSe_2 layer (orange) and encapsulated between two hBN flakes (pale blue and green); (b) Band diagram of the p–n junction. Source: Massicotte et al. [142]. https://www.nature.com/articles/s41467-018-03864-y. Licensed under CC BY 4.0.

electrical contacts, are positioned at both ends of the monolayer flake. A stable dielectric environment is obtained by hexagonal boron nitride (hBN) encapsulation, providing a smooth and disorder-free substrate to minimize the non-radiative processes and acting as insulator layer [143]. Many functionalities can be realized by such a lateral homojunction device: when irradiated by light, the device outputs current or voltage and can therefore act as a solar cell, photodiode, or photodetector. The corresponding device characterization is shown in some devices based on WSe_2 [143–145] and MoS_2 [12, 146, 147]. A challenge [4] for lateral devices is the simultaneous realization of a wide depletion region (the gap between two bottom electrodes, 500 nm [148]) to capture excitons, and a high built-in electric field to separate the electrons and holes.

When lateral devices are tuned by in-plane strain, the photocurrent and photoresponse can increase greatly, up to 26 times [149]. The piezoelectric polarization charges at p–n homojunction interfaces or metal–semiconductor interfaces will modulate the band structure and affect the carrier generation, separation, transport, or recombination. The reason behind this lies in two aspects, the depletion zone broadening and barrier height decrease. The piezophototronic effects and their application are reported in strain-gated adaptive photodetection. Specifically, strain was applied via the elastic substrate-bending technique [149, 150].

Similar phenomena were also reported in heterojunctions/heterostructures, such as MoS_2/WSe_2 [151] or MoS_2/CuO [152]. An example of such a device is shown in Figure 14.14.

Compared to monolayer TMD devices, heterobilayers allow the separation of electrons and holes into the two different layers of the vertical device, with the gap between the two layers acting as the depletion region. Therefore a very high built-in electric field can be generated. This kind of device shows rectification characteristics, and its properties can be tuned by changing the gate voltage [145]. Next to that, also the photovoltaic effect has been characterized [154].

TMDs show further potential for applications in optics: Coupling 2D TMDs to microcavities was first shown by Englund and coworkers [155] for MoS_2. Significant enhancement of the emission was observed [155–160] and Purcell factors up to 70 were reported. The existence of single-photon emitters (based on defects or local strain) has been shown in WSe_2 [61, 161–163] and the single-photon character confirmed, e.g. in WSe_2-bilayers on a patterned substrate.

Piezoresistive devices are typically used as strain sensors. The gauge factor (describing the change of resistance in response to strain) in some 2D TMDs is higher than for conventional Si-based sensors [164]. Besides, the higher mechanical

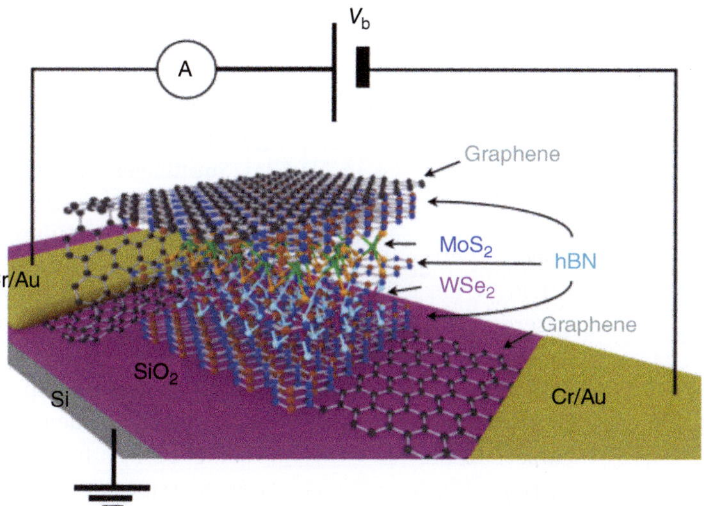

Figure 14.14 Schematic of a 2D $MoSe_2/WSe_2$ heterojunction vertical device. Graphene is used as top and bottom electrode. Source: Binder et al. [153]. https://www.nature.com/articles/s41467-019-10323-9. Licensed under CC BY 4.0.

strength enables TMD-based devices to withstand higher strain, as reported by using AFM tips [78, 165] or elastic substrate deformation [166, 167]. The mechanical flexibility of the devices is interesting for flexible electronics applications such as transparent electronic skin based on MoS_2 as reported by Park and coworkers [168]. TMD-based nanogenerators for application in energy harvesting were also reported [169, 170]. The output voltage of the monolayer (bilayer) WSe_2 nanogenerators was reported to depend on strain, yielding 90 mV (85 mV) at 0.64% (0.95%) strain.

14.6 Summary and Outlook

In this chapter, the properties of 2D TMDs have been introduced and their modification by applying strain fields discussed. Different configurations of strain have been described and their effect on the materials was summarized. Strain modifies the structural properties such as interatomic distances and angles, atomic arrangements and symmetry. Therefore various physical quantities can be engineered, such as the bandgap and the effective masses of charge carriers. These changes lead to a modified PL intensity, PL polarization, and changes in the SHG or Raman signal. Strain-tuning techniques such as elastic substrate deformation, substrate patterning, AFM tips, DAC, piezoelectric actuators, and others were reviewed, followed by a closer look at practical applications of strain-tuned 2D TMDs such as p–n junction devices or piezoresistive strain sensors.

Beyond the reported efforts, many more possibilities and challenges exist for the strain-tuning of 2D TMDs. For many fundamental studies and practical applications, high magnitudes of strain are required in combination with versatile and well controllable tuning technique (e.g. electrically driven and allowing low- and room-temperature measurements) allowing for the dynamic generation of different types of strain. A promising avenue is the use of micropatterned piezoelectric substrates to gain complete control over the in-plane strain tensor in 2D TMDs. Although the resulting anisotropic strain fields have been considered by means of *ab initio* calculations, an experimental demonstration remains yet outstanding.

Next to interesting practical applications, strain-tuning holds great promise for gaining a deeper understanding of the physics in 2D TMD multilayer structures. With the very long lifetime and valley depolarization time of interlayer excitons, the valley pseudospin of interlayer excitons can be exploited as carrier of (quantum) information. A complete picture of the underlying mechanism for the charge transfer process in TMDs heterostructures has yet to be found. Strain-tuning can help to shed light on the underlying mechanisms and therefore constitutes a valuable tool in the ongoing research on 2D TMDs [95, 113].

References

1 Novoselov, K.S., Geim, A.K., Morozov, S.V. et al. (2005). Two-dimensional gas of massless Dirac fermions in graphene. *Nature* 438 (7065): 197–200.

2 Tan, C. and Zhang, H. (2015). Two-dimensional transition metal dichalcogenide nanosheet-based composites. *Chem. Soc. Rev.* 44 (9): 2713–2731.

3 Roldán, R., Castellanos-Gomez, A., Cappelluti, E., and Guinea, F. (2015). Strain engineering in semiconducting two-dimensional crystals. *J. Phys. Condens. Matter* 27 (31): 313201.

4 Mak, K.F. and Shan, J. (2016). Photonics and optoelectronics of 2D semiconductor transition metal dichalcogenides. *Nat. Photonics* 10 (4): 216–226.

5 Choi, W., Choudhary, N., Han, G.H. et al. (2017). Recent development of two-dimensional transition metal dichalcogenides and their applications. *Mater. Today* 20 (3): 116–130.

6 Bao, W., Cai, X., Kim, D. et al. (2013). High mobility ambipolar MoS_2 field-effect transistors: substrate and dielectric effects. *Appl. Phys. Lett.* 102 (4): 042104.

7 Frey, G. and Elani, S. (1998). Optical-absorption spectra of inorganic fullerene-like MS_2 (M = Mo, W). *Phys. Rev. B: Condens. Matter* 57 (11): 6666–6671.

8 Choudhary, N., Islam, M.R., Kang, N. et al. (2016). Two-dimensional lateral heterojunction through bandgap engineering of MoS_2 via oxygen plasma. *J. Phys. Condens. Matter* 28 (36): 364002.

9 Gao, Y.P., Wu, X., Huang, K.J. et al. (2017). Two-dimensional transition metal diseleniums for energy storage application: a review of recent developments. *CrystEngComm* 19 (3): 404–418.

10 Feng, J., Liu, K., Bulushev, R.D. et al. (2015). Identification of single nucleotides in MoS_2 nanopores. *Nat. Nanotechnol.* 10 (12): 1070–1076.

11 Kumara, A. and Ahluwalia, P.K. (2012). Electronic structure of transition metal dichalcogenides monolayers $^1H\text{-}MX_2$ (M = Mo, W; X = S, Se, Te) from ab-initio theory: new direct band gap semiconductors. *Eur. Phys. J. B* 85 (6): 18–22.

12 Voiry, D., Mohite, A., and Chhowalla, M. (2015). Phase engineering of transition metal dichalcogenides. *Chem. Soc. Rev.* 44 (9): 2702–2712.

13 Duerloo, K.A.N., Li, Y., and Reed, E.J. (2014). Structural phase transitions in two-dimensional Mo- and W-dichalcogenide monolayers. *Nat. Commun.* 5 (May): 1–9.

14 Durán Retamal, J.R., Periyanagounder, D., Ke, J.J. et al. (2018). Charge carrier injection and transport engineering in two-dimensional transition metal dichalcogenides. *Chem. Sci.* 9 (40): 7727–7745.

15 Yu, H. and Yao, W. (2017). Valley-spin physics in 2D semiconducting transition metal dichalcogenides. In: *2D Materials: Properties and Devices* (ed. P. Avouris, T.F. Heinz and T. Low), 279–294. Cambridge: Cambridge University Press.

16 Chaves, A., Azadani, J.G., Alsalman, H. et al. (2020). Bandgap engineering of two-dimensional semiconductor materials. *npj 2D Mater. Appl.* 4: 29.

17 Yoshikawa, N., Nagai, K., Uchida, K. et al. (2019). Interband resonant high-harmonic generation by valley polarized electron–hole pairs. *Nat. Commun.* 10 (1): 1–7.

18 Zhang, L., Zhang, Z., Wu, F. et al. (2020). Twist-angle dependence of moiré excitons in $WS_2/MoSe_2$ heterobilayers. *Nat. Commun.* 11 (1): 1–8.

19 Stier, A.V., McCreary, K.M., Jonker, B.T. et al. (2016). Exciton diamagnetic shifts and valley Zeeman effects in monolayer WS_2 and MoS_2 to 65 Tesla. *Nat. Commun.* 7: 1–8.

20 Premasiri, K. and Gao, X.P.A. (2019). Tuning spin-orbit coupling in 2D materials for spintronics: a topical review. *J. Phys. Condens. Matter* 31 (19): 93001.

21 Xie, S., Faeth, B.D., Tang, Y. et al. (2020). Direct observation of distinct minibands in moiré superlattices. *arXiv* 2: 1–18.

22 Yu, H., Liu, G.-B., and Yao, W. (2018). Brightened spin-triplet interlayer excitons and optical selection rules in van der Waals heterobilayers. *2D Mater.* 5 (3): 035021.

23 Britnell, L., Ribeiro, R.M., Eckmann, A. et al. (2013). Strong light–matter interactions thin films. *Science* 340 (June): 1311–1314.

24 Chiu, M.H., Zhang, C., Shiu, H.W. et al. (2015). Determination of band alignment in the single-layer MoS_2 WSe_2 heterojunction. *Nat. Commun.* 6 (May): 1–6.

25 Seyler, K.L., Rivera, P., Yu, H. et al. (2019). Signatures of moiré-trapped valley excitons in $MoSe_2/WSe_2$ heterobilayers. *Nature* 567 (7746): 66–70.

26 Yu, J., Kuang, X., Zhong, J. et al. (2020). Observation of double indirect interlayer exciton in WSe_2/WS_2 heterostructure. *Opt. Express* 28 (9): 13260.

27 Mouri, S., Zhang, W., Kozawa, D. et al. (2017). Thermal dissociation of inter-layer excitons in $MoS_2/MoSe_2$ hetero-bilayers. *Nanoscale* 9 (20): 6674–6679.

28 Unuchek, D., Ciarrocchi, A., Avsar, A. et al. (2018). Room-temperature electrical control of exciton flux in a van der Waals heterostructure. *Nature* 560 (7718): 340–344.

29 Hill, H.M., Rigosi, A.F., Rim, K.T. et al. (2016). Band alignment in MoS_2/WS_2 transition metal dichalcogenide heterostructures probed by scanning tunneling microscopy and spectroscopy. *Nano Lett.* 16 (8): 4831–4837.

30 Ceballos, F., Bellus, M.Z., Chiu, H.Y., and Zhao, H. (2014). Ultrafast charge separation and indirect exciton formation in a MoS_2–$MoSe_2$ van der Waals heterostructure. *ACS Nano* 8 (12): 12717–12724.

31 Rivera, P., Seyler, K.L., Yu, H. et al. (2016). Supplementary material for valley-polarized exciton dynamics in a 2D semiconductor heterostructure. *Science* 351 (688): 688.

32 Rivera, P., Yu, H., Seyler, K.L. et al. (2018). Interlayer valley excitons in heterobilayers of transition metal dichalcogenides. *Nat. Nanotechnol.* 13 (11): 1004–1015.

33 Scalise, E., Houssa, M., Pourtois, G. et al. (2014). First-principles study of strained 2D MoS_2. *Physica E* 56: 416–421.

34 Jones, A.M., Yu, H., Ghimire, N.J. et al. (2013). Optical generation of excitonic valley coherence in monolayer WSe_2. *Nat. Nanotechnol.* 8 (9): 634–638.

35 Wang, G., Robert, C., Glazov, M.M. et al. (2017). In-plane propagation of light in transition metal dichalcogenide monolayers: optical selection rules. *Phys. Rev. Lett.* 119 (4): 1–7.

36 Mak, K.F., McGill, K.L., Park, J., and McEuen, P.L. (2014). Valleytronics. The valley Hall effect in MoS_2 transistors. *Science* 344 (6191): 1489–1492.

37 Xu, X., Yao, W., Xiao, D., and Heinz, T.F. (2014). Spin and pseudospins in layered transition metal dichalcogenides. *Nat. Phys.* 10 (5): 343–350.

38 Tran, K., Moody, G., Wu, F. et al. (2019). Evidence for moiré excitons in van der Waals heterostructures. *Nature* 567 (7746): 71–75.
39 Yu, H., Liu, G.B., Tang, J. et al. (2017). Moiré excitons: from programmable quantum emitter arrays to spin-orbit coupled artificial lattices. *Sci. Adv.* 3 (11): 1–8.
40 Cudazzo, P., Tokatly, I.V., and Rubio, A. (2011). Dielectric screening in two-dimensional insulators: implications for excitonic and impurity states in graphane. *Phys. Rev. B: Condens. Matter* 84 (8): 1–7.
41 Stier, A.V., Wilson, N.P., Clark, G. et al. (2016). Probing the influence of dielectric environment on excitons in monolayer WSe_2: insight from high magnetic fields. *Nano Lett.* 16 (11): 7054–7060.
42 Zhang, X.X., You, Y., Zhao, S.Y.F., and Heinz, T.F. (2015). Experimental evidence for dark excitons in monolayer WSe_2. *Phys. Rev. Lett.* 115 (25): 1–6.
43 Dufferwiel, S., Schwarz, S., Withers, F. et al. (2015). Exciton-polaritons in van der Waals heterostructures embedded in tunable microcavities. *Nat. Commun.* 6 (May): 1–7.
44 Wang, G., Bouet, L., Lagarde, D. et al. (2014). Valley dynamics probed through charged and neutral exciton emission in monolayer WSe_2. *Phys. Rev. B: Condens. Matter* 90 (7): 1–6.
45 Schaibley, J.R., Yu, H., Clark, G. et al. (2016). Valleytronics in 2D materials. *Nat. Rev. Mater.* 1 (11): 16055.
46 Yu, H., Liu, G.B., Gong, P. et al. (2014). Dirac cones and Dirac saddle points of bright excitons in monolayer transition metal dichalcogenides. *Nat. Commun.* 5 (May): 1–7.
47 You, Y., Zhang, X.X., Berkelbach, T.C. et al. (2015). Observation of biexcitons in monolayer WSe_2. *Nat. Phys.* 11 (6): 477–481.
48 Ross, J.S., Wu, S., Yu, H. et al. (2013). Electrical control of neutral and charged excitons in a monolayer semiconductor. *Nat. Commun.* 4: 1–6.
49 Bernardi, M., Palummo, M., and Grossman, J.C. (2013). Extraordinary sunlight absorption and one nanometer thick photovoltaics using two-dimensional monolayer materials. *Nano Lett.* 13 (8): 3664–3670.
50 Rivera, P., Schaibley, J.R., Jones, A.M. et al. (2015). Observation of long-lived interlayer excitons in monolayer $MoSe_2$–WSe_2 heterostructures. *Nat. Commun.* 6: 4–9.
51 Yu, H., Wang, Y., Tong, Q. et al. (2015). Anomalous light cones and valley optical selection rules of interlayer excitons in twisted heterobilayers. *Phys. Rev. Lett.* 115 (18): 1–5.
52 Terrones, H., Del Corro, E., Feng, S. et al. (2014). New first order Raman-active modes in few layered transition metal dichalcogenides. *Sci. Rep.* 4: 1–9.
53 Wurstbauer, U., Miller, B., Parzinger, E., and Holleitner, A.W. (2017). Light–matter interaction in transition metal dichalcogenides and their heterostructures. *J. Phys. D: Appl. Phys.* 50 (17): 173001.
54 Shahzad, R., Kim, T.W., and Kang, S.W. (2017). Effects of temperature and pressure on sulfurization of molybdenum nano-sheets for MoS_2 synthesis. *Thin Solid Films* 641 (November): 79–86.

55 Luo, X., Zhao, Y., Zhang, J. et al. (2013). Anomalous frequency trends in MoS_2 thin films attributed to surface effects. *Phys. Rev. B: Condens. Matter* 88 (7): 1–13.
56 Grzeszczyk, M., Gołasa, K., Molas, M.R. et al. (2018). Raman scattering from the bulk inactive out-of-plane $B^1{}_{2g}$mode in few-layer $MoTe_2$. *Sci. Rep.* 8 (1): 1–9.
57 Luo, X., Zhao, Y., Zhang, J. et al. (2013). Effects of lower symmetry and dimensionality on Raman spectra in two-dimensional WSe_2. *Phys. Rev. B: Condens. Matter* 88 (19): 1–7.
58 Zeng, H., Liu, G.B., Dai, J. et al. (2013). Optical signature of symmetry variations and spin-valley coupling in atomically thin tungsten dichalcogenides. *Sci. Rep.* 3 (4): 4908–4916.
59 Huang, J.K., Pu, J., Hsu, C.L. et al. (2014). Large-area synthesis of highly crystalline WSe_2 monolayers and device applications. *ACS Nano* 8 (1): 923–930.
60 Li, H., Lu, G., Wang, Y. et al. (2013). Mechanical exfoliation and characterization of single- and few-layer nanosheets of WSe_2, TaS_2, and $TaSe_2$. *Small* 9 (11): 1974–1981.
61 Koperski, M., Nogajewski, K., Arora, A. et al. (2015). Single photon emitters in exfoliated WSe_2 structures. *Nat. Nanotechnol.* 10 (6): 503–506.
62 Brotons-Gisbert, M., Baek, H., Campbell, A. et al. (2021). Moiré-trapped interlayer trions in a charge-tunable $WSe_2/MoSe_2$ heterobilayer. *Phys. Rev.* 11 (3): 29–31.
63 Hong, J., Hu, Z., Probert, M. et al. (2015). Exploring atomic defects in molybdenum disulphide monolayers. *Nat. Commun.* 6: 1–8.
64 Qiu, H., Xu, T., Wang, Z. et al. (2013). Hopping transport through defect-induced localized states in molybdenum disulphide. *Nat. Commun.* 4: 1–6.
65 Griffith, A.A. (1920). The phenomena of rupture and flow in solids. *Philos. Trans. R. Soc. London, Ser. A* 221: 163–198.
66 Ding, F., Ji, H., Chen, Y. et al. (2010). Stretchable graphene: a close look at fundamental parameters through biaxial straining. *Nano Lett.* 10 (9): 3453–3458.
67 Bi, Z., Yuan, N.F.Q., and Fu, L. (2019). Designing flat bands by strain. *Phys. Rev. B: Condens. Matter* 100 (3): 1–9.
68 Mennel, L., Furchi, M.M., Wachter, S. et al. (2018). Optical imaging of strain in two-dimensional crystals. *Nat. Commun.* 9: 516.
69 Zhang, L., Bampoulis, P., Rudenko, A.N. et al. (2016). Structural and electronic properties of germanene on MoS_2. *Phys. Rev. Lett.* 116 (25): 1–6.
70 Zeng, F., Zhang, W.B., and Tang, B.Y. (2015). Electronic structures and elastic properties of monolayer and bilayer transition metal dichalcogenides MX_2 ($M = Mo, W; X = O, S, Se, Te$): a comparative first-principles study. *Chin. Phys. B* 24 (9): 097103.
71 Li, J., Medhekar, N.V., and Shenoy, V.B. (2013). Bonding charge density and ultimate strength of monolayer transition metal dichalcogenides. *J. Phys. Chem. C* 117 (30): 15842–15848.
72 Radisavljevic, B., Radenovic, A., Brivio, J. et al. (2011). Single-layer MoS_2 transistors. *Nat. Nanotechnol.* 6 (3): 147–150.

73 Liu, K., Yan, Q., Chen, M. et al. (2014). Elastic properties of chemical-vapor-deposited monolayer MoS_2, WS_2, and their bilayer heterostructures. *Nano Lett.* 14 (9): 5097–5103.

74 Yang, Y., Li, X., Wen, M. et al. (2017). Brittle fracture of 2D $MoSe_2$. *Adv. Mater.* 29 (2): 1604201.

75 Iberi, V., Liang, L., Ievlev, A.V. et al. (2016). Nanoforging single layer $MoSe_2$ through defect engineering with focused helium ion beams. *Sci. Rep.* 6: 30481.

76 Zhang, R., Koutsos, V., and Cheung, R. (2016). Elastic properties of suspended multilayer WSe_2. *Appl. Phys. Lett.* 108 (4): 042104.

77 Liu, W., Kang, J., Sarkar, D. et al. (2013). Role of metal contacts in designing high-performance monolayer n-type WSe_2 field effect transistors. *Nano Lett.* 13 (5): 1983–1990.

78 Manzeli, S., Allain, A., Ghadimi, A., and Kis, A. (2015). Piezoresistivity and strain-induced band gap tuning in atomically thin MoS_2. *Nano Lett.* 15 (8): 5330–5335.

79 Deng, S., Sumant, A.V., and Berry, V. (2018). Strain engineering in two-dimensional nanomaterials beyond graphene. *Nano Today* 22: 14–35.

80 Branny, A., Kumar, S., Proux, R., and Gerardot, B.D. (2017). Deterministic strain-induced arrays of quantum emitters in a two-dimensional semiconductor. *Nat. Commun.* 8 (May): 1–7.

81 Zhai, D. and Yao, W. (2020). Layer pseudospin dynamics and genuine non-Abelian Berry phase in inhomogeneously strained moiré pattern. *Phys. Rev. Lett.* 125 (26): 266404.

82 Johari, P. and Shenoy, V.B. (2012). Tuning the electronic properties of semiconducting transition metal dichalcogenides by applying mechanical strains. *ACS Nano* 6 (6): 5449–5456.

83 Lu, P., Wu, X., Guo, W., and Zeng, X.C. (2012). Strain-dependent electronic and magnetic properties of MoS_2 monolayer, bilayer, nanoribbons and nanotubes. *Phys. Chem. Chem. Phys.* 14 (37): 13035–13040.

84 Desai, S.B., Seol, G., Kang, J.S. et al. (2014). Strain-induced indirect to direct bandgap transition in multilayer WSe_2. *Nano Lett.* 14 (8): 4592–4597.

85 Horzum, S., Sahin, H., Cahangirov, S. et al. (2013). Phonon softening and direct to indirect band gap crossover in strained single-layer $MoSe_2$. *Phys. Rev. B: Condens. Matter* 87 (12): 1–5.

86 Ghorbani-Asl, M., Borini, S., Kuc, A., and Heine, T. (2013). Strain-dependent modulation of conductivity in single-layer transition-metal dichalcogenides. *Phys. Rev. B: Condens. Matter* 87 (23): 1–6.

87 Zhu, C.R., Wang, G., Liu, B.L. et al. (2013). Strain tuning of optical emission energy and polarization in monolayer and bilayer MoS_2. *Phys. Rev. B: Condens. Matter* 88 (12): 1–5.

88 Niehues, I., Schmidt, R., Drüppel, M. et al. (2018). Strain control of exciton–phonon coupling in atomically thin semiconductors. *Nano Lett.* 18 (3): 1751–1757.

89 Yun, W.S., Han, S.W., Hong, S.C. et al. (2012). Thickness and strain effects on electronic structures of transition metal dichalcogenides: 2H-MX_2 semiconductors (M = Mo, W; X = S, Se, Te). *Phys. Rev. B: Condens. Matter* 85 (3): 1–5.

90 Yu, S., Xiong, H.D., Eshun, K. et al. (2015). Phase transition, effective mass and carrier mobility of MoS_2 monolayer under tensile strain. *Appl. Surf. Sci.* 325 (C): 27–32.

91 Dong, L., Dongare, A.M., Namburu, R.R. et al. (2014). Theoretical study on strain induced variations in electronic properties of 2H-MoS_2 bilayer sheets. *Appl. Phys. Lett.* 104 (5): 053107.

92 Shi, H., Pan, H., Zhang, Y.W., and Yakobson, B.I. (2013). Quasiparticle band structures and optical properties of strained monolayer MoS_2 and WS_2. *Phys. Rev. B: Condens. Matter* 87 (15): 1–8.

93 Bhattacharyya, S. and Singh, A.K. (2012). Semiconductor–metal transition in semiconducting bilayer sheets of transition-metal dichalcogenides. *Phys. Rev. B: Condens. Matter* 86 (7): 1–7.

94 Pak, S., Lee, J., Lee, Y.W. et al. (2017). Strain-mediated interlayer coupling effects on the excitonic behaviors in an epitaxially grown MoS_2/WS_2 van der Waals heterobilayer. *Nano Lett.* 17 (9): 5634–5640.

95 Bai, Y., Zhou, L., Wang, J. et al. (2020). Excitons in strain-induced one-dimensional moiré potentials at transition metal dichalcogenide heterojunctions. *Nat. Mater.* 19 (October): 1068–1073.

96 Aivazian, G., Gong, Z., Jones, A.M. et al. (2015). Magnetic control of valley pseudospin in monolayer WSe_2. *Nat. Phys.* 11 (2): 148–152.

97 Zollner, K., Junior, P.E.F., and Fabian, J. (2019). Strain-tunable orbital, spin-orbit, and optical properties of monolayer transition-metal dichalcogenides. *Phys. Rev. B: Condens. Matter* 100 (19): 1–10.

98 Conley, H.J., Wang, B., Ziegler, J.I. et al. (2013). Bandgap engineering of strained monolayer and bilayer MoS_2. *Nano Lett.* 13 (8): 3626–3630.

99 Liang, J., Zhang, J., Li, Z. et al. (2017). Monitoring local strain vector in atomic-layered $MoSe_2$ by second-harmonic generation. *Nano Lett.* 17 (12): 7539–7543.

100 Mennel, L., Paur, M., and Mueller, T. (2019). Second harmonic generation in strained transition metal dichalcogenide monolayers: MoS_2, $MoSe_2$, WS_2, and WSe_2. *APL Photonics* 4 (3): 034404.

101 Zhang, Q., Chang, Z., Xu, G. et al. (2016). Strain relaxation of monolayer WS_2 on plastic substrate. *Adv. Funct. Mater.* 26 (47): 8707–8714.

102 Plechinger, G., Castellanos-Gomez, A., Buscema, M. et al. (2015). Control of biaxial strain in single-layer molybdenite using local thermal expansion of the substrate. *2D Mater.* 2 (1): 015006.

103 Frisenda, R., Drüppel, M., Schmidt, R. et al. (2017). Biaxial strain tuning of the optical properties of single-layer transition metal dichalcogenides. *npj 2D Mater. Appl.* 1 (1): 1–7.

104 Lloyd, D., Liu, X., Christopher, J.W. et al. (2016). Band gap engineering with ultralarge biaxial strains in suspended monolayer MoS_2. *Nano Lett.* 16 (9): 5836–5841.

105 Yang, R., Lee, J., Ghosh, S. et al. (2017). Tuning optical signatures of single- and few-layer MoS_2 by blown-bubble bulge straining up to fracture. *Nano Lett.* 17 (8): 4568–4575.

106 Wang, Y., Cong, C., Yang, W. et al. (2015). Strain-induced direct–indirect bandgap transition and phonon modulation in monolayer WS_2. *Nano Res.* 8 (8): 2562–2572.

107 Sichert, J.A., Hemmerling, A., Cardenas-Daw, C. et al. (2015). Tuning the optical, magnetic, and electrical properties of $ReSe_2$ by nanoscale strain engineering. *Nano Lett.* 15 (3): 1660–1666.

108 Castellanos-Gomez, A., Roldán, R., Cappelluti, E. et al. (2013). Local strain engineering in atomically thin MoS_2. *Nano Lett.* 13 (11): 5361–5366.

109 Song, S., Keum, D.H., Cho, S. et al. (2016). Room temperature semiconductor–metal transition of $MoTe_2$ thin films engineered by strain. *Nano Lett.* 16 (1): 188–193.

110 Fu, X., Li, F., Lin, J.F. et al. (2017). Pressure-dependent light emission of charged and neutral excitons in monolayer $MoSe_2$. *J. Phys. Chem. Lett.* 8 (15): 3556–3563.

111 Nayak, A.P., Bhattacharyya, S., Zhu, J. et al. (2014). Pressure-induced semiconducting to metallic transition in multilayered molybdenum disulphide. *Nat. Commun.* 5 (May): 1–9.

112 Fu, X., Li, F., Lin, J.F. et al. (2018). Coupling-assisted renormalization of excitons and vibrations in compressed $MoSe_2$–WSe_2 heterostructure. *J. Phys. Chem. C* 122 (10): 5820–5828.

113 Xia, J., Yan, J., Wang, Z. et al. (2021). Strong coupling and pressure engineering in WSe_2–$MoSe_2$ heterobilayers. *Nat. Phys.* 17 (1): 92–98.

114 Nayak, A.P., Yuan, Z., Cao, B. et al. (2015). Carrier density, and mobility of multilayered tungsten disulfide. *ACS Nano* 9 (9): 9117–9123.

115 Nayak, A.P., Pandey, T., Voiry, D. et al. (2015). Pressure-dependent optical and vibrational properties of monolayer molybdenum disulfide. *Nano Lett.* 15 (1): 346–353.

116 Kim, J.-S., Moran, S.T., Nayak, A.P. et al. (2016). High pressure Raman study of layered $Mo_{0.5}W_{0.5}S_2$ ternary compound. *2D Mater.* 3 (2): 025003.

117 Nayak, A.P., Yuan, Z., Cao, B. et al. (2015). Pressure-modulated conductivity, carrier density, and mobility of multilayered tungsten disulfide. *ACS Nano* 9 (9): 9117–9123.

118 Zhao, X.M., Liu, H.Y., Goncharov, A.F. et al. (2019). Pressure effect on the electronic, structural, and vibrational properties of layered 2H-$MoTe_2$. *Phys. Rev. B: Condens. Matter* 99 (2): 2–7.

119 Hui, Y.Y., Liu, X., Jie, W. et al. (2013). Exceptional tunability of band energy in a compressively strained trilayer MoS_2 sheet. *ACS Nano* 7 (8): 7126–7131.

120 Castellanos-Gomez, A., Poot, M., Steele, G.A. et al. (2012). Elastic properties of freely suspended MoS_2 nanosheets. *Adv. Mater.* 24 (6): 772–775.

121 Wang, R., Yu, Y., Zhou, S. et al. (2018). Strategies on phase control in transition metal dichalcogenides. *Adv. Funct. Mater.* 28 (47): 1–17.

122 Lee, J., Hu, X., Voevodin, A.A. et al. (2018). Effect of substrate support on dynamic graphene/metal electrical contacts. *Micromachines* 9 (4): 1–9.

123 Aslan, O.B., Datye, I.M., Mleczko, M.J. et al. (2018). Probing the optical properties and strain-tuning of ultrathin $Mo_{1-x}W_xTe_2$. *Nano Lett.* 18 (4): 2485–2491.

124 He, K., Poole, C., Mak, K.F., and Shan, J. (2013). Experimental demonstration of continuous electronic structure tuning via strain in atomically thin MoS_2. *Nano Lett.* 13 (6): 2931–2936.

125 Shen, T., Penumatcha, A.V., and Appenzeller, J. (2016). Strain engineering for transition metal dichalcogenides based field effect transistors. *ACS Nano* 10 (4): 4712–4718.

126 Pérez Garza, H.H., Kievit, E.W., Schneider, G.F., and Staufer, U. (2014). Controlled, reversible, and nondestructive generation of uniaxial extreme strains (>10%) in graphene. *Nano Lett.* 14 (7): 4107–4113.

127 Li, Z., Lv, Y., Ren, L. et al. (2020). Efficient strain modulation of 2D materials via polymer encapsulation. *Nat. Commun.* 11 (1): 1–8.

128 Island, J.O., Kuc, A., Diependaal, E.H. et al. (2016). Precise and reversible band gap tuning in single-layer $MoSe_2$ by uniaxial strain. *Nanoscale* 8 (5): 2589–2593.

129 Koenig, S.P., Boddeti, N.G., Dunn, M.L., and Bunch, J.S. (2011). Ultrastrong adhesion of graphene membranes. *Nat. Nanotechnol.* 6 (9): 543–546.

130 Tripathi, L.N., Iff, O., Betzold, S. et al. (2018). Spontaneous emission enhancement in strain-induced WSe_2 monolayer-based quantum light sources on metallic surfaces. *ACS Photonics* 5 (5): 1919–1926.

131 Palacios-Berraquero, C., Kara, D.M., Montblanch, A.R.P. et al. (2017). Large-scale quantum-emitter arrays in atomically thin semiconductors. *Nat. Commun.* 8 (May): 1–6.

132 Mirõ, P., Ghorbani-Asl, M., and Heine, T. (2013). Spontaneous ripple formation in MoS_2 monolayers: electronic structure and transport effects. *Adv. Mater.* 25 (38): 5473–5475.

133 Castellanos-Gomez, A., Buscema, M., Molenaar, R. et al. (2014). Deterministic transfer of two-dimensional materials by all-dry viscoelastic stamping. *2D Mater.* 1 (1): 011002.

134 Li, F., Cabral, M.J., Xu, B. et al. (2019). Giant piezoelectricity of Sm-doped $Pb(Mg_{1/3}Nb_{2/3})O_3$–$PbTiO_3$ single crystals. *Science* 364 (6437): 264–268.

135 Chen, Y., Zhang, J., Zopf, M. et al. (2016). Wavelength-tunable entangled photons from silicon-integrated III–V quantum dots. *Nat. Commun.* 7: 1–7.

136 Chakraborty, C., Mukherjee, A., Moon, H. et al. (2020). Strain tuning of the emission axis of quantum emitters in an atomically thin semiconductor. *Optica* 7 (6): 580.

137 Hou, W., Azizimanesh, A., Sewaket, A. et al. (2019). Strain-based room-temperature non-volatile $MoTe_2$ ferroelectric phase change transistor. *Nat. Nanotechnol.* 14 (July): 668–673.

138 Goldsche, M., Sonntag, J., Khodkov, T. et al. (2018). Tailoring mechanically tunable strain fields in graphene. *Nano Lett.* 18 (3): 1707–1713.

139 Cao, K., Feng, S., Han, Y. et al. (2020). Elastic straining of free-standing monolayer graphene. *Nat. Commun.* 11 (1): 1–7.

140 Fu, X., Liao, Z.M., Liu, R. et al. (2015). Strain loading mode dependent bandgap deformation potential in ZnO micro/nanowires. *ACS Nano* 9 (12): 11960–11967.

141 Chen, H., Zhuo, Z., Huang, J. et al. (2020). Correlation between exciton polarized lifetime and fine structure splitting in InAs/GaAs quantum dots. *Appl. Phys. Lett.* 116 (8): 082101.

142 Massicotte, M., Vialla, F., Schmidt, P. et al. (2018). Dissociation of two-dimensional excitons in monolayer WSe_2. *Nat. Commun.* 9 (1): 1–7.

143 Ross, J.S., Klement, P., Jones, A.M. et al. (2014). Electrically tunable excitonic light-emitting diodes based on monolayer WSe_2 p–n junctions. *Nat. Nanotechnol.* 9 (4): 268–272.

144 Pospischil, A., Furchi, M.M., and Mueller, T. (2014). Solar-energy conversion and light emission in an atomic monolayer p–n diode. *Nat. Nanotechnol.* 9 (4): 257–261.

145 Baugher, B.W.H., Churchill, H.O.H., Yang, Y., and Jarillo-Herrero, P. (2014). Optoelectronic devices based on electrically tunable p–n diodes in a monolayer dichalcogenide. *Nat. Nanotechnol.* 9 (4): 262–267.

146 Sundaram, R.S., Engel, M., Lombardo, A. et al. (2013). Electroluminescence in single layer MoS_2. *Nano Lett.* 13 (4): 1416–1421.

147 Zhang, Y.J., Ye, J.T., Yomogida, Y. et al. (2013). Formation of a stable p–n junction in a liquid-gated MoS_2 ambipolar transistor. *Nano Lett.* 13 (7): 3023–3028.

148 Wang, R., Ruzicka, B.A., Kumar, N. et al. (2012). Ultrafast and spatially resolved studies of charge carriers in atomically thin molybdenum disulfide. *Phys. Rev. B: Condens. Matter* 86 (4): 1–5.

149 Wu, W., Wang, L., Yu, R. et al. (2016). Piezophototronic effect in single-atomic-layer MoS_2 for strain-gated flexible optoelectronics. *Adv. Mater.* 28 (38): 8463–8468.

150 Zhang, K., Zhai, J., and Wang, Z.L. (2018). A monolayer MoS_2 p–n homogenous photodiode with enhanced photoresponse by piezo-phototronic effect. *2D Mater.* 5 (3): 035038.

151 Lin, P., Zhu, L., Li, D. et al. (2018). Piezo-phototronic effect for enhanced flexible MoS_2/WSe_2 van der Waals photodiodes. *Adv. Funct. Mater.* 28 (35): 1–8.

152 Zhang, K., Peng, M., Wu, W. et al. (2017). A flexible p-CuO/n-MoS_2 heterojunction photodetector with enhanced photoresponse by the piezo-phototronic effect. *Mater. Horiz.* 4 (2): 274–280.

153 Binder, J., Howarth, J., Withers, F. et al. (2019). Upconverted electroluminescence via Auger scattering of interlayer excitons in van der Waals heterostructures. *Nat. Commun.* 10 (1): 1–7.

154 Lee, C.H., Lee, G.H., Van Der Zande, A.M. et al. (2014). Atomically thin p–n junctions with van der Waals heterointerfaces. *Nat. Nanotechnol.* 9 (9): 676–681.

155 Gan, X., Gao, Y., Fai Mak, K. et al. (2013). Controlling the spontaneous emission rate of monolayer MoS_2 in a photonic crystal nanocavity. *Appl. Phys. Lett.* 103 (18): 1–5.

156 Schwarz, S., Dufferwiel, S., Walker, P.M. et al. (2014). Two-dimensional metal-chalcogenide films in tunable optical microcavities. *Nano Lett.* 14 (12): 7003–7008.

157 Ye, Y., Wong, Z.J., Lu, X. et al. (2015). Monolayer excitonic laser. *Nat. Photonics* 9 (11): 733–737.

158 Wu, S., Buckley, S., Schaibley, J.R. et al. (2015). Monolayer semiconductor nanocavity lasers with ultralow thresholds. *Nature* 520 (7545): 69–72.

159 Hou, S., Tong Edwin, T.H., Birowosuto, M.D., and Wang, H. (2019). Simultaneous inhibition and redistribution of spontaneous emission from perovskite photonic crystals. In: *2019 Conference on Lasers and Electro-optics*, San Jose, California United States (5–10 May 2019), CLEO: Science and Innovations 2019, 1296–1299. Optical Society of America.

160 Englund, D., Fattal, D., Waks, E. et al. (2005). Controlling the spontaneous emission rate of single quantum dots in a two-dimensional photonic crystal. *Phys. Rev. Lett.* 95 (1): 2–5.

161 Srivastava, A., Sidler, M., Allain, A.V. et al. (2015). Optically active quantum dots in monolayer WSe_2. *Nat. Nanotechnol.* 10 (6): 491–496.

162 He, Y.M., Clark, G., Schaibley, J.R. et al. (2015). Single quantum emitters in monolayer semiconductors. *Nat. Nanotechnol.* 10 (6): 497–502.

163 Chakraborty, C., Kinnischtzke, L., Goodfellow, K.M. et al. (2015). Voltage-controlled quantum light from an atomically thin semiconductor. *Nat. Nanotechnol.* 10 (6): 507–511.

164 Li, F., Shen, T., Wang, C. et al. (2020). *Recent Advances in Strain-Induced Piezoelectric and Piezoresistive Effect-Engineered 2D Semiconductors for Adaptive Electronics and Optoelectronics*. Singapore: Springer.

165 Qi, J., Lan, Y.W., Stieg, A.Z. et al. (2015). Piezoelectric effect in chemical vapour deposition-grown atomic-monolayer triangular molybdenum disulfide piezotronics. *Nat. Commun.* 6 (May): 1–8.

166 Zhu, M., Sakamoto, K., Li, J. et al. (2019). Piezoresistive strain sensor based on monolayer molybdenum disulfide continuous film deposited by chemical vapor deposition. *J. Micromech. Microeng.* 29 (5): 055002.

167 Neri, I. and López-Suárez, M. (2018). Electronic transport modulation on suspended few-layer MoS_2 under strain. *Phys. Rev. B: Condens. Matter* 97 (24): 1–5.

168 Park, M., Park, Y.J., Chen, X. et al. (2016). MoS_2-based tactile sensor for electronic skin applications. *Adv. Mater.* 28 (13): 2556–2562.

169 Zhou, Y., Liu, W., Huang, X. et al. (2016). Theoretical study on two-dimensional MoS_2 piezoelectric nanogenerators. *Nano Res.* 9 (3): 800–807.

170 Lee, J.H., Park, J.Y., Cho, E.B. et al. (2017). Reliable piezoelectricity in bilayer WSe_2 for piezoelectric nanogenerators. *Adv. Mater.* 29 (29): 1–7.

Index

Nanomembranes: Materials, Properties, and Applications, First Edition.
Edited by Yongfeng Mei, Gaoshan Huang, and Xiuling Li.
© 2022 WILEY-VCH GmbH. Published 2022 by WILEY-VCH GmbH.

n

t